FORTRESS NASHVILLE
Pioneers, Engineers, Mechanics, Contrabands & U.S. Colored Troops

Mark Zimmerman
Author of Guide to Civil War Nashville and Mud, Blood & Cold Steel

How the Heavily Fortified Logistics and Transportation Hub
Became the Key to Victory in the Western Theater

Fortress Nashville:
Pioneers, Engineers, Mechanics, Contrabands & U.S. Colored Troops
Copyright © 2022 Mark Zimmerman
Zimco Publications LLC
Website: zimcopubs.com
Email: info@zimcopubs.com

ISBN 978-0-578-37936-4

All rights reserved.
All Text, Photographs, and Maps by the Author Unless Otherwise Noted.
Original artwork by Philip Duer, David Meagher, Andy Thomas, John Paul Strain, and Rick Reeves used with permission.

No part of this book may be reproduced or transmitted in any form or by any means — electronic or manual, including photocopy, scanner, email, CD or other information storage and retreival system — without written permission from the author, except for personal use or as provided to the news media and book sellers.

Printed in the United States of America.

The text in this book, which is published to inform and entertain, should be used for general information and not as the ultimate source of educational or travel information. Every effort has been made to ensure the accuracy and relevence of information in this book but the author and publisher do not assume responsibility for any errors, inaccuracies, omissions or inconsistencies within. Any slights of people, places, or organizations are strictly unintentional.

The names of organizations and destinations mentioned in this book may be trade names or trademarks of their owners. The author and publisher disclaims any connection with, sponsorship by or endorsement of such owners.

Also by Mark Zimmerman:

Guide to Civil War Nashville, 2nd Edition

God, Guns, Guitars & Whiskey: An Illustrated Guide to Historic Nashville, Tennessee, 2nd Edition

Gone Under: Historic Cemeteries of Nashville, Tennessee, 2nd Edition

Iron Maidens and the Devil's Daughters: US Navy Gunboats versus Confederate Gunners and Cavalry on the Tennessee and Cumberland Rivers, 1861-65

Land of the Free, Home of the Brave: Our Founding Documents & Concise History of the USA

The Zone: True Tales from the Heartland

Content related to this publication can be found on the publisher's website at:
www.zimcopubs.com

Table of Contents

Introduction 5

Frontier Stations 8
Cumberland Settlements, Fort Nashborough, Buchanan's Station, Mansker's Station, Rock Castle, Cragfont, Sevier Station

The River Forts 21
The Federal Gunboat Flotilla 22
Building the River Forts 32
Capture of Fort Henry 37
Battle of Fort Donelson 47
Capture of Clarksville and Nashville 55

Defenses of Nashville 62
Nashville 1864 72
Fort Andrew Johnson 74
Fortified Railroad Bridge 79
Pontoon Bridges 80
Fort Negley 82
Fort Morton 126
Blockhouse Casino 128
Fort Houston 130
Redoubt for Hill 210 132
Fort Gillem 134
Fort Whipple 136
Fort Garesché 138
Railroad Redoubt 140
Magazine Granger 142
Brentwood Stockade 144

Table of Contents

Middle Tennessee Infrastructure 148
U.S. Army Corps of Engineers 150
U.S. Signal Corps 151
U.S. Military Hospitals 152
Fort Granger and Triune Works 154
The Pioneer Brigade 164
Fortress Rosecrans 170
U.S. Military Railroads and River Freighters 177
First Michigan Engineers & Mechanics 209
Engineer, Quartermaster Uniforms & Buttons 211
Guerrillas, Gunboats & Convoys 214
Johnsonville and Nashville & Northwestern RR 222
Federal Garrison Towns 238
Columbia, Gallatin, Sumner County, Tullahoma, Shelbyville, Bridgeport, Stevenson, Paducah, Decatur, Bowling Green, Pulaski
Federal General/Engineer Also Spymaster 250

The Battle of Nashville 254
The Confederate Redoubts 259
U.S. Colored Troops 271
Granbury's Lunette 281
Peach Orchard Hill 289
Battle of Nashville Trust 296

Addendum A: Glossary of Fortification Terms 297
Addendum B: Timeline of Events 305
Addendum C: Inspection Reports on Defenses 311
Acknowledgements & Notes on Sources 329
Bibliography and Suggested Reading 330
About the Author / Zimco Publications LLC 336

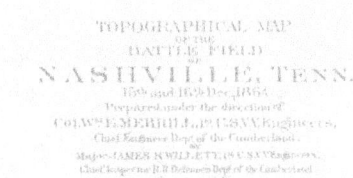

The saga of NASHVILLE during the Civil War
played out in three acts:

I. Secession and unsuccessful fortification of the Tennessee and Cumberland Rivers, leading to the joint Federal army-navy invasion of Middle Tennessee and capture of Nashville, the first Confederate state capital to fall.

II. The Federal occupation of Nashville, the building of fortifications to repel enemy attacks, and the massive development of transportation and logistics facilities to support further invasion into the heartland.

III. Hood's 1864 invasion of Tennessee, culminating in the battles at Franklin and Nashville, and the rout of the Confederate army. After the end of the war, Federal troops occupied the city until 1872 and then abandoned it.

Along the way, black men in Middle Tennessee
escaped their chains of bondage,
labored for the National army,
donned the two-tone blue uniform,
and fought valiantly for their freedom.

At the same time, the roles of the pioneer, the engineer,
the quartermaster, the pilot, and the mechanic
proved vital to the success
of the Federal armed forces.

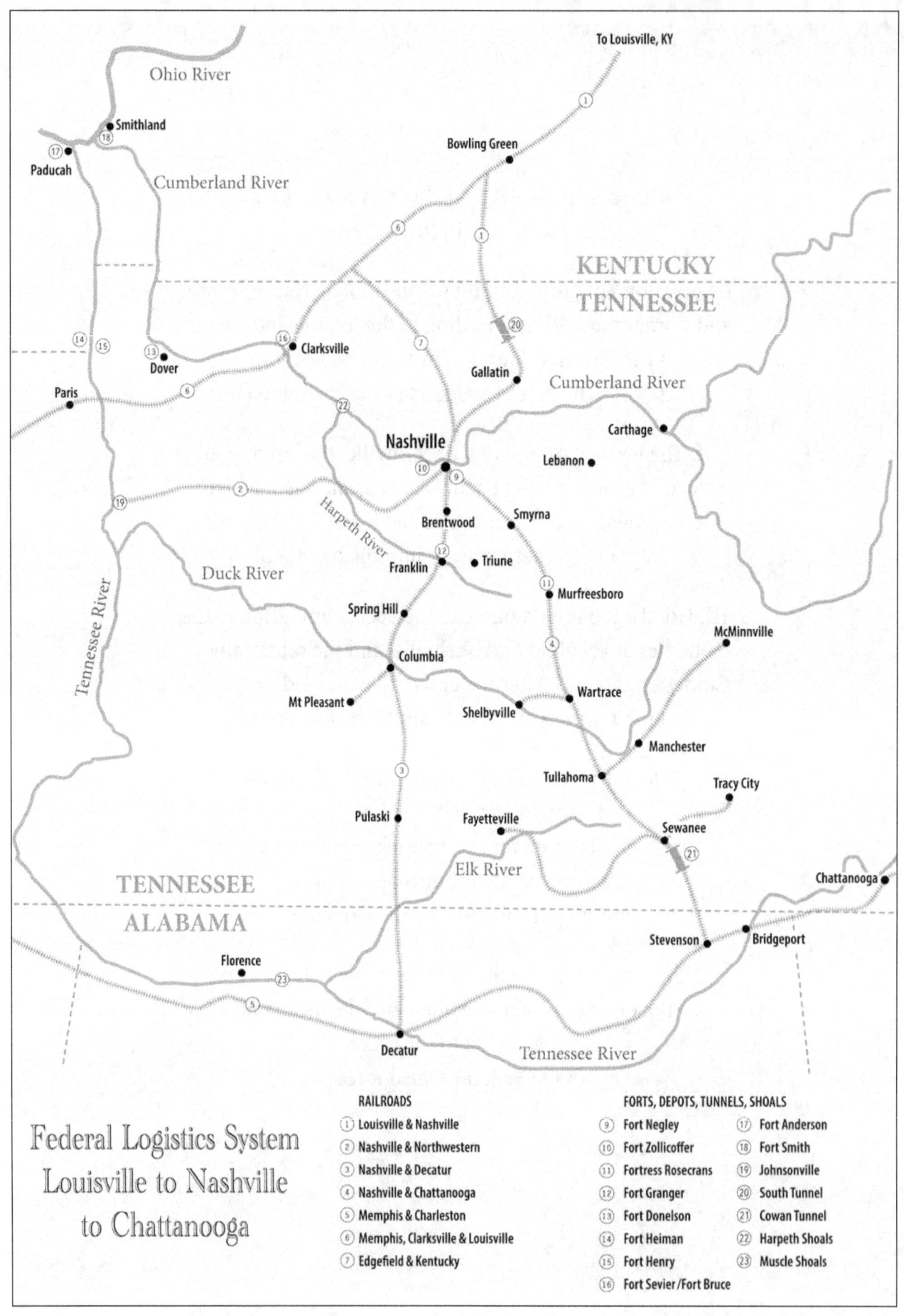

Introduction

Through the centuries, humans have used fortifications as places of refuge against wild animals and the elements but mostly against hostile human enemies. In Europe, forts took the form of castles and walled cities. In America, indigenous tribes erected stockades made of timber and walls constructed of stacked limestone, a building material which could be found in abundance. The early European hunters made use of caves and even hollow trees to hide from hostiles. Early settlers erected timber stockades in which to gather when threatened. These structures also served as domiciles and trading posts. Some wealthy settlers from back East built their homes of thick, cut stone. These settlements and homes were called stations. Around the turn of the century, relations between the European settlers and the local tribes improved (with a few minor exceptions), and the settlement on the banks of the Cumberland River became a village and eventually a town, Nashville, and fortifications were no longer needed.

In the early 1860s, when Tennessee seceded from the Union, local volunteers were charged with building forts or artillery batteries on the Tennessee and Cumberland Rivers to thwart invasion by the Federal army and navy. In this role, the Confederate forts failed, due mostly to poor siting and lack of strong leadership. Several fortifications were simply abandoned. In early 1862, the Federal army captured Nashville without firing a shot and proceeded to develop the city into a major transportation and logistics hub.

The captor of Nashville, General D.C. Buell, began to make slow but steady progress toward capturing Chattanooga, proceeding roughly along Tennessee's southern border. Confederate cavalry raids by the likes of John Hunt Morgan and Nathan Bedford Forrest and the invasion of Kentucky by Braxton Bragg and Edmund Kirby Smith pulled Buell's army back through Nashville and into the Bluegrass State, leaving Nashville with few defenses. At this time, Buell ordered his brilliant chief of engineers, Capt. James St. Clair Morton, to erect fortifications around Nashville to protect the city and the Federal infrastructure. The result was Morton's stone masterpiece, Fort Negley, finished by the beginning of 1863.

> **for·ti·fi·ca·tion** | fôrdəfə′kāSH(ə)n |
> noun
> a defensive wall or other reinforcement built to strengthen a place against attack: *the building and maintenance of fortifications | a medieval fortification.*
> • the action of fortifying or process of being fortified: *the fortification of the frontiers.*
> ORIGIN
> late Middle English: via French from late Latin *fortificatio(n-),* from *fortificare.*
> New Oxford American Dictionary

Construction of the other Nashville forts lagged behind, suffering from the lack of labor, the rocky terrain (Nashville was known as Rock City), their complicated, elaborate designs, and other priorities. Guns in the forts and other locations were positioned so they could be turned upon the city itself in the case of a rebellion by the pro-Confederate populace.

The Federal fortifications were built mostly by slaves, impressed freedmen, and contrabands (fugitive slaves). Over time, the Federal installations became a place of refuge for runaway slaves and their families although the living conditions were usually quite squalid. Freedom, protection, and opportunity were sought by these formerly enslaved people. First used as laborers, they later were recruited into the Federal army as U.S. Colored Troops. The USCT units were stationed at blockhouses along the railroads — the vital arteries of supply for the advancing Union armies — to protect vulnerable trestles and bridges. Many officers and troops desperately wanted to exchange garrison duty for fighting in the field, a wish that was granted at the Battle of Nashville in December 1864. Inexperienced and engaged in assaults upon strong field fortifications, the USCT soldiers fought valiantly and proved their mettle in battle. Prior to the battle, the Confederate Army of Tennessee built several redoubts and a lunette to strengthen their positions south of the city. The lunette defended its position, while the redoubts were overwhelmed by superior numbers and firepower.

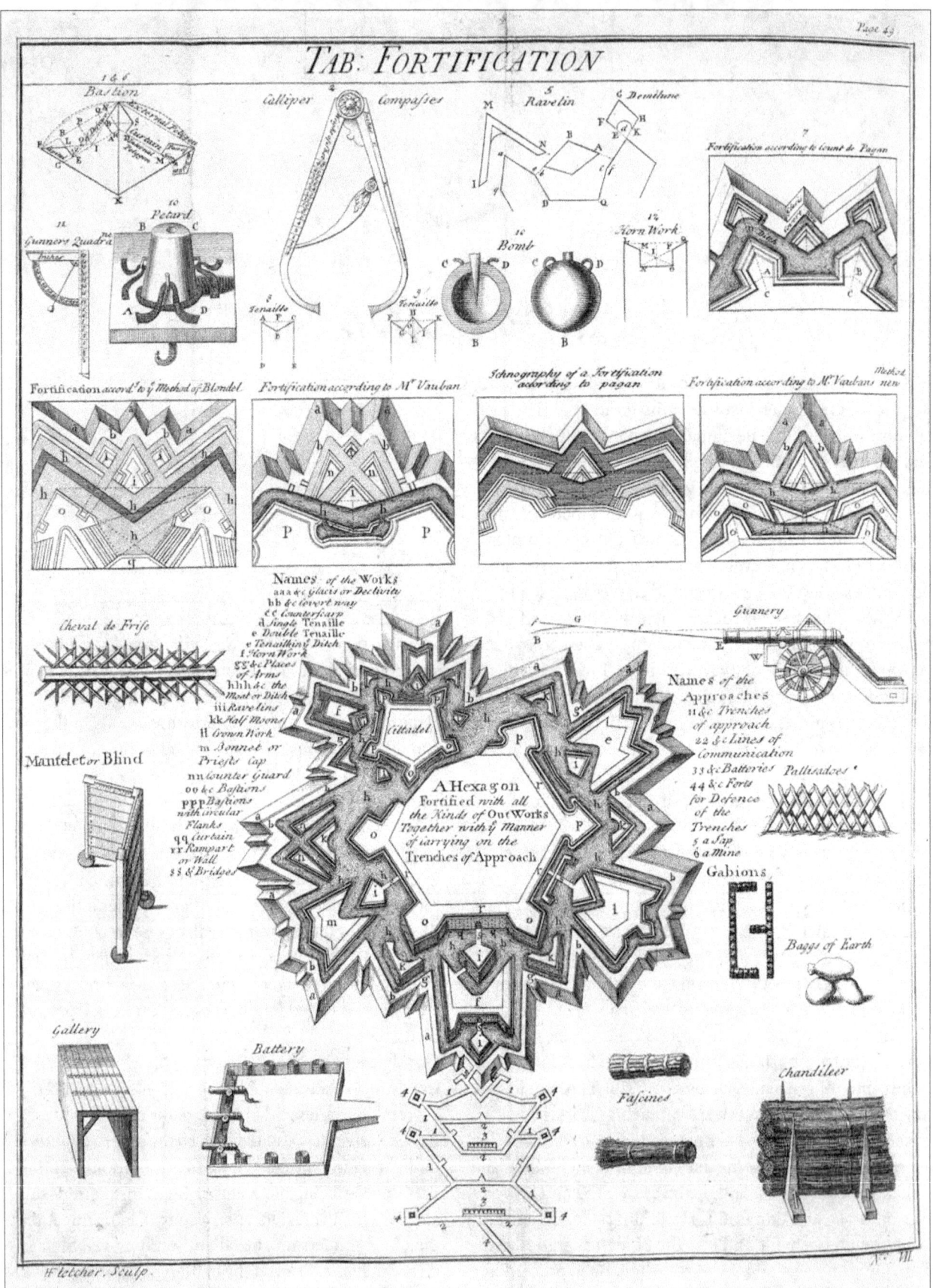

Fortifications were designed with mathematical and geometric precision, as shown in the *Cyclopædia: An Universal Dictionary of Arts and Sciences* (1792). The construction and maintenance of fortifications were a bit more haphazard.

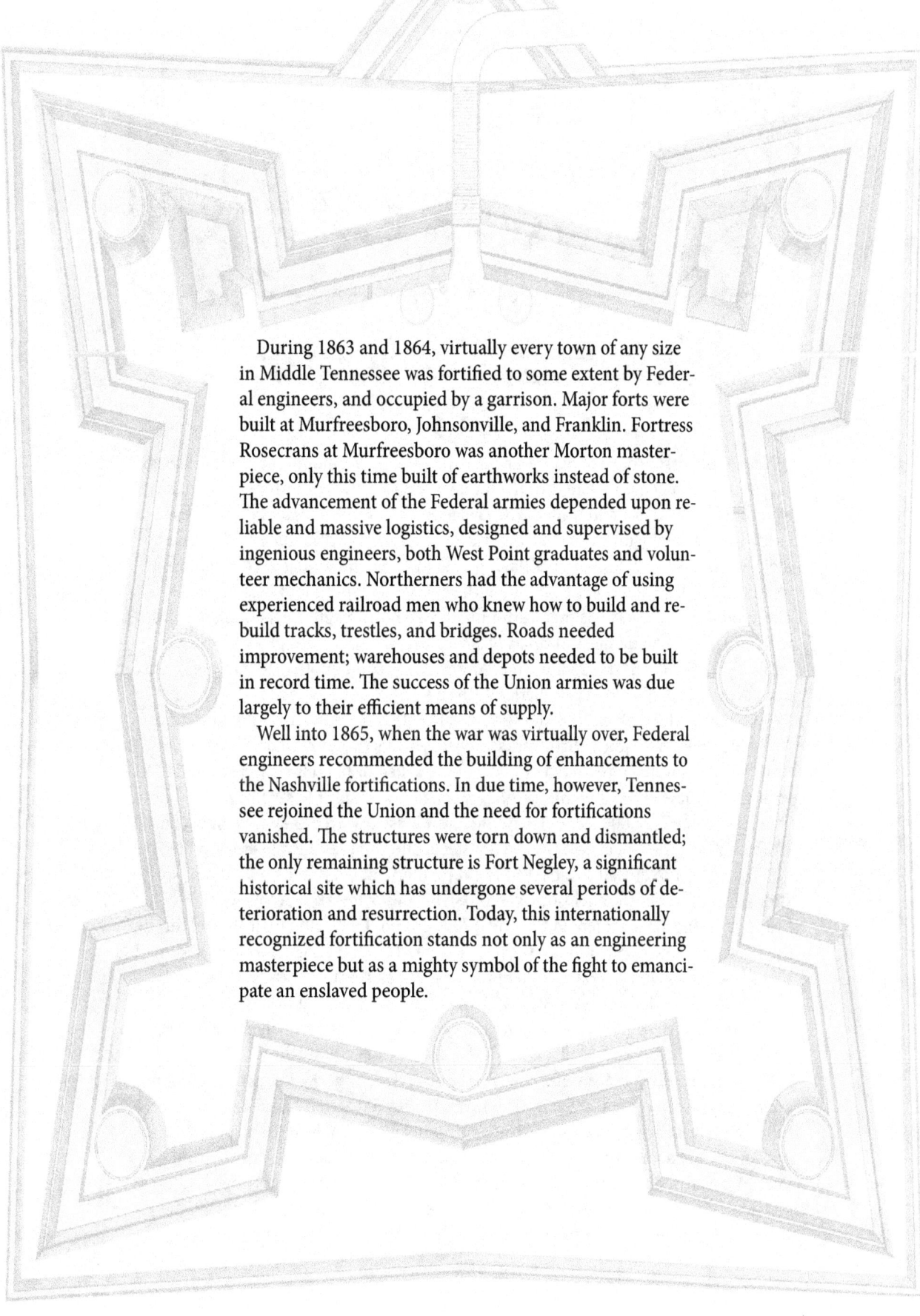

During 1863 and 1864, virtually every town of any size in Middle Tennessee was fortified to some extent by Federal engineers, and occupied by a garrison. Major forts were built at Murfreesboro, Johnsonville, and Franklin. Fortress Rosecrans at Murfreesboro was another Morton masterpiece, only this time built of earthworks instead of stone. The advancement of the Federal armies depended upon reliable and massive logistics, designed and supervised by ingenious engineers, both West Point graduates and volunteer mechanics. Northerners had the advantage of using experienced railroad men who knew how to build and rebuild tracks, trestles, and bridges. Roads needed improvement; warehouses and depots needed to be built in record time. The success of the Union armies was due largely to their efficient means of supply.

Well into 1865, when the war was virtually over, Federal engineers recommended the building of enhancements to the Nashville fortifications. In due time, however, Tennessee rejoined the Union and the need for fortifications vanished. The structures were torn down and dismantled; the only remaining structure is Fort Negley, a significant historical site which has undergone several periods of deterioration and resurrection. Today, this internationally recognized fortification stands not only as an engineering masterpiece but as a mighty symbol of the fight to emancipate an enslaved people.

Frontier Stations

One of the first groups of Europeans to explore the "overmountain" territory west of the Appalachians were the long hunters, small groups of intrepid frontiersmen who sought wild game and traded with friendly natives for hides and pelts. These hardy men, armed with the deadly accurate Pennsylvania or Kentucky rifle, lived by their wits and their wiles, at times having to fend off hostile Indians. They followed waterways and animal trails, roaming in the late 1750s throughout what is now Middle Tennessee and what was then communal tribal hunting grounds populated with vast herds of bison and deer. In times of strife, Thomas Sharpe "Bigfoot" Spencer found refuge inside a large hollow tree. Another pioneer, Timothy Demonbreun, lived in a cave on the cliffs of the Cumberland River.

At the forks of the Duck River on terrain called The Barrens, early white hunters stumbled upon prehistoric earthworks and stone walls that enclosed a 50-acre plateau. The walls were constructed of undressed stacked or piled limestone covered with earth. These embankments, originally ranging from four to six feet in height, would have totaled 4,600 feet in length if continuous; however, they were only constructed where the stream bluffs were not steep. The complex entranceway to the "Old Stone Fort" featured parallel walls that pointed toward the sun at summer solstice. The interior of the enclosure was grassland with a few trees, otherwise featureless. The natives there in the mid-1700s did not know who built the old fort or the relatively large earthen mounds that also dotted the landscape. Pioneers assumed that some unknown civilization had built these structures, perhaps Welsh or Norse, or Spanish troops led by conquistador Hernando de Soto during the 1500s. Some more imaginative souls speculated that the earthworks may have been built by extraterrestrials.

Not until the early 1960s were these myths dispelled for good. Using carbon dating, archaeologists from the University of Tennessee proved that the Old Stone Fort was built 2,000 years ago by prehistoric Native American groups. And the structure was most likely not a fort but some sort of celestial observatory, although the true purpose of the ancient structure remains a mystery. The Old Stone Fort is now a state archaeological park located at Manchester. These days, each year, celebrants gather nearby for the multi-day Bonnaroo music festival. As for ancient structures, the

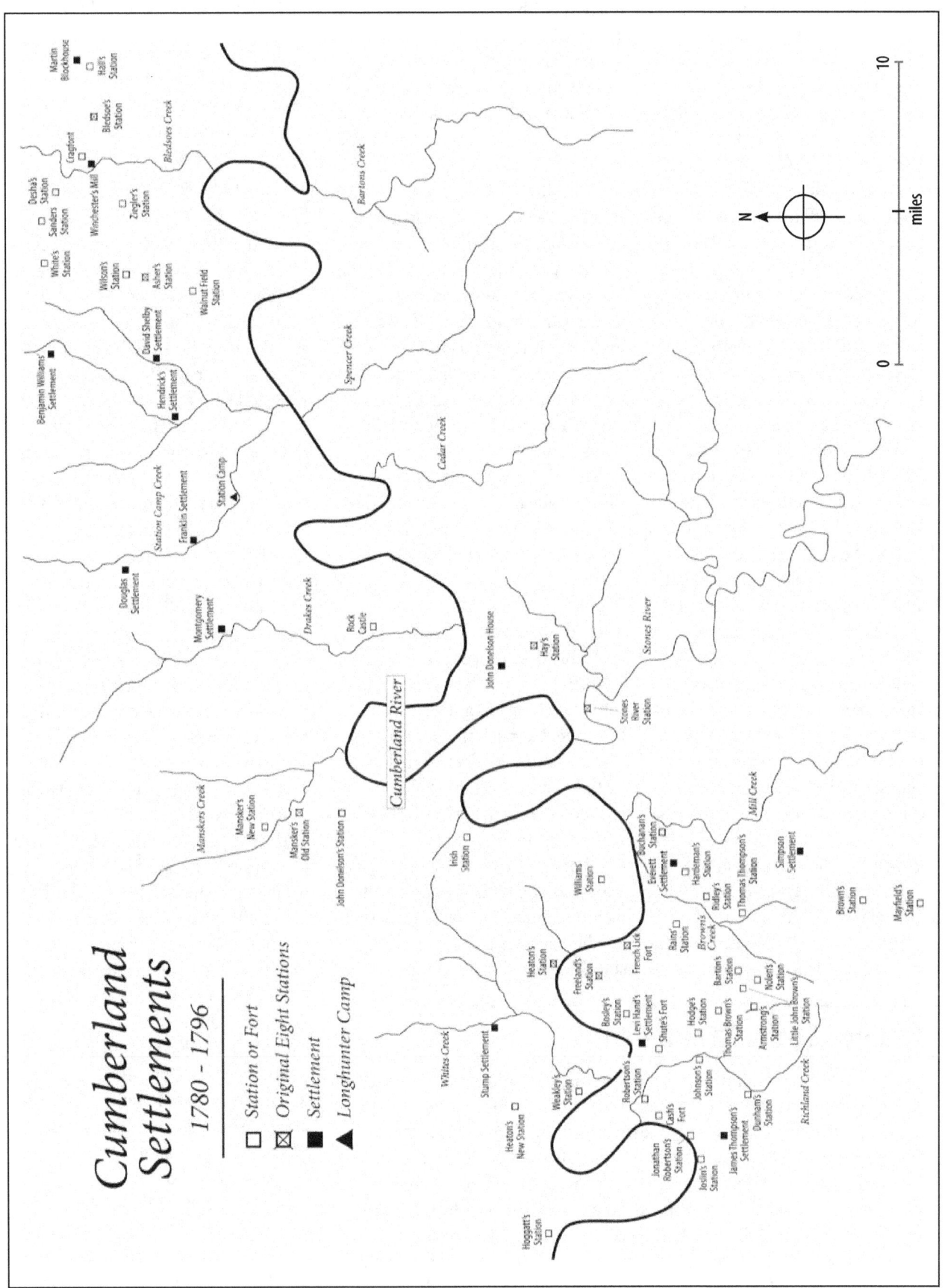

Middle Tennessee area is also inundated with ceremonial and burial mounds built by the Indians of the Mississippian Era (1,000 years ago).

The first significant fortification built by Europeans west of the Allegheny Mountains and east of the Mississippi River was Fort Loudoun, erected by the British military in 1756-57 in an effort to prevent Cherokee tribes from uniting with French forces. The timber fort was named for John Campbell, the Earl of Loudoun, commander-in-chief of the British forces in North America. Fort Loudoun featured a large log palisade, inside of which were a row of barracks, a powder magazine, a blacksmith shop (that also served as a meeting house or chapel), two corn houses, a guardroom, and various storehouses. By 1760, however, during the Seven Years War, aka the French and Indian War, relations between the British colonists and the Cherokee deteriorated. The Cherokee laid siege to the fort, and in August 1760 the British surrendered. The British were allowed to march to the South Carolina settlements but were attacked en route by Cherokee warriors. Several whites were killed, the others were taken prisoner, later to be ransomed. The site is now commemorated at the Fort Loudoun State Historic Area, which features a replica of the fort as it existed in 1759-60.

Later, in 1794-95, the Tellico Blockhouse was constructed by the fledgling U.S. government to maintain relations with the Cherokee and serve as a trading post. In 1807 the Indian agency was moved to another location and the blockhouse was abandoned.

The first white settlements in what is now Tennessee were in the valleys of the upper tributaries of the Tennessee River. In 1777 (during the War for Independence), at Fort Watauga in present-day Elizabethton, former North Carolina judge and land speculator Richard Henderson signed a treaty with a large group of Cherokee granting him a vast territory of nearly one million acres in what is now Kentucky. Disgruntled young Cherokees led by Dragging Canoe opposed the Transylvania Purchase and vowed violence. His faction, called the Chickamaugan Cherokee, battled with white settlers up until shortly before Tennessee was granted statehood in 1796.

The Cumberland Settlements and Fort Nashborough

North Carolina voided the Transylvania Purchase but granted Henderson a much smaller tract as compensation. He determined to settle that region, beginning at a particular place. Long hunters had described the area along the Cumberland River near a salt spring known as French Lick as a land of plenty, teaming with wild game such as whitetail deer and bison. Henderson chose two distinguished men to lead expeditions to the French Lick to establish settlements — James Robertson and John Donelson. Robertson journeyed overland to the French Lick and determined a suitable site for settlement. Later, in late 1779, Robertson returned with a group of men to prepare a suitable site for friends and relatives who planned to join them in a few months. The men arrived on Christmas Day and drove their cattle across the frozen Cumberland River. Crude cabins were erected for immediate winter housing, and a fort was built atop the bluff along the river. The fort was called the Bluff Station and later became known as Fort Nashborough, in honor of Francis Nash, who had fought alongside Robertson at the Battle of Alamance in 1771. Author, historian, and president Teddy Roosevelt called this migration the "Great Leap Westward."

The early settlers, comprising about 60 families, built stockades and fortifications in which to live and work. By May 1780 there were eight forts or stations in the Cumberland settlements — French Lick or the Bluff (later known as Fort Nashborough); Freeland's to the north; Gasper's at Goodlettsville; Fort Union, about six miles upriver; Eaton's, three miles downriver; Stones River, on that tributary; Asher's, near Gallatin; and Bledsoe's, at Castalian Springs in what is now Sumner County.

On Dec. 22nd, 1779, a flotilla of flatboats, led by Donelson in his craft *Adventure,* left Fort Patrick Henry on the Holston River to join Robertson's men. The flotilla included many women, children, and slaves. One of

They Do Not Pity Me

The Great Being above gave us the land, but the white people seem to want to drive us from it. I pity the white people, but the white people do not pity me.

— Cherokee Chief Atakullakulla

Ceding Land to Pay Off Debt

As much as anything, the Cherokee land cessions were a matter of debt. First and foremost, British colonists often bartered with alcohol, rather than necessary goods like guns and ammunition, when trading with the Cherokee for deerskins. This led to a large deficit between traders and the Cherokee. Thus, the cession of land became the only valued commodity to pay off what they owed. Second, by the 1770s the Cherokee favored their trade alliances with the English colonists over their relationship with other native tribes, especially the Iroquois. In their world, the value of English and colonial trade goods took precedence.

— Fort Nashborough Interpretive Master Plan, 2013

Donelson's children was his 13-year-old daughter Rachel, who later would become the wife of Andrew Jackson.

The boats, which could only flow with the current or be poled, were attacked many times by hostile Indians. The winter weather was frightfully cold and severe. The rivers were dangerous, with many shoals, rapids, and impediments. The flotilla floated down the Tennessee River and then north to the Ohio. Then upriver to the mouth of the Cumberland and upriver to the French Lick. Some of the travelers stopped for good at the present site of Clarksville. The main party reached the bluffs on April 24, 1780, four months after they started.

On May 13th, representatives from each station, 256 in total, made their marks on what's become known as the Cumberland Compact, an organization of government formed not only in the overmountain territory but many years prior to the U.S. Constitution.

Up until the summer of 1780, the indigenous peoples had not troubled the European settlers. Caution resulting from previous encounters and the harsh winter weather kept the tribal warriors at bay. Also, one of the Donelson flatboats, kept to the rear of the flotilla, bore a family infected with smallpox. Easy prey, the boat was captured and its occupants killed by the Indians, who contracted the highly contagious disease and died by the hundreds. However, by that summer,

A Dark and Bloody Ground

You have bought a fair land, but there is a cloud hanging over it. You will find its settlement dark and bloody.
— Chickamaugan Cherokee warrior Dragging Canoe, referring to the Transylvania Purchase.

Drawing based on Andrew Castleman's memory of Fort Nashborough

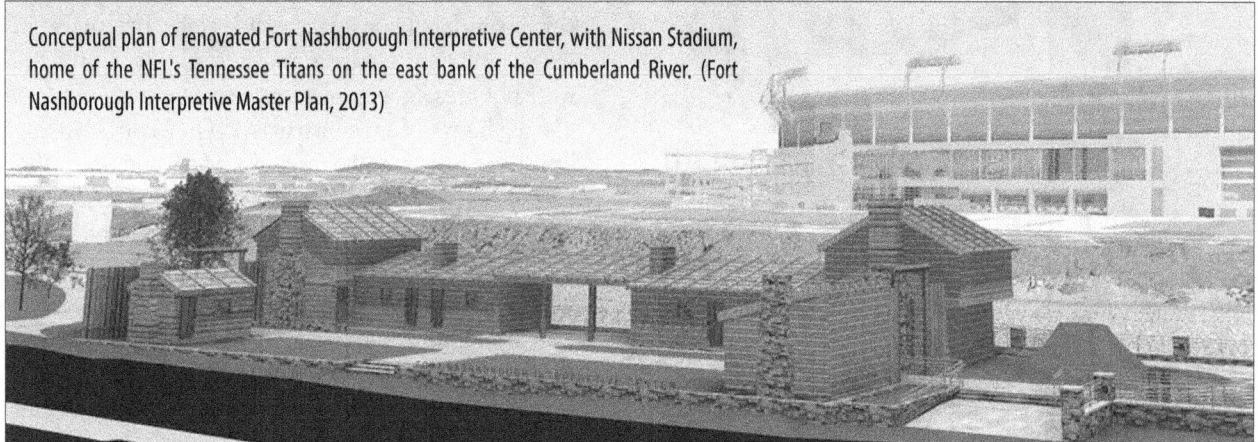

Conceptual plan of renovated Fort Nashborough Interpretive Center, with Nissan Stadium, home of the NFL's Tennessee Titans on the east bank of the Cumberland River. (Fort Nashborough Interpretive Master Plan, 2013)

settlers in small groups or alone found themselves set upon by Indians, warfare that would not abate until 1792, when Bledsoe's and Ziegler's stations north of Nashville were overrun by Indians. By the end of the first year of settlement, only two stations remained occupied due to hostile relations with the Indians — Heaton's Station and the Bluff Station.

James Robertson was 38 when he arrived at the Bluffs. He stood six feet tall, with dark hair and blue eyes, a man with inner composure and an even temper. He and his wife, Charlotte Reeves, had 13 children, two who died in infancy. Two of his sons were killed by Indians and another was scalped. His two brothers were slain by hostile natives. He settled in the upper Holston Valley near the Watauga River. He led the Watauga Association and commanded Fort Watauga. Robertson died in 1814; he and his family are buried at the Old City Cemetery in Nashville in the shadow of Fort Negley.

Fort Nashborough, as it is now known, was a stockaded collection of timber structures built along a high bluff of the Cumberland River near a

spring. The Bluff Station was roughly 248 feet long by 124 feet wide, or a little more than two-thirds of an acre. The structure consisted of a series of log houses laid out in an elongated pattern and supported at the corners by blockhouses. Stockade walls, twelve feet in height, were erected in areas where log structures were not present, completing the enclosed station. The log houses were constructed of hewn logs with a limestone hearth and chimney, and a wood floor. The logs were 8 to 12 inches in diameter and used a composite of mud, rock, animal hair, or other bonding materials for chinking. The structures were covered by a gable roof with split-shake shingles held in place by a grid of weighted poles. Gates were located on the northeast (riverbank) and the northwest. The spring was 50 feet beyond the north wall of the stockade and ran parallel to the fort wall before flowing over the

Robertson (left) and Donelson

side of the bluff and down to the river.

On April 2nd, 1781, Creek Indians launched a surprise attack against the Bluff Station. Several Indians taunted the settlers from a distance outside the fort. Robertson knew the tactic was an ambush, but he was admonished by James Leeper, who accused Robertson of being too cautious if not outright cowardly. Twenty of the men rode horses outside the station to a creek about a mile away to the southwest. There they were ambushed by 250 Indians. As the settlers hastily retreated they were cut off from the station by another 250 warriors. Fortunately the Indians became distracted while rustling some of the settlers' riderless horses. There was also another threat. Robertson's wife, Charlotte, let loose the hounds from the fort. These large dogs were trained to attack Indians.

"The dogs have been trained to hostility against Indians, and made a furious onset upon them," according to 19th-Century historian John Haywood, "disabling them from doing more than defending themselves. The retreating Whites passed near them, through the interval made by those who had gone in pursuit of horses. Had it not been for these circumstances, the whites could never have returned to the fort."

The Indians lost about 40 men. Two of the settlers were killed outside the fort. Three died later at the fort. Leeper was wounded and died two weeks later. One of the settlers, Edward Swanson, wrote about the aftermath of the Battle of the Bluffs:

"George Kennedy died on the ground and was scalped. Peter Gill escaped to the riverbank after being wounded, and died immediately afterwards at the canoe landing. Isaac Lucas, besides being wounded very badly in the thigh, was also scalped all over the head, and being brought in, died the third or fourth day afterwards. One of the 10 reinforcements, Zachariah White, aged about 50, was

shot through the bowels, but escaped into the fort and died the same day about sunset. Alex Buchanan came in, also wounded through the bowels, and died within a few minutes of White's death. James Leeper came into the fort wounded in the back, and died after several weeks. Jonas Menifee was badly wounded in the hip and Kasper Mansker was wounded in the left shoulder, but both recovered."

The Creeks laid siege to the fort but departed after a day or two, only to be replaced by a large party of Cherokee. This band was fended off by firing the four-pounder swivel gun, loaded with musket balls, mounted inside the fort. The victory by Robertson's party saved the settlements from total destruction, and possibly changed the course of westward expansion in the region, according to some historians.

Nashville was officially chartered in 1784. By 1800, Nashville had grown to a population of 345. Almost 40 percent of those were black, and all but three of those were slaves. Of the 191 white residents, only 12 were older than 45.

A small-scale replica of Fort Nashborough was built in 1930 along the river just south of the actual site. Over the years, like many of Nashville's historic sites, the public attraction fell into disrepair and neglect. In recent years, however, the replica has been reconstructed by the city and offers self-guided tours and great views of the river. The Founders Statue of Robertson and Donelson stands nearby, as does a statue of Timothy Demonbreun.

Buchanan's Station

The family of John and Jean Trindle Buchanan was among the earliest permanent settlers in the Cumberland. They began construction of a fortified settlement, or station, on Mill Creek about four miles east of Nashville in the spring of 1784. During those early years, many settlers, including John Buchanan Sr. and his son Samuel Buchanan, were killed at the station by Indians. By 1787 the settlement was led by the oldest surviving son, John Buchanan, who later became known as Major John Buchanan. He was a farmer, land speculator, surveyor, businessman, and writer. In 1781, he authored a manuscript known as *Buchanan's Arithmetic*. On Sunday, Sept. 30th, 1792, several hundred Creek,

Cherokee, and Shawnee warriors attacked the station en route to Fort Nashborough. At that time, the station was occupied by only about 20 men. They killed and severely wounded a number of Indian leaders, including Kiachatallee and Cheeseekau, older brother of Tecumseh, during the battle. Sarah (Sally) Ridley Buchanan, the wife of Major John Buchanan, kept the men supplied with black powder and musket balls during the attack, continually encouraging them to keep up the fight. Eighteen years old at the time, she was pregnant with the first of her 13 children. The child was born only 11 days after the battle. Nineteenth-Century historian J.G.M. Ramsey called the settler victory "a feat of bravery which has scarcely been surpassed in all the annals of border warfare." John Sugden, an English historian, noted that "the assault on Buchanan's Station was not a simple raid, but an attempt to wipe out the Nashville settlements entirely, backed by Spanish arms and supplies secured in Pensacola." Members of the Buchanan family and others lived at Buchanan's Station until the property was sold in 1841 (Major John Buchanan died in 1832). Today, the site of the station is marked by a family cemetery and signage, a small plot along the creek nestled among businesses in a light-industry park near Elm Hill Pike and Massman Drive.

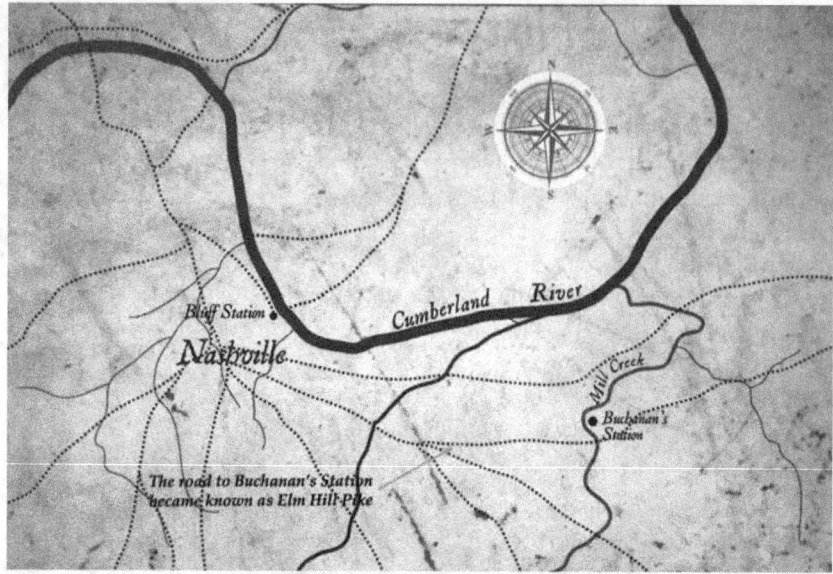

The road to Buchanan's Station became known as Elm Hill Pike

Buchanan Family Cemetery

Mansker's Station

Kasper Mansker, a German born onboard a ship en route to America, became a long hunter and in 1780 built a station of timber near a creek and salt lick in what is now Goodlettsville, a suburb of Nashville. The settlers left Mansker Station during the winter of 1780-81; Indians burned the abandoned fort. In 1782-83, Mansker built a new fort about one mile north of the first one. Mansker lived there the rest of his life. Boarders included Andrew Jackson, John Overton, and French botanist André Michaux. Mansker was a signer of the Cumberland Compact. In 1787, Mansker was elected major of the Sumner County militia and served on the first county grand jury. When the Southwest Territory was established in 1790, Territorial Governor William Blount appointed Mansker lieutenant colonel of the Sumner County militia. Mansker later participated as a volunteer in the 1794 campaign against the Chickamauga villages. He fought with Jackson against the British at the Battle of New Orleans at age 62. He died in 1820. His simple grave is located at a park in Goodlettsville. An impressive reproduction of Mansker's Station sits in Moss-Wright Park near the Bowen-Campbell house (1787), the oldest brick structure in Middle Tennessee. Living-history events with a variety of re-enactors are often conducted at the station.

Rock Castle

In 1784, General Daniel Smith (1748-1818) began work on his home on a land grant of more than 3,000 acres along Drakes Creek. Today, the historic site, Rock Castle, is preserved in Hendersonville, Sumner County, northeast of Nashville. Smith was a

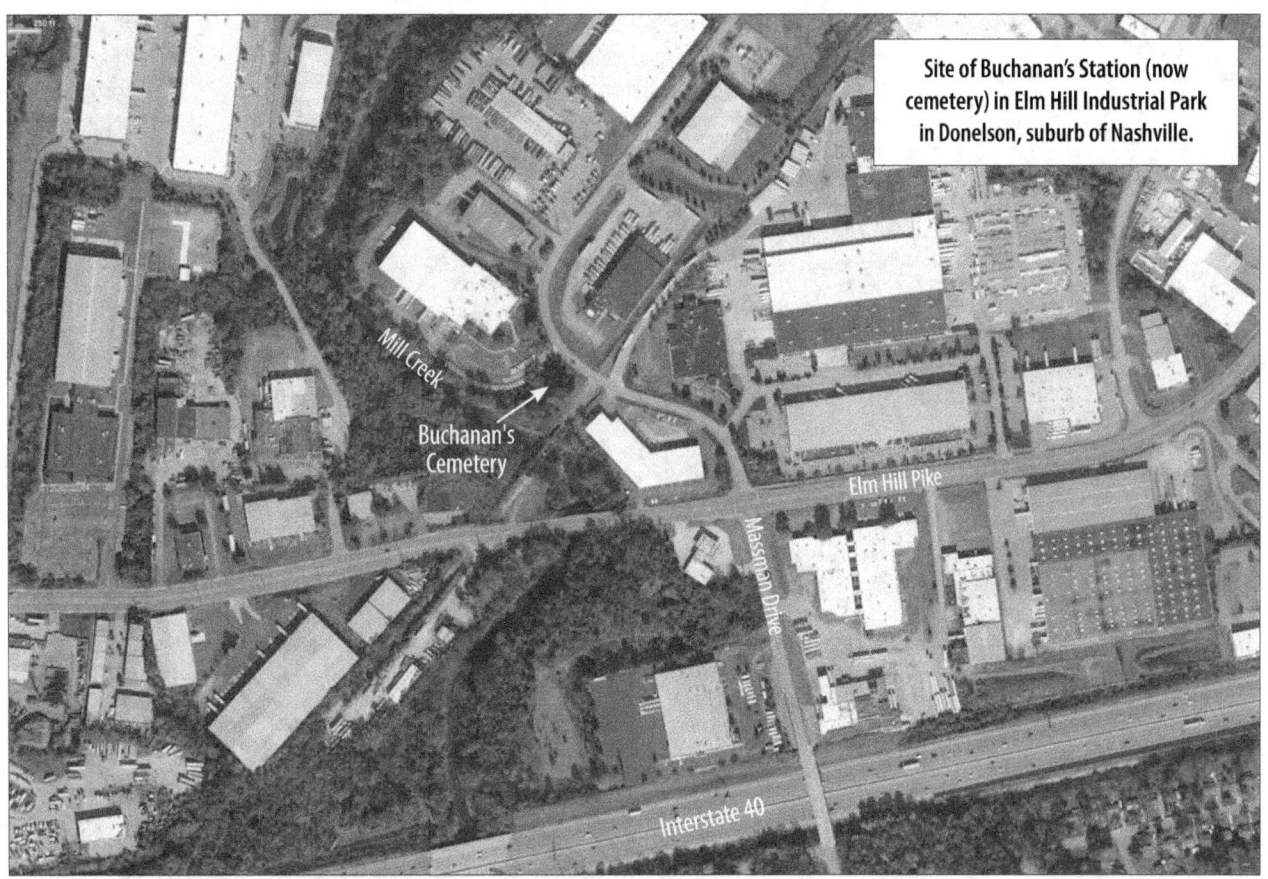

Site of Buchanan's Station (now cemetery) in Elm Hill Industrial Park in Donelson, suburb of Nashville.

Mansker's Station

Mansker's Station replica fort with gardens and Interpretive Center (upper right) at Moss-Wright Park in Goodlettsville.

Bowen-Campbell House (right)
Interior of Mansker's Station (below)

Rock Castle

highly educated man who attended the College of William and Mary. A captain in the Revolutionary War, Smith was a licensed surveyor who is famous for producing the first map of the Tennessee territory. Smith, his wife Sarah, and their two children, lived in a log house while Rock Castle was being built of limestone quarried on site. When their log cabin burned down, they moved to the two rooms that had been completed at Rock Castle. The house, completed in 1796, was more of a small fortress than a castle. "Built largely by nephews Peter and Smith Hansborough under the supervision of Sarah Smith during her husband's frequent absences on military assignment, Rock Castle shows a degree of architectural mastery rarely found on the Tennessee frontier," according to Arlene F. Young, writing in the *Tennessee Encyclopedia of History and Culture*. "Although it strives for Georgian symmetry and balance, its two-story portico, which was added during the Greek Revival period, is not quite centered on the front facade. The interior has the common central-hall plan of the time, and the carpenters added unique fireplaces that feature central-mantel, over-mantel, and floor-to-ceiling side cabinets placed within an integrated and painted black walnut-paneled wall."

General Smith served as the Commander of the Mero District, Secretary of the Southwest Territory, chairman of the committee that drafted the Tennessee State Constitution, U.S. Senator, and U.S. Commissioner for Indian negotiations.

Smith's daughter Polly was 15 when it was decided she would attend a boarding school in Philadelphia. However, she had fallen in love with Samuel Donelson, 22, Rachel Jackson's brother and Andrew Jackson's partner-in-law. General Smith was against their proposed marriage. One night, Samuel and Andrew sneaked over to Rock Castle, assisted Polly down a ladder from her bedroom window, and took her to the Jackson home, where the two were married. The general was enraged and basically disowned his daughter. A year later, they reconciled when she gave the general his first grandchild.

Today, Historic Rock Castle, consisting of 18 acres, the house, family cemetery, and visitor's center, is owned by the state and operated by the Friends of Rock Castle. The public site opened in 1981.

Cragfont

A native of Maryland and a veteran

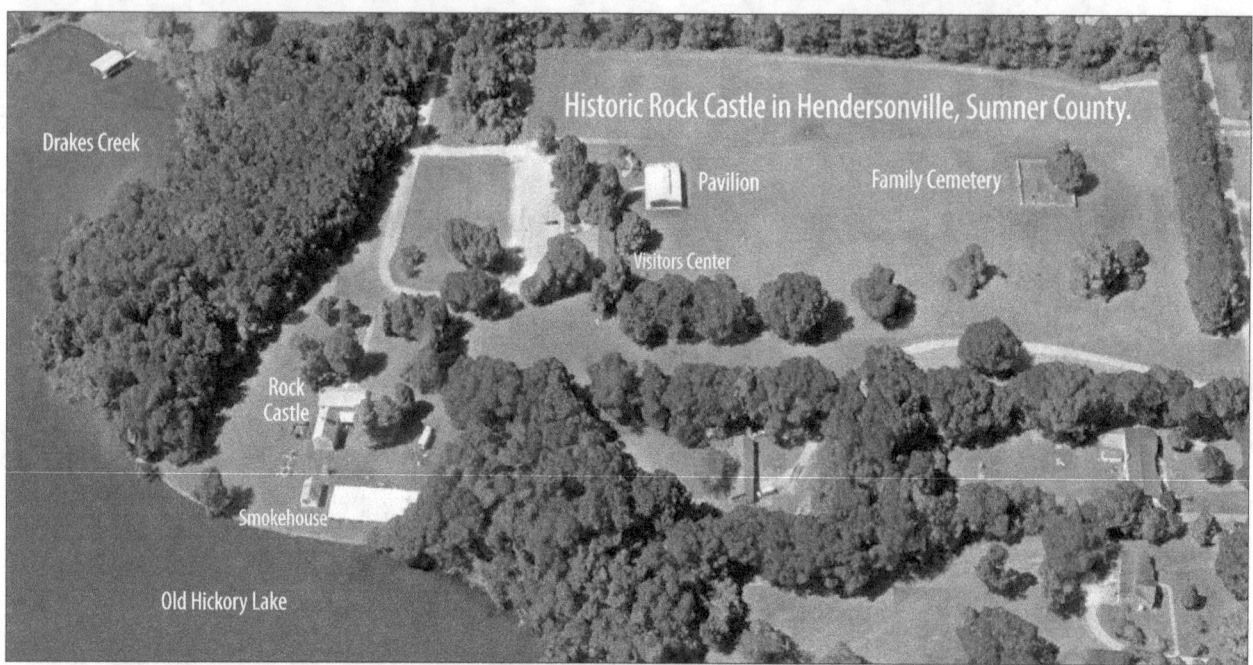

Historic Rock Castle in Hendersonville, Sumner County.
Drakes Creek
Pavilion
Family Cemetery
Visitors Center
Rock Castle
Smokehouse
Old Hickory Lake

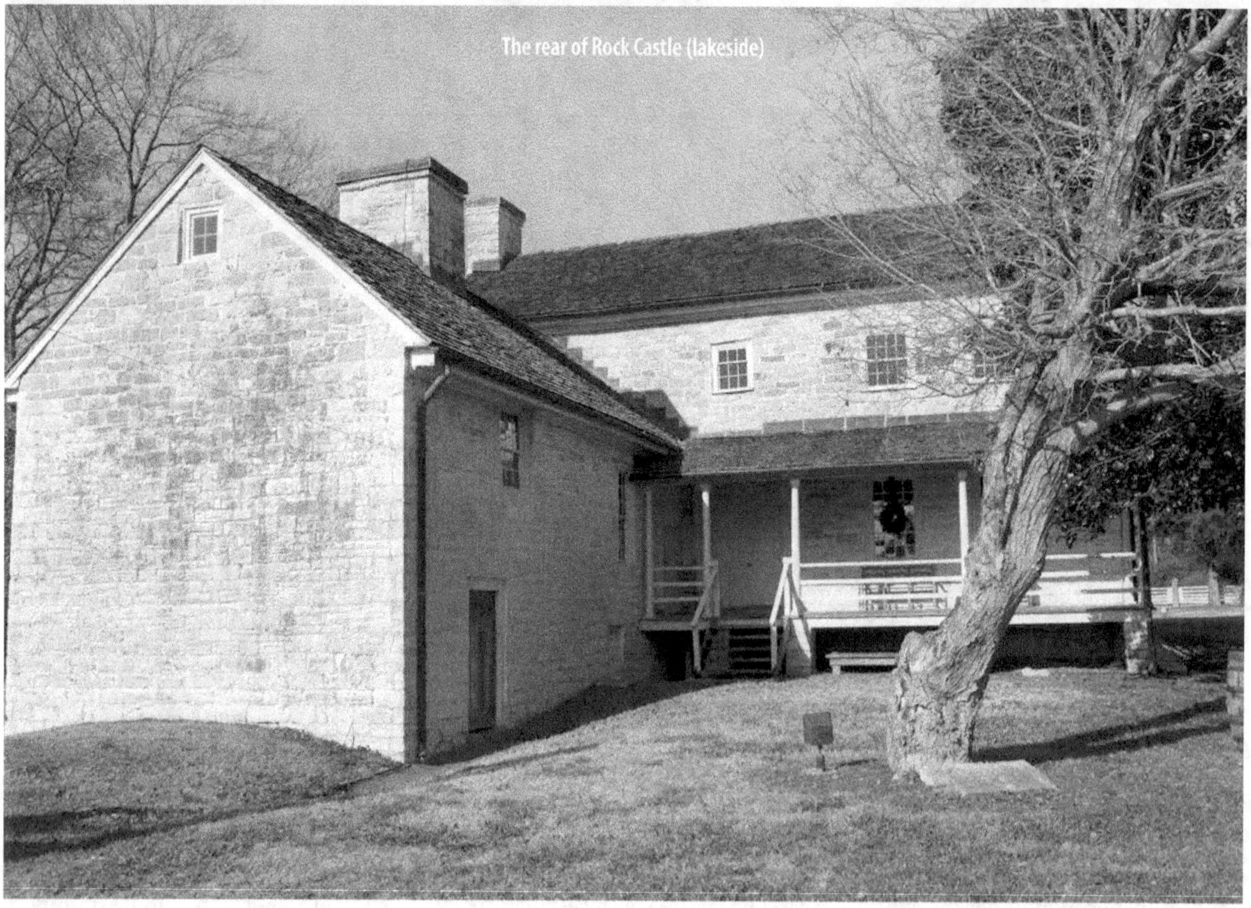
The rear of Rock Castle (lakeside)

of the Revolutionary War, James Winchester was a pioneer, entrepreneur, military commander, shipbuilder, and a founder of Memphis. In 1785, he settled on Bledsoe's Creek in Sumner County, near the present-day city of Gallatin. There he built a magnificent stone mansion or station called Cragfont. The structure enclosed many rooms, including a large, second-story ballroom. It is said the attic timbers resemble the structure of a ship, which isn't surprising since Winchester hired carpenters and shipbuilders from Philadelphia to build his home. It was the most elegant residence on the Tennessee frontier. The front of Cragfont shows several decorative metal stars, which are actually the bolts for structural metal rods running through the house, installed after the Great Earthquakes of 1812-13 hundreds of miles away. A man of many talents, Winchester constructed two oceangoing schooners near his mill in 1806. Winchester purchased much land, including a tract along the Mississippi River which was developed as the city of Memphis. His son Marcus became the first mayor. The basic planning of the city was done around the dining room table at Cragfont. Winchester, a rival of Andrew Jackson and William Henry Harrison, was a hardluck soldier and military officer, having been captured in both the Revolutionary War and the War of 1812. That he is not better known is mostly due to the fact that he didn't have a good biographer, according to the late Walter Durham, State Historian. In 1958 the Tennessee Historical Commission acquired Cragfont, which is listed on the National Register of Historic Places and is open to the public.

Sevier Station

In Clarksville near the confluence of the Cumberland and Red Rivers, Valentine Sevier (1747-1800), the younger brother of Tennessee's first governor, John Sevier, established about 1790 a small station made of limestone and featuring a fireplace and chimney in the center of the structure and loopholes for the use of rifles in defense. The simple building stands today, close to the Fort Defiance Civil War site. Sevier's estate consisted of 640 acres, several buildings, and homes for several other families. In 1792, three of Sevier's sons were killed by a Chickamauga raiding party. On Nov. 11th, 1794, the station was attacked by 40 Creek Indi-

VALENTINE SEVIER MEMORIAL
Col. Valentine Sevier, defender of the early settlers of this community, on July 11th, 1792 purchased from George Cook, for the sum of 100 pounds, 640 acres, lying between this point and Cumberland and Red Rivers, known as Red Paint Hill, hunting ground of the Cherokee Indians, it became the site of Sevier Station.
Erected by Captain Wm. Edmiston Chapter D.A.R. 1936

ans and many of the settlers were killed. Sevier was a sergeant, and one of the spies, at the Battle of Point Pleasant, where "he was distinguished for vigilance, activity, and bravery." He subsequently served in the Indian wars in East Tennessee, and commanded a company at Thicketty Fort, Cedar Springs, Musgrove's Mill, and King's Mountain. He was the first sheriff of Washington County, a justice of the court, and rose in the militia to the rank of colonel.

The River Forts
Federal Gunboats Invade Middle Tennessee

Federal ironclad gunboats shelling Fort Henry on the Tennessee River. (Library of Congress)

The importance of the river system in the Western Theater as a route of invasion into the heartland of the Confederacy and as an avenue of supplying forward bases was self-evident to any military or government official who could read a map. The Federal military would need to create a river gunboat flotilla to lead the invasion. The Confederates would need to build river forts to thwart that invasion.

The Federal Gunboat Flotilla

USS Cairo City Class Ironclad Gunboat
(The Photographic History of the Civil War)

Just ten days after the fall of Fort Sumter in April 1861, Federal forces occupied the small river port of Cairo in a region of southern Illinois known as Little Egypt. The city featured ample wharfs on the banks of the Ohio River (where the current was slower than on the Mississippi River) and served as the terminus of the Illinois Central Railroad. Cairo would be the headquarters and staging area for the Federal fight on the Mississippi and home port for what would be called the Western Gunboat Flotilla.

At the outset of the Civil War, the business of steamboating on the inland rivers dropped considerably. Most of the transport fleet, built and owned by Northern interests, were moored and unused. There were plenty of boats for the Federal military to purchase and convert into gunboats, which was faster and easier than building gunboats from scratch. Following inspections, Brigadier General Joseph G. Totten advised that there were 400 steamers available on the rivers for transporting troops, 400 coal barges, and 200 freight barges.

Boatbuilding facilities were available at Pittsburg, Pa.; Wheeling, Va.; Cincinnati; Madison and New Albany, Ind.; and Mound City, Ill.

After several fits and starts, and trips between St. Louis and Washington, renowned and wealthy marine engineer James Buchanan Eads was directed to consult with Major General George McClellan in Cincinnati about forming an inland navy, with the assistance of Navy Commander John Rodgers. Esteemed naval designer Samuel M. Pook was directed to examine the proposal. It was determined that the naval equipment for the boats, plus armament and officers and crews, should be fulfilled by the U.S. Navy. It was most likely the largest naval building project since the construction in the early 1800s of the first U.S. naval frigates under President Thomas Jefferson.

On June 8, 1861, Rodgers purchased three steamers at Cincinnati for conversion into gunboats and charged the expense to the navy, generating a rebuke from Navy Secretary Gideon Welles. He sternly advised that the army would foot the bill; that the navy would supply only guns and crews. This dictate put the U.S. Army squarely in charge of what would become the Western Gunboat Flotilla. This "unified" command would remain in effect until September 1862, when the U.S. Navy took over all riverboat flotilla operations.

USS Conestoga Timberclad Gunboat
(US Naval Historical Center)

This was the beginning of a unique, sometimes contentious, cooperation between the army and navy which would reap many dividends. The joint command also necessitated the cooperation of skilled civilians, including that of the river pilots required on each gunboat to navigate the tricky Western rivers.

The first three gunboats were the timberclads *USS Conestoga*, *USS Lexington,* and *USS Tyler*. These were river steamboats converted into lightly armored (five inches of oak planking) and heavily armed gunboats. They were sidewheelers, powered by coal-burning steam engines, and recognizable by their tall twin chimneys (smokestacks).

The initial conversion of the timberclads left much to be desired. Rogers said the project "was more like the work of Irish laborers than mechanics." More work was required, and repairs on the timberclads would hinder their operations for many months.

On Aug. 12th, 1861, the *USS Conestoga* completed its maiden trip down the Ohio and anchored at Cairo. Her captain, Lt. Commander Seth Ledyard Phelps, shook hands with Col. Richard Oglesby of the 8th Illinois Regiment, the first face-to-face encounter between the leaders of U.S. Army forces and the U.S. Navy's Western Gunboat Flotilla.

The *Conestoga* was the smallest, least expensive, and the fastest of the three timberclads, and would serve as the timberclad flotilla's flagboat. The *Lexington* was the newest, the most expensive, the slowest, and would become the best-known due to her extensive wartime service. The *Tyler* was the oldest, largest, and heaviest but one knot faster than the *Lexington*. Earlier, in January 1861, as a transport, she had been fired upon at Vicksburg, sunk, and then salvaged.

Marine designer John Lenthall drew up plans for what a Western river gunboat would look like—a flat-bottomed sidewheeler about 170 feet long and 28 feet wide that was double-ended, so that the boat could change directions without turning about. Totten suggested that ten to twenty gunboats be built at an average cost of $20,000 each, without public bidding so as not to alert the Confederates as to their intentions.

General Winfield Scott approved the report and forwarded it to Secretary of War Cameron on June 10, 1861, recommending that 16 gunboats be built and be ready for service by September 20 of that year. Cameron turned the proposal over to Quartermaster General Montgomery Meigs, who requested that Commander Rodgers have naval designer Pook and his engineers study the proposed designs for suitability of production and operation of gunboats on the Western waters.

Pook consulted builders, engineers, and river captains in the Cincinnati area and then revised Lenthall's design extensively. The gunboats would be 175 feet long and 50 feet wide and draw six feet of water. The flat bottom would have three keels. The double-ender concept was abandoned. The hull would support a casemate (superstructure) that would feature sloped sides and house 20 naval guns. The sides of the casemate and around the steam engines and boilers would be clad in iron plate. Specifications for the steam engines, which would be powered from coal-fired tubular boilers, were drawn up by A. Thomas Merritt and approved by chief naval engineer B.F. Isherwood. A proposal for a gunboat using a screw propeller system was rejected as impractical for the shallow, debris-ridden inland rivers. The sidewheeler concept was rejected in favor of a single enclosed paddlewheel at the stern of the boat. Named for their designer and the ap-

(Continued on Page 27)

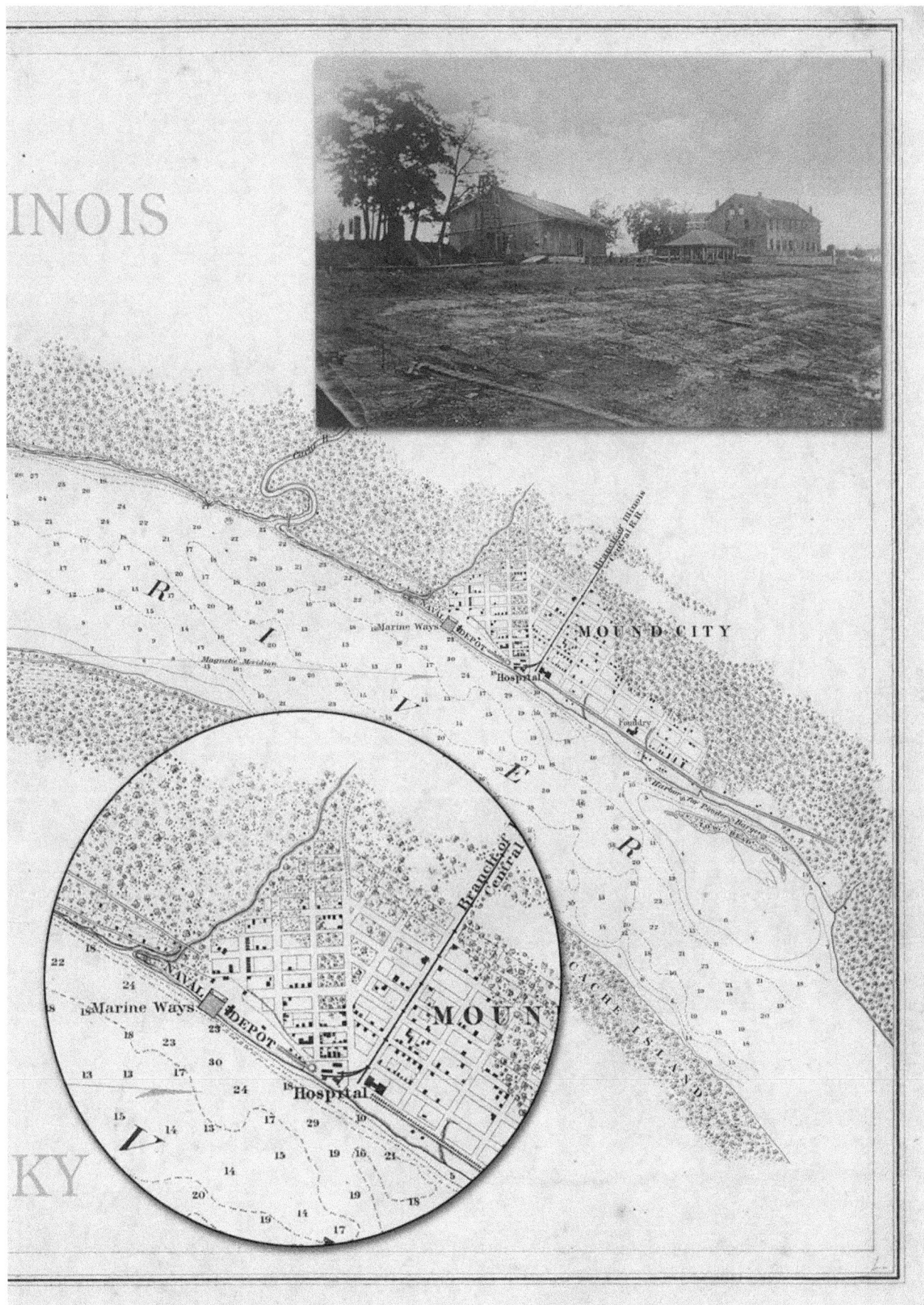

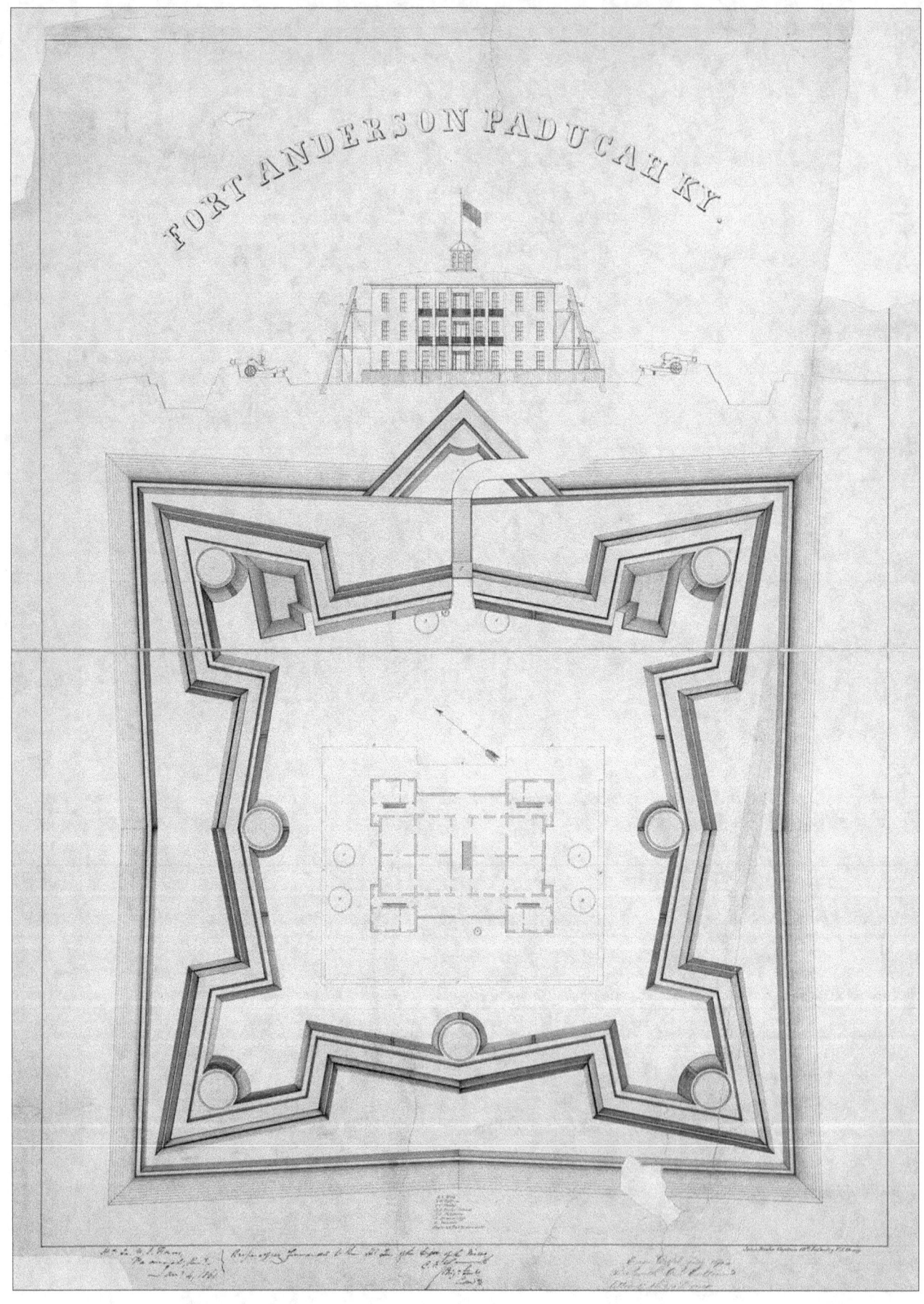

Fort Anderson

(Continued from Page 23)
pearance of the sloped walls of their casemates and low-lying hulls, these ironclad gunboats would be called "Pook's turtles."

On July 15, 1861, three prominent men from St. Louis wrote to Meigs that their city would be a favorable site for the construction of the proposed gunboats—drydocks were in place; machinery could be manufactured locally; skilled mechanics were available; and its location on the Mississippi allowed for the greatest usage relative to river levels. Three days later, Meigs advertised proposals for bids for the construction of the gunboats. When the seven bids were opened on Aug. 5, 1861, the low bid was by James Eads, who proposed building four to 16 gunboats at $89,600 apiece and delivering them to the government at Cairo in 59 days (by October 5). Subsequently, Eads signed a contract to build seven gunboats for $89,600 each, on or before October 10.

The specifications called for casemate ports for 20 guns, an enclosed paddlewheel, 75 tons of protective iron plating, five boilers 25 feet long, a firebox lined with firebrick, and two chimneys 44 inches in diameter and 28 feet high. At some point, the armament requirement was lowered to 13 guns per boat (three bow, four starboard and port, and two astern). The boat's two steam engines would have cast-iron cylinders inclined at a 15-degree angle with a bore of 22 inches and a stroke of six feet. The piston would be attached to a driving rod four inches in diameter and nine feet long. Hartupee and Company of Pittsburg was subcontracted to build and supply the gunboat engines and boilers.

Eads leased the Carondelet Marine Railway and Drydock Co., just south of St. Louis, for the construction site. Meigs named engine designer Merritt as overall superintendent of the project; Rodgers named John Litherbury to oversee the construction of the gunboats at Carondelet.

Commander Rodgers wrote to the chief of naval ordnance that he needed 35 rifled 42-pounders and 70 nine-inch Dahlgren smoothbores to arm the gunboats, a request that was tentatively denied. The Secretary of the Navy recalled that the War Department had promised to provide 35 rifled 42-pounders for the gunboats and requested they be sent to St. Louis posthaste. Eagle Iron Works of Cincinnati won the contract for providing the wooden gun carriages and implements.

Due to time constraints, Eads decided to build the remaining three gunboats at the Marine Railway and Ship Yard owned by Captain William Hambleton and located at Mound City, Illinois, on the Ohio River. By the end of August, an additional 130 carpenters and workers had been hired, and construction at Mound City commenced immediately. Eventually, 800 men and 13 sawmills in five states would be working on the boats. Two large iron rolling mills in Cincinnati worked day and night for two weeks to produce the iron cladding. Gaylord, Son and Company in Portsmouth, Ohio, and Newport, Ky. rolled 700 tons of iron into 2.5-inch-thick plates for the cladding. In building the boats, Eads determined that each boat would require not the 75 tons of ironcladding specified, but actually 122 tons, an increase of 63 percent in weight.

By the second week of September, the engines and boilers were almost ready for delivery, but Eads had run into a problem—the government had not been paying him for the work completed as agreed upon in the contract. He had spent $700,000 on the project so far and had not received a penny from the government. Sub-contractors were pressuring him for payment. The government forked over about $44,000, which was not nearly enough, Eads said, to prevent him from "being cramped and annoyed for money."

As work progressed, the builders became curious as to the strength of the ironcladding and set up a test range on the sandy Mississippi shore opposite the Carondelet boatyards. Lt. Albert R. Buffington of the St. Louis Arsenal oversaw the firing of two 10-pounder Parrott rifles into sheets of gunboat iron securely fastened at a 35-degree angle to oak blocks 16 inches thick. At 800 yards it was difficult to hit the target, but one shot

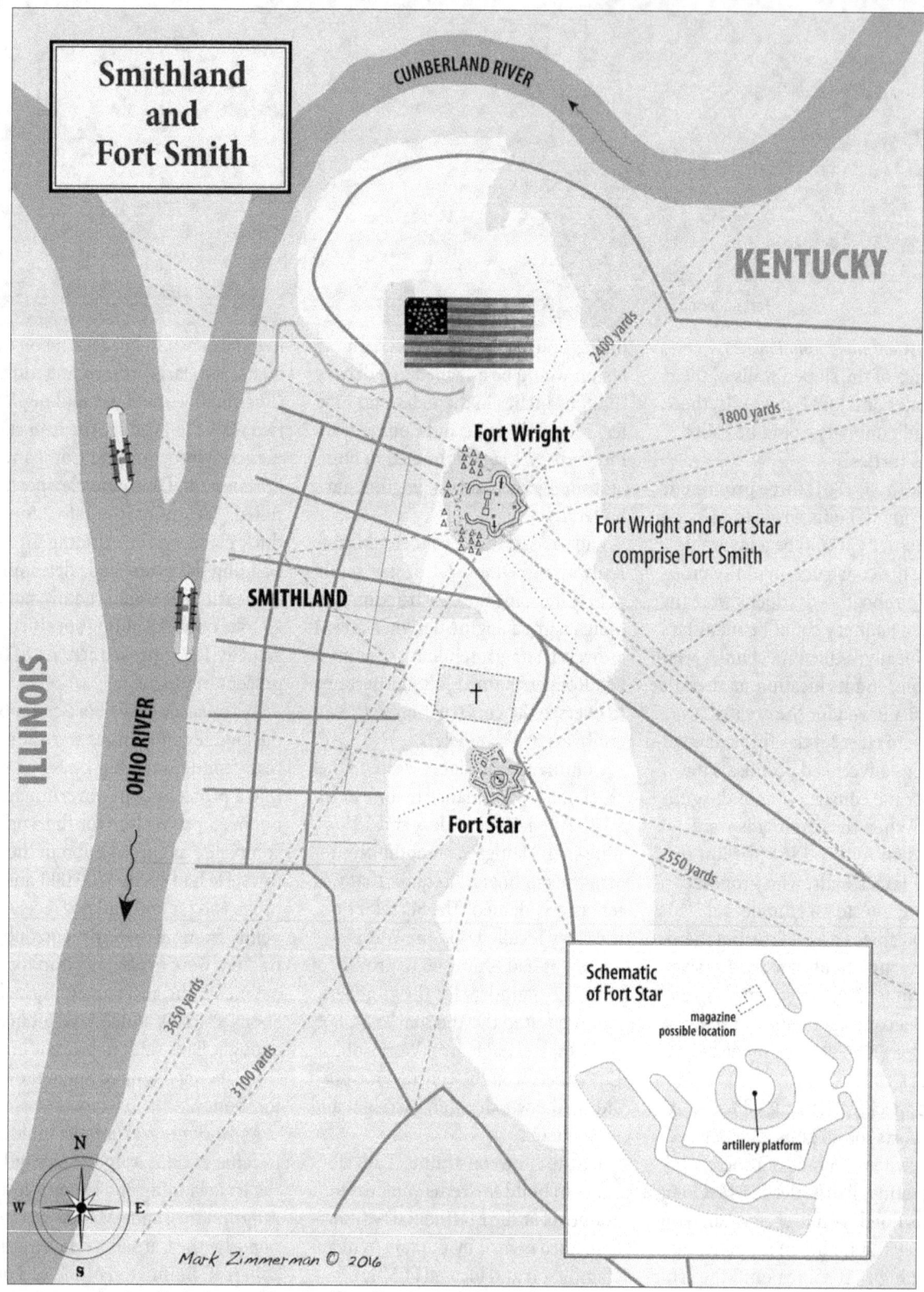

(Author's artwork based on Map of Fort Star prepared by Andie Kellie of Murray State University.)

dislodged a bolt fastener and made a "raking" mark. At 500 yards, the solid shot made a one-inch indentation. Moving ever closer, at 200 yards, the shot made a deep indentation but no sign of cracking or breaking. Finally, at 100 yards (point-blank range) and the iron set up perpendicular, the projectile knocked the target down and splintered it into a thousand pieces. Commander Rodgers concluded, "The iron resisted beyond all expectations, and proved to be of a very superior quality." The performance of the iron in actual combat, they would later learn, would be a different experience.

On Sept. 4, 1861, the same day that Brigadier General U.S. Grant established his command at Cairo, the Confederates made a stupendous strategic blunder. Gen. Leonidas Polk, the "fighting bishop," directed a 1,500-man rebel force to occupy the Kentucky town of Columbus on the Mississippi River. This movement, in effect, violated the state of Kentucky's tenuous neutrality (the state would remain in the Union but had two state capitals during the war). The high cliffs at Columbus were subsequently fortified with so many big guns that it was called the Gibralter of the West. Realizing what was happening, Grant immediately moved to occupy Paducah and Smithland, at the mouths of the Tennessee and Cumberland rivers, respectively. Grant's rapid response would prove propitious.

Work progressed on the ironclads until at last the glorious day arrived, on Saturday afternoon, Oct. 12th, 1861 (two days past the deadline) when the first ironclad gunboat in the Western Hemisphere was launched at Carondelet with so much care and consideration that, according to newspaper reports, "not even a lady was frightened." This first gunboat would be named the *Carondelet*. The second vessel, the *St. Louis*, was launched three days later, and the other two, the *Louisville* and the *Pittsburg*, would slide into the river the third week of October. Thus were born the City Class of ironclad "casemate" gunboats, named after ports on the rivers.

USS *Essex* Ironclad Gunboat

The *St. Louis* would be the first of the gunboats to be fitted out and, by the end of 1861, ready for combat. When the flotilla was turned over to the U.S. Navy in October 1862 the name of the USS *St. Louis* was changed to USS *Baron deKalb* because the Navy already had a ship named *St. Louis*. The ironclad was named for Baron Johann DeKalb, a German officer and major general in George Washington's Revolutionary War army. The other ironclads were not commissioned until January 1862. These included *Mound City*, *Cairo*, and *Cincinnati*, all built at the Mound City works.

On Sept. 6, 1861, command of the Western Flotilla was taken from Rodgers and assigned to Captain Andrew Hull Foote, a naval veteran who was soon elevated to the rank of Flag Officer. It was Foote who rejected the original names for the ironclads (Eads suggested naming the ironclads for generals and prominent officials) and named them for cities along the river. Foote was a competent and able seaman, highly disciplined, religious, and vehement in his disgust for liquor. While other sailors might be issued a well-earned dram of whiskey, Foote would not allow alcohol to be consumed in his command. Hardworking, Foote died during the war, in June of 1863.

One of Foote's most immediate problems was finding able-bodied crews for the boats, a mission that would plague the naval command throughout the war. Duty on the gunboats was generally believed to be too dangerous and not nearly as glamorous as working on a seafaring ship. By the middle of October 1861, only 100 men had volunteered to serve. A month later, the navy department sent 500 seamen from Washington to Cairo. When mustered a week later, it was found that 58 men had simply disappeared enroute. The naval department warned that few men could be spared due to the increase in the seafaring saltwater fleet. Foote claimed he needed an additional 1,100 men. The navy shifted the burden of supplying men to the army. General McClellan ordered General Henry W. Halleck (who had succeeded Fremont as commander in St. Louis) to detail 1,100 soldiers stationed in St. Louis to man the gunboats. The army insisted, however, that the army men, labelled "marines," would answer only to army officers. By January the now-commissioned gunboats averaged only 60 men onboard, one-third of the actual crew needed. Freedmen of color were allowed to serve on the gunboats much sooner than they were allowed to enlist in the U.S. Colored Troops infantry. They drew the same pay as whites, and were employed as gunners, cooks, and coal tenders.

Foote complained about the need for manning the gunboats, noting that he had enough crew for only seven of the existing eleven gunboats. Grant suggested converting men from the brig into sailors, and Foote did not object. Infantrymen who were trouble for their commanders were offered up for gunboat service along with men such as the 4th Illinois Cavalry, mostly Norwegians and Germans who could barely speak English. Also, one man accused of trying to kill his lieutenant who chose gunboat duty instead of a court-martial.

On Jan. 15, 1862, following the required inspections, the seven City Class ironclads were accepted into service by Flag Officer Foote.

Meanwhile, Eads was petitioning the government for the $150,000 in payments owed to him according to the contract. During construction, the government had not kept up with its partial payments. Quartermaster Meigs reminded Eads that the gunboats had not been completed on time; the government was due damages. Eads wrote: "The Govt. failed to pay me according to agreement, and I failed to build the boats in time—Question, am I liable to forfeiture?"

As it would happen, the Union military attacked Fort Henry in February 1862 using gunboats that it legally did not own. Meigs assigned a staff member to investigate all claims and his subsequent report was generally favorable to Eads, who had not been paid promptly and had to deal with changes in specifications and additional workload. The Federal victory at Fort Henry also aided Eads' cause. In mid-February, $234,781 in government bonds were issued to Eads as final payment.

Late in the fall of 1861, itching to take action, Grant and Foote devised a plan to attack the Confederate camp on the Mississippi River at Belmont, Missouri, opposite Columbus. On Nov. 7, 1861, the timberclads *Tyler* and *Lexington* escorted five troop transports to a landing site just upstream from Belmont. While the U.S. troops attacked the Confederate camp, the gunboats engaged in a shooting match with the Columbus batteries, evading hits as moving targets. Grant's troops drove the rebels from the camp but then began looting the campsite. The rebels circled around and, with newly arrived reinforcements from Columbus, nearly surrounded Grant's men. Grant pushed his men through the entanglement and hightailed it back to the transports just in time to avoid a terrible defeat. The men were hastily evacuated under the protection of the timberclads. Grant's expedition wasn't much of a victory, but he did demonstrate the usefulness of joint army-navy operations, the use of transports and gunboat escorts for unopposed amphibious landings of troops, and his willingness to meet the enemy and engage. It was Grant's first taste of combat in the Civil War.

City Class Ironclads

The City Class ironclads were capable of only six knots and consumed one ton of coal per hour. Often, the ironclads had to be towed or lashed to another vessel. Some towed their own coal barges as sources of fuel. Despite the enormous number of bushels of coal that had to be loaded aboard each ironclad, in addition to the bags of gunpowder, artillery shells, rations, and numerous other types of gear, the gunboats did not have a hatch—all supplies had to be loaded through the gunports!

Each ironclad was armed with 13 big guns—four on each side, three in the bow, and two in the stern. Although variable, a typical configuration would be:
- Three 42-pound army rifles
- Three 64-pound navy smoothbores
- Six 32-pound navy smoothbores
- One 32-pound Parrott gun

The crew of a City Class ironclad numbered 175, composed of the commander and 16 officers, 27 petty officers, 111 seamen, 4 or 5 landsmen, 1 or 2 apprentices, 12 firemen, and 4 coal heavers.

One or two civilian pilots were required on each boat to navigate the treacherous rivers. It should be noted that the familiar octagonal ironclad pilothouses were added to the gunboats only after the battle at Fort Donelson.

The ironclads were 175 feet long and 50 feet wide with a draft of only six feet. They displaced 888 tons. The flat bottom had three keels. The estimated weight of the hull and casemate was 350 tons, plus 122 tons of charcoal iron plating (armor) at 2.5 inches thick. The casemate was sloped 35 degrees on the sides and 45 degrees on the ends. The exterior was painted black and the interior white-washed.

The seven City Class ironclads were built to identical specifications; the only noticeable difference was the color of bands painted near the top of the chimneys. The colors were *USS Cairo* (gray), *USS Carondelet* (red), *USS St. Louis* (yellow), *USS Louisville* (green), *USS Mound City* (orange), *USS Cincinnati* (blue), and *USS Pittsburg* (brown).

The first convoy of transports filled with blueclad soldiers which reached Nashville on Feb. 25, 1862 was led by the *USS Cairo*. On Dec. 12th of that year, on the Yazoo River in Mississippi, the *Cairo* became the first boat to be sunk by an electrically detonated underwater torpedo (mine). The sunken boat was discovered in 1956, and the pilothouse, cannon, and other artifacts were recovered in 1960. Much, but not all, of the structure was raised in 1964, and the restored ship was put on permanent display in 1977 at the Vicksburg National Military Park, where it can be seen today.

The *Carondelet* was the first City Class gunboat launched. She was the most celebrated of the ironclads but also known as the slowest. She participated heavily in operations at Bell's Bend near Nashville in December 1864, and helped turn Hood's right flank during the Battle of Nashville.

Built in 1856 in New Albany, Indiana, as the steam ferry *New Era*, the new vessel was converted into the 355-ton timberclad gunboat *New Era* and took part in the Cumberland River expedition in November 1861. Then she was converted into a 1,000-ton ironclad and renamed *USS Essex*. Nearly 200 feet long and 58 feet wide, she drafted 6 feet, 10 inches and made 5.5 knots. She was armed with one 10-inch and three 11-inch Dahlgren smoothbores, one 12-pound howitzer, and one 32-pound rifle. Known as the *S.X.*, she was a hardluck ironclad. The ironclad participated at Fort Henry and sustained heavy damage. She was commanded by Commander William "Dirty Bill" Porter.

Foote's gunboats at the Battle of Fort Henry were all three timberclads, three City Class ironclads — the *USS Cincinnati* (flagboat), *USS Carondelet*, and *USS St. Louis* — and the ironclad gunboat *USS Essex*.

Participating at the Battle of Fort Donelson were two timberclads — *USS Conestoga* and *USS Tyler* — and four City Class ironclads *USS St. Louis* (flagboat), *USS Carondelet*, *USS Pittsburg*, and *USS Louisville*.

One additional ironclad, the powerful *USS Benton*, a converted snag boat, plied the inland rivers, bristling with 16 guns.

Fort Anderson

Following Grant's occupation of the Ohio River ports of Paducah at the foot of the Tennessee River and Smithland at the foot of the Cumberland, forts were built at both locations. In order to protect Paducah, an earthen fortification was constructed by General Charles F. Smith's command in the fall of 1861. Fort Anderson was originally meant to be a quartermaster supply base, but the 400 foot by 160 foot fort was eventually armed with 32-pounder artillery pieces, which faced the Ohio River. An imposing fifty foot, water-filled ditch surrounded the fort. The interior of the fort held the Marine Hospital building, which burned in 1863.

For much of the war, Fort Anderson served as a garrison and communications center. But, on March 24, 1864, Confederate cavalry under the command of General Nathan Bedford Forrest attempted to capture Paducah. Fort Anderson was under the command of Colonel Stephen G. Hicks. Forrest first demanded Fort Anderson's surrender, but Hicks, having the support of gunboats on the Ohio River, refused to surrender. After two unsuccessful assaults, the Confederates occupied homes in Paducah and fired on the fort. The Southerners burned the town's railroad depot, a steamboat, and some military supplies and then withdrew toward Mayfield. The greatest property damage, however, probably occurred on March 25, when Hicks ordered the burning of a number of Paducah homes that were used by the Confederates for cover during the engagement.

One unit that was garrisoning Fort Anderson at the time of Forrest's attack was the 8th United States Colored Heavy Artillery. This unit consisted largely of formerly enslaved men from McCracken and surrounding counties and was one of the first black regiments enlisted in Kentucky.

Fort Smith

Fort Smith, an encampment and fortification complex, was constructed in September 1861 and would serve its purpose until the last regiment there was mustered out of service in November 1865. The star-shaped fortifications at Smithland — one of the finest examples of earthen fortifications in Kentucky — were separated by a half mile of rolling river bluffs. One gun emplacement guarded the confluence of the Ohio and Cumberland Rivers while the other guarded the floodplain that surrounded them.

The northern gun emplacement was protected by both a 32- and 64-pound cannon and housed the Union encampment and rifle pits. The southern gun emplacement was a 32-pounder that covered the southern and western flanks and the roads to Tilene and Gilbertsville. It was located on Cemetery Hill and behind what is now Livingston Central High School.

The fort became an important supply depot and staging area for Grant's upcoming campaigns. It was garrisoned throughout the war by various regiments, including the 13th U.S. Colored Heavy Artillery.

On Sept. 5, 1861, four companies from the Illinois 12th Regiment Infantry under the leadership of Gen. C.F. Smith were the first to arrive at Smithland. Others from their group were stationed at Fort Anderson. On Sept. 8, Companies B and I separated from the 41st Regiment Infantry at Paducah and were stationed at Fort Smith.

On Jan. 31, 1862, Smith received orders from Grant to take all of the command from Smithland, except the 52nd Illinois, to Fort Henry. After the victories at Forts Henry and Donelson, captured Confederates were to be sent north to Chicago. One of the first stops on the route north was Smithland. What Federal General Lew Wallace described as a mostly quiet town was enraged at the sight of the tattered and torn prisoners in gray.

On July 8, 1864, defense of Fort Smith was turned over to the 13th Heavy Artillery Colored Regiment, which had garrisoned with others at Camp Nelson and other points in Kentucky. They remained at Fort Smith until mustered out of service on November 18, 1865.

Once a bustling port, Smithland is reputed to be the oldest incorporated town on the Ohio River.

Target: Nashville

Nashville Gazette
Aug. 2, 1860

About 10 o'clock last night the city was brilliantly illumined with a meteor that passed from southeast to northwest. The streets were so completely lit up from the reflection of the meteor that a pin could have been readily discovered. We were in the house at the time, and thought for the moment that it was the most brilliant lightning we had ever seen.

Cincinnati Gazette
Dec. 20, 1861

A reader who has just traveled from Tennessee: There is no place between Bowling Green and Nashville that admits of defence. At Nashville they are making preparations to resist the anticipated attack, and ... if we wait on them till next year, they will probably be able to make a successful defence ... [but] the progress is very slow. On Capitol Hill a few cannon have been mounted, but now there are no defences that would more than momentarily delay our army.

The New York Times
Dec. 21, 1861

The only fortification on the Tennessee River, of much importance, is Fort Henry...The armament of the fort consists of eight 32-pounders, four 12-pounders, and two 6-pounders... At Dover, about a day's march from Fort Henry (eastward), is the principal fortification on the Cumberland, below Clarkesville. It mounts twelve 32-pounders. Some 3,000 troops are reported to be at this point, with some field artillery...Steps are also being taken for the erection of fortifications near Nashville; but not much has yet been done.

Tennessee State Capitol
Nov. 10, 1861

Governor Harris asks the people of Tennessee to donate "every double-barrel shotgun and rifle they have, to arm the troops now offering their services."

Tennessee and Cumberland Rivers
Building the River Forts

At the time of the river invasion, the Confederacy boasted six fortifications on the rivers downstream from Nashville, most unfinished. One was captured easily by the Federal navy, one was victorious over the Lincoln gunboats, and the four others were abandoned and/or destroyed. Why the abysmal record? Why did it take so long to build the forts and why were they basically useless?

Much of the reason derives from the status of Kentucky, a slave-holding border state of divided loyalties (as was Tennessee) that wished to remain neutral. As it turned out, neutrality was impossible and Kentucky stayed in the Union, but most of the political and military officials in Tennessee were banking on the state as an invasion buffer.

Also, it was a matter of priorities. The Confederacy was concentrating on the territory between Washington, D.C. and Richmond. And most of the priority in the Western Theater was occupying and fortifying the Mississippi River Valley. And besides, most thought the war would be over in 90 days.

On May 7, 1861, Tennessee joined the Confederacy, after Governor Isham Harris rejected Lincoln's call for volunteers to put down the rebellion. Nashville was a prosperous riverport, second in the South only to New Orleans in its vitality. But the aristocracy in Nashville, the "Nashville Gods" who lived at Belle Meade, Belmont, Two Rivers, Clover Bottom, and Burlington, were not interested in building defenses for their city, according to historian Thomas Connelly. Lethargy was due to the "self assured, snobbish attitude of the influential Nashvillians," most of whom would not lend out their slave labor to build fortifications. Harris was more interested in defending the Mississippi River, according to Connelly. "The governor's lack of interest in providing defenses for the inland rivers and Nashville was probably his most serious blunder as commander of the state army. He believed in the neutrality of Kentucky." Also in this frame of mind was the second-in-command of the Tennessee state army, Gideon Pillow, a vain, ambitious politician-general.

In May, Harris ordered Adna Anderson, chief engineer of the Edgefield & Kentucky Railroad, to select sites on the Tennessee and Cumberland Rivers for fortifications to be built. He was assisted by Major Wilbur Fisk Foster of the 10th Tennessee, who would later become chief engineer in A.P. Stewart's corps. At the time, Kentucky was neutral so no site could be chosen inside that state. (The best site for the forts would have been Birmingham, Ky., where the rivers were three miles apart, 30 miles north of Dover.)

Anderson chose a high bluff a mile downstream from the village of Dover for the Cumberland River site. On the Tennessee River, he selected a site slightly below Standing Rock Creek directly across from Big Sandy Creek. Both forts would lie between the rivers so as to support each other.

The governor wanted the opinion of a military man so he sent General Bushrod Johnson, newly appointed chief engineer of the Army of Tennessee, to examine the sites. Johnson affirmed the Dover site but disliked

Anderson's Tennessee River site. He choose Old Kirkman's Landing instead, 12 miles due west of the Cumberland River. Many historic accounts have Brigadier General Daniel S. Donelson, the newly appointed adjutant general, making the selection of the Fort Henry site. According to researchers M. Todd Cathey and Ricky Robnett, however, the only primary source for Donelson's involvement in the selection of the Fort Henry site was by A.S. Johnston's son in his father's biography. Cathey and Robnett discount Donelson's involvement entirely. Donelson did use the Tennessee River site to recruit men into the new Tennessee army; that site became known as Fort Donelson.

Foster thought Kirkman's Landing was a bad choice and protested but to no avail. Ordered to do so, he surveyed the site and work began on June 14th by the 10th Tennessee under Colonel Adolphus Heiman. A native of Prussia, Heiman came to America, settled in Nashville, and earned the reputation as a skilled architect and builder. He was a Mexican War hero. He would be captured at Fort Donelson and died in 1862 as a prisoner of war.

Fort Henry was named for state senator and Clarksville resident Gustavus Henry. The first gun (32-pounder smoothbore) was mounted and fired a blank cartridge on July 12th. Foster then left the forts and rejoined his regiment in West Virginia. In August, Heiman would be put in command of the garrisons of both forts.

On July 10th, the Rock City Guards, formed two years earlier in Nashville, and the First Tennessee shipped out to Virginia and fought at the Battle of Bull Run.

Work on Fort Donelson was sporadic at best. In June, about 40 untrained, unarmed men worked on the fort but then abandoned the site until October. Local slaveowners refused to lend out their hands, and most white laborers were working at Columbus.

In July, the department was put in command of Major General Leonidas Polk, a West Pointer, Episcopal bishop, and personal friend of Confederate President Jefferson Davis. Fixated only on defense of the Mississippi River, Polk ordered the seizure of Columbus, Ky. for the site of a massive fortification. Kentucky's neutrality was officially over. Three days later, a little-known brigadier named U.S. Grant had the wherewithal to seize the Ohio River ports of Paducah and Smithland, at the mouths of the Tennessee and Cumberland Rivers respectively. Federal command, encampments, and marine ways were established at Cairo and Mound City, Illinois.

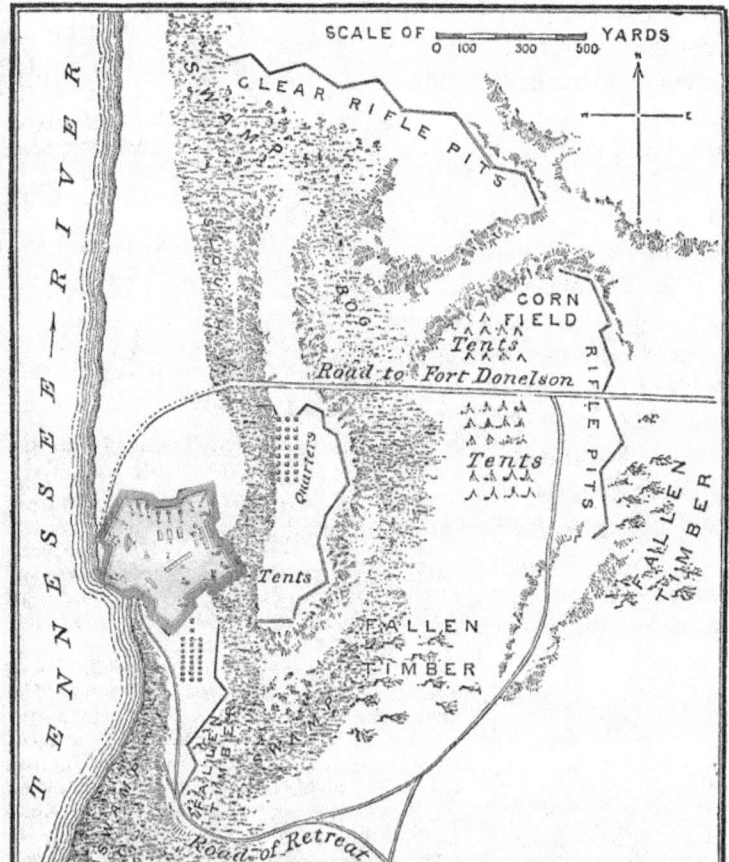

Jeremy Gilmer   Wilbur F. Foster   Adna Anderson

On September 15th, General Albert Sidney Johnston arrived to take command of Confederate forces in the Western Theater. Three days later, rebel forces occupied Bowling Green, Ky. Charged with defending a thousand-mile front, Johnston spread his force of 50,000 men too thin. Bushrod Johnson reported to Johnston that Fort Henry was bastioned, armed with eight artillery pieces, and required 1,000 men to garrison.

About this time, Lt. Col. Milton Haynes sent Capt. Jesse Taylor of the 1st Tennessee Artillery to Fort Henry to train the gunners there. Taylor noted that "extraordinarily bad judgement, or worse" had been used to site the fort, which stood in a floodplain flanked by hills to the east and also across the river. Talking to local residents and noting high-water marks on trees, Taylor discovered that normal

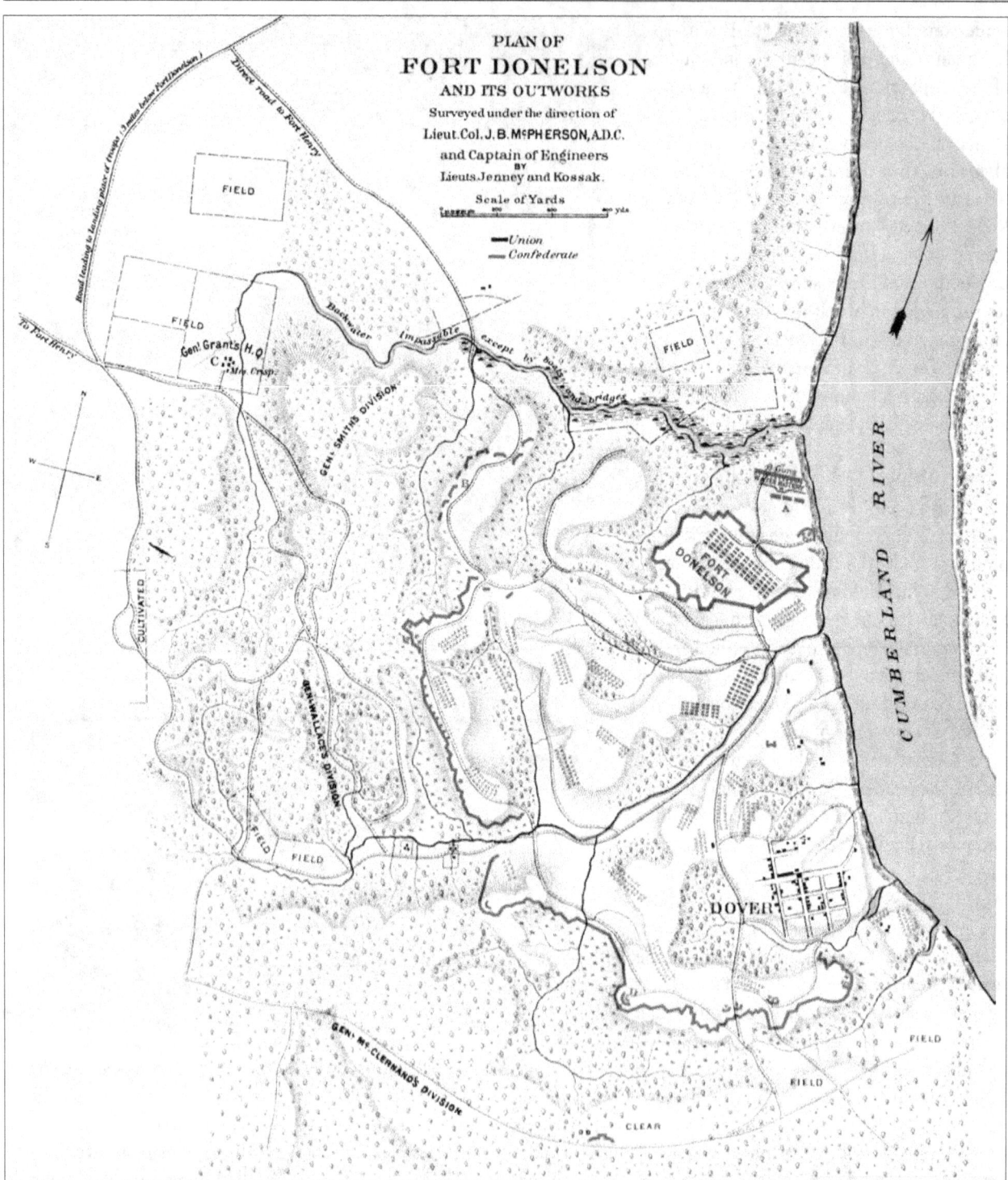

February river levels would leave the fort two feet underwater. He reported his findings to Haynes and Harris, who asserted that the site had been recommended by competent engineers. Harris referred him to Polk, who passed him on to Johnston, who sent Major Jeremy Gilmer, his chief of engineers, to investigate Taylor's concerns. Gilmer was also placed in charge of building all the river defenses, including sites at Clarksville and Nashville.

Gilmer was more than qualified for the task. He graduated 4th out of 31 from West Point in 1839 and was commissioned 2nd lieutenant of engineers. He served as assistant professor of engineering at West Point (1839-40), assistant engineer building Fort Schuyler at New York harbor (1840-44), assistant engineer at Washington, D.C. (1844-46), and chief engineer of the Army of the West during the Mexican War (1846-48). He supervised the construction of Fort Jackson and Fort Pulaski in Savannah, Ga. In October 1861, Gilmer was commissioned major of engineers and posted to the Army of Tennessee as chief engineer on the staff of General Albert Sidney Johnston.

The problem was Gilmer's attitude.

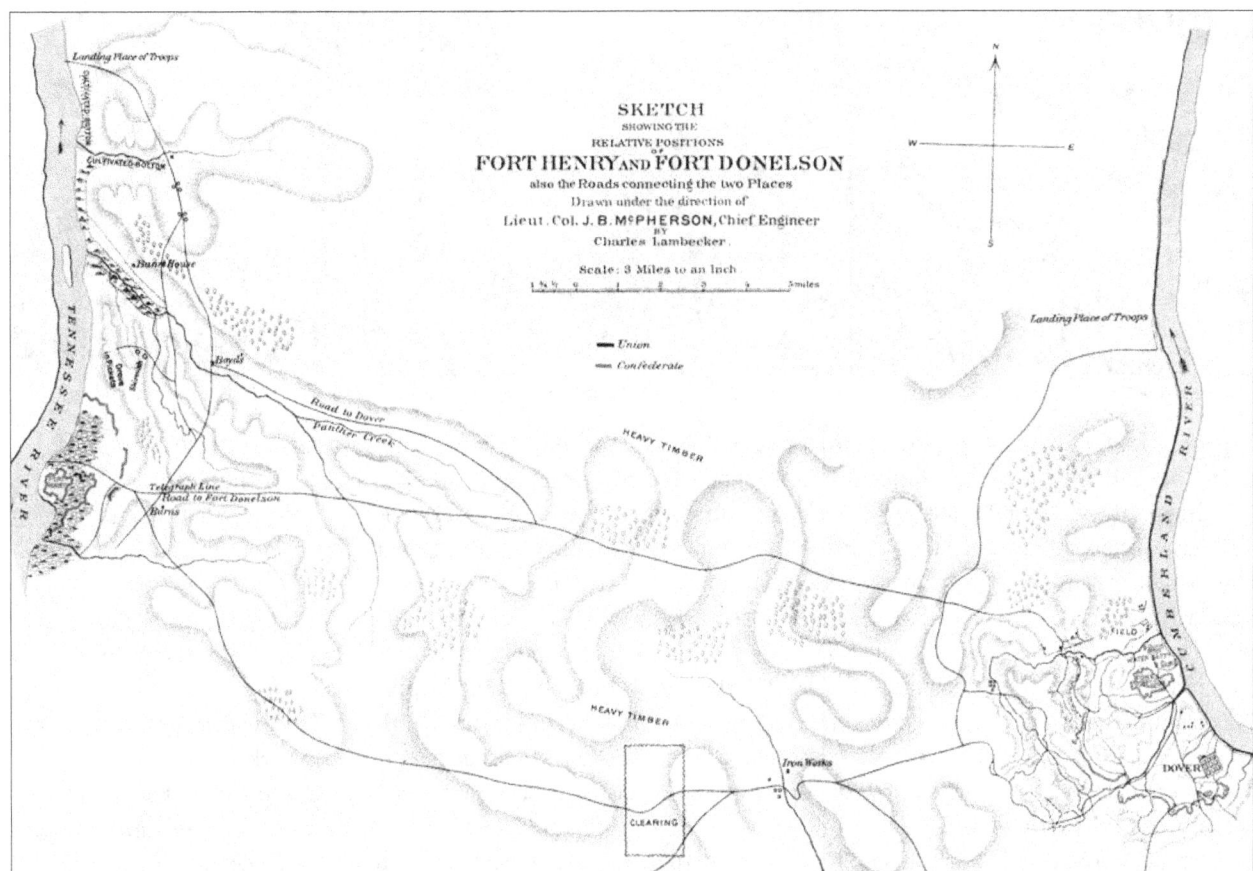

He really wanted to be in Savannah with his family and friends and the warm weather. He thought he deserved higher rank. He failed to communicate with Johnston, who was led to believe that defensive fortifications were more than adequate. Also, Gilmer believed that the Yankees would not advance until spring 1862 and by that time Johnston would have advanced upon Louisville. "Gilmer simply began his assignment with an indifferent, unpleasant attitude..." noted historian Connelly.

On October 7th, 1st Lt. W. Ormsby Watts of Taylor's Battery transferred from Fort Henry to serve as Ordnance Officer at Fort Donelson. There he found two 32-pounder howitzers, four bronze 6-pounder guns, and two 9-pounder guns.

The next day, Johnston transferred Lt. Joseph Dixon from Columbus to Fort Donelson to supervise the placement of heavy guns. He reported to Lt. Col. Randall W. McGavock of the 10th Tennessee, who commanded the

Engineer Officer's Button (CSA)

garrison, which would later become the 50th Tennessee.

The Confederates were not alone. As early as October 12th, Federal gunboats were reconnoitering up the Tennessee River as far as Fort Henry.

Gilmer spent November surveying the terrain between Nashville and Clarksville. He proposed earthworks north of Edgefield and a river fort just downstream from Nashville. This would become Fort Zollicoffer, named for the Nashville Whig politician and newspaper editor who was killed at the Battle of Fishing Creek as leader of Confederate forces. A submarine battery was planned, even a chain across the river.

Gilmer sited a fort in Clarksville at the confluence of the Red River and ordered local railroad engineer Edward Sayers to design the fort and supervise its construction. While Gilmer worked at Clarksville, Dixon was left to supervise work at Fort Donelson, but he lacked the manpower and equipment he needed. "There are not more than 200 troops here fit for duty; all the rest are sick or have leave of absence," he reported to Gilmer.

By October, Heiman commanded 870 men at Fort Henry and McGavock had 300 men at Fort Donelson, although they were untrained and unarmed.

On October 18th, Col. Heiman reported to Gen. Polk that Fort Henry was armed with six 32-pounders, two 12-pounders, and one 6-pounder and manned by the 10th Tennessee (850 men) and fifty officers and men of Taylor's Tennessee battery. He said more gunners were needed; also four more 12-pounders to prevent a land attack. Heiman noted the fort's future

Bushrod Johnson   Adolphus Heiman   Daniel Donelson   Gustavus Henry   Randal McGavock   Albert Sidney Johnston

depended on the skill of its gunners. Extra powder charges would be required due to the inferiority of the gunpowder.

Heiman said he did not think Fort Donelson could repel a Federal gunboat attack, that the defense of the Cumberland had been "almost entirely overlooked." Fort Donelson had two 32-pounder carronades in the upper battery which were basically useless and was awaiting four 32-pounder smoothbores railed from Memphis to Clarksville. He said Fort Donelson was garrisoned by three newly organized companies of Col. McGavock armed with shotguns. He said he detached Lt. W.O. Watts of Taylor's battery to train the gunners at Fort Donelson.

Attempts were made to deny the gunboats access to the river. From Oct. 26 to Nov. 1st, Lt. Dixon and Capt. Frank Maney's four-gun battery plus two companies under Capt. Young and D.C. Kelley's troopers at Cadiz sank eight barges full of stone to obstruct the Cumberland River (two at Line Island and six at Ingram Shoals). Dixon reported that it would be impossible for gunboats to pass Ingram Shoals, an assessment echoed by Gilmer. A few days later, Federal gunboats did, in fact, navigate the shoals.

On November 5th, Johnston ordered Lt. Hugh Lawson Bedford from Bowling Green to train artillerymen at Fort Donelson. Bedford reported "very little in the way of defense had been accomplished," with only five guns in position.

On November 8th, Johnston reported to Richmond that the river obstructions were viable, that Fort Donelson had sufficient men and guns, and that Fort Henry was "a strong work."

Two weeks later, Brig. Gen. Lloyd Tilghman, a native of Paducah, was ordered to take command of Fort Henry and Fort Donelson.

Labor was still a major problem. By late December, planters still refused to lend their slaves, white volunteers were few, and Gilmer's plans remained on the sketch board," noted Connelly. Many of the men under MacGavock were disabled or killed by the measles.

Johnston sent 1st Lt. Jacob Culbertson, chief of artillery at Bowling Green, to Fort Donelson to assist with the heavy guns. On Dec. 11th, the Ordnance Department in Richmond informed Johnston that two 10-inch columbiad smoothbores, one 6.5-inch rifled columbiad, and two more 32-pounders would be delivered to the fort via Nashville (only one of the big smoothbores actually arrived).

On Christmas Day, Gen. Johnston, who never personally inspected the river forts, assured Gov. Harris that Nashville could be properly defended.

At the beginning of 1862, Tilghman reported that only one-third of the planned earthworks were under construction and none completed, the guns at Fort Henry remained unmounted, there were no trained men in the river batteries, and half of his 4,600 men were unarmed. Work on a small fort across the river from Fort Henry on higher ground—Fort Heiman—was progressing slowly. Gilmer clashed with Tilghman, who complained to headquarters, but Gilmer had more pull with Johnston.

Gilmer planned to place even more obstructions in the Cumberland River at Ingram Shoals downstream (north) of Fort Donelson and put civil engineer S.P. Glenn in charge. Tilghman ordered Glenn to cease work immediately. Glenn complained to Gilmer, who ordered him to continue and complained to Johnston that Tilghman was interfering. Tilghman tried to justify the move, but was informed by Johnston that he was not to interfere with Gilmer.

On Jan. 15th, Lt. Col. Milton Haynes of the Tennessee Artillery Corps was ordered from the Columbus fortifications to Fort Henry to serve under Tilghman. He organized Maney's Battery and the Beaumont and Bidwell infantry companies into a provisional artillery battalion. Lts. George Martin and J.J. McDaniel were also sent as artillery instructors.

On February 6th (the same day as the battle at Fort Henry), Pillow reported to Johnston that none of the Clarksville defenses were complete. That same day, the big columbiad at Fort Donelson was test-fired for the first time. Newly arrived men worked feverishly to finish the parapets and magazines at Fort Donelson and trained on the guns that would face the Federal gunboat flotilla. Time was running out for the raw-recruit gunners.

First Significant US Victory of the War
Capture of Fort Henry

On Jan. 27th, 1862, President Lincoln issued General War Order No. 1, which required all of his generals to show movement against the enemy by February 22nd. General U.S. Grant had already been pressing his superior, General Henry "Old Brains" Halleck, to allow him to move up the twin rivers. Flag Officer Andrew H. Foote assured Halleck that his ironclad gunboats could defeat any Confederate river fort. On January 30th, Halleck wired Grant: "Make your preparations to take and hold Fort Henry."

On the last day of January 1862, the timberclads *Conestoga* and *Lexington* moved up the Tennessee River ever closer to Fort Henry, trying to get one last look before the upcoming attack. Accompanying Lt. Commander Seth Ledyard Phelps was landlubber Brigadier General Lew Wallace, the future author of the bestselling novel *Ben Hur*, who took notes of his trip. The boats had anchored for the night in mid-channel, planning to conduct their reconnaissance first thing in the morning. Then the baying of hounds was heard on the shore, getting louder and ever closer. Emerging from a riverside cornfield was their target, an escaped slave. The black man was running toward the boats. A landing party was sent. The sailors used their paddles to drive away the hounds and rescued the fugitive slave, a contraband. He was taken back to the gunboat.

At the crack of dawn, the gunboats paddled slowly upstream, with two men in the bow on either side, staring intently into the waters ahead. They were searching for torpedoes, as submerged naval mines were called in those days. At least a dozen torpe-

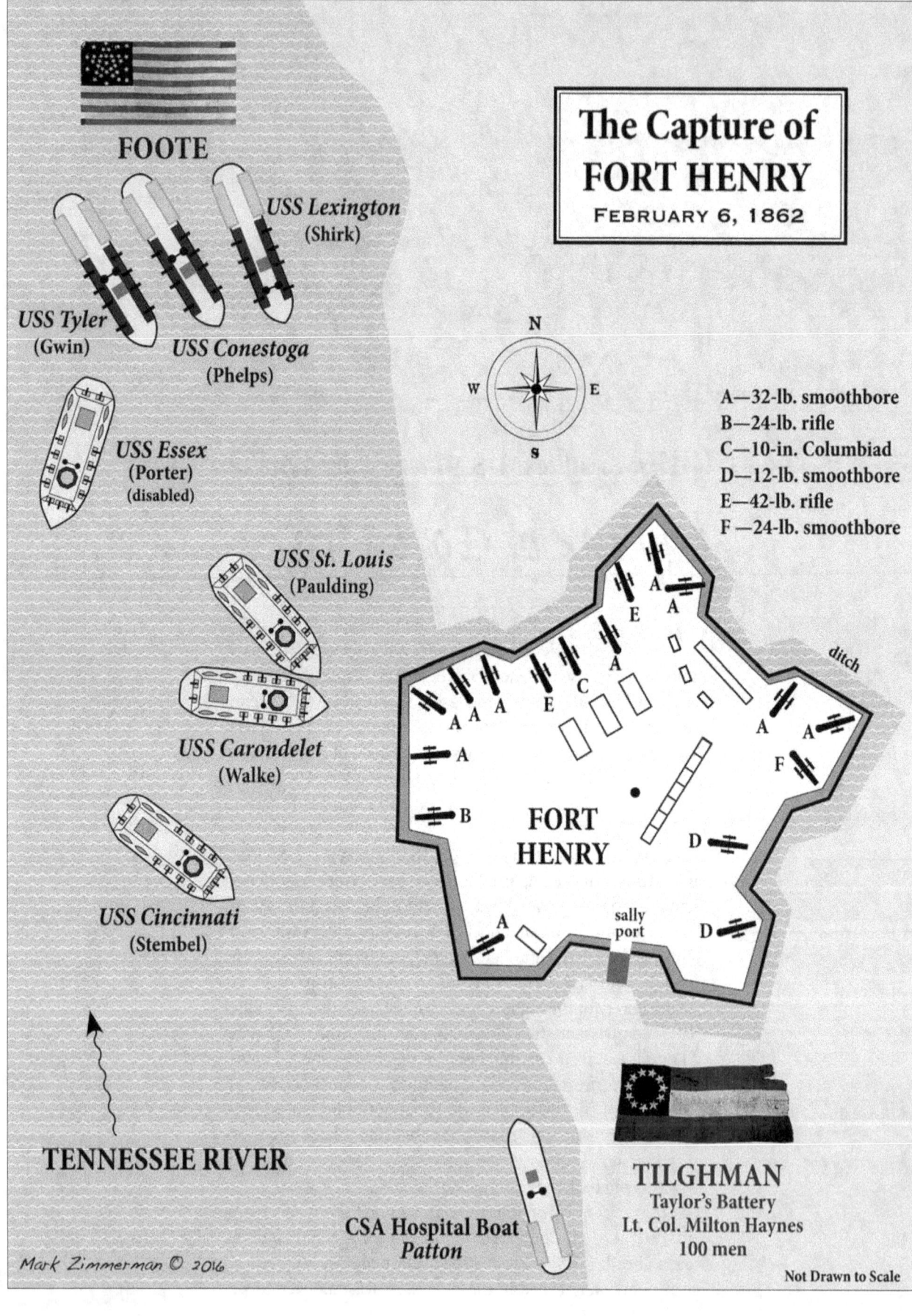

The 24-pound rifle bursts at the Fort Henry parapet, killing and injuring several Confederate gunners.

does had been anchored in the western chute of the river around Panther Island, which was located about two miles downstream from Fort Henry. The torpedoes were sheet-metal cylinders about six feet long, a foot in diameter, filled with 70 pounds of black powder, and sporting a protruding prong, which would trigger an explosion if tripped by the bottom of a passing boat. It was flood season and the river level was rising, the currents strong. As it turned out, due to the floodwaters, most of the torpedoes had been swept away and sank or rendered inert due to damage or seepage.

The flagboat *Conestoga* cleared Panther Island and boldly moved into the main channel, in full view of the fort's occupants, who were ready and waiting. What Phelps and Wallace saw was a bastioned fort "built squat on low ground" and covering ten acres, with one of its bastions extending out into the river and holding three heavy guns. They counted 17 guns in all, 11 trained on the river. A water-filled ditch circled the fort, as did rifle pits. Behind the fort on slightly higher ground were trees bearing watermarks, indicating that at flood stage the fortification would be underwater. On the opposite side of the river, further upstream and in Kentucky, was the unfinished fortification known as Fort Heiman, much higher in elevation than Fort Henry. Engineers had been ordered to build Fort Heiman back in November, but it remained unfinished.

Fort Henry had been poorly sited, as noted by several engineers, but it was built at Kirkman's Landing nonetheless. In June 1861, at the time of the fort's construction, Kentucky was still neutral, so the fort had to be situated on Tennessee soil. The water level was low during the summer of 1861 and nearly everybody assumed the war would be a short and decisive one. One factor in Fort Henry's favor was that it was only 12 miles west of Fort Donelson on the Cumberland River. The forts could reinforce each other, as it was correctly assumed that both forts would not be attacked at the same time.

Fort Henry was commanded by Brigadier General Lloyd Tilghman, a native of Paducah. The fort bristled with 17 heavy guns, ten of them 32-pounders, with eleven trained on the river, including a 10-inch columbiad, a 42-pounder rifle, and two 42-pounder smoothbores (these last two were inoperable due to lack of suitable ammunition). During training, the Confederate gunners were having trouble with the recoil of the big columbiad, as the length of the chassis was too short and the massive tube threatened to dislodge itself.

Where did the Confederates obtain such large siege artillery pieces? The columbiads were cast at the famous Tredegar ironworks in Richmond, Va., but the other guns came from the 1,200 heavy artillery pieces confiscated by Virginia state troops when they captured the Federal naval yard at Gosport/Norfolk in April 1861. The

An Unfinished Fort. Fort Heiman. Artist Andy Thomas. Used with permission.

The View From Fort Heiman. Artist Andy Thomas. Used with permission.

Battle of Fort Henry. Artist Andy Thomas. Used with permission.

Confederates certainly would have been ill-equipped without the bounty seized from that early raid.

Fort Henry was garrisoned by 1,885 men and Fort Heiman by 1,100 men, although sickness had reduced the effective total strength to 2,600. Many were not properly trained or equipped, the exceptions being the 10th Tennessee and the 4th Mississippi regiments.

In a classic case of too little too late, Fort Heiman had been built on the high west banks of the river, overlooking Fort Henry. This bastion was named for its commander, Colonel Adolphus Heiman, a Prussian architect who had emigrated to America in 1834 and eventually settled in Nashville. A hero of the Mexican War, Heiman was the colonel of the 10th Tennessee Infantry Regiment. "Construction began in December 1861 with the arrival of the Twenty-seventh Alabama and Fifteenth Arkansas infantry regiments who, along with some 500 slaves, were tasked with building the works," stated National Park Service (NPS) Historian Timothy Parsons. "Its suitable defensive position — protected by 150-foot bluffs in front and impassible roads and rough terrain in the rear — stood in marked contrast to the poor placement of Fort Henry. On the morning of February 3rd, General Tilghman made an inspection of the incomplete works at Fort Heiman."

On February 3rd, a convoy of steam transports carrying Federal troops left the Ohio River port of Paducah and made its way up the Tennessee River. The transports were escorted by the ironclad gunboats *Essex* and *St. Louis*. A few hours later, the remainder of the gunboats, the City Class ironclads *Cincinnati* and *Carondelet,* and the three timberclads departed for Fort Henry, all under the command of Flag Officer Foote. The army troops, under the overall command of Grant, disembarked on the east bank of the river several miles below the fort, planning to advance upon the Confederates in a joint attack with the naval forces. Another small Federal force put ashore on the west bank and advanced upon Fort Heiman. The weather had been rainy and the river was swollen, making advancement by the infantry difficult due to the muddy conditions.

Reports were reaching Confederate fort commander Tilghman that transports were landing thousands of U.S. soldiers below the fort, and the gunboats were shelling the riverbanks as Confederate cavalry patrols skirmished with the advancing forces. On February 4th at noon, three gunboats moved to within two miles of the fort and began shelling it. The boats were out of range of the 32-pounders so the rebel gunners discharged the columbiad and 24-pounder rifle. Several shots from the columbiad threatened to dismount it, so it was abandoned. Soon the range had closed to a point where the 32-pounders were ordered to open fire. After half an hour, with none of their shells landing in the fort, the gunboats withdrew.

The next day the Confederate garrison at Fort Heiman was ferried across the river to bolster Fort Henry. Tilghman's chief of artillery, Lt. Col. Milton Haynes, thought Fort Henry also should be abandoned due to the rising level of the river and his estimation that the Confederate forces were vastly outnumbered. They agreed that the fight at Fort Henry would be a holding action. Tilghman ordered Heiman to remove all the remaining troops from Fort Henry except for Taylor's Battery, which would remain to man the guns.

On the morning of Thurs., Feb. 6th, 1862, the Federal gunboat flotilla pulled away from Bailey's Landing at 10:50 am and headed for the fort. The four ironclads lined up abreast with their bows to the enemy (the *St. Louis* positioned closest to the east bank, then the *Carondelet, Cincinnati,* and *Essex*). The vulnerable timberclads (*Conestoga, Lexington,* and *Tyler*) would stay behind the ironclads and provide artillery support. When the range had closed to 1,700 yards, Flag Officer Foote ordered the captain of the *Cincinnati*, Commander Roger Stembel, to open fire. This was the signal for all of the gunboats to commence firing. At 12:34 pm, the bottle-shaped 8-inch Dahlgren in the bow of the *Cincinnati* roared to life. The commander of the *Essex,* William "Dirty Bill" Porter, cautioned his gunners to watch the flight of the shells from the *Cincinnati* before firing. As the first shells fell short, the gunners raised the elevation of the guns. The No. 2 port bow gun of the *Essex* spoke with authority and landed the first hit on the fort's earthworks, exploding "handsomely" with a scattering of dirt and evoking a cheer from the sailors. The percussive pounding of the big guns, accompanied by the flames and bursts of gray smoke, lighter in color than the black sooty discharges of the chimneys, echoed back and forth from the riverbanks. It seemed as if all hell had broke loose. "The gunboat-men enjoyed the terror they inspired," Wallace duly noted.

Despite all this, the Confederate

The capture of Fort Henry as depicted in *Harper's Weekly*, sketches by Alexander Simplot.

gunners held their fire. Then the columbiad opened the ball, followed by the 24-pounder, and finally the 32-pounder smoothbores. All this time Tilghman was receiving reports that troops were moving upon Fort Heiman and the outer rifle pits of Fort Henry itself, which was now about one-third underwater.

As the officers aboard the *Essex*, including Commander Porter, prepared to go below, a solid shot hit the pilothouse, sending deadly wooden splinters through the air. The pilot was killed instantly while his assistant fell to the deck below and died within minutes. On the gundeck, Porter ordered the changing of gun crews, one in relief of the other, and at that time another solid shot hit a porthole and tore its way through the iron plating and oak backing and struck the middle boiler, causing a terrific blast of scalding steam to escape. The resulting scene was "almost indescribable," stated Second Master James Laning. One of the sailors was kneeling while handing a new shell to the gunner and caught the scalding steam right in the face. The steam blast forced many gunners to leap out the gunports and into the river, some managing to cling to the casemate. Others weren't so lucky. The acting master's mate was killed instantly and Porter was scalded badly. He tried to jump into the river, but a crewman caught him and carried him astern. The first master took over the commander's duties as the *Essex* drifted downriver, away from the fort, "a number of her officers and crew dead at their posts, whilst many others were writhing in their last agony." Porter managed to stay conscious until a crewman told him that the fort had surrendered. He called for three cheers but by the second cheer he collapsed, exhausted. He would survive his injuries and eventually report back for duty.

Ten crewmen on the *Essex* died, 23 were wounded, and five went missing. The gunboat had taken 15 hits while discharging 72 shells from her three 9-inch Dahlgrens.

All in all, the Confederate gunners at Fort Henry hit the gunboats 59 times, with more than half aimed at the *Cincinnati*, in an attempt to disable Foote's flagboat. The gunners also remarked that the *Cincinnati* seemed to be dispensing more accurate fire than the other gunboats. The rebel gunners managed to shred the spardeck, chimneys, after-cabin, and small boats, while a direct strike on the pilothouse dented it but bounced away. On the gundeck one round tore

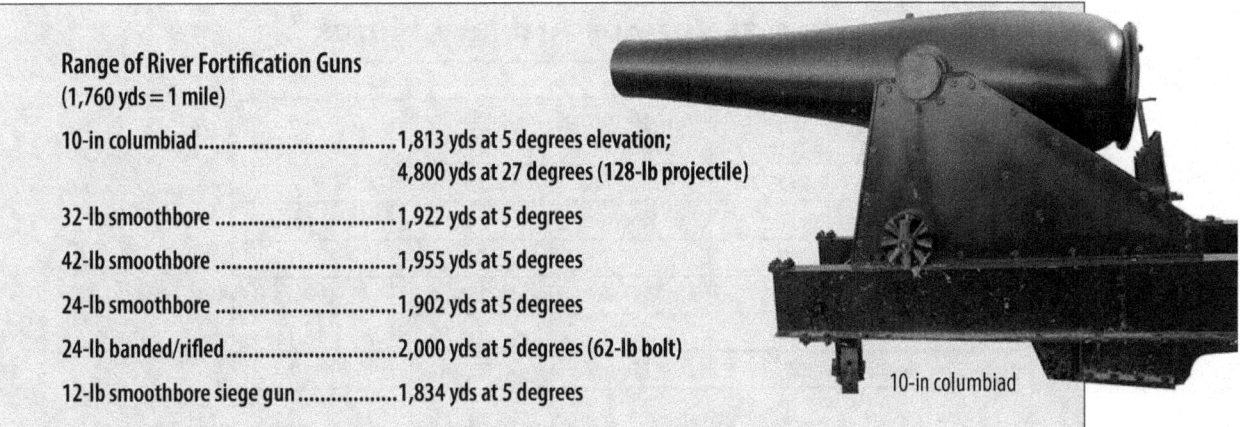

Range of River Fortification Guns
(1,760 yds = 1 mile)

10-in columbiad	1,813 yds at 5 degrees elevation; 4,800 yds at 27 degrees (128-lb projectile)
32-lb smoothbore	1,922 yds at 5 degrees
42-lb smoothbore	1,955 yds at 5 degrees
24-lb smoothbore	1,902 yds at 5 degrees
24-lb banded/rifled	2,000 yds at 5 degrees (62-lb bolt)
12-lb smoothbore siege gun	1,834 yds at 5 degrees

10-in columbiad

through the seam between the front and port side of the casemate, decapitating a gunner. Another shell crashed through and splintered the woodwork and hit the paddlewheel. One 8-inch Dahlgren was hit and one 32-pounder smoothbore disabled. Nine casualties were suffered in all. The *Cincinnati* fired 112 rounds at the fort.

Elsewhere, the *Carondelet* and the *St. Louis* collided, interlocked, and remained mated for a large portion of the battle. The *Carondelet,* commanded by Henry Walke, fired 107 projectiles and was struck nine or ten times. One shot from the *Carondelet* went through the upper deck of the *Patton,* a Confederate hospital boat trying to escape upriver. The Federal gunners realized their mistake when they finally were able to see the boat's yellow flag. Nobody on the hospital ship was injured.

The *St. Louis* fired 116 times, the most of any gunboat, and was hit seven times but suffered little damage and no casualties.

Inside the fort, Captain Jesse Taylor was directing the gunners, telling each crew to concentrate on one gunboat. Then he took charge of the big 24-pounder rifle, while Capt. Charles Hayden manned the 10-inch columbiad. Shortly after the *Essex* was disabled, the 24-pounder burst, killing a sergeant and disabling the crew. Then, the priming wire got stuck in the breech of the columbiad and could not be cleared, rendering the big gun unusable. As the gunboats plowed to within a thousand yards of the fort, two of the 32-pounder guns in the fort were struck and put out of action. A premature discharge at another 32-pounder killed two of its gunners. By the time the gunboats were within 600 yards, the fort had only seven functioning guns. A few minutes later, only four guns remained serviceable. The cannoneers were exhausted from an hour's worth of tremendous effort loading and firing the heavy guns.

Colonel Heiman, who had led the majority of the fort's defenders on a 20-mile march to Fort Donelson, returned to Fort Henry and asked Tilghman how to proceed. Heiman agreed with Col. Gilmer that the time had come to surrender, but Tilghman refused. The general threw off his coat and manned one of the 32-pounders himself. Heiman was ordered to fetch 50 men back to the fort to man the guns. By this time, the three ironclads had closed to within 600 yards of the fort, which was now being ripped apart. Finally, Tilghman agreed to surrender but not right away. He tied a white flag to a staff and waved it atop a parapet but the smoke was too dense for the flag to be seen. After another conference with his officers, Tilghman ordered the colors struck. While Taylor struck the colors at the fort, Haynes, who had not been consulted, ordered the Confederate flag raised again and directed that any man who tried to interfere be shot. Haynes sought out Tilghman and became indignant when told of the pending surrender. Haynes wanted no part of it and left the fort, headed to Fort Donelson.

The battle lasted 75 minutes. Two Confederate officers rowed a small boat to the *Cincinnati*. Stembel and Phelps were ordered by Foote to take possession of the fort. After raising the U.S. flag over the half-submerged fortification, Stembel accompanied Tilghman back to the *Cincinnati*. The Federal sailors went crazy with jubilant emotion, so much so that Foote "had to run among the men and knock them on the head to restore order."

Fort Henry's garrison suffered 99 casualties (five dead, 11 wounded, five missing, and 78 captured) during the holding action. Seventeen pieces of heavy ordnance were captured or destroyed. The hospital boat *Patton* was captured. General Grant reached the fort about an hour after its capture. Entering the fort, Union officers described what they saw. "The effect of the fire on the fortifications here was terrible — guns dismounted — earthworks torn up and the evidences of carnage meet the eye on every hand." Another remarked that "the devastation astonished me all the more when I recalled the short time in which it had been accomplished." One sailor noted, "On every side lay the lifeless bodies of the victims, in reckless confusion, intermingled with shattered implements of war."

Flag Officer Foote worried about

Attributes of Various Artillery Pieces

	Smoothbore		Rifled	
	12-pounder Howitzer	12-pounder Napoleon	10-pounder Parrott	3-inch Ordnance
Bore Diameter	5.62"	4.62"	2.9"	3"
Tube Material	Bronze	Bronze	Iron	Wrought Iron
Tube Length	53"	66"	78"	73"
Tube Weight	788 lbs	1,227 lbs	890 lbs	816 lbs
Powder Charge	1 lb	2.5 lbs	1 lb	1 lb
Range (5 degrees)	1,077 yds	1,680 lbs	1,950 lbs	1,950 lbs

the damage to his fleet. He vowed "never again will I go into a fight half-prepared." He also thought that "we have made the narrowest escape possible with our boats and our lives." The gunboats were ordered back to Paducah and Cairo for repairs.

The fall of Fort Henry to the Federal gunboats made a heavy impression upon Confederate theater commander Albert Sidney Johnston, who noted that the battle "indicates that the best open earth works are not reliable to meet successfully a vigorous attack of iron clad gun boats." Johnston began to intimate that he might pull his entire army out of Kentucky and southward into Tennessee.

On Fort Henry, the Union's first significant victory of the war, historian Kendall D. Gott wrote: "One of the poorest lessons the Confederate troops learned from their officers was that surrender or fleeing the battlefield was somehow tolerable. Colonel Heiman, Lt. Col. Haynes, and Major Gilmer fled from the scene even though they were inside the fort at the time of surrender...the precedent had been set that a fortified position was not necessarily defended to the bitter end."

Grant telegraphed headquarters, "Fort Henry is ours." Halleck wired Washington: "The flag of the Union is re-established on the soil of Tennessee. It will never be removed." Grant also told Halleck, "I shall take and destroy Fort Donelson on the 8th and return to Fort Henry."

Historian Benjamin F. Cooling wrote, "Fort Henry demonstrated that the Civil War in the West would be fought largely for control of the rivers — antebellum commercial arteries that became wartime barriers to effective Confederate unity and Union avenues for military, political, and economic reconstruction."

The significance of the capture of Fort Henry was stated plainly by *New York Times* correspondent Franc B. Wilkie, who wrote: "The value of the victory at Fort Henry will, in view of the shortness of the battle and the small number of prisoners, be very generally regarded as of no great consequence. In truth, the result is quite the contrary … The surrender of the fort breaks up the rebel line of fortification, which extended from Columbus on the west to Bowling Green on the east, and opens to our forces a highway easy to be traversed and along which operations can be carried nearly to the Gulf without opposition. By its taking, our troops can reach the very heart of the South without formidable hindrance — indeed there are no works of any kind between Fort Henry and any point further north than Tuscumbia, in Alabama."

Immediately following the capture of Fort Henry, by prior instructions, Lt. Commander Phelps took his fleet of three timberclads on a six-day expedition up the Tennessee River as far as Florence and Muscle Shoals in northern Alabama, meeting little Confederate resistance. They disabled the Memphis, Clarksville & Louisville Railroad bridge at Danville, 25 miles upriver (south) of the fort. Overall, the raid forced the scuttling of six transports and their cargoes, caused a seventh vessel to be destroyed trying to escape, and captured the *CSS Eastport* ironclad and two other vessels. The raid had rattled the Confederate authorities and frightened Southern civilians while encouraging those who remained loyal to the Union.

Following the fall of Fort Henry, the Tennessee River completely inundated the fortification, and although the river would recede in the following weeks, the fortress was ruined and never used again. (Today it sits submerged under the waters of Kentucky Lake.) Earthworks at Fort Heiman are extant. According to a report by the National Park Service, "The task of occupying Fort Heiman fell to Colonel W. W. Lowe and the Iowa Fifth Cavalry regiment, also known as the Curtis Horse. Although no battles or skirmishes were fought at Fort Heiman during their occupation of the post, the Curtis Horse's time in Kentucky and Tennessee was not uneventful. Federal soldiers at Heiman were often bothered by Confederate sympathizing bushwhackers and partisans, not to mention regular Confederate cavalry, while on patrol. And, several times Union forces engaged assembled Confederate troops in Paris, Tennessee, usually suffering numerous casualties. During one particularly

(Continued on Page 46)

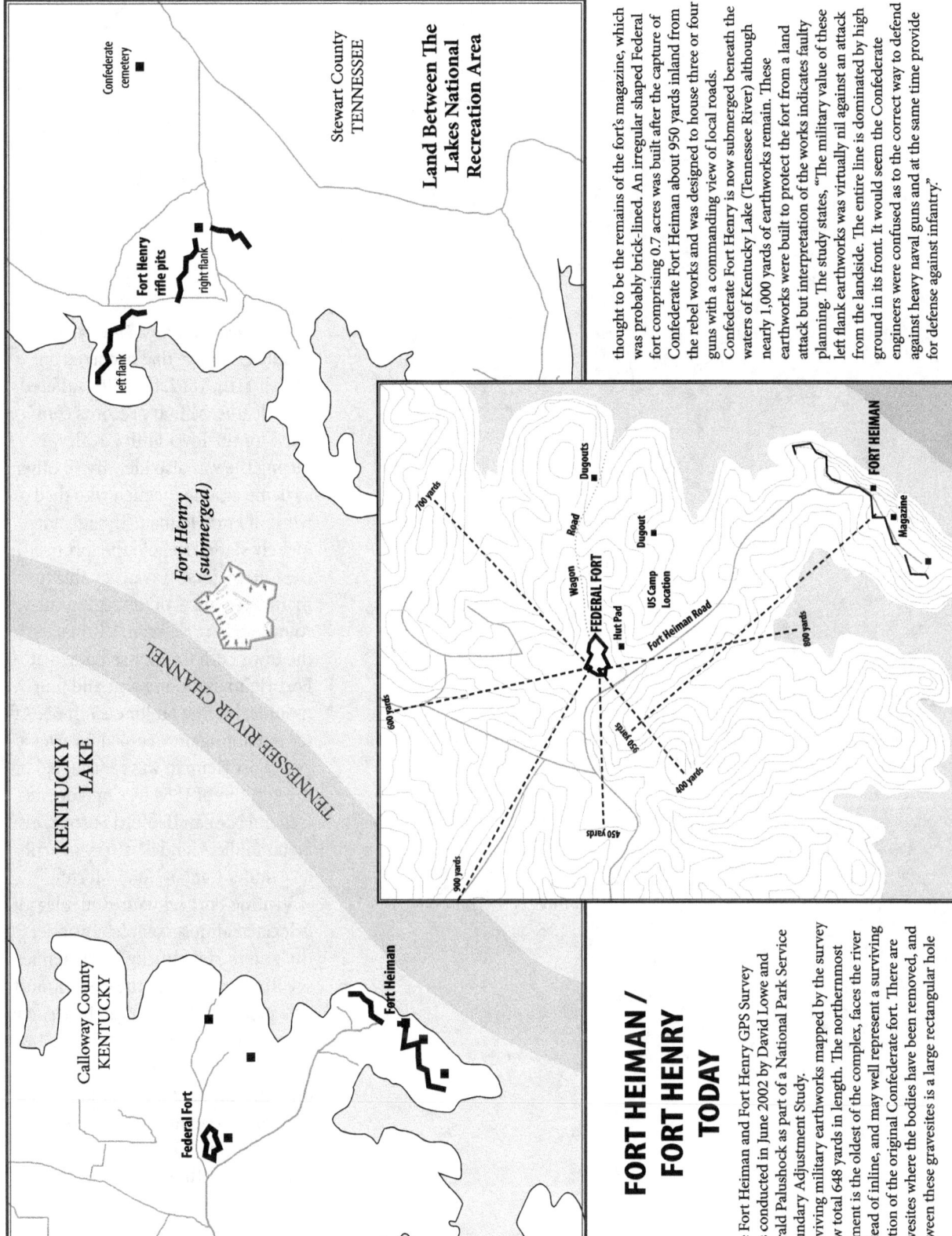

thought to be the remains of the fort's magazine, which was probably brick-lined. An irregular shaped Federal fort comprising 0.7 acres was built after the capture of Confederate Fort Heiman about 950 yards inland from the rebel works and was designed to house three or four guns with a commanding view of local roads. Confederate Fort Henry is now submerged beneath the waters of Kentucky Lake (Tennessee River) although nearly 1,000 yards of earthworks remain. These earthworks were built to protect the fort from a land attack but interpretation of the works indicates faulty planning. The study states, "The military value of these left flank earthworks was virtually nil against an attack from the landside. The entire line is dominated by high ground in its front. It would seem the Confederate engineers were confused as to the correct way to defend against heavy naval guns and at the same time provide for defense against infantry."

FORT HEIMAN / FORT HENRY TODAY

The Fort Heiman and Fort Henry GPS Survey was conducted in June 2002 by David Lowe and Gerald Palushock as part of a National Park Service Boundary Adjustment Study.

Surviving military earthworks mapped by the survey crew total 648 yards in length. The northernmost segment is the oldest of the complex, faces the river instead of inline, and may well represent a surviving portion of the original Confederate fort. There are gravesites where the bodies have been removed, and between these gravesites is a large rectangular hole

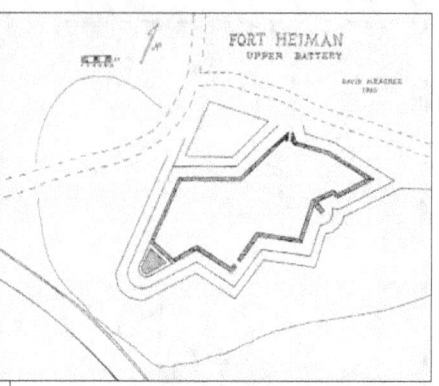

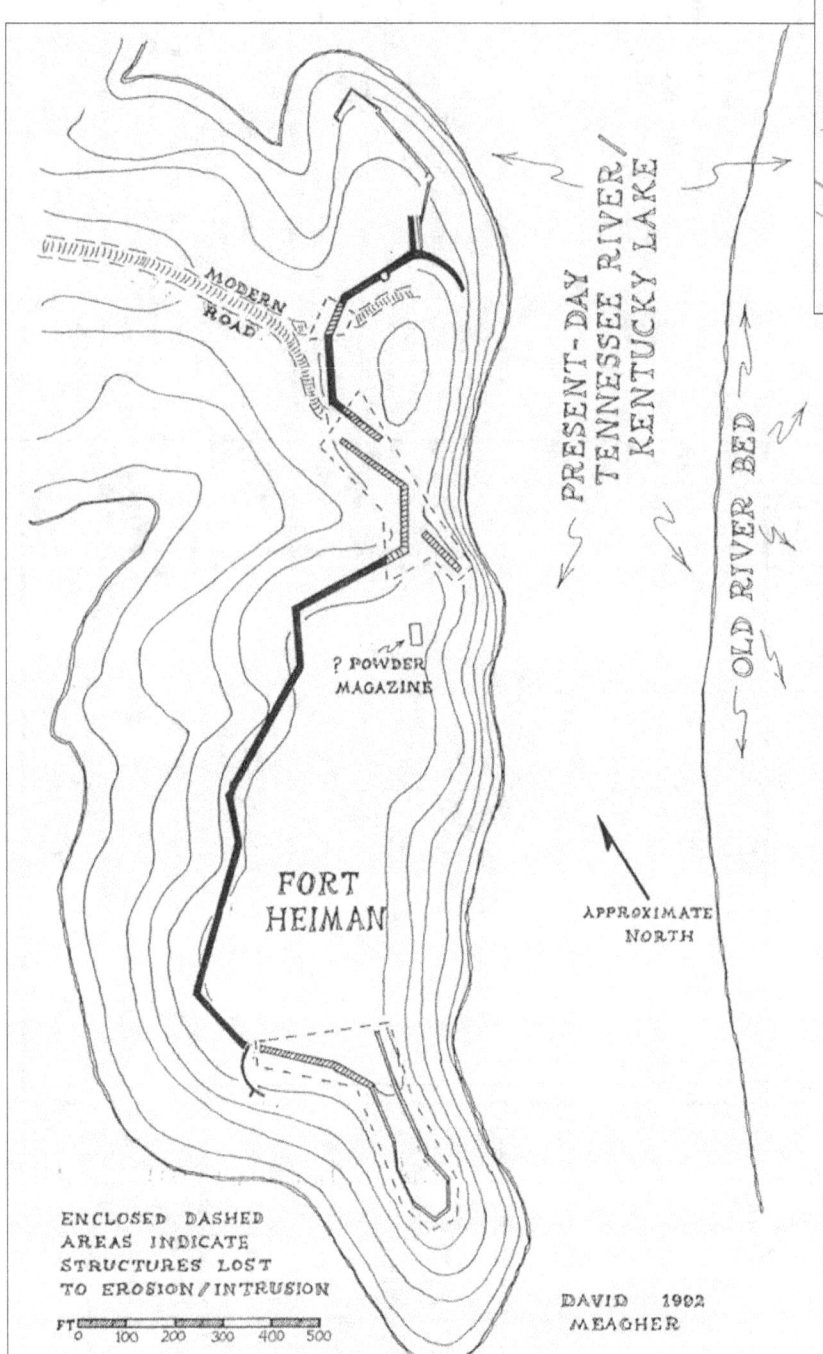

(Continued from Page 44) deadly exchange that occurred on March 11th, 1862, the Fifth suffered nine deaths. Military records compiled for the Iowa Fifth Cavalry during the war also identify 14 others as being among the men who died while at Fort Heiman. Though adequately staffed to hold the fort itself, the Curtis Horse was never able to maintain control of the region surrounding Fort Heiman. Ultimately, the Iowa Fifth Cavalry remained at Fort Heiman for one year and four months, leaving on June 25, 1863. After remaining unoccupied for over a year, Fort Heiman was reoccupied in the autumn of 1864 by Confederate General Nathan Bedford Forrest, his 3,500 soldiers, and a battery of artillery under Captain John Morton."

Visiting Fort Henry is difficult as it is located almost completely under the waters of Kentucky Lake (Tennessee River) except for the easternmost rifle pits. Fort Heiman can be visited; many of the earthworks remain. The fort was linear in shape on top of a bluff overlooking the river and has two sets of remaining earthworks totaling 648 yards in length and between 8 to 10 feet deep.

Battle of Fort Donelson

General Grant's troops spent Feb. 11-13th, 1862 investing the sprawling 554-acre Fort Donelson, their 12-mile march from Fort Henry having been hotly contested by Colonel Forrest's troopers. Flag Officer Foote tended to his damaged fleet at the Cairo dockyards and prepared for another naval expedition. This time the attacking fleet would consist of the City Class ironclads *Carondelet* and *St. Louis,* plus the *Louisville* (Commander Benjamin Dove) and the *Pittsburg* (Egbert Thompson), and two of the timberclads, the *Conestoga* and the *Tyler.* Meanwhile, General Albert S. Johnston, commander of Confederate forces in the Western Theater, began to withdraw from his base at Bowling Green, Kentucky, and headed to Murfreesboro, Tennessee, declining to reinforce the troops at Fort Donelson or defend Nashville. Due to several factors, he reacted to the fall of Fort Henry by professing apprehension of the Federal gunboat fleet. He did not believe that Fort Donelson could withstand a naval assault. He was alarmed at the thought of being caught between Grant's army and that of Union General Don Carlos Buell, moving southward from Louisville, Ky.

Unlike Fort Henry, the river batteries at Fort Donelson, located between the earthen fort itself and the river and just downstream from the village of Dover, were located significantly above the surface of the Cumberland River. The elevated site would allow plunging fire from the Confederate guns to hit the sloped sides of the ironclads at right angles, with much more force, instead of glancing off. The ironclads were more lightly armored on the top deck and around the rear of the casemate. The rudders in the stern were also vulnerable.

Fort Donelson's lower battery, the one furthest downstream and closest to the advancing gunboats, was 20 feet above the river's surface and featured a 10-inch columbiad that fired a 128-lb. shell, and eight 32-pounder smoothbores. The upper battery was 150 feet upstream and 50 feet above the river and featured a 6.5-inch columbiad rifle that fired a 68-lb. shell and two 32-pounder naval carronades, which had a much more limited range. Both batteries were commanded by Colonel Milton Haynes and directed by Lieutenant Joseph Dixon, with 200 men of Captain Reuben Ross' Maury Artillery (part of the 1st Tennessee Artillery) manning the guns. Lt. Hugh Bedford directed the columbiad in the lower battery, while Capt. Ross commanded the rifled columbiad in the upper battery.

Around noon on Wed., Feb. 12th, the U.S. ironclad *Carondelet,* commanded by Henry Walke, arrived in the fort's vicinity under tow from the steamer *Alps.* Approaching the fort alone, the *Carondelet* opened fire to provoke a response and signal to Grant that the gunboat flotilla had arrived. The Confederates held their fire. The gunboat retreated.

Fort Donelson. Artist Andy Thomas. Used with permission.

At 10:00 am the next morning, the *Carondelet* again pushed upstream alone into position about a mile from the imposing river batteries and opened fire from her three bow guns. This time, Dixon ordered the two columbiads to return fire. Their third shot zeroed in on the gunboat but caused little damage. At the end of the 90-minute duel, however, disaster struck, as a projectile from the 6.5-inch rifled columbiad hit the corner where the port side of the *Carondelet's* casemate joined the bow protection. "Tearing through the exterior," wrote historian Jack Smith, "it ricocheted over the temporary log barricade Walke had placed around the boilers, jumped over the steam-drum, hit upper deck beams, carried away the railing around the engine room, and burst the steam-heater before bouncing back to rest in the engine room."

One of the assistant engineers later remarked that the shot seemed to "bound after the men like a wild beast pursuing its prey." Twelve sailors were wounded, seven severely, mostly by flying wooden splinters of various sizes.

On the other hand, one of the 139 shells launched by the *Carondelet* that morning hit the carriage of a 32-pounder and dismounted it, wounding four and killing two men, including Capt. Dixon, who was struck in the side of the head by a dislodged piece of hardware and killed instantly. The shell was a Dyer from a 42-pounder army rifle. Major Chris Robertson said, "The gun at which Dixon was standing had not fired a single time, when one of the cheeks of the carriage was struck by a shot from the enemy, dismounting the gun and throwing bolts and splinters in every direction. One of the iron bolt-heads struck Capt. Dixon carrying away the left side of his face and head, killing him instantly."

Replacing Dixon was Captain Jacob Culbertson (although Ross had more seniority).

The Yankee ironclad withdrew from the skirmish only to return in the afternoon to play a cat-and-mouse game with the gunners. She nearly depleted her magazine by expending 45 more shells, causing more damage to the fort that day than the following day's major naval battle involving six gunboats.

Friday, Feb. 14th found Foote's gunboat fleet exchanging "iron valentines" with the two river batteries at Fort Donelson. At dawn the temperature was ten degrees and there was two inches of snow on the ground. At 2:38 pm, the rifled columbiad opened fire on the four ironclads a mile-and-a-half away, the two timberclads a thousand yards further back. At 2:50 pm, the guns of the flagboat *St. Louis* opened fire within range of the fort's 32-pounders. During the exchange, the *Carondelet* began firing so rapidly that the methodical Foote hailed Walke with his speaking trumpet and told him to slow down. The smoke from the guns and the belching chimneys plus the thunderous explosions of the guns, the paddling of the wheels against the strong current, and the clanking of shells against iron cladding thrilled the shorebound audience of blueclad Federal troops, the Confederates in the fort, and the villagers of Dover. Within 45 minutes, the gunboats had approached to 400 yards of the lower battery.

Confederate cavalry commander Colonel Nathan Bedford Forrest, at the time purely a spectator, was definitely concerned about the shelling. He watched the shot of the 32-pounders "roll off the boats like water off a duck's back." The heaviest gun in the rebel battery was knocked off its carriage, leaving only a single heavy gun (the 32-pounders remained in service). Not a man to be easily alarmed or intimidated, Forrest asked his fighting parson, Major David Campbell Kelley pointedly, "Are you praying? Pray, Major, only God almighty can save us. We have no power to look to but God and one gun."

At this point, however, many of the shells from the gunboats were flying over the river batteries and into the ranks of Federal troops beyond, those who had invested the earthen fort.

Then a Confederate gunner at the rifled columbiad exclaimed to his fellows that he was going to take away a chimney. He pulled the lanyard and his shot did exactly that — one of the chimneys on the *Carondelet* was carried away. "Come on, you cowardly scoundrels; you are not at Fort Henry!" the gunner bellowed at the

One of the two carronades in the Upper Battery and a 32-pounder smoothbore in the Lower Battery.

sailors. The *Carondelet* was being literally shot to pieces. One of the shells hit the metal pilothouse and an iron splinter mortally wounded the civilian pilot. Another shell wounded the second pilot. Planking and plating were being ripped away "as lightning tears the bark from a tree," according to one witness.

At one point in the frenzy, the premature firing of the *Carondelet's* port-bow 42-pounder resulted in an explosion that rocked the boat, wounding more than a dozen men. The big gun burst into four large pieces, one of which tumbled through the port and into the river. Amazingly, no one was killed — the crew had recognized the impending danger and had time to back away before the explosion.

The flagboat *St. Louis* was hit 59 times and had her rudder chains severed. The pilot was killed in his house and Flag Officer Foote himself wounded, in the foot. The *Louisville* was hit numerous times, one shot decapitating three sailors and splattering Commander Dove with their brains.

The *Pittsburg* sustained 30 hits and was hit below the hull and took on water. She turned and crashed into the stern of the *Carondelet,* disabling her starboard rudder.

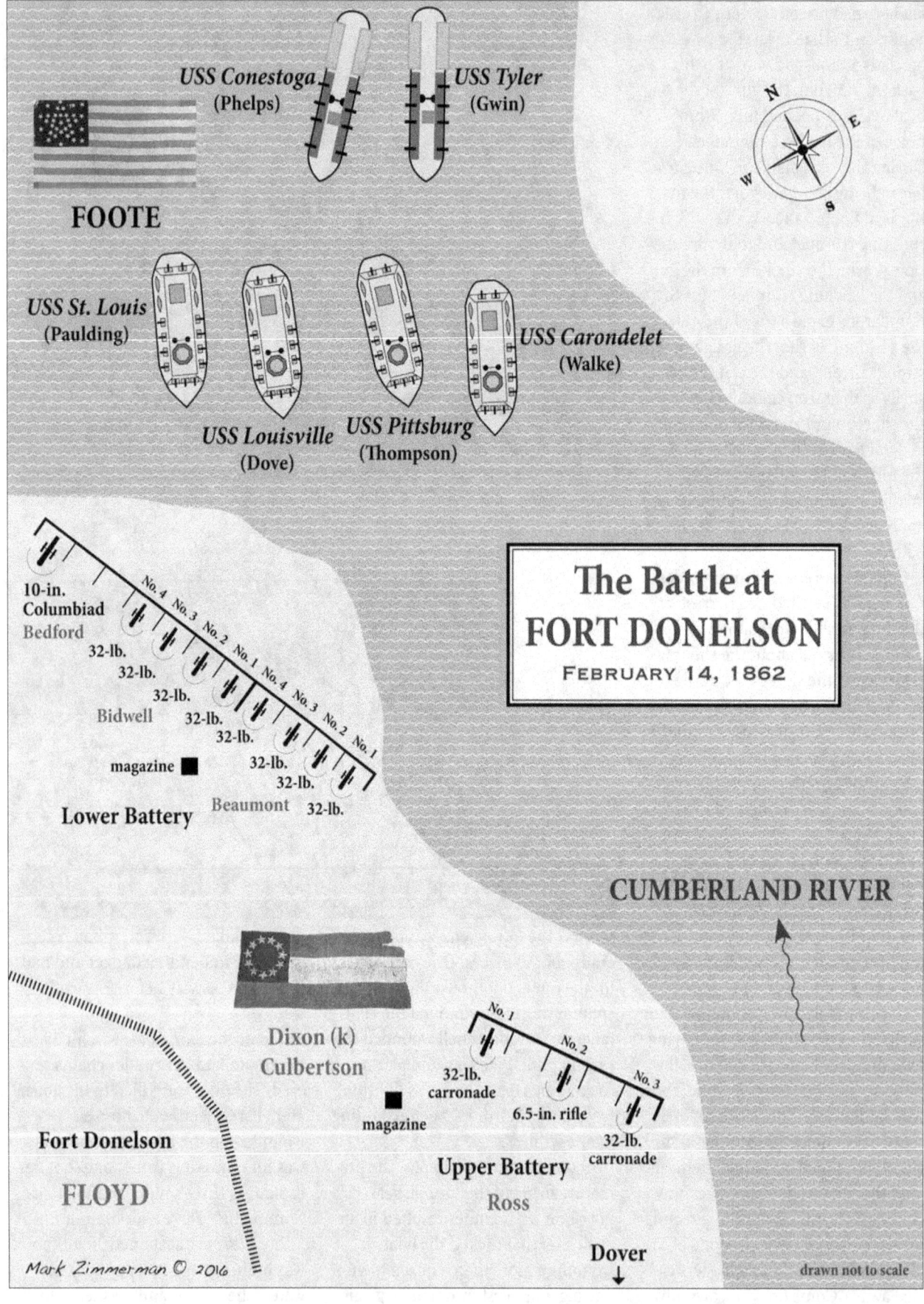

Exchanging Iron Valentines. Artist Andy Thomas. Used with permission.

The gunboats were all drifting downriver away from the fort, the *Carondelet* exposing only her bow. The Confederate gunners began skipping their shots across the water like flat stones. One rattled into and through a bow port but caused no damage. Another shot glanced off a bow gun and decapitated two sailors who refused to follow the "Down!" command. Signal Quartermaster Matthew Arther of the *Carondelet* later received the Navy Medal of Honor for carrying out his duties as signal master and captain of the rifled bow gun.

Shells from the timberclads back in the rear of the formation began to hit the ironclads. An 8-inch shell from the *Tyler* hit the stern of the *Carondelet;* a piece of shrapnel was later dug out of the wood and kept by Commander Walke, no fan of the timberclads, as a souvenir. All in all the *Carondelet* was hit 54 times, with four killed and 29 injured.

The battle ended as the disabled Union ironclads were carried downriver by the current. During the 70-minute shootout, the ironclads were hit 180 times, resulting in nine dead and 45 wounded.

The Confederate gunners had fired 370 shot and shells and did not sustain a single casualty. The gunners in the batteries and soldiers in the rifle pits put up a great celebration, shouting until they were hoarse. "That cheer was the most direful sound," said one soldier from Iowa. The dreaded gunboats, once thought invincible, had been beaten back. One Confederate colonel ordered that the victorious gunners receive a round of good Tennessee whiskey.

During the next two days, the Confederates in the fort pushed the surrounding Federal forces aside in an attempted break-out which was eventually thwarted by indecision within the Confederate high command. The two highest-ranking generals fled for their lives, yet another decided to surrender; Grant was accorded his demand for unconditional surrender; and 13,000 Confederate soldiers became prisoners of war. Only Colonel Forrest and his cavalry of 1,200 troopers managed to escape the enclosure, fording icy streams, and finding their way to Nashville to fight another day (another 1,000 escaped by basically walking through the lines). Many of the Confederates taken prisoner would die in Northern POW camps, while those who survived were exchanged late in 1862. The surrender of so many soldiers was unprecedented, and the Federal officials were woefully unprepared to take care of them.

Grant was hailed throughout the North for his victory and his demand for unconditional surrender, but it wasn't originally his idea. The demand had been suggested previously by Flag Officer Foote and General Charles Ferguson Smith.

Fort Donelson National Battlefield Historian Jim Jobe speculates that one major reason for the Federal naval defeat on February 14 was that the gunboats needlessly approached too close to the rebel batteries. The gunboats could have stopped at 1,600 yards and shelled the batteries and fort almost with impunity, he noted.

General Johnston was wrong about his fear of the gunboats. Historian Kendall Gott wrote: "When Fort Henry fell to a purely gunboat attack in but two hours, General Johnston himself became convinced that Fort Donelson had no hope of standing up to them. From that point, it seems that Johnston lost his will to put up a determined resistance. Had Johnston seen for himself just how poorly Fort Henry was constructed, he would have realized that its fall was not due simply to the ironclads, and that though they were powerful, they were not invincible."

On Feb. 26th, Grant issued Order No. 14: "Such slaves as were within the lines at the time of the capture of Fort Donelson…will be employed in the quartermaster's department, for the benefit of Government." Freed men were used to build the new Union Fort Donelson between the old fort and the village of Dover (half of the site of the new fort would be used after the war for the National Cemetery). By March 1863, some 300 contrabands (fugitive slaves) lived at the freedman's camp known as Free State, which included houses, a school, and a church. The encampment existed until the 1880s. Benevolent societies, such as the Western Freedmen's Aid Society, assisted the Federal Army by providing clothing and teachers, and administering religious and medical services.

In the months following the fall of Fort Donelson, the occupying Union forces abandoned the old Confederate works and fortified the nearby riverfront village of Dover, manned by four companies of the 71st Ohio, who arrived in April 1862. The 83rd Ohio arrived in September. The Union garrison of 600 men was composed of the 83rd Illinois infantry, commanded by Colonel Abner C. Harding, a banker by occupation, and the 5th Iowa Cavalry. Stationed at the town square in

After-Action Report
Gunboats Rendered Wholly Unmanageable

FLAG-SHIP ST. LOUIS,
Near Fort Donelson, Cumberland River, February 15, 1862
Major-General HALLECK,
Commanding Army of the West, Saint Louis, Mo.

SIR: I have the honor to report that, as you regarded the movement as a military necessity, although not in my judgment properly prepared, I made an attack on Fort Donelson yesterday, the 14th instant, at 3 o'clock pm with four iron clad and two wooden gunboats, the St. Louis, Carondelet, Louisville, and Pittsburg, with the Tyler and Conestoga, and after a severe fight of an hour and a half, being in the latter part of the action less than 400 yards from the fort, the wheel of this vessel, by a shot through her pilot-house, was carried away, and the tiller-ropes of the Louisville also disabled by a shot, which rendered the two boats wholly unmanageable. They then drifted down the river, the relieving tackles not being able to steer or control them in the rapid current. The two remaining boats, the Pittsburg and Carondelet, were also greatly damaged between wind and water, and soon followed us, as the enemy rapidly renewed the fire as we drifted helplessly down the river. This vessel, the St. Louis, alone received 59 shots, 4 between wind and water and one in the pilot-house, mortally wounding the pilot and others, requiring some time to put her in repair. There were 54 killed and wounded in this attack, which, notwithstanding our disadvantages, we have every reason to suppose would in fifteen minutes more, could the action have been continued, have resulted in the capture of the two forts bearing upon us, as the enemy's fire materially slackened and he was running from his batteries when the two gunboats helplessly drifted down the river from disabled steering apparatus, as the relieving tackles could not control the helm in the strong current, when the fleeing enemy returned to their guns and again boldly reopened fire upon us from the river battery, which we had silenced.

Flag Officer Andrew Foote

The enemy must have brought over twenty heavy guns to bear upon our boats from the water batteries and the main fort on the side of the hill, while we could only return the fire with twelve bow guns from the four boats. One rifled gun aboard the Carondelet burst during the action. The officers and men in this hotly-contested but unequal fight behaved with the greatest gallantry and determination, all deploring the accident rendering two gunboats suddenly helpless in the narrow river and swift current.

On consultation with General Grant and my own officers, as my services here, until we can repair damages by bringing up a competent force from Cairo to attack the fort, are much less required than they are at Cairo, I shall proceed to that point with two of the disabled boats, leaving the two others here to protect the transports, and with all dispatch prepare the mortar boats and Benton, with other boats, to make an effectual attack upon Fort Donelson. I have sent the Tyler to the Tennessee River to render impassable the bridge, so as to prevent the rebels at Columbus re-enforcing their army at Fort Donelson.

I transmit herewith a list of casualties. I am informed that the rebels were served by the best gunners from Columbus.

Very respectfully, your obedient servant,
A. H. FOOTE,
Flag-Officer, Comdg. U. S. Naval Forces on the Western Waters.

the ruins of the old county courthouse was the impressive swivel gun, a 32-pounder confiscated from the former Confederate river battery, and four 12-pounders under the supervision of Capt. James H. Flood's Battery C, 2nd Illinois Artillery. The siege gun was likely serviced by Pvt. Virgil Earp of the 83rd Illinois. Earp would become famous in 1881 when he, as marshal of Tombstone, Arizona, along with his three younger brothers and friend Doc Holliday, fought at the O.K. Corral. Half of the 83rd Illinois, commanded by Lt. Col. A.A. Smith, and two field pieces, were positioned at fortifications behind the old village graveyard several hundred yards west of the town square.

In August 1862, the brigade of Confederate Lt. Colonel Thomas G. Woodward, consisting of the Kentucky Battalion and Adam Rankin Johnson's 10th Kentucky Partisan Rangers, raided Dover after recapturing Clarksville. Based at nearby Cumberland Furnace, Woodward had been harassing Federal communications in the Dover area for quite some time. Battling the 71st Ohio under Major Hart, Woodward left the village of Dover "mostly destroyed by fire" (the Federals burned 17 buildings to clear their field of fire).

On Feb. 3rd, 1863, Confederate cavalry forces under General Joseph Wheeler, including those of Gen. Forrest and Col. John Wharton, attacked the village of Dover occupied by the 83rd Ohio and were handsomely repulsed by the Union infantry and river gunboats, due to faulty maneuvering and lack of ammunition. After the debacle, Forrest vowed never to serve under Wheeler again.

After the raid, the Federal forces built a new fort between the old Confederate fort and Dover, where the National Cemetery is now located.

In November 1863, Fort Donelson became a USCT recruiting station, enlisting freedmen into the 8th Heavy Artillery, and the 13th, 16th, and 17th infantry regiments.

Fort Donelson National Battlefield was dedicated in July 1932. The 368-acre tract includes a Visitors Center (currently under much-needed renovation), the National Cemetery (670 interments, most unknown), earthworks, soldier hut replicas, the river batteries with authentic and replica guns, various monuments and gun batteries, the Surrender House (Dover Hotel), and a self-guided driving tour.

The Fall of Fort Donelson

Headquarters, Army in the Field, Camp near Fort Donelson, Feb. 16, 1862,
General S.B. Buckner, Confederate Army:
Sir: Yours of this date, proposing an armistice and appointment of Commissioners to settle terms of capitulation, is just received. No terms except unconditional and immediate surrender can be accepted. I propose to move immediately upon your works.
— I am, sir, very respectfully, your obedient servant,
U.S. Grant, Brigadier General Commanding.

U.S. Grant and CSA Gen. Simon Bolivar Buckner, who surrendered Fort Donelson.

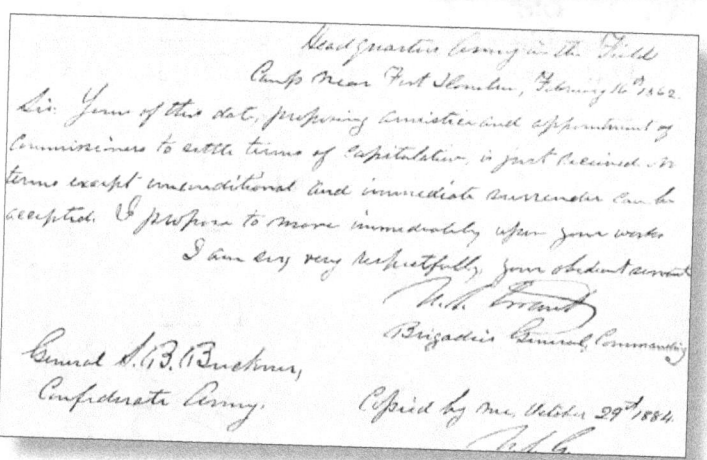

Grant's handwritten terms of surrender (Internet Archive Book Images).

Disease Ravages Fort Donelson Garrison

I have just come in from visiting the sick of our Regt. They have increased very rapidly. The first attacks are diarrhea, which reduces the men very fast. Then fevers chills & fevers in many of the cases follow…We have about forty of our Co at this place who are sick & unfit for duty but I think none of them dangerous.

— Capt. John McClanahan, age 68, Co. B, 83rd Illinois Volunteer Infantry Regiment, stationed at Fort Donelson, October 3, 1862, in a letter to his wife. The captain was wounded during the Battle of Dover on Feb. 3, 1863 and died three weeks later.

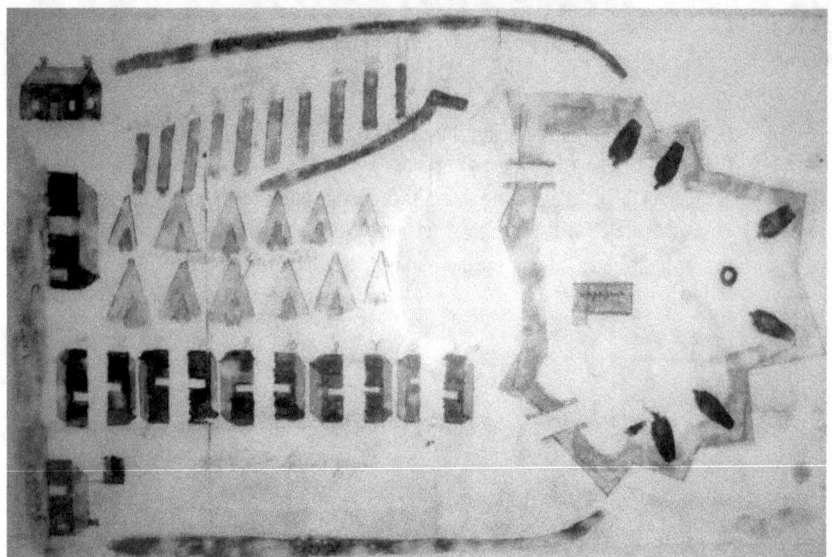

Union Fort Donelson

Map below based on a watercolor rendering of Union Fort Donelson in 1863 by D.N. Graham, a Federal soldier in the garrison, which consisted of two companies of the 13th Wisconsin and 83rd Illinois. This fort sat where the National Cemetery is located today. The photo was taken from an interpretive panel displayed at the cemetery office.

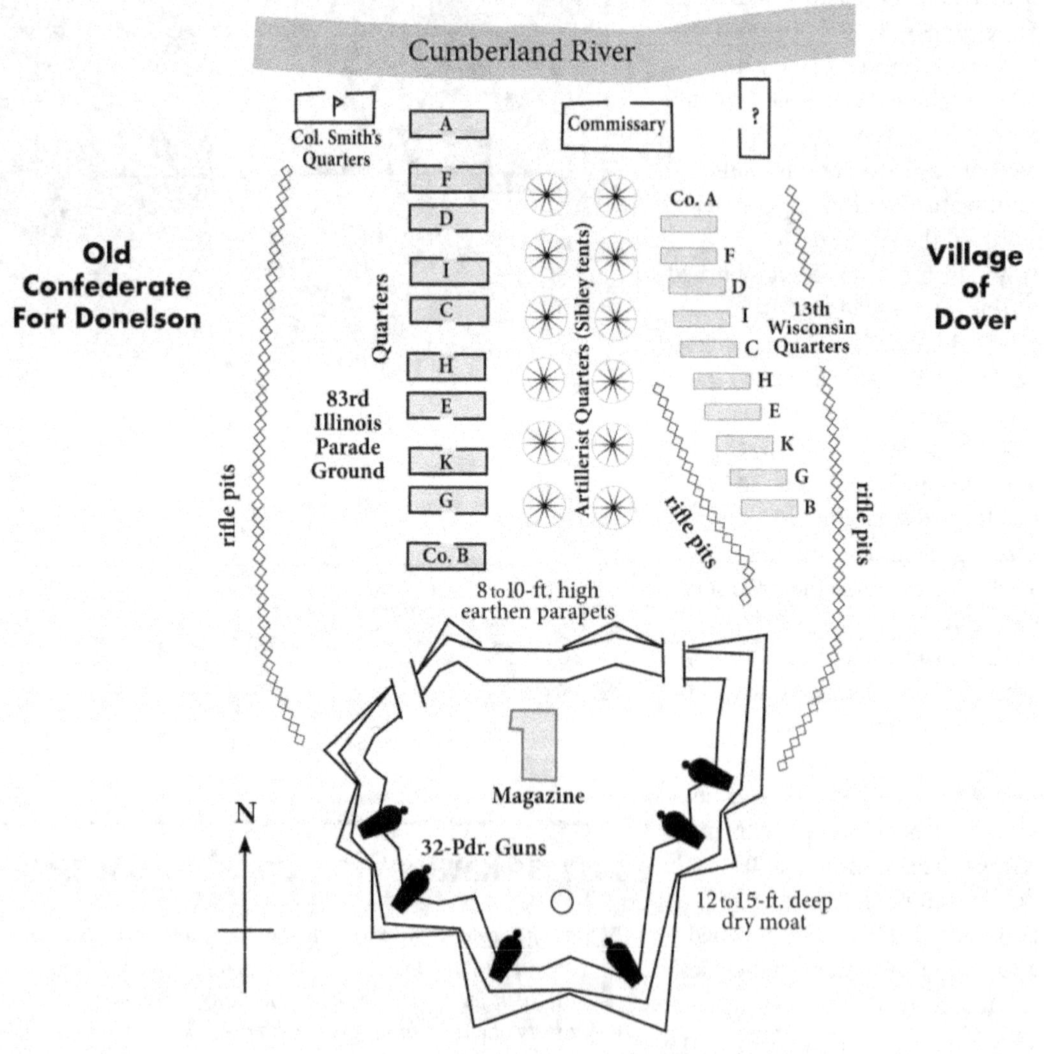

UNION FORT DONELSON

Capture of Clarksville & Nashville

At 2:00 am on Sunday, Feb. 16th, 1862, General John Floyd telegraphed Major General Albert Sidney Johnston in Edgefield that he had defeated U.S. Grant at Fort Donelson. Three hours later, Johnston was awakened by a messenger who told him that Floyd and other generals had agreed to surrender the river fortification and then fled the scene.

Church services were suspended in Nashville that morning when the shocking and unexpected news of the surrender of Fort Donelson arrived. Unsure what to do, and certain that the dreaded gunboats would reduce the city to rubble, the citizens launched into what has been described as the "Great Panic." Depositers rushed the banks, crowds mobbed the warehouses, and southbound trains were filled to overflowing, all while the retreating Confederate army of General Johnston tramped through the city on their journey from Bowling Green to Murfreesboro. Johnston subsequently announced that the army would not defend the city.

"A perfect panic reigned throughout the whole city," said a local Union loyalist. "The streets were thronged with people wild with excitement."

Much to the chagrin of the citizenry, Johnston abandoned Nashville and moved southward to Columbia. (The Confederate forces would eventually arrive and converge at Corinth, Mississippi.)

Nashville Mayor R.B. Cheatham, after consulting with Johnston, addressed a crowd at the courthouse, begging for calm and asking his citizens not to burn the city. He said he would surrender the city to General D.C. Buell when he and his troops arrived from Kentucky. The First Missouri Regiment was assigned to guard the city against vandalism and looting.

Dr. John B. Lindsley of the University of Nashville wrote in his diary: "Sunday: (Albert Sidney) Johnston's army passing by the University from 10 am until after dark, camped out near Mill Creek. Light of campfires very bright at night. The army was in rapid retreat the men disliked bitterly giving up Nashville without a struggle. The southern army, however, was too small to make a stand against the overwhelmingly superior numbers of union troops. During all Sunday from about 10:00 am when the news of the fall of Fort Donelson reached here, the wildest excitement prevailed in the city. Very many persons left the city in vehicles many on the (rail) cars — the government and Legislature decamped — Nashville was a panic stricken city."

John Miller McKee with the *Nashville Union and American* reported: "Every available vehicle was chartered, and even drays were called into requisition, to remove people and their plunder, either to the country or to the depots, and the trains went off crowded to their utmost capacity, even the tops of the cars being literally covered with human beings. A large number of citizens left the city from fear of fire. They had been led to believe that the town would be shelled during the afternoon or night at farthest, and reduced to heap of ruins. These went only a short distance into the country, and returned as soon as they felt they could do so with safety."

Railroad cars were being used to evacuate sick soldiers from Kentucky. Spring freshets had washed bridges away. The Nashville & Chattanooga Railroad suffered from 1,200 broken rails. V.K. Stevenson, the "father of Tennessee railroads" and superintendent of supplies at Nashville caught a train to Chattanooga. A train of 100 wagons was stranded at Clarksville. Thirty-five tons of bacon was abandoned at the Nashville wharf (125 tons were saved). An estimated 50 percent of the supplies at Nashville were lost during the panic.

On Feb. 18th, 1862, the *USS St. Louis* destroyed the Cumberland Iron Works, just upstream from Dover, which was partially owned by former U.S. Senator John Bell. Two of Bell's partners were taken prisoner. Bellwood Furnace was also destroyed.

Three days after the fort's surrender, on Feb. 19th, 1862, Flag Officer Foote proceeded from Fort Donelson up the Cumberland River toward Clarksville (pop. 5,000) with the timberclad *Conestoga* and the ironclad *USS Cairo* (captained by Lt. Commander Nathaniel Bryant). At mid-afternoon they reached Linwood Landing, a mile below Clarksville and under the high bluff occupied by Fort Sevier (later called Fort Bruce and Fort Defiance). The fort displayed a white flag, having been abandoned the day before. Although formidably gunned, the fort had been sited too high on the hillside (by CSA engineer Jeremy Gilmer) and at a bad angle to the river. Shortly thereafter, the gunboat crew sighted Fort Clark, housing three guns, at the confluence of the Red River. Apparently a storm had blown down the white flag there; a muddy and dirty white flag was raised up the standard, only to be quickly replaced with the Stars and Stripes by a landing party. Fort Clark had been sited on a flood plain and was partly underwater.

Three prominent Clarksville citizens, including the mayor, met with Flag Officer Foote and advised that the Confederates had abandoned the city after hearing of the gunboats' approach but not before torching the railroad bridge over the Cumberland River (it was saved) and the road bridge over the Red River. Two-thirds of the panic-striken population had fled the town. Foote, who tagged the hilltop structure as Fort Defiance, assured the delegation that "we came not to destroy anything but forts, military stores, and army equipment."

Landing parties at Clarksville found a message in the telegraph office that the Confederate commander in the town had sent ahead to Nashville advising that the gunboats were coming. The Federal officers were thus convinced that Nashville was in a panic and would surrender to the gunboats without firing a shot, which is what eventually did happen.

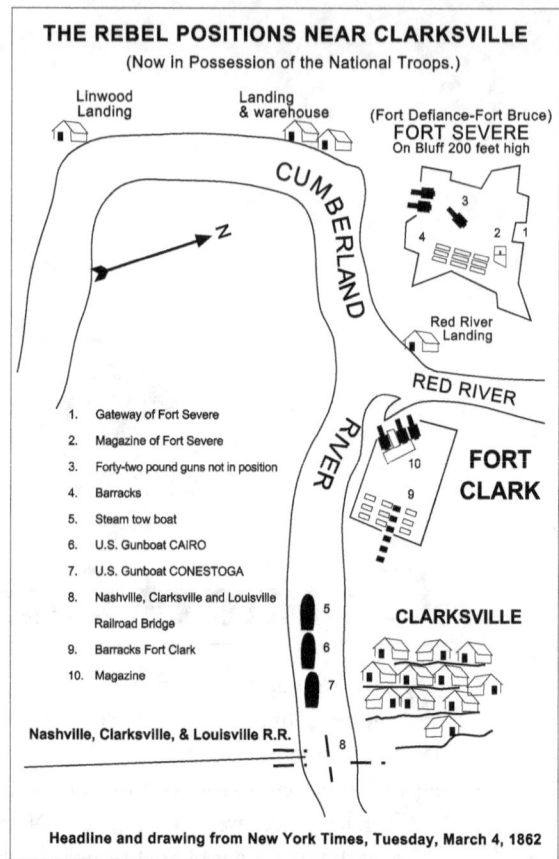

Headline and drawing from New York Times, Tuesday, March 4, 1862

Clarksville was left with a small garrison of Federal troops, consisting of the 7th Illinois and 9th Illinois under Gen. C.F. Smith. In April 1862, a regiment led by Col. Rodney Mason took over garrison duties.

The subsequent history of Clarksville during the early years of the war was a turbulent one. During July-August 1862, there was an increase in guerrilla activity around Clarksville. On August 18th, Clarksville was re-captured by Confederate cavalry under Col. Thomas Woodward. Col. Mason was cashiered for surrendering Clarksville without much of a fight. In September, Federal troops were sent from Fort Donelson to retake Clarksville under command of Col William Lowe of the 5th Iowa Cavalry, to include four companies of the 71st Ohio, the 13th Wisconsin, the 11th Illinois, and two sections of Illinois artillery. Skirmishes were fought at Riggins Hill and then New Providence on September 7th. The town and forts were ungarrisoned until reoccupied by Federal troops on Christmas Day 1862. Col. Sanders D. Bruce was placed in command with 2,300 men, and Fort Sevier was renamed Fort Bruce. Clarksville was occupied by Federal troops for the remainder of the war.

Fort Sevier was the major Confederate fort sited at Clarksville by Confederate military engineer Jeremy Gilmer to protect the bend in the Cumberland River.

The *USS Cairo* ironclad gunboat and *USS Conestoga* timberclad gunboat on reconnaissance up the Cumberland River reported a white flag at Fort Sevier on Feb. 17th, 1862 and aided in the capture of Clarksville as Confederate forces withdrew. From this perspective, the fort appears larger than in reality. Original artwork by Philip Duer, used with permission.

Gilmer hired civilian engineer Edward Sayers to build the fort. Local slaves were leased, and Confederate soldiers were ordered to build the fort. Continuous labor shortages caused construction to lag. Sayers laid out another fort, on ground higher than Fort Sevier, which was never built. Several factors made Fort Sevier untenable. The hill's height required a downward angle that was too steep for the cannons to effectively fire down upon that portion of the river. The shape of the fort and the structural core did not comply with military engineering guidelines. Fort Clark was sited on the opposite side of the Red River, much closer to the Cumberland, and, in fact, in a floodplain. Fort Sevier and Fort Clark were initially garrisoned by the Montgomery Heavy Artillery, CSA. Fort Terry, a third fort, was sited inland to protect the Memphis, Clarksville & Louisville Railroad.

After Union forces took control of Clarksville and New Providence in early 1863, Fort Sevier was renamed Fort Bruce in honor of the city's commander Colonel Sanders Bruce. At some point later, it became known as Fort Defiance. Many escaped slaves and a few uprooted white Unionists came here for protection and assistance. They were housed in tobacco warehouses and in camps near fortified posts such as Fort Defiance/Fort Bruce. Called "contrabands," some were hired by the army as civilian labor.

Fourteen infantry regiments and eight artillery units were organized with 24,000 men from Tennessee beginning in October 1863. The 16th USCT (United States Colored Troops) and one company of the 9th US Colored Heavy Artillery were raised in Clarksville. While in Clarksville, these soldiers engaged in duties including checking civilian passes, driving wagons, handling supplies, and guarding contraband camps (later known as Freedmen's camps).

All USCT units were transferred from Clarksville to other parts of the state for similar duties. Men from Clarksville serving in the 12th USCT and 13th USCT completed the Nashville & Northwestern Railroad from Kingston Springs to Johnsonville, Tenn.

Fort Defiance was abandoned after the war and sat many decades overgrown with vegetation and undeveloped. In 1982, in preparation for Clarksville's Bicentenary, Judge and Mrs. Sam E. Boaz, who owned the fort property, deeded it to the city. Faculty and students from Austin Peay State University cleared the site, and Mayor Ted Crozier arranged for its maintenance as a city park. In 2002, Mayor John Piper established the Fort Defiance Commission to devise a plan for developing the site. With a $2.2 million federal grant and funding awarded by the city in 2008, a modern interpretive center and other amenities were built for tourism.

On February 20th, Foote returned to Dover to assemble a task force to move on Nashville only to learn that General Halleck had ordered the gunboats to go no further than Clarksville. Many of the gunboats in the task force would be used instead to assault the Confederate works at Columbus, Kentucky, on the Mississippi River. Halleck was paranoid about Grant's accomplishments overshadowing his own contributions. Halleck decided that General Buell, then stationed at Bowling Green, should have the honor of capturing Nashville.

On February 21st, four regiments from Gen. Charles Ferguson Smith's division were transported up the river from Dover to Clarksville, under cover of the *USS Cairo*. Landing parties found flour and bacon in the city's warehouses and coal reserves at the steamboat landing. They dismantled the batteries at the forts and completed the destruction of the railroad bridge. The soldiers set up their camps at Fort Sevier and Fort Clark.

Private John King of the 92nd Illinois noted in his diary that Clarksville was a pretty town but notorious for "making the world nastier and filthier by raising and shipping large quantities of that weed called tobacco" which "boys love to eat, chew, and smoke and they learn to swear and become lazy, filthy loafers."

A delegation of Nashville citizens chartered the steamer *Iatan* on February 23rd and met with Union officials

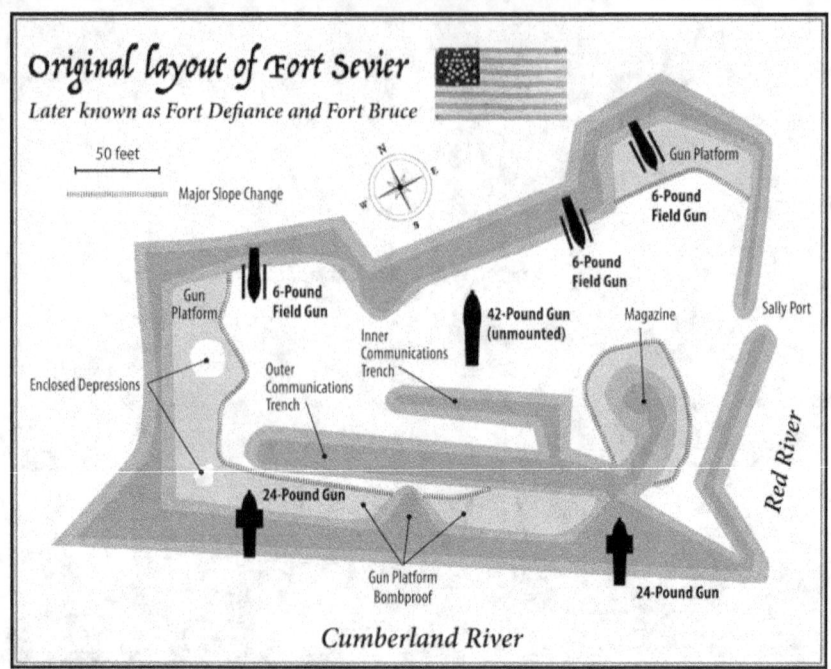

at Clarksville under a white flag. They brought surgeons to attend to Confederate soldiers wounded at Fort Donelson and asked for assurances that property would be protected once the army reached Nashville.

The *Iatan* was accompanied by the *Cairo* back to Nashville. At this time, two hours behind them, 16 steamers were carrying the 10,000 troops of Brigadier General William "Bull" Nelson up the river. They had boarded the boats at West Point, Ky. five days earlier, originally intending to fortify the army at Fort Donelson. Buell had dispatched Brigadier General O.M. Mitchel and his 9,000 men from Bowling Green to march on Nashville. They would reach Edgefield on the east bank of the river opposite Nashville and would need to be ferried across. The Cumberland River suspension bridge at Nashville had been torched the night of February 19th and the cables cut. At 3:00 am the next morning the Louisville & Nashville railroad bridge was destroyed, except for the stone piers. At this time, the Cumberland River was in full flood. En route to Nashville, the transports would leave the main channel and pass directly over farms and cruise by houses with families living on the second floor, the ground floor inundated with water.

A five-gun battery was spotted on an island at Harpeth Shoals but the troop transports did not stop, Nelson knowing that the *Cairo* had already passed. At 9:30 pm, 15 river miles downstream from Nashville, the gunboats and transports tied up to the bank for the night.

Earlier that month, Confederate soldiers had erected Fort Zollicoffer six miles below Nashville on a high bluff 130 feet above the river and boasting eight heavy guns. One was a 6-inch rifle weighing 9,490 pounds and made at Tredegar Iron Works in Richmond. One of the others was a six-pounder made by Ellis & Co. of Nashville. The fort was named for General Felix Zollicoffer, a Nashville newspaper editor and Confederate commander who had been killed at the Battle of Mill Springs (Fishing Creek) far up the Cumberland River in Kentucky on Jan. 19th, 1862.

On the sunny morning of Tuesday, Feb. 25th, the boats sighted the guns of Fort Zollicoffer at the top of a high cliff. A landing party of Col. Jacob Ammen's 36th Indiana found the fort in shambles. According to historian Ed Bearss: "Several of the oak bar-

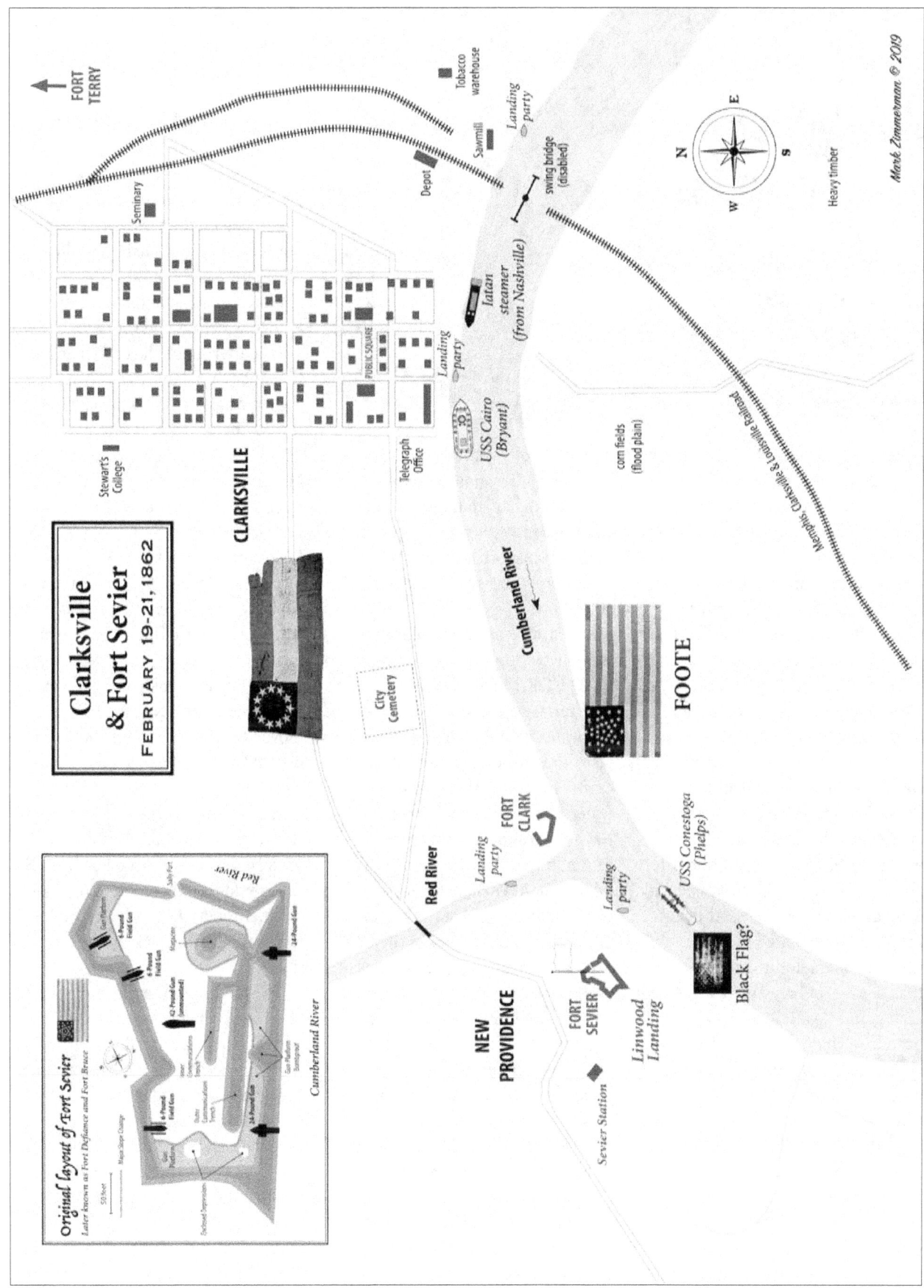

Field artillery piece at Fort Defiance Interpretive Center, Clarksville.

bette carriages were still smouldering. [The] guns, ranging in size from 32-pounder smoothbores to 6.5-inch rifles, had been dismounted or spiked. Scattered about the stronghold's interior were thousands of bars of railroad iron, many of them badly twisted by the terrific explosion that had shattered the magazine that they were intended to shield. Cotton bales used to reinforce the parapets had been burned."

A special correspondent from the *New York Times* described the fort: "Six miles below Nashville we reached Fort Zollicoffer. It is located on the west bank of the river, some sixty feet above the water, and is mounted with eight guns—32s and 64s. Although the guns are mounted the fort is unfinished, being nothing more as yet than a series of breastworks— one for each gun. Two additional guns had been thrown down the bank and lay close to the water's edge—one or two others are supposed to have been thrown in the river, while the balance are indifferently well spiked. The rebels who constructed the fort evidently knew but little of the existence of the gunboats, or else they would have placed the pieces in quite a different position. The guns stand very nearly on a line parallel with the river, thus exposing them to our enfilading fire from the gunboats. The gallant Commodore Foote, with his fleet, would have swept the whole battery out of existence in half an hour; but they were evidently intended to operate against transports carrying troops, in which case they would have answered admirably."

Reboarding and getting underway, the convoy came around a bend and saw the taller buildings and spires of the capital city, several white flags and yellow hospital flags flying but no Stars and Stripes, and none atop the capitol building. Aboard the flagboat *Diana*, Gen. Nelson, a former US Navy ship captain also known as Old Buster, ordered the flotilla ahead at full speed. At 9:00 am blueclad soldiers walked ashore at the wharf under the watchful guns of the *Cairo* ironclad. Even then, small patrols of Confederate cavalry were just leaving the city and lurking on the outskirts. The color sergeant of Company C, Sixth Ohio Volunteer Infantry, carried shore the blue Guthrie Grey flag, followed by five companies of the regiment marching up Front Street to a band playing "Dixie" and then to Cedar, where a group of slaves applauded as the Federals advanced to "Yankee Doodle." Some local citizens were cursing General Johnston for abandoning the city without a fight. The bluecoats marched from the courthouse up Cedar Street to the magnificent capitol building with its Greek choragic tower. At 8:45 a.m., Gen. Nelson hoisted the regimental flag of the 6th Ohio over the Capitol. The 6th Ohio Volunteer Regiment then assisted retired sea captain William Driver, a noted resident of Nashville, in hoisting his old ship's Union flag up the mast over the main portico, a flag he called "Old Glory." The name caught on with the occupying forces, and soon the U.S. flag was known widely as Old Glory.

That afternoon, Mayor Richard B. Cheatham and ten citizens crossed the river by steamer to Edgefield on the east bank and officially surrendered the city to General Buell, as Mitchel and Nelson witnessed. Soon Buell's men were across the river as regimental bands from Indiana and Ohio entertained the curious crowd by playing "Dixie." A Northern newspaper correspondent stated that "there is but little Union sentiment expressed here; in fact, far less than I had anticipated. All hands appear to hate us cordially." Nelson's troops moved two miles down the Murfreesboro Pike and set up Camp Andrew Jackson.

The next day, as the Federals continued to occupy the city, a dozen of John Hunt Morgan's troopers brashly snuck into the city and set fire to a steamer moored near the city waterworks well within sight of Federal troops.

On March 1st, Gen. Buell moved into the St. Cloud Hotel. He paid his respects to Mrs. Polk, the widow of President James K. Polk, at her home, Polk Place, a precedent followed by all succeeding Federal commanders. Mrs. Polk always made plain her allegiance to the Confederacy. Later, Buell selected the house of George W. Cunningham at 13 North High Street as his residence, two blocks south of the Capitol.

On March 3rd, President Lincoln appointed Sen. Andrew Johnson of Greeneville, Tenn. to be military gov-

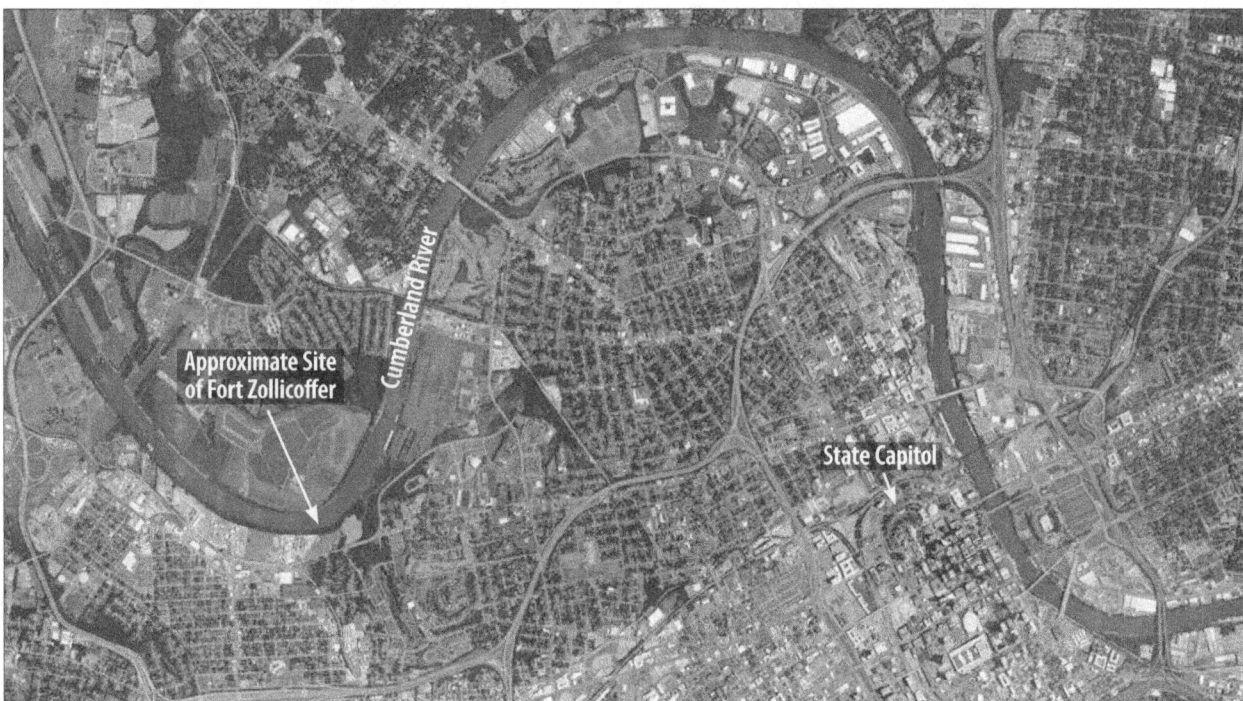

ernor of Tennessee, a position he would hold for the next 34 months. Late on the night of March 12th, Johnson entered Nashville "without display" and checked into the St. Cloud Hotel.

On March 5th, Nelson assembled the soldiers of his Fourth Division to witness the execution by firing squad of Private Michael Connell, who had been sentenced to death for firing upon a sentry while trying to sneak back into camp with applejack. Many rank-and-file soldiers considered the shooting an "official murder."

On March 12th, Bull Nelson marched his men to the Hermitage plantation in honor of President Andrew Jackson's 95th birthday (Jackson died in 1845). Five days later, Nelson's division was marching to the southwest to join the Federal troops converging on the Tennessee River at Pittsburg Landing, in preparation for attacking and capturing the railroad junction at Corinth, Mississippi.

Confederates Save Stores During The Great Panic

Testimony by Col. N.B. Forrest on his activities during the Great Panic: "On Tuesday morning I was ordered by General Floyd to take command of the city, and attempted to drive the mob from the doors of the departments, which mob was composed of straggling soldiers and citizens of all grades. The mob had taken possession of the city to that extent that every species of property was unsafe. Houses were closed, carriages and wagons were concealed to prevent the mob from taking possession of them. Houses were being seized everywhere. I had to call out my cavalry, and, after every other means failed, charge the mob before I could get it so dispersed as to get wagons to the doors of the departments to load up the stores for transportation. After the mob was partially dispersed and quiet restored a number of citizens furnished wagons and assisted in loading them. I was busily engaged in this work on Friday, Saturday, and Sunday. I transported 700 large boxes of clothing to the Nashville & Chattanooga Railroad depot, several hundred bales of Osnaburg and other military goods from the Quartermaster's Department, most, if not all, of the shoes having been seized by the mob. I removed about 700 or 800 wagon loads of meat. The high water having destroyed the bridges so as to stop the transportation over the Nashville & Chattanooga Railroad, I had large amounts of this meat taken over the Tennessee & Alabama Railroad. By examination on Sunday morning I found a large amount of fixed ammunition in the shape of cartridges and ammunition for light artillery in the magazine, which, with the assistance of General Harding, I conveyed over 7 miles on the Tennessee & Alabama Railroad in wagons, to the amount of 30 odd wagon loads, after the enemy had reached the river. A portion was sent on to Murfreesborough in wagons. The quartermaster's stores which had not already fallen into the hands of the mob were all removed, save a lot of rope, loose shoes, and a large number of tents. The mob had already possessed themselves of a large amount of these stores. A large quantity of meat was left in store and on the river bank and some at the Nashville & Chattanooga Railroad depot, on account of the break in the railroad. I cannot estimate the amount, as several store-houses had not been opened up to the time of my leaving. All stores left fell into the hands of the enemy, except forty pieces of light artillery, which were burned and spiked by order of General Floyd, as were the guns at Fort Zollicoffer. My proposition to remove these stores, made by telegraph to Murfreesborough, had the sanction of General A. S. Johnston."

Defenses of Nashville

Nashville, a vital city which had barely been fortified by the Confederates in 1861, became, under Federal occupation, the most heavily fortified city in North America, second only to Washington, D.C. Nestled in a bend of the Cumberland River, major portions of Nashville were protected by this natural barrier and the ironclad Federal gunboats which patrolled it. During the course of the war, Nashville became a major transportation hub and logistics center. Holding Nashville in Federal hands was imperative, the military governor arguing that the city be burned to the ground rather than fall into Confederate hands again.

Eventually, Nashville's defenses would consist of three major forts, a fortified state capitol building, fortified railroad bridge over the Cumberland River, and 21 minor installations along 30 miles of earthworks configured into inner and outer lines around the city, plus patrols by a variety of Federal gunboats. On occasion, field artillery batteries would be stationed on the pikes outside the city.

Shortly after the city's capture, the Union forces placed batteries at barricades controlling the eight roads into the city—the Lebanon, Murfreesboro, Nolensville, Franklin, Granny White, Hillsboro, Harding and Charlotte turnpikes. The guns were placed so that they could easily be turned against the city itself in case of a civilian uprising.

The fortified railroad bridge and the State Capitol each were supplied with four artillery pieces and six companies of infantry. The guns at the bridge were on the south or west side of the river. The building at the corner of Broad and Spruce will garrison two companies and overlook a position holding one cavalry regiment and one battery of horse artillery.

On Aug. 6th, 1862, General Don Carlos Buell ordered Capt. James St. Clair Morton of the US Army Corps of Engineers to Nashville to prepare fortifications. "For the present," said Buell, "I only propose to throw up small works to hold from four to six companies and from two to four pieces of artillery. They should be in the edge of the city, to command the principal thoroughfares and other prominent points...See Governor Johnson, and if he approves, devise some defenses also around the capitol; devise also some defenses for the bridge."

About this time, Capt. Morton placed earthen parapets, a log stockade, and bales of cotton at the State Capitol at the urging of Military Gov. Andrew Johnson. Fort Johnson, as it came to be known, was armed with 15 heavy guns and a regiment of infantry.

Morton began work on three main forts—Fort Negley on St. Cloud Hill, Fort Morton, and Fort Houston.

Using impressed labor, slaves, and mules, Fort Negley was built in 1862-64 as an impressive defensive work, the largest inland stone fort built during the war. Late in the war, Fort Negley was renamed Fort Harker in honor of a general killed in action (the new name didn't stick).

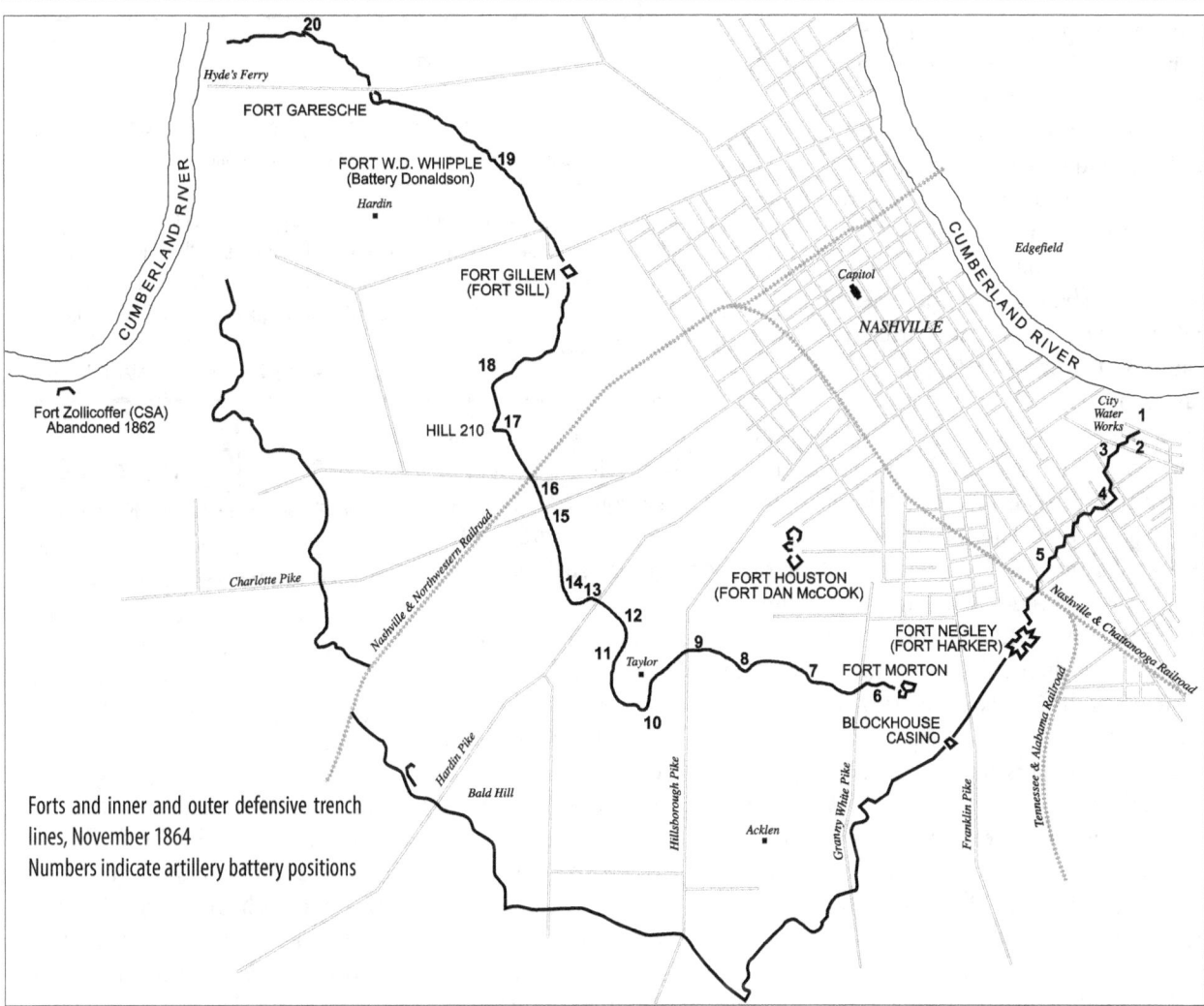

Forts and inner and outer defensive trench lines, November 1864
Numbers indicate artillery battery positions

Across Franklin Pike was Fort Morton, another large and complicated structure situated on the solid limestone of Curry Hill (current location of Rose Park). The site had been occupied by the house of Dr. William P. Jones, superintendent of the Insane Asylum and prominent Unionist, which was blown up and cleared.

Further south was Blockhouse Casino on Kirkpatrick Hill, now occupied by the City Reservoir. A large cross-shaped blockhouse was constructed there which could be defended, if necessary, by guns at Fort Negley and Fort Morton.

The third fortification, Fort Houston, was built on high ground near the current Music Row traffic roundabout. The home of prominent Union loyalist Russell Houston was destroyed to built the fort named after him. Later it was renamed Fort Dan McCook.

Work progressed slowly at Forts Morton and Houston due to the lack of labor, the complex design of the structures, the rocky terrain, and other more timely priorities.

About this time, Morton placed earthen parapets and a log stockade at the State Capitol at the urging of Military Gov. Andrew Johnson. Fort Johnson, as it came to be known, was armed with 15 heavy guns and a regiment of infantry.

The Louisville & Nashville railroad bridge across the Cumberland River was fortified with guardhouses and stockades with loopholes so that infantry on the bridge could shoot at any attackers. The modifications required 40,000 board feet of heavy timbers. Nine heavy guns were set on the bank of the river for defense of the eastern part of the city.

"Each prominent street is barricaded, and beautiful homes have been relieved of their roofs and turned into forts or rifle pits, and the fort on St. Cloud Hill, which commands the town and surrounding country has been made of such strength that it can scarcely be taken," wrote a Louisville newspaper correspondent.

By late 1862, artillery, usually consisting of one 32-pounder Parrott rifle captured from the Confederates, was stationed at various sites around the city, including the reservoir near the river, Lebanon Pike, the end of Summer Street, General Palmer's headquarters, the railroad tunnel, the old Lunatic Asylum, and the pontoon river bridge. There were also 28 other captured guns at the ordnance depot, only four of which were considered safe to use.

In May 1864, a grand depot magazine was completed at the center of an eight-acre field where the city hospital had been located prior to burning down in early 1863. The rectangular underground structure measured 65 feet by 200 feet and was covered with earth eight feet thick. Magazine Granger was supplied by a branch railroad line built on a trestle. The depot was well-ventilated and virtually waterproof. Citizens were relieved to have gunpowder stored outside the city limits.

In October 1864, under the direction of Gen. Zealous B. Tower, Army engineer, revised plans for the defensive fortifications of Nashville were submitted to Washington. Original plans were simplified so that work could resume on the forts. With the approach of the Confederate Army of Tennessee in November 1864, work by laborers, soldiers, and quartermaster employees commenced on both the inner and outer defensive lines and continued around the clock.

The inner defensive line was seven miles long and contained 20 batteries of heavy guns. Encompassing the city and all significant Federal works, the line would be manned by a garrison of 3,000 soldiers supported by 2,000 mobile troops plus 7,000 quartermaster employees and could repel a force of up to 30,000 men.

The inner works began at the Cumberland River at the water works (current site of the old General Hospital), and ran across University Hill to Fort Negley. From Fort Morton it ran westward to a strong salient at the Taylor farm (current site of Vanderbilt University). Then it ran northerly (along the eastern boundary of the current Centennial Park) to a fortification known as Hill 210 (current site of Washington School). Here was placed two bastion fronts for 15 cannons, supported by rifle pits.

The next large fort was Fort Gillem (later renamed Fort Sill) located at the current site of Jubilee Hall on the Fisk University campus. It was built by the 10th Tennessee Regiment, commanded by Gen. Alvan Gillem. It was a redoubt 120 feet square, with six-foot-high stone walls, embrasures for 13 guns, two service magazines, and a blockhouse.

A mile north of Fort Gillem was Fort Garesché, built by the 2nd Ohio Volunteers and housing 14 guns and three magazines. It was located at the present intersection of Buchanan Street and 26th Avenue. Between Fort Gillem and Fort Garesché was Battery Donaldson (aka Fort W.D. Whipple), a small battery with an octagonal bombproof blockhouse.

Beginning at Blockhouse Casino and diverting from the inner lines was the outer line of defense, also running west to the bend in the Cumberland River. The distance between the inner and outer lines ranged from half-a-mile to a mile.

The outer line ran south along Granny White Pike to the hilly main salient south of the Acklen estate. This was the pivot point around which the Union armies marched against the enemy left flank on Dec. 15th, 1864.

The outer line then ran northwest to Bald Hill (currently Love Circle), where a strong battery was located. The line then turned north, with a short offset along the railroad, and then north to the river where Tennessee State University is now located.

By mid-1863, Nashville was the supply center for armies in the Western Theater, handling massive amounts of materiel. Supplies manufactured or warehoused in Louisville

The March of the Armies

The movement of troops in the West is on a grand scale, and you can hardly conceive — though you have seen much in New-York — with what energy and activity new armies are gathering for the field, from Ohio, Indiana, Illinois and Missouri. The reserves are moving forward. As the army advances through another State its place will be filled up and its communications maintained by new divisions. As place after place is taken in the South, they will be garrisoned and maintained, till the whole territory of rebellion is held and crushed — like a dwarf in the hands of a giant. Such is the programme of the campaign going on, and I doubt not it will be carried out to the least particular.

— The New York Times, March 2, 1862

Corps heavily fortified D.C.

Nashville was second to Washington, D.C. in scale of fortifications. In the District of Columbia, by April 1865, the Corps of Engineers had overseen the use of $1.4 million to construct 20 miles of rifle pits and 30 miles of military roads serving more than 1,400 gun emplacements in 68 forts and 93 batteries.

and Cincinnati were transported down the Ohio to the Cumberland River at Smithland, Ky., then upstream to Nashville. During March 1864, 213 steamboats delivered and unloaded 62,666 tons of cargo at the city's two river levies (a third levy was later added). During the summer and early fall of 1864, the quartermaster's corps handled 2,000 tons of supplies each day for a period of 150 consecutive days. It is interesting to note that although a comprehensive system of fortifications was planned for Nashville in 1862-63, it wasn't until the approach of the Confederate Army of Tennessee in November 1864 that much of the works were completed, albeit with less elaborate design. The Confederates did not attack the works, knowing they were formidable and that they were outnumbered and outpowered.

Some of the Nashville works were not complete in early 1865, and were recommended to be completed despite the fact that the war was clearly winding down. By the summer of 1865, it became evident that work on

(Continued on Page 71)

Fortress Nashville: Pioneers, Engineers, Mechanics, Contrabands & U.S. Colored Troops

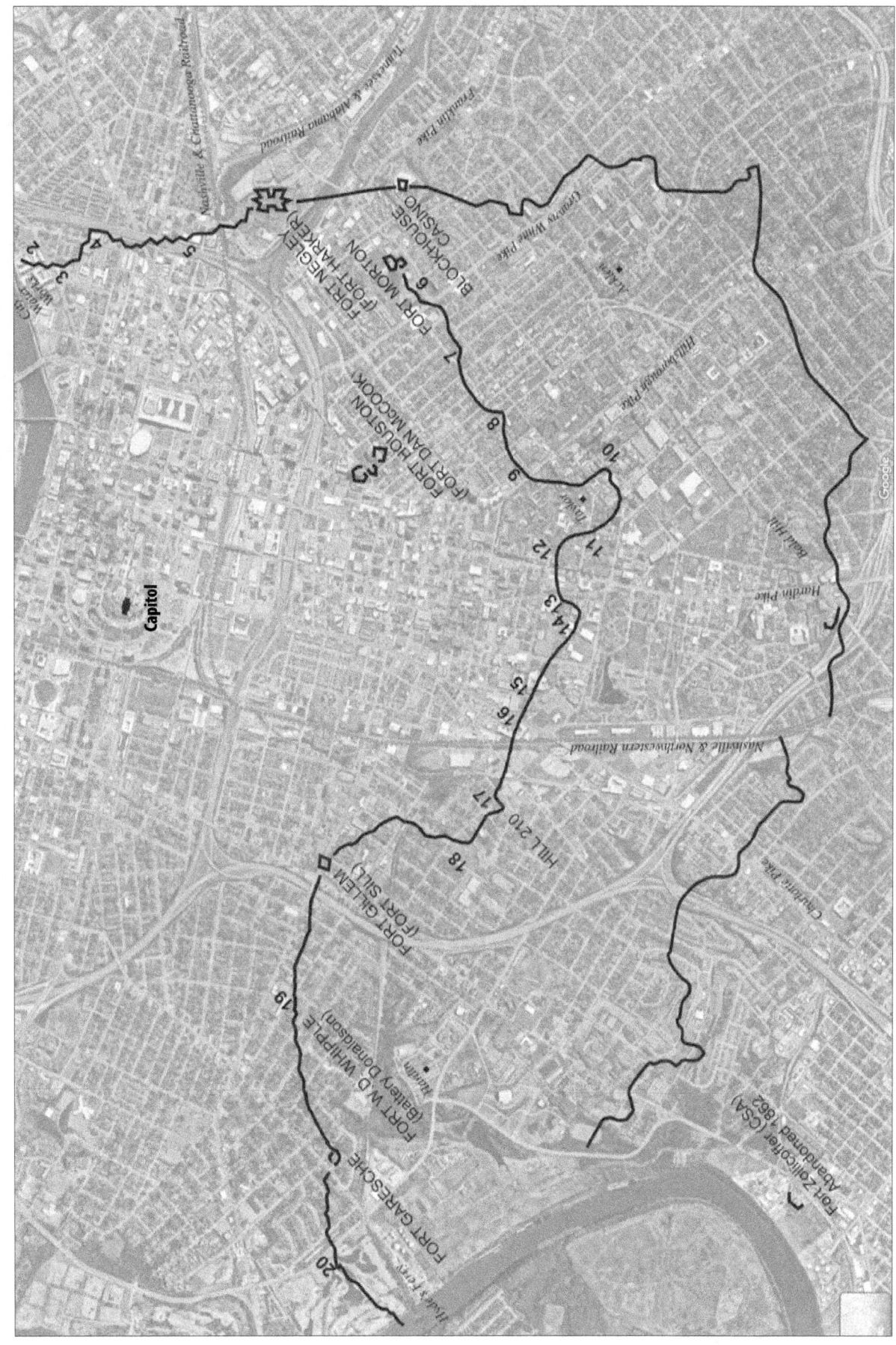

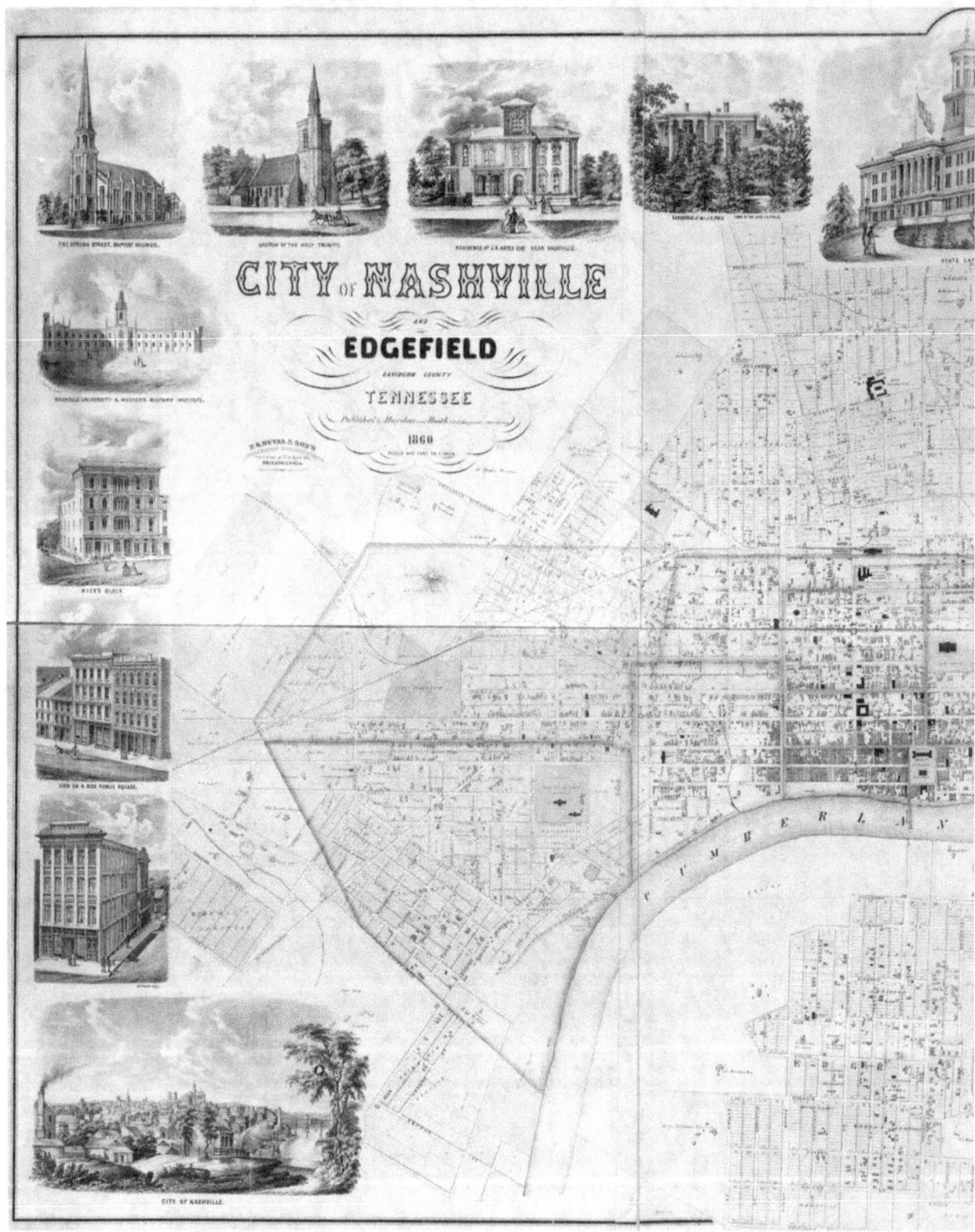

Detailed map of Nashville in 1860 and some of the city's significant architecture, drawn and published by Haydon & Booth engineers and surveyors.

Fortress Nashville: Pioneers, Engineers, Mechanics, Contrabands & U.S. Colored Troops

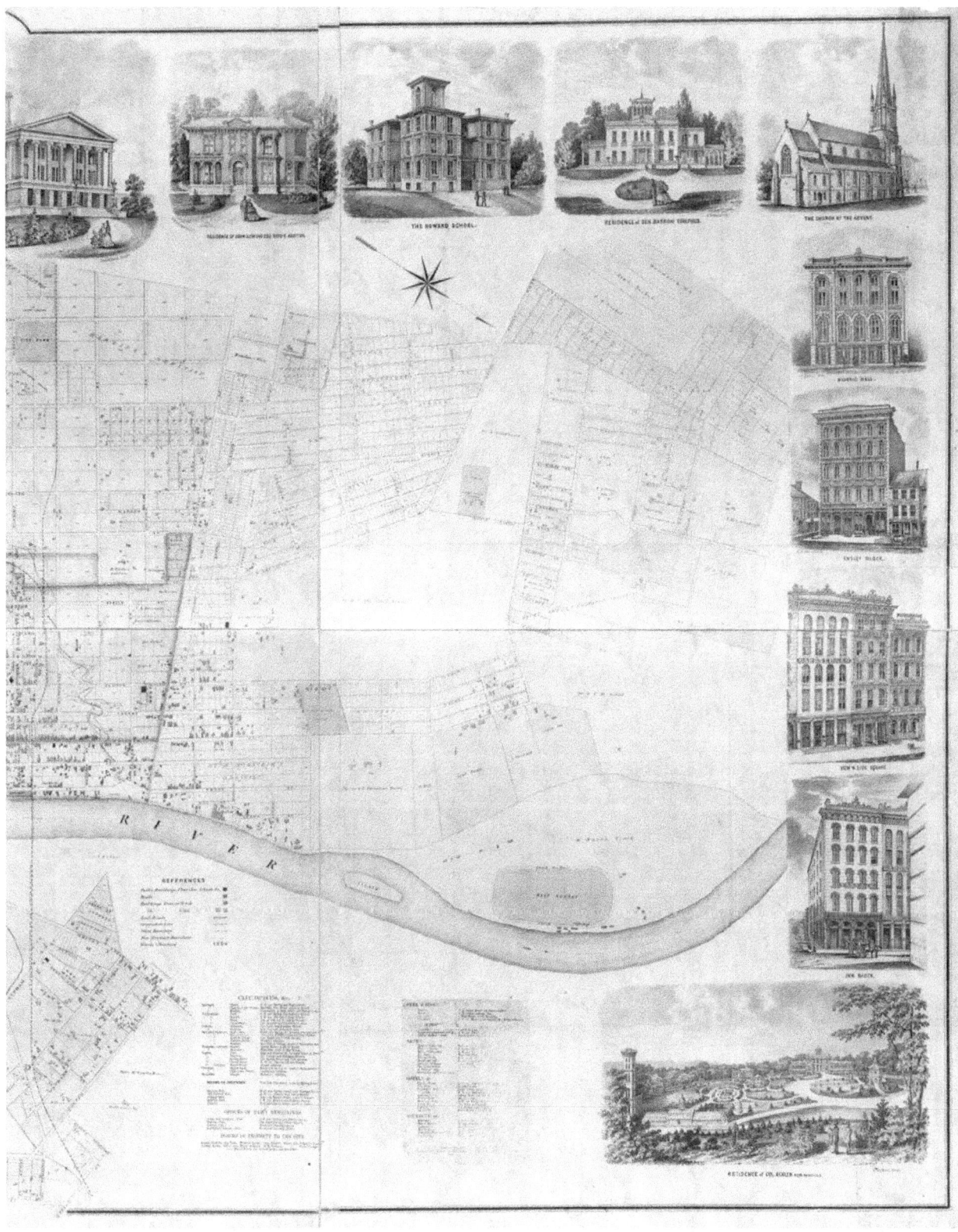

Fortress Nashville: Pioneers, Engineers, Mechanics, Contrabands & U.S. Colored Troops

NASHVILLE, TENN., FROM THE SOUTH-EAST.
SHOWING THE STATE CAPITOL.

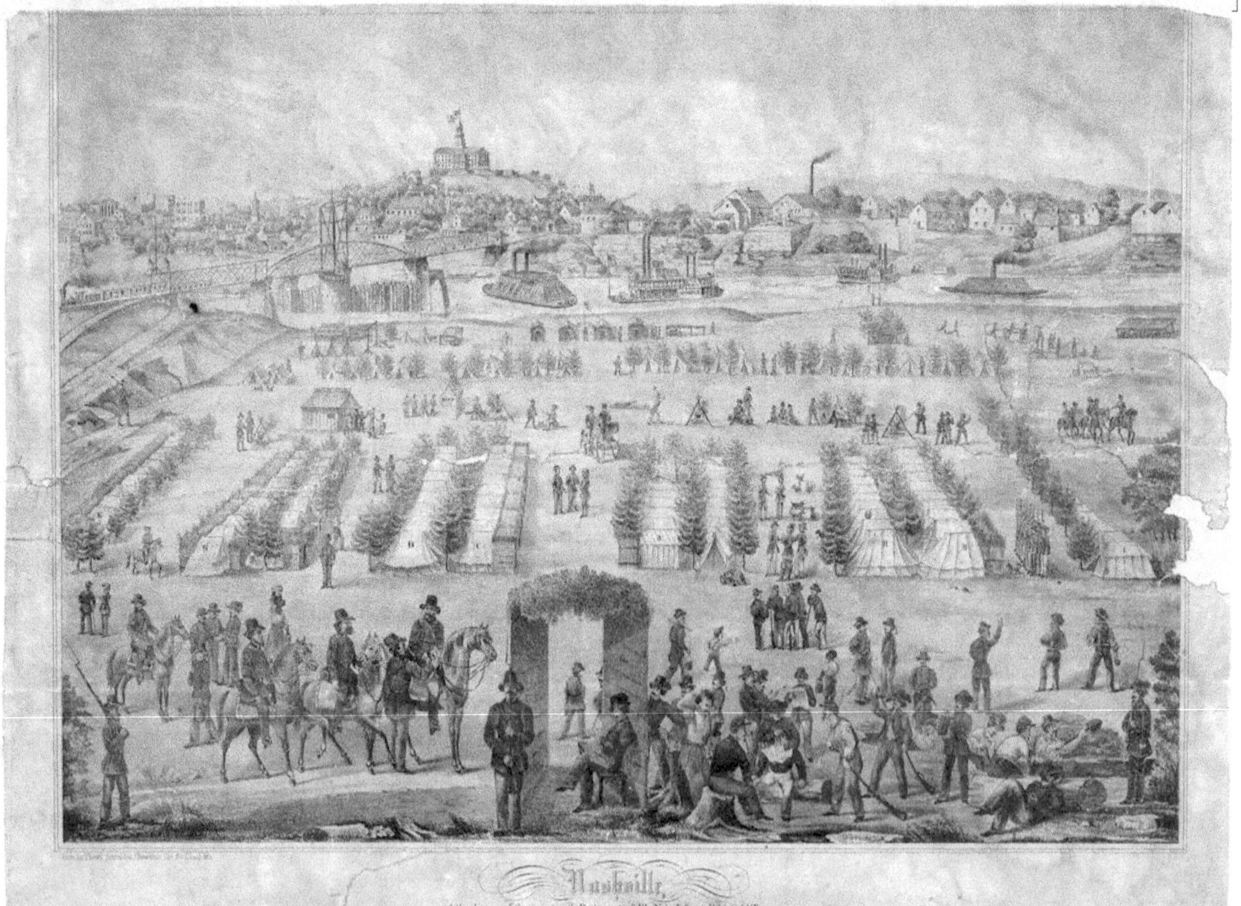

Camp of the 16th Illinois Volunteer Infantry Regiment at Edgefield, across the Cumberland River from Nashville.
The lithograph was created by Theodore Schrader of St. Louis, Missouri.

ARMY OF THE CUMBERLAND—NASHVILLE AND ITS FORTIFICATIONS.
(Library of Congress)

Topographical Map of Nashville (1880s) Showing Railroad Lines and Major Hills

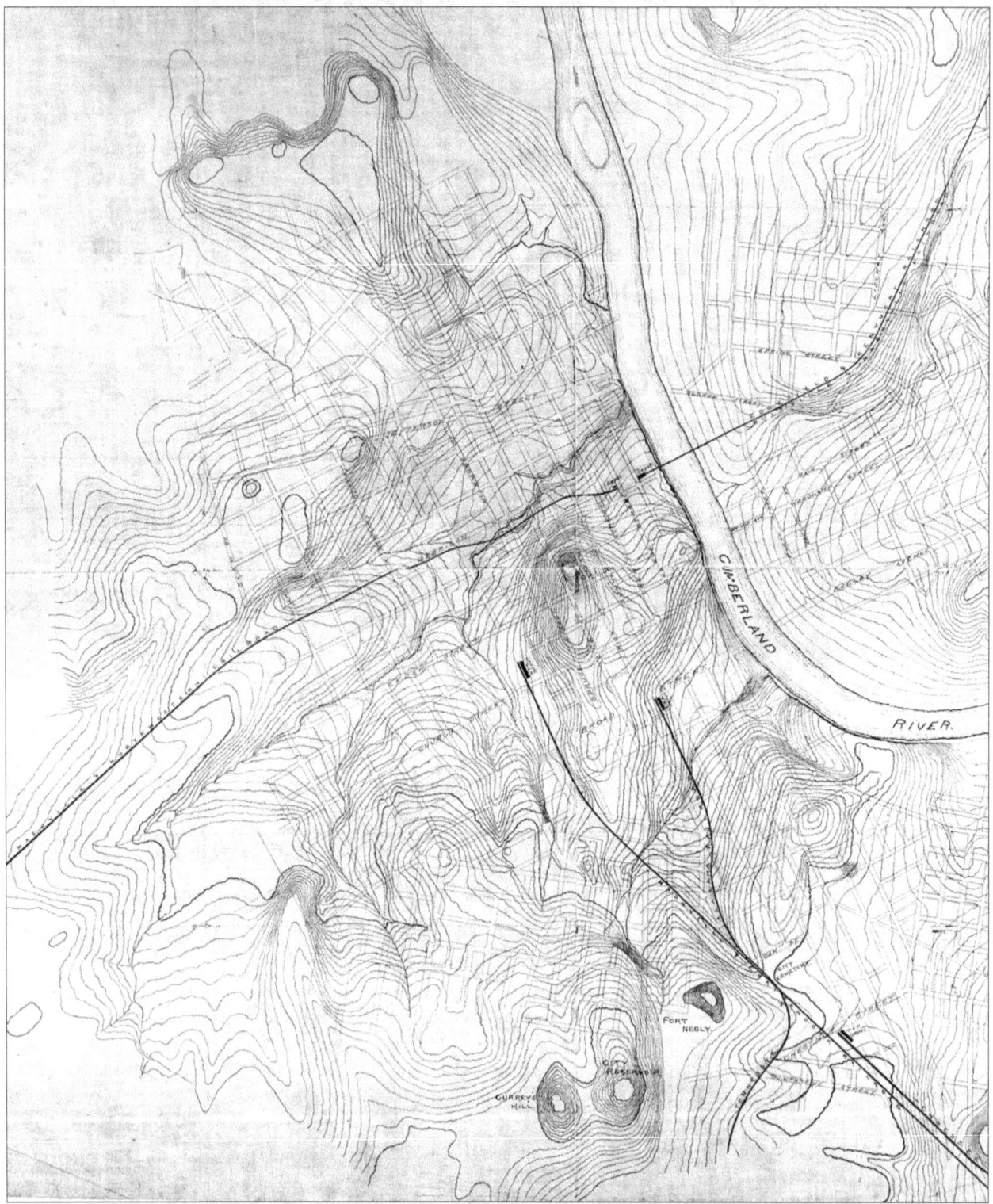

(Tennessee State Library and Archives)

The modern and everchanging skyline of Nashville as seen from the sally port of Fort Negley atop Saint Cloud Hill.

(Continued from Page 64)
the fortifications should cease.

In 1865, General Z.B. Tower explained: "Col. Merrill gave little attention to the defenses of this depot, being principally occupied with those at Chattanooga. For so important a place, held so long by our troops, the Nashville defenses certainly were not pushed forward as much as they should have been. Little aid is given by commanding officers of posts when those posts are not in the front or constantly exposed. In such positions building redoubts is the first operation, while far back on the line of communication it is very difficult to get a detail to throw up lines. Every other labor takes precedence. Capt. Barlow took immediate charge of the works around Nashville the 13th of November, under my direction, and has performed his duties faithfully and intelligently. Capt. Jenney gave me much assistance, superintending at Forts Houston and Gillem and upon the lines.

Majors Dickson, Powell, and Willett assisted in the construction of the entrenchments around the city, which were mostly executed the first week in December, 1864, by the quartermaster and railroad employees."

Nashville was transformed over time. According to historian Flagel, soldiers of the Army of the Cumberland found the capital exceedingly unpleasant to the senses. "The war had mutated the vibrant center of commerce and higher education into an odiferous, crowded, military base of cheerless earthworks, careworn edifices, and morbid hospitals. Especially for the rural boys, it all had the feel of a dark satanic mill. During his brief stay, Cpl. John Leek of the 92nd Illinois wrote, 'Mother, this is a forsaken looking country. It don't look like it was worth fighting for. The country is stripped of everything but the hills'."

Most of the fortifications were abandoned and taken down (usable materials sold at auction for roughly $30 million) during 1865-66. The Federal army left Nashville in 1872.

Nashville's Rock City Guards Volunteered for Confederacy

By the end of April 1861 there were 16 companies of volunteers drilling in Nashville, fearful they would not be ready in time to participate in the upcoming glorious, yet brief war against Northern aggression. There was Beauregard's Light Infantry, the Hickory Guards, the Hermitage Guards, Tennessee Rangers, Tennessee Rifles, Cheatham Rifles, and Harris Guards.

Formed two years prior and recognized by the Legislature as the city's official volunteer militia was the Rock City Guards, which boasted three companies of 100 men each. The eager volunteers drilled in the City Square, near the company headquarters at the Market House, and in Edgefield across the river.

In April, the ladies of Nashville fashioned a flag for the Rock City Guards and presented it with these words, in part: "We place in your hand this flag that its folds may wave over you in the dark hour of conflict and recall to your minds the recollection that your mothers, your sisters and your children are praying at your firesides for your triumph and your safe return."

On May 3rd at the State Capitol, the Rock City Guards were mustered into service as part of the First Tennessee Infantry Regiment of Volunteers.

After military training and drilling at Camp Harris in Allisonia and at Camp Cheatham in Springfield, the Rock City Guards returned to Nashville and camped on the lawn of the Nashville Female Academy and were greeted by wellwishers the next day at the train station.

On July 10th, the Guards and the First Tennessee shipped out to Virginia and then saw action in most of the battles of the Western Theater. They returned to Nashville in December 1864 for their final desperate fighting.

Some of the Rock City Guards survived the war, but it was written in October 1863 that of the 300 men enlisting in the Rock City Guards more than two years earlier only about 25 were still alive.

Map of City of Nashville - 1864

1. Hynes High School, part of Hospital No. 15, for soldiers with venereal disease.
2. Adelphi Theatre. Built in 1850 by local architect Adolphus Heiman with second largest stage in America.
3. Hicks Building, Morgan Building, and Douglas Building, on Public Square. Dry goods and readymade clothing companies; used by Federals to store ordnance. Also Union Hotel.
4. Inn Block. Site of old Nashville Inn, which burned in 1856. Clothing merchants, druggists, grocer, liquor store and book bindery located here. Used as medical and commissary storage.
5. Southern Methodist Publishing House (1854). Used to print army forms and reports.
6. City Hotel.
7. Watson House hotel. Part of Hospital No. 19.
8. Ensley Building. Five-story Italianate building used as part of Hospital No. 3.
9. Morris Stratton & Co. Wholesale grocers; used as part of Hospital No. 19.
10. French & Co. Three-story brick building used as part of Hospital No. 19.
11. Polk Place (1818-20) was home of the widow of President James K. Polk and site of the President's tomb, designed by William Strickland. The tomb is now located on the State Capitol grounds.
12. Felix DeMoville House (1857), served as Gen. Rousseau's HQ.
13. George W. Cunningham House. Used by commanding generals as their headquarters, including Grant, Sherman, Thomas, Buell, and Rosecrans.
14. New Theatre. Located in the Odd Fellows Grand Lodge.
15. Zollicoffer House. Townhouse on High Street, used as Provost Marshal's Office.
16. Three separate buildings — St. Mary's Catholic Church (1844-47) designed by Heiman or Strickland. The oldest standing church in Nashville. Next door was architect Adolphus Heiman's townhouse (1850). Also, Planter's Hotel; used as Officers' Hospital No. 17 and the Soldiers' Home by U.S. Sanitary Commission.
17. First Baptist Church (1841). Part of Hospital No. 15.
18. Bank of Tennessee. Built 1853 by Francis Strickland. Used as paymaster's department building.
19. Maxwell House Hotel. Begun in 1859 by John Overton, Jr. Used by Confederates as Zollicoffer Barracks and later by Federals as a prison for captured Confederates.
20. St. Cloud Hotel. One of the city's finest hotels and temporary quarters of Union generals.
21. Christ Church Episcopal. Built 1829-31 by Hugh Roland.
22. McKendree Methodist Church. Site of famous 1850 convention on Southern secession. Used as part of Hospital No. 21.
23. Cumberland Presbyterian Church. Part of Hospital No. 8.
24. First Presbyterian Church (Downtown Presbyterian Church). Egyptian Revival architecture. Used as part of Hospital No. 8. National Historic Landmark. Across street is Masonic Hall (1860), also part of Hospital No. 8.
25. E.H. Ewing & Co. wholesale grocers used as carpenter's shop.
26. Four commercial buildings used as Hospital No. 16, for "coloreds."
27. Hume School (1855). City's first public school building. Federals quartered railroad employees in it.
28. Dr. John Shelby's Medical College (1858) closed during the war and never reopened. An indigent hospital adjoined the structure, which included a teaching hospital.
29. Refugee camp, with wash station and kitchen.
30. Stables for RR depot drayage (16 bldgs. in neighborhood).
31. Commercial buildings and house used as part of Hospital No. 3.
32. Broad Street Fire Company No. 2. Firehall and engine house featuring a bell tower topped by statue of fireman.
33. University Medical Department. Classes were held here throughout the war. Used in conjunction with Hospital No. 10.
34. Factory (teamsters' quarters). Three-story brick building with wooden shingle roof and openable skylights.
35. Elm Street Methodist Church.
36. Cherry Street Baptist Church, used as the post hospital.
37. Primitive Baptist Church, part of Hospital No. 1.
38. Third Presbyterian Church and College Hill Armory, used as Hospital No. 1.

Extant Buildings: Tennessee State Capitol; St. Mary's Catholic Church; First Presbyterian Church (Downtown Presbyterian Church); Elm Street Methodist Church (office building); Fort Negley (Fort Negley Park); University of Nashville Literary Building (city government office building), Primitive Baptist Church.

Street Names 1864	Today	Street Names 1864	Today
Broad Street	Broadway	Locust St.	Jo Johnston St.
Castleman St.	Peabody	Market St.	Second Ave.
Cedar St.	Charlotte Ave.	McLemore St.	Ninth Ave.
Cherry St.	Fourth Ave.	Mulberry St.	Sixteenth Ave.
Clark St.	Bank St.	Priestley St.	Peabody
College St.	Third Ave.	Spring St.	Church St.
Cumberland Alley	Commerce St.	Spruce St.	Eighth Ave.
Front St.	First Ave.	Summer St.	Fifth Ave.
Guthrie St.	No longer exists	Union Alley	Union St.
High St.	Sixth Avenue	Vine St.	Seventh Ave.
Line St.	Jo Johnston St.	Williams St.	Seventeenth Ave.

Fortress Nashville: Pioneers, Engineers, Mechanics, Contrabands & U.S. Colored Troops

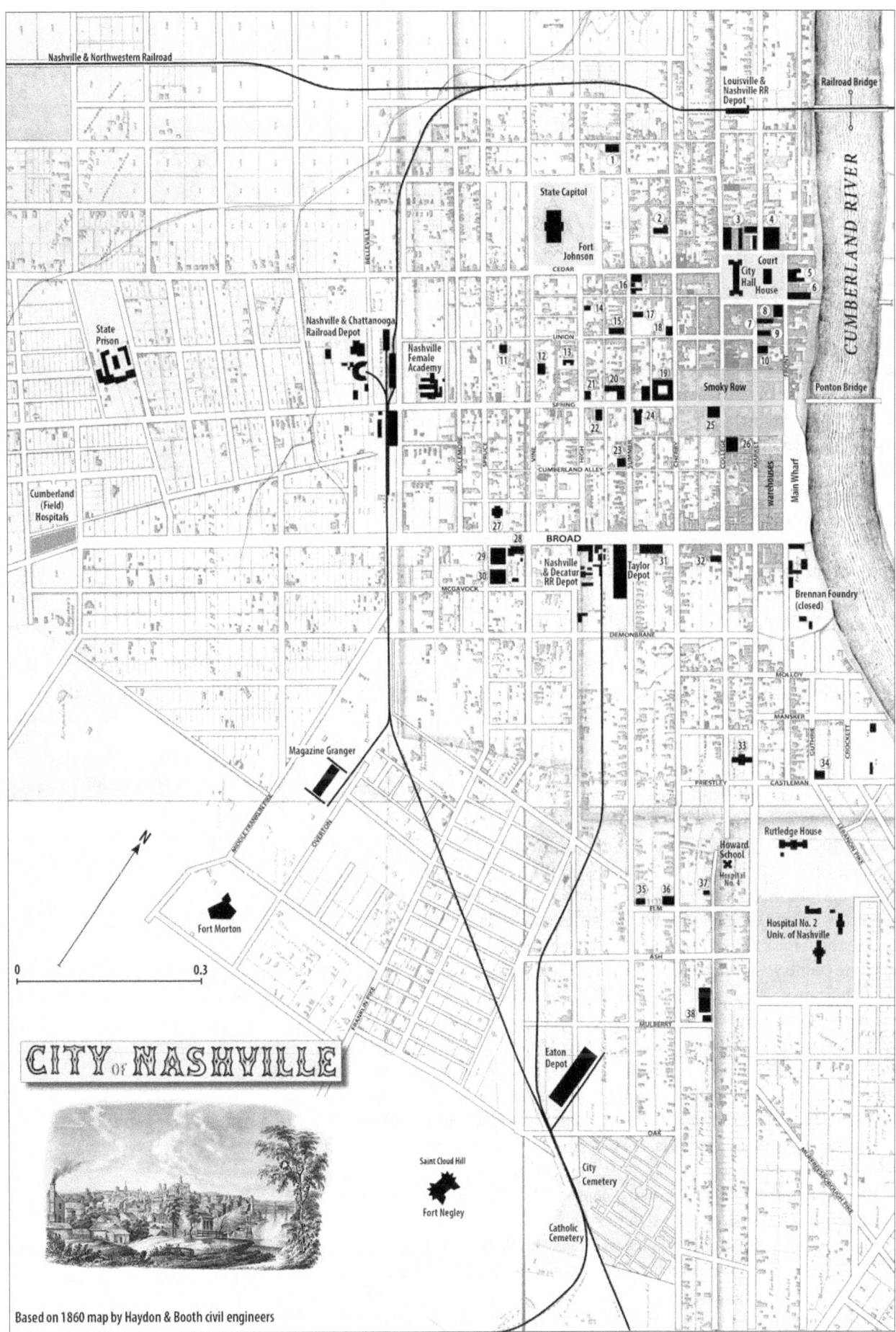

Fort Andrew Johnson

State capitol fortified to protect military governor

Andrew Johnson statue at the Tennessee State Capitol

Fort Andrew Johnson was the name given to the fortification of the newly completed Tennessee State Capitol atop one of Nashville's highest elevations, Cedar Knob or Campbell Hill. It was named for Andrew Johnson, military governor of Tennessee, who was concerned about Confederate cavalry raids and acts of sabotage. It was also known as Camp Andy Johnson and Capitol Redoubt.

Captain James St. Clair Morton of the U.S. Army Corps of Engineers converted the capitol building into a strong fortification with a stockade of cedar logs surrounding the building, reinforced bales of cotton and earthen parapets. Fifteen pieces of heavy artillery were emplaced at strategic points around the Capitol. Several companies of infantry and artillery garrisoned the fort. Fort Johnson became the headquarters of the command.

The big guns were never used in anger, but were fired several times in honor of Union victories or elections. For example, on April 3, 1865, when news of the fall of Richmond arrived, business was canceled and at noon the First Missouri Battery manned the guns and fired 100 rounds. Like many buildings in Nashville, the Capitol was temporarily used as a military hospital following the battle at Murfreesboro, Tenn. (Stones River), which ended on Jan. 2, 1863.

An inspection report dated May 25, 1865 by Brigadier General Zealous B. Tower, Inspector General of Fortifications, Military Division of the Mississippi, stated the following: "Capitol Hill.-Gen. Morton built some earth parapets and stockades around the capitol building large enough to mount fifteen guns and to give room for a regiment of infantry. The position has a good command over the country around, and, thus strengthened, was a good keep for the north portion of the city. No longer needed, the stockade is being removed at the request of the Legislature and by direction of the commanding general. Gen. Morton's line of defense successfully resisted Morgan's and Forrest's attacks during Buell's march into Kentucky. Afterward, Nashville became a great depot, and public buildings, as hospitals, store-houses, and corrals, extended far beyond the limits of the city and necessitated a much longer defensive line."

Johnson told U.S. Army officials that before surrendering the city he would have it burned to the ground.

The fort was abandoned by Union troops in 1867 after the end of the war and after Tennessee had returned to the Union.

The four-acre site, originally known as Cedar Knob, was the home of attorney George W. Campbell. Although Tennessee became the 16th state in 1796, Nashville wasn't designated the permanent capital until 1843. Knoxville, Kingston, and Murfreesboro served at times as the state capital, and in Nash-

The Tennessee State Capitol, aka Fort Johnson, sat atop Campbell's Knob, the highest point in Nashville, in this view from the southeast. The grounds are in obvious disarray and include military tents, arbors, and a construction office along with visible stockades. (Library of Congress)

ville the legislature met at the Davidson County Courthouse.

Representing the city, Mayor Powhattan Maxey bought the tract for $39,000 and offered it to the state for free, a major inducement for locating the capital in Nashville. Campbell's house was moved to south of Cedar Street (now Charlotte Avenue) and served as home to Military Governor Johnson during the Union occupation of Nashville. Along with his appointment as military governor, Johnson, who had no military training, was appointed a brigadier general in the Federal army. Johnson was shot and wounded by a drunken Federal soldier while walking from the Capitol to his executive residence across the street. The wound was not serious, and Johnson later pardoned the assailant.

A native of Raleigh, N.C., Johnson was a self-made man from humble origins, working as a tailor in East Tennessee when he first embraced politics. A Democrat, Johnson rose from local politics to become a Congressman and Senator, Tennessee Governor, and the military governor of Tennessee during the Union occupation (1862-65). He was the only U.S. Senator in the South to refuse to resign when his state seceded.

As military governor, he dealt harshly with secessionists, demanding they take an oath of allegiance to the Union and imprisoning many who did not cooperate. Lincoln chose him as his running mate in the 1864 election. He served as Vice-President until the assassination of Lincoln in the spring of 1865. He then served as the country's 17th President until 1869. Politics were bitter following the war, and Johnson became the first President to be impeached. Afterward, he served briefly again in the U.S. Senate, the only President to do so. He died in 1875. He is buried in Greeneville, Tenn., at the Andrew Johnson National Historic Site. A statue of Johnson, an exact copy of one at Greeneville, was erected at the southeast corner of the State Capitol in 1995.

Tennessee's is one of the few state capitols topped by a tower (forty others have domes). The Capitol is also a National Historic Civil Engineering Landmark for being the first to use structural iron roof trusses.

The Tennessee State Capitol is a

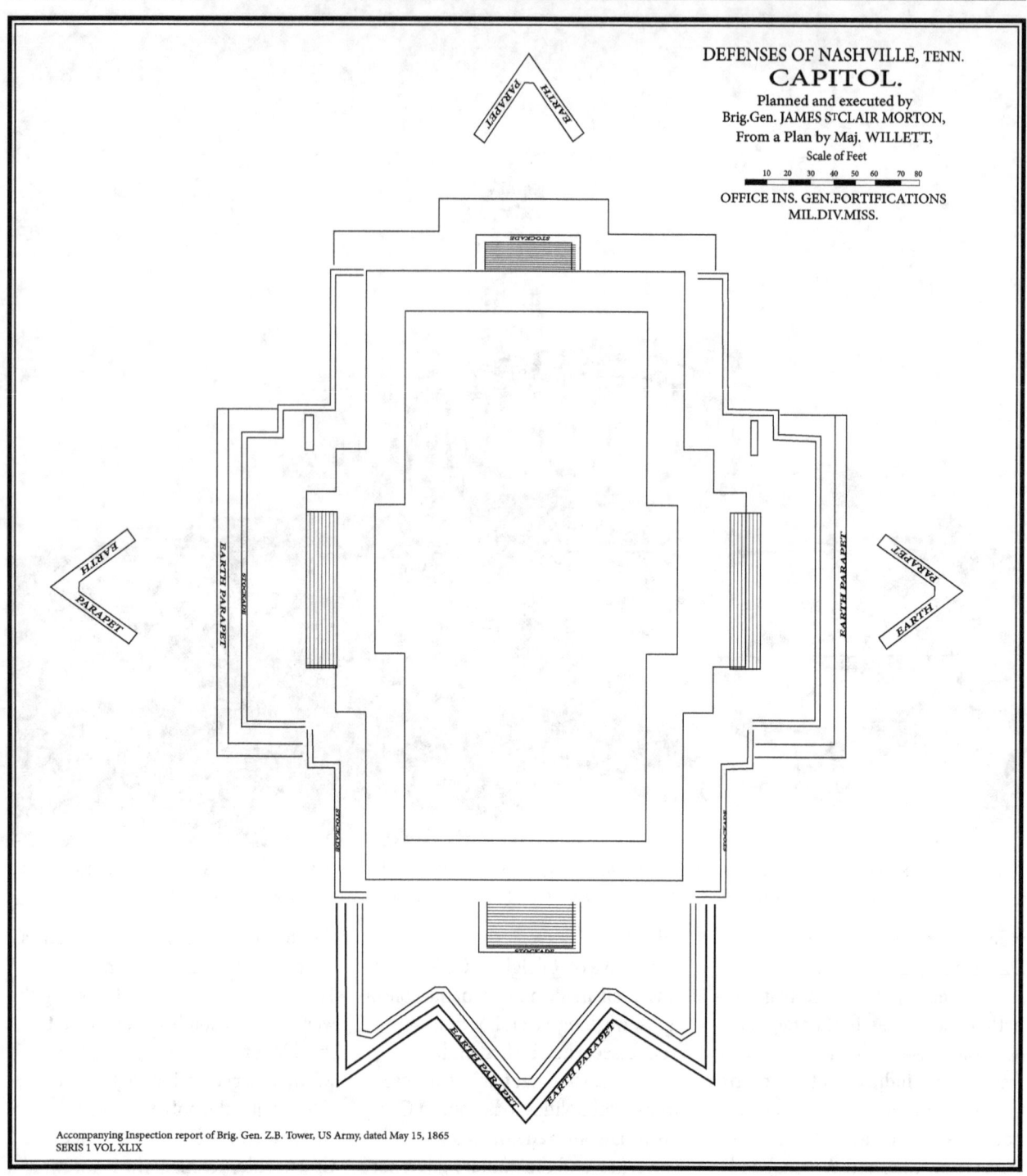

Accompanying Inspection report of Brig. Gen. Z.B. Tower, US Army, dated May 15, 1865
SERIS 1 VOL XLIX

Greek Ionic temple by design, 236 feet by 109 feet, with a unique tower bringing its height to 206 feet. There are porticoes at each of the four main facades. The east facade, facing the river, is the historic main entrance. The east and west porticoes each feature six Ionic stone columns. The larger north and south porticoes each feature eight columns. Each carved stone column is 36 feet tall and 4.5 feet in diameter.

The square rusticated base of the tower is 42 feet tall and the slender circular tower is 37 feet tall. The tower is based on the Choragic Monument built in Athens, Greece about 334 B.C. by Lysicrates, a great choral leader. The Tennessee tower is twice the size of the original Greek monument, also known as the Lantern of Diogenes. The columns of the tower are of the Corinthian order, with acanthus leaves. The Capitol was built entirely of stone. Efforts during construction to hold down costs by substituting brick interior work were rejected. The interior walls and columns were made from East Tennessee marble. Architect William Strickland noted that there are "no examples of any buildings in the United States, either public or private, in which the walls are con-

View from Tennessee State Capitol looking northeastward and showing the heavy siege guns, earthworks, and a portion of stockade around the structure also known as Fort Johnson. In the distance can be seen the Cumberland River and a tent encampment. Photo taken by George N. Barnard. (The J. Paul Getty Museum, Los Angeles. Used with permission.)

structed of rubbed or polished stones on the interior, indeed, there are very few buildings in Europe of this handsome and permanent class." Strickland died during the construction and is entombed in the Capitol.

In July 1863, Unionists from across the state met at the Capitol and pressed for state elections, resolving that only Unionists be able to vote, as well as Tennesseans in the U.S. Army. In September, Lincoln instructed Johnson to re-inaugurate a loyal state government, confident that Tennessee was clear of "armed insurrectionists."

In December, Lincoln announced his Amnesty Oath, designed to encourage Southerners to return to their homes. Radicals thought the oath too lenient. In the first three months of 1864, more than 1,300 took the amnesty oath at Nashville. However, it was not uncommon to find guerillas who had signed copies of loyalty oaths on their persons.

By the beginning of 1864 Johnson was mentioned as a candidate for higher federal office, even as Lincoln's vice-presidential running mate in the 1864 election.

In late January, however, Johnson shocked many by proclaiming that all voters in the March county election would subscribe to a stringent loyalty oath which included the extension of the Emancipation Proclamation to Tennessee. Johnson's oath was more stringent than Lincoln's amnesty oath. Thousands refused to take Johnson's "Damnesty Oath." In an election that newspaper editors called a farce, the Union candidates were elected 1,220 to 420.

Johnson still had much work to do before he left Nashville for Washington in late February 1865 to be inaugurated as U.S. Vice-President. In January a Unionist convention proposed amendments to the state constitution, the abolishment of slavery, and nullified the state legislative acts of 1861. Also, the convention denied a request by local blacks seeking the right to vote. In February a referendum ratified those amendments overwhelmingly since only Unionists were allowed to vote. The vote in Nashville was 1,416 to 3.

On March 4, 1865, William "Parson" Brownlow was elected Governor

FRANK LESLIE'S ILLUSTRATED NEWSPAPER. [SEPT. 24, 1864.

THE CAPITOL, NASHVILLE.

Our Special Artist at Nashville some time since, when the rebels were menacing that important city in hopes of diverting Sherman from Atlanta, was struck with the scene at the State Capitol. The building itself, of immense proportions and of magnificent design, was entirely surrounded by earthworks and stockades of great strength; all the staircases and other approaches being thus defended. The strange association of the beautiful in art with the grim weapons of war is as sad as it is picturesque. The view of the country beyond, with tents scattered far off in the distance, and the city around formed a scene seldom equalled and not easily forgotten.

VIEW OF THE STOCKADES AND FORTIFICATIONS AROUND THE CAPITOL, NASHVILLE, TENN.—FROM A SKETCH BY FRED. B. SCHELL.

along with the entire Union slate of candidates.

On April 3rd, the state General Assembly convened in Nashville. News was received that Richmond, Va. had fallen to Union forces. Business was canceled. At noon the First Missouri Battery manned the guns at the Capitol and fired 100 rounds. The festivities continued into the night.

Two days later, Brownlow was inaugurated as Tennessee's civilian governor. The legislature ratified the 13th Amendment to the U.S. Constitution abolishing slavery.

In 1866 Tennessee was formally readmitted into the Union.

The Capitol played a significant role in the story of the American flag becoming known as Old Glory.

Captain William Driver (1803-1886), a veteran merchant ship's captain from New England who had sailed around the world twice, retired to Nashville in 1837 after the death of his wife. In 1824 his wife had sewn a large 10x17 American flag to fly on his ship. When he saw it for the first time, he called it Old Glory.

In Nashville, Capt. Driver, a staunch Unionist, flew his beloved flag over the street at his home on holidays and his birthday, March 17th. Originally it bore 24 stars but ten more were added in 1860, reflecting the growth of the nation, along with an anchor sewn on the canton. When the Confederates took over Nashville, he hid the flag within a quilt.

On Feb. 25, 1862, Union soldiers captured Nashville without firing a shot and marched up Cedar Street to the State Capitol. They raised the Stars and Stripes above the building. Driver uncovered his cherished flag and climbed above the east entrance and hoisted Old Glory up the flag mast. Soldiers and civilians then began to refer to the American flag as Old Glory, and the name stuck.

Play Ball!

In a field north of Capitol Hill, in what is now Bicentennial Capitol Mall State Park, encamped Federal soldiers played baseball, a sport new to Nashville. At the time of the war, baseball was a far different game than the one we know today. The ball was softer; outfielders and even some infielders played without gloves. A runner was out when he was hit by a thrown or batted ball, or if a fielder caught the ball in the air or on a bounce. Scores were most always in the double digits—sometimes more than 60 runs scored. Home runs were aces.

Union officers encouraged baseball games among their soldiers because they reduced the boredom of camp life and kept the men in a competitive spirit. In 1870 the field became known as Athletic Park, and baseball continued to be played here until the final season of the Nashville Vols in 1963. Today, a new stadium, home of the minor-league Nashville Sounds, sits nearby.

Fortified Railroad Bridge

The Louisville & Nashville Railroad bridge over the Cumberland River was rebuilt and fortified by Federal authorities during the occupation to discourage Confederate raiding parties. The swing-span bridge was built during the 1850s. The center span could be swung 90 degrees on a central pivot to allow steamboats with tall chimneys to pass. In this view looking west toward downtown Nashville, the wooden bulwarks with loopholes for riflemen, turret observations posts, swinging door gates, and aerial telegraph lines can be seen. This bridge, replaced in the 1900s with a metal structure, is still used today; the original stone piers are still in use. (Library of Congress)

Pontoon Bridges

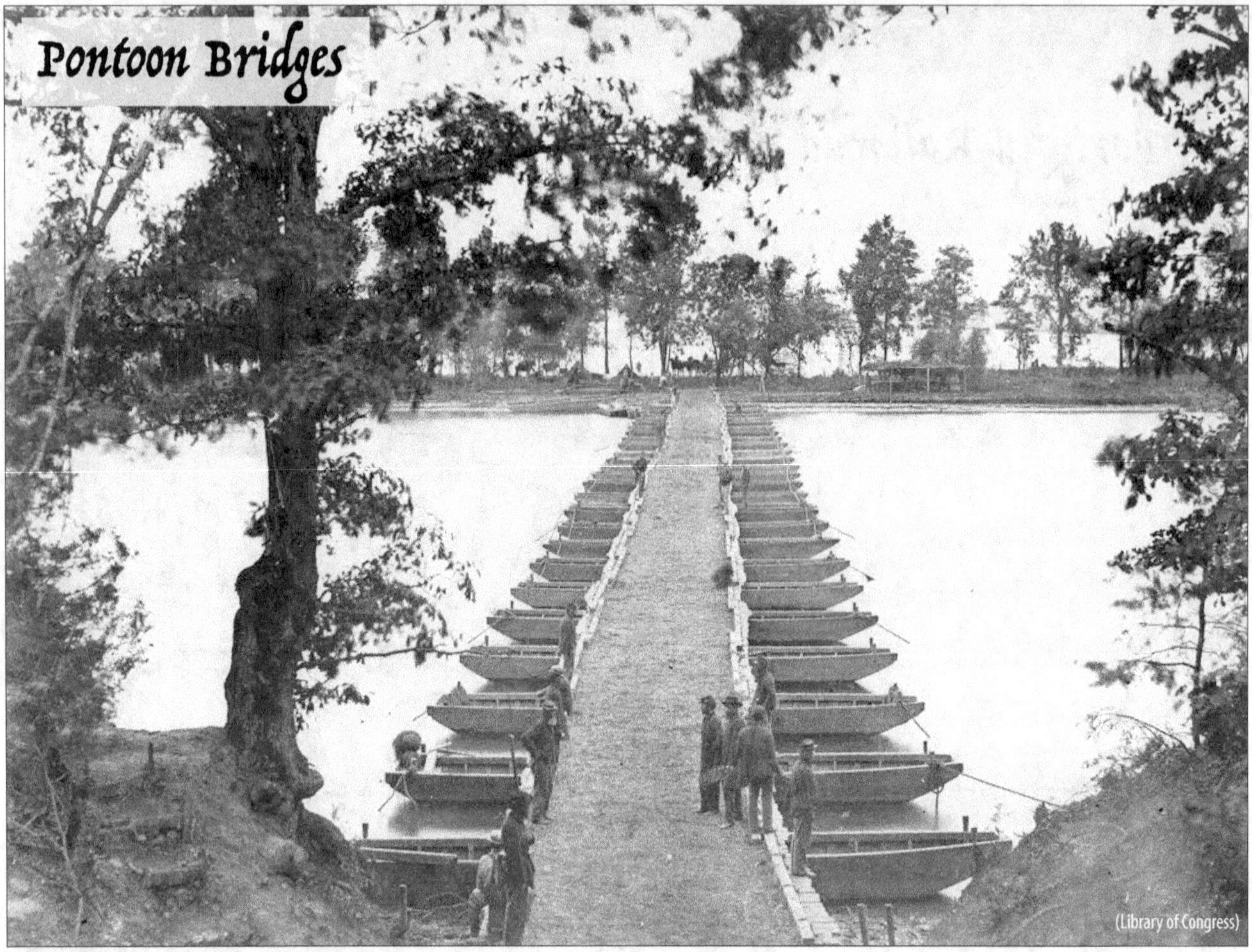

(Library of Congress)

One of the most useful skills of Civil War engineers was the transportation, assembly and dismantling of portable, floating bridges — pontoon bridges, sometimes written as ponton bridges. Such a bridge spanned the Cumberland River at Nashville. Pictured above is a pontoon bridge over the James River in Virginia. Pontoon boats, which came in several versions, had their own special carriages, as shown below. The bridges were built in a methodical fashion, with various pieces numbered to facilitate construction.

(Library of Congress)

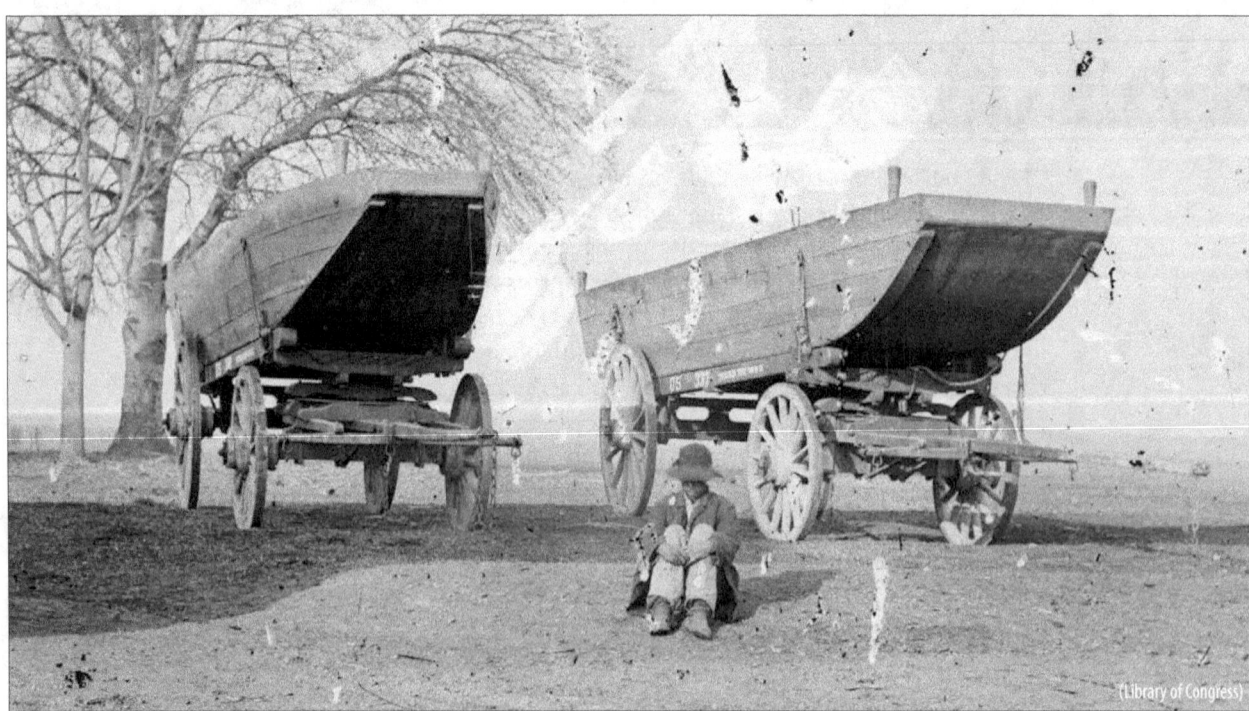

Construction of Pontoon Bridge

riverbank

Side rails, 27 ft. long, 5 x 5

Chess 13 ft. long

Balks 27 ft. long, 5 x 5

Pontoon-boats, 26 ft. long, 20 ft. apart center to center

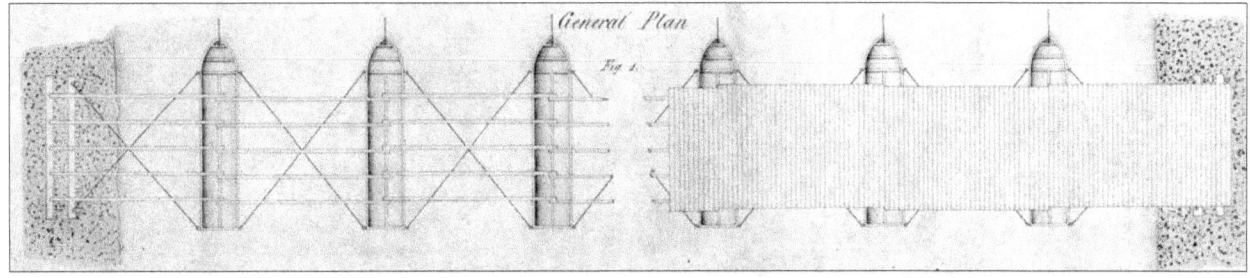

(Atlas to Accompany the Official Records of the Union and Confederate Armies)

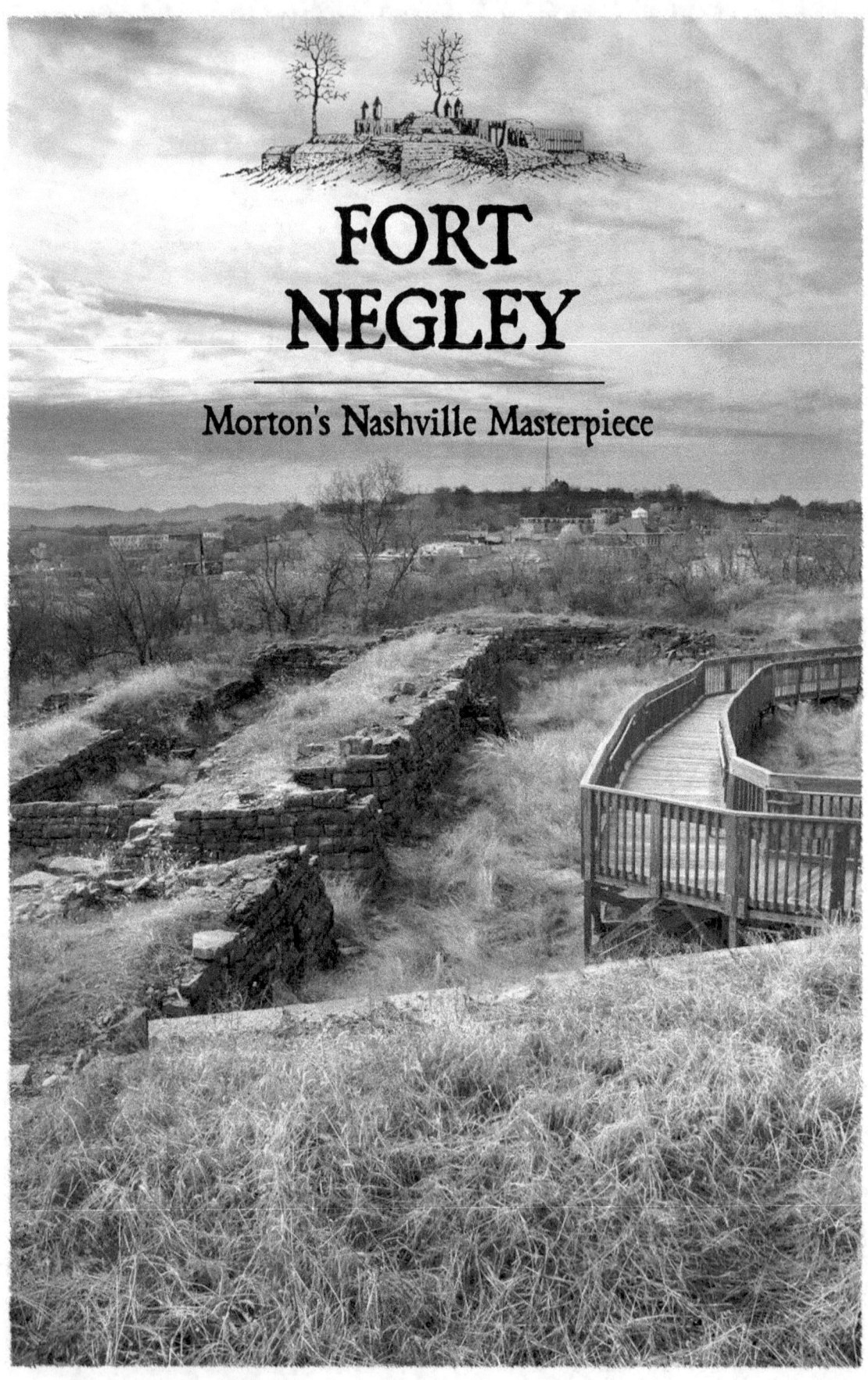

Fort Negley was built by Federal forces during the Civil War to anchor their fortifications around Nashville, the capital of Tennessee. In addition to its unique star-bastion design, Fort Negley was the largest inland stone fortification built during the war. Nashville eventually became the most fortified city in North America, second only to Washington, D.C.

Fort Negley Park, with its modern visitors center, is now a major Civil War attraction, having twice been rescued from its steady deterioration. Too risky and expensive to reconstruct, the structure atop Saint Cloud Hill is "interpreted as ruins." Other than the Tennessee State Capitol, it is perhaps the most significant local structure built of native limestone, the masterpiece of brilliant military engineer James St. Clair Morton. The fortress is a testament to the spirit of mankind, having been built by slaves, impressed freedmen, and the contrabands of war. These ragtag men, hundreds of whom died during construction due to appalling living conditions, raised their shovels and picks to defend the fort. Later, these proud men, many of them fugitive slaves, would bear arms as recruits in the newly formed U.S. Colored Troops. At Nashville, in late 1864, many would die in battle defending their newfound freedom.

The renovation and reopening of Fort Negley is due in large part to one of the largest municipal appropriations for Civil War interpretation ever.

Fort Negley was so impressive and imposing that it was never attacked during the war. The fort was rebuilt expertly during the Great Depression as a federal works project and then allowed to deteriorate again. Only in recent decades has it been reclaimed — the fort you see today is mostly that from the 1930s, not the Civil War.

The survival of the fort over the past 150 years has been linked to providence, and also irony because the father of its designer was the founder of scientific racism, the theory that humans consist of different species ranked by cranial capacity.

"Fort Negley, long hidden, is a map to Nashville black history," stated Learotha Williams Jr., PhD., professor of African-American studies at Tennessee State University. The contraband camp at the fort during the war grew into a viable black community in the decades following the war until disrupted by interstate highway construction in the 1960s.

The significance of Fort Negley and St. Cloud Hill to African-American history has been finally recognized in recent years. In 2019, Fort Negley was among the first four "sites of memory" associated with UNESCO's Slave Route Project Sites in the U.S., along with Fort Mose in St. Augustine, Florida; Freedmen's Town Historic District in Houston, Texas; and The Middle Passage Ceremonies and Port Markers

James St. Clair Morton

was the brilliant architect of Fort Negley and most of the fortifications around Nashville. He was a native of Philadelphia, Pa. and the son of renowned physician Samuel Morton, the founder of craniometry (measurement of human skulls) and scientific racism (the belief in several human species and the intellectual ranking thereof). There's little evidence that the father's racism rubbed off on the son. James Morton attended the University of Pennsylvania and graduated 2nd in his class at West Point Military Academy in 1851.

An intelligent yet quirky individual, Morton has been described as dutiful, outspoken, confrontational, energetic, brash, opinionated, meticulous, and merciless.

As a 2nd Lieutenant in the Corps of Engineers, Morton served as assistant engineer in the construction of the defenses of Charleston harbor and the building of Fort Delaware, and then as assistant professor of engineering at West Point for two years. He was a student of the famous professor Dennis Hart Mahan, and he challenged current conventional thinking about fortifications. Morton believed that earthworks would stand up to modern rifled shelling better than the masonry forts of the time, and he was proven correct. Ironically, his most famous work—Fort Negley in Nashville—was built of limestone instead of dirt due to the rocky terrain.

(Continued on Page 85)

Project. Fort Monroe in Virginia was added to the project in 2021.

Fort Negley has fought many battles due to neglect, deterioration, invasive vegetation, abuse, squatters, and even developers of recreational and residential projects. That Fort Negley survives today is a minor miracle. Owned and maintained by the Metro Nashville Department of Parks and Recreation since 1928, the public site is listed on the National Register of Historic Places.

Saint Cloud Hill is basically an knob of white limestone, 260 feet above the level of the Cumberland River, formed millions of years ago from the tropical sea that lay where Middle Tennessee is now. Many types of fossils (cephalopods, crinoids, and trilobites) can be found at the site today. Removal of fossils formed naturally within the park boundaries is prohibited, but visitors are encouraged to explore the fossil pit donated by Vulcan Materials Company located near the visitors center. During the first half of the 19th Century, Saint Cloud Hill, part of the Judge John Overton estate, was a picnic spot for Nashvillians, who enjoyed the grove of ancient oak and chesnut trees and the panoramic views. All that changed when the Federal army captured and occupied Nashville in February 1862. They would not leave until well after the war. In the meantime, the Federal military converted the city into a huge transportation hub and supply depot that needed to be protected from Confederate attack.

Military Governor Andrew Johnson, fearful of enemy attack, insisted that critical locations, including the State Capitol in which he worked, be fortified and protected. He warned that the city would be burned to the ground rather than fall into Confederate hands again. The occupiers placed gun batteries at barricades controlling the eight roads into the city — Lebanon, Murfreesboro, Nolensville,

(Continued on Page 94)

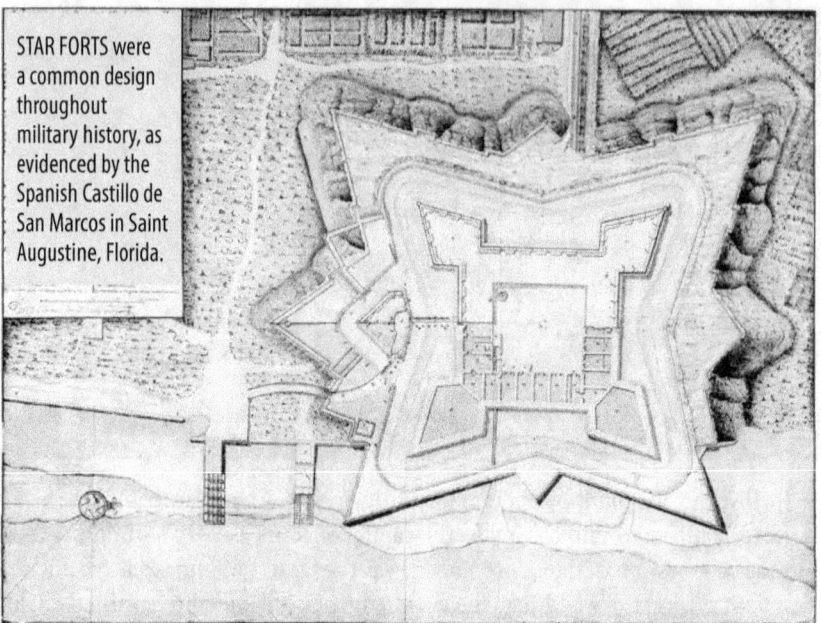

STAR FORTS were a common design throughout military history, as evidenced by the Spanish Castillo de San Marcos in Saint Augustine, Florida.

Negley defended Nashville, later exonerated of cowardice

Major General James Scott Negley (1826-1901), a native of Pennsylvania, was "heavyset, twinkling eyes, cherubic face, he cared deeply for his soldiers," according to historian Peter Cozzens. He dropped out of college at age 19 to enlist in the 1st Pennsylvania Regiment and fought in the Mevxican War. He served as a brigadier general in the state militia. His first love was gardening; he was a brilliant horticulturalist. *New York Herald* correspondent William Shanks stated, "when in the field of war his leisure hours were devoted to the study of various fruits, flowers and shrubs in which the southern fields and woods abounded. Many a march, long, tedious, exhausting, has been rendered delightful to his staff by his interesting descriptive illustrations of the hidden beauties and virtues of fragrant flowers and repulsive weeds." Negley served in Gen. Don Carlos Buell's Army of the Ohio and was posted in Nashville to defend the city while Buell chased Bragg into Kentucky in late 1862. He served well at the Battle of Stones River and in the Tullahoma Campaign. In September 1863 at Chickamauga he led the 2nd Division of Gen. George H. Thomas' Corps and decided to leave the field, based partially on false rumors, with desperately needed infantry and 22 cannons. Gen. Thomas J. Wood denounced Negley to deflect criticism from himself, and was joined by Gen. John Brannan. Commander William Rosecrans relieved Negley, but after he read his report he admitted that Negley had acted in his best judgment under the circumstances. A court of inquiry, held in June 1864, cleared Negley and reprimanded Wood. Negley resigned from the army in January 1865. He claimed that he had been discriminated against by West Pointers. He served several terms in Congress. He is buried at Allegheny Cemetery, Pittsburgh.

This sketch of Fort Negley shows the steepness of Saint Cloud Hill, the encampments of Federal troops, the barren landscape, the sally port, the big gun casemate, the turreted stockade, the flag pole, and trees within the enclosure used as lookout posts, and the interconnecting trenchworks.

Captain James St. Clair Morton

(Continued from Page 83)

Other projects that Morton supervised or worked on prior to the war included Sandy Hook Fort, East Coast Lighthouse District 3, Potomac Waterworks, Washington Aqueduct, Fort Jefferson, and Fort Mifflin. He also engineered the Chiriqui Expedition to Central America in 1860 in search of a passage between the oceans. After adequately recovering from malaria, Morton was assigned as chief engineer to General Don Carlos Buell's Army of the Ohio. When Buell was forced to chase Confederate General Braxton Bragg into Kentucky, he assigned Morton to begin the construction of defenses in Nashville, which had been occupied by Federal troops since February 1862.

On Oct. 24th, 1862, Major General William S. Rosecrans replaced Buell as commander of the now-renamed Army of the Cumberland. A former engineer, Rosecrans created the Pioneer Brigade, led by Morton, to build roads and fortifications and to repair railroads. Their first role was as combat engineers at the Battle of Stones River. Commanding the brigade along with Stokes' Chicago Board of Trade Battery, Morton successfully held off several Confederate assaults and protected the Nashville Pike from capture. He was promoted to lieutenant colonel for his heroics. Following the battle, Morton supervised the building of a massive depot and fortification at Murfreesboro named Fortress Rosecrans, this time composed of earthworks.

Neither Fortress Rosecrans nor Fort Negley were ever directly attacked by Confederate forces, their commanders wisely deciding that discretion was the better part of valor.

During the Tullahoma Campaign, the lack of discipline within the Pioneer Brigade became a problem, and Morton was berated by his benefactor, General Rosecrans. After being wounded at Chickamauga and building fortifications at Chattanooga, Morton was mustered out of the volunteer service. Morton requested a transfer, which was denied. He then made the rare request that his rank be reduced from brigadier general of volunteers to his Regular Army rank of major of engineers, the only instance during the Civil War of a general voluntarily reducing his rank.

After briefly returning to Nashville to work on defenses, he was assigned to work as assistant to the Chief Engineer in Washington, D.C. In May 1864 he found himself chief engineer to General Ambrose Burnside of the IX Corps near Petersburg, Virginia. On June 17th, 1864, at the age of 35, Morton was shot in the head and killed while reconnoitering the front lines without escort. His death was witnessed by the brother of poet Walt Whitman. Morton was brevetted brigadier general and buried with military honors at Laurel Hill Cemetery in Philadelphia.

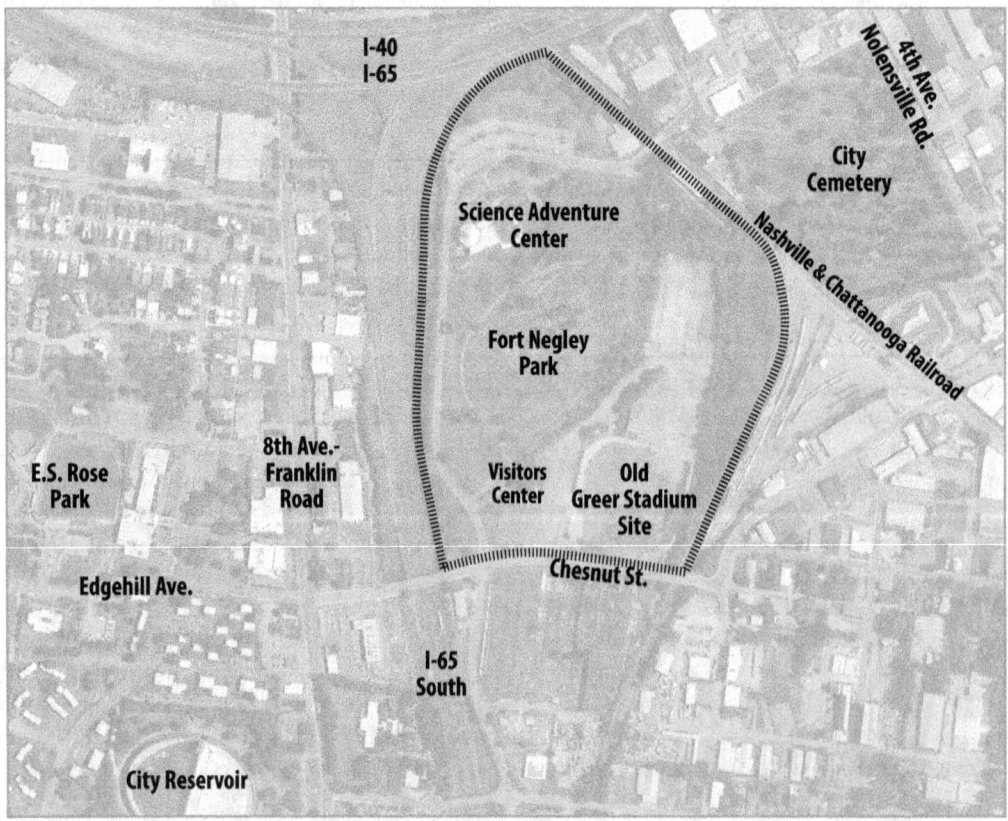

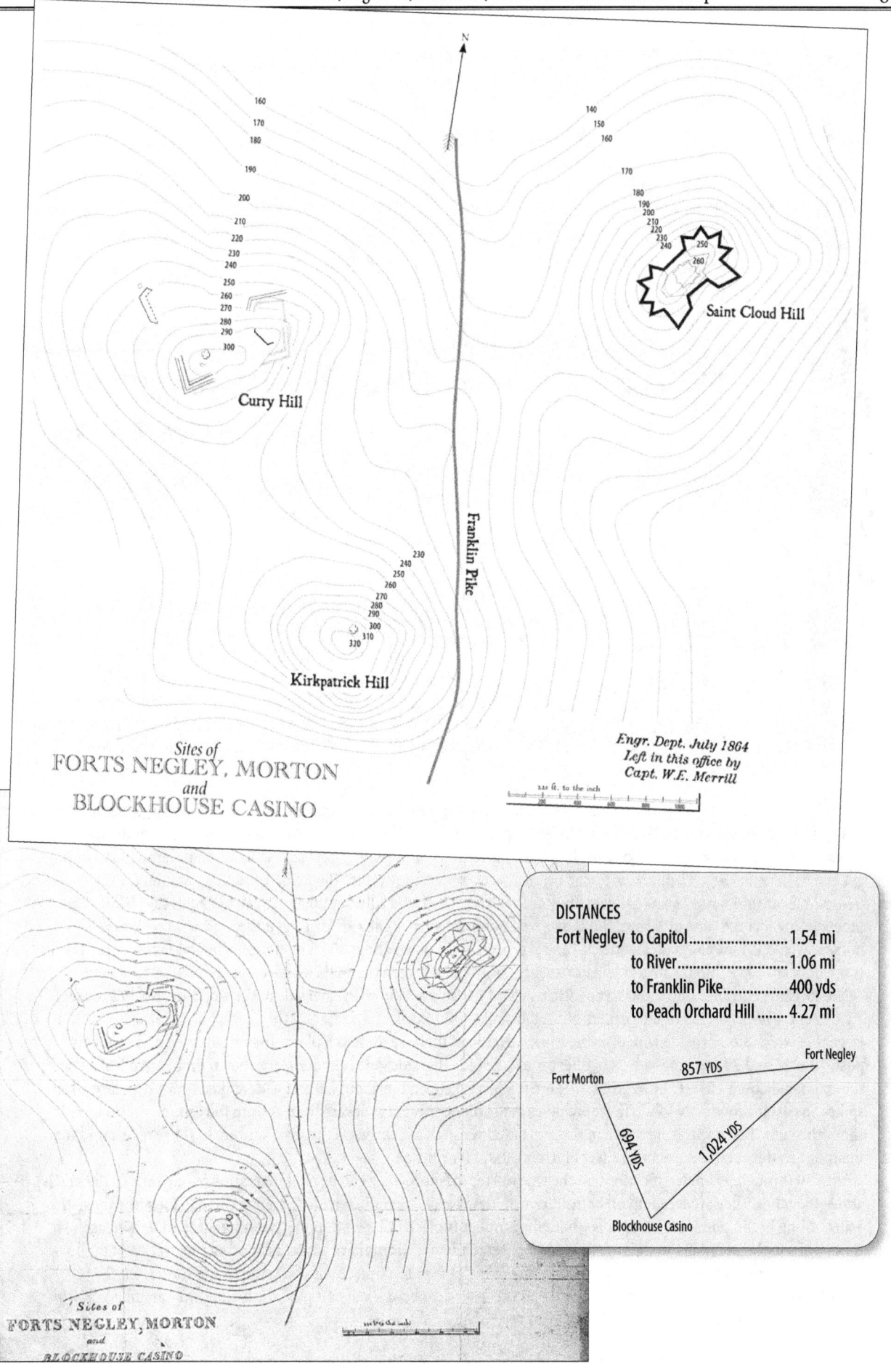

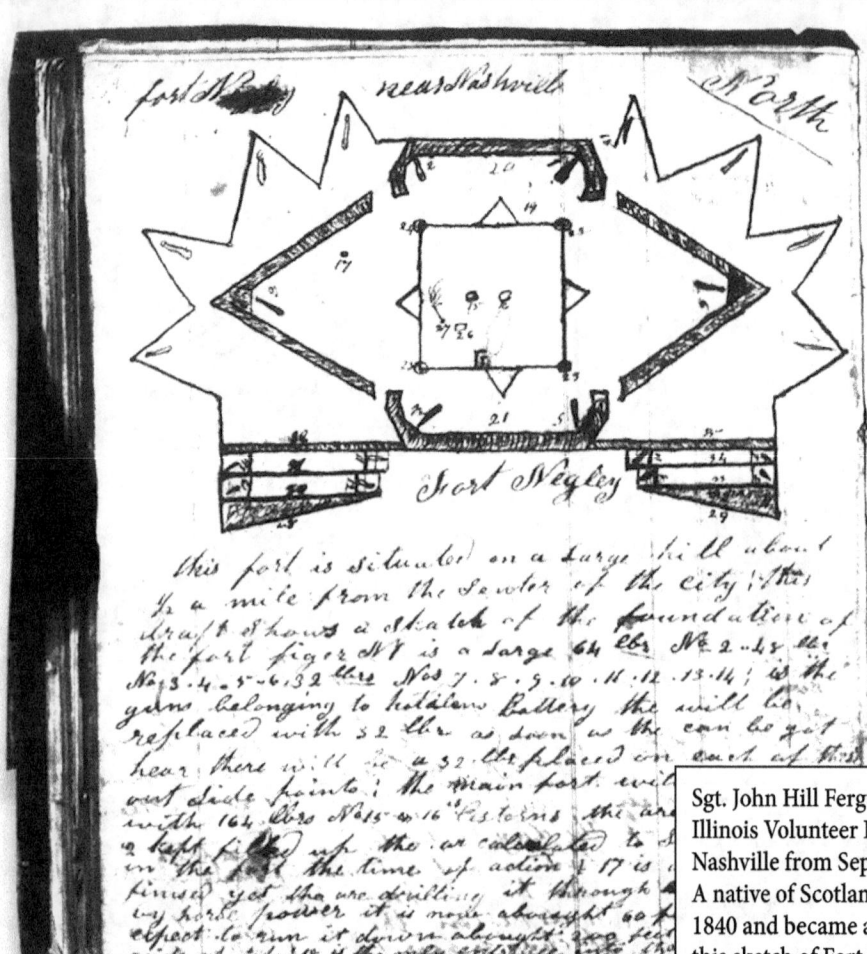

(Fort Negley Archives)

Sgt. Ferguson's Sketch of Union Fort Negley and Diary Description

Sgt. John Hill Ferguson, 34, of Co. G, 10th Illinois Volunteer Infantry served in Nashville from September 1862 to July 1863. A native of Scotland, he came to the U.S. in 1840 and became a citizen in 1856. He drew this sketch of Fort Negley and wrote in his diary November 4, 1862 (edited for clarity):

This fort is situated on a large hill about half a mile from the center of the city. This draft shows a sketch of the foundation of the fort. Figure 1 is a large 64-pounder. No. 2 is 48-pounder. Nos. 3-4-5-6 are 32 pounders. Nos. 7 through 14 are guns belonging to Houghtailing's Battery. They will be replaced with 32-pounders as soon as they get here. There will be a 32-pounder placed on each of these outside points. The main fort will be mounted with 64-pounders. No. 15 and 16 are cisterns. They are very large and kept filled up; they are calculated to supply the regiments in the fort the time of action. No. 17 is a well not finished yet. They are drilling it through solid rock by horsepower. It is now about 60 feet deep. They expect to run it down about 200 feet. It is 5 inches wide at the top. No. 18 is the only entrance into the fort. It will have a large iron gate between the walls when finished.

No. 19 is the entrance into the stockade. Nos. 20 and 21 are magazines on each side of the stockade. Nos. 22 through 25 are sentry boxes on top of the corners of the stockade. No. 26 is the tent wherein the telegraph operates. No. 27 is a large tree which supports the wire over the works. On top of this tree there is a platform built. It is used as a lookout post. Nos. 28 and 29 are wings where artillery may be used. This stockade is built of large hewn timber two feet square set up on end about 12 feet above ground. There is a large plate on top about 2.5 feet wide spiked down with large iron spikes so that it is perfectly solid. There are holes cut through these logs about five feet from the ground for infantry to shoot through. Those wings are made in the same manner so as to command the main entrance. The walls around the main fort in the outside are about 12 feet high and about four inside.

Nos. 30 through 35 are fortifications on the side hill for infantry. They can be all in operation at one time as the one is above the other. They are faced up on both sides with hewn rock, then filled with dirt some three to four feet above the walls. Along in the center of these walls where the number is placed there is a tunnel running through under each of these walls to the main fort so that the infantry can get to them or from there without exposing themselves to the enemy. These outside points are about 14 feet high on the extreme points and about six on the inside corners on the outside. The fall of the hill makes the difference as the walls are about level on top on the inside. They are filled up with rock and dirt within four or 4.5 feet of the top all around.

(Library of Congress)

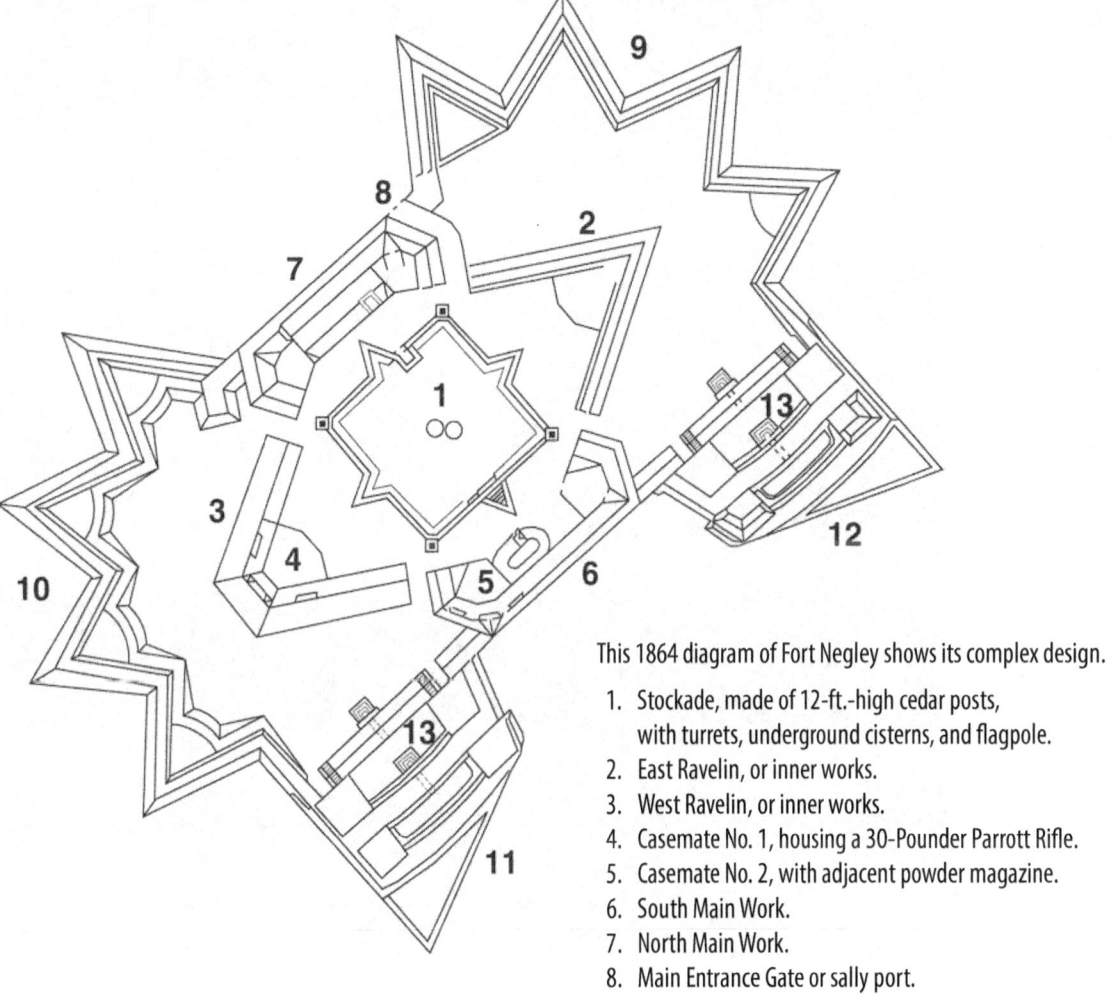

This 1864 diagram of Fort Negley shows its complex design.

1. Stockade, made of 12-ft.-high cedar posts, with turrets, underground cisterns, and flagpole.
2. East Ravelin, or inner works.
3. West Ravelin, or inner works.
4. Casemate No. 1, housing a 30-Pounder Parrott Rifle.
5. Casemate No. 2, with adjacent powder magazine.
6. South Main Work.
7. North Main Work.
8. Main Entrance Gate or sally port.
9. East Outer Parapets, with redans allowing crossfire.
10. West Outer Parapets, with redans allowing crossfire.
11. Southwest Bastion, a bombproof structure.
12. Southeast Bastion, a bombproof structure.
13. Bastion tunnels.

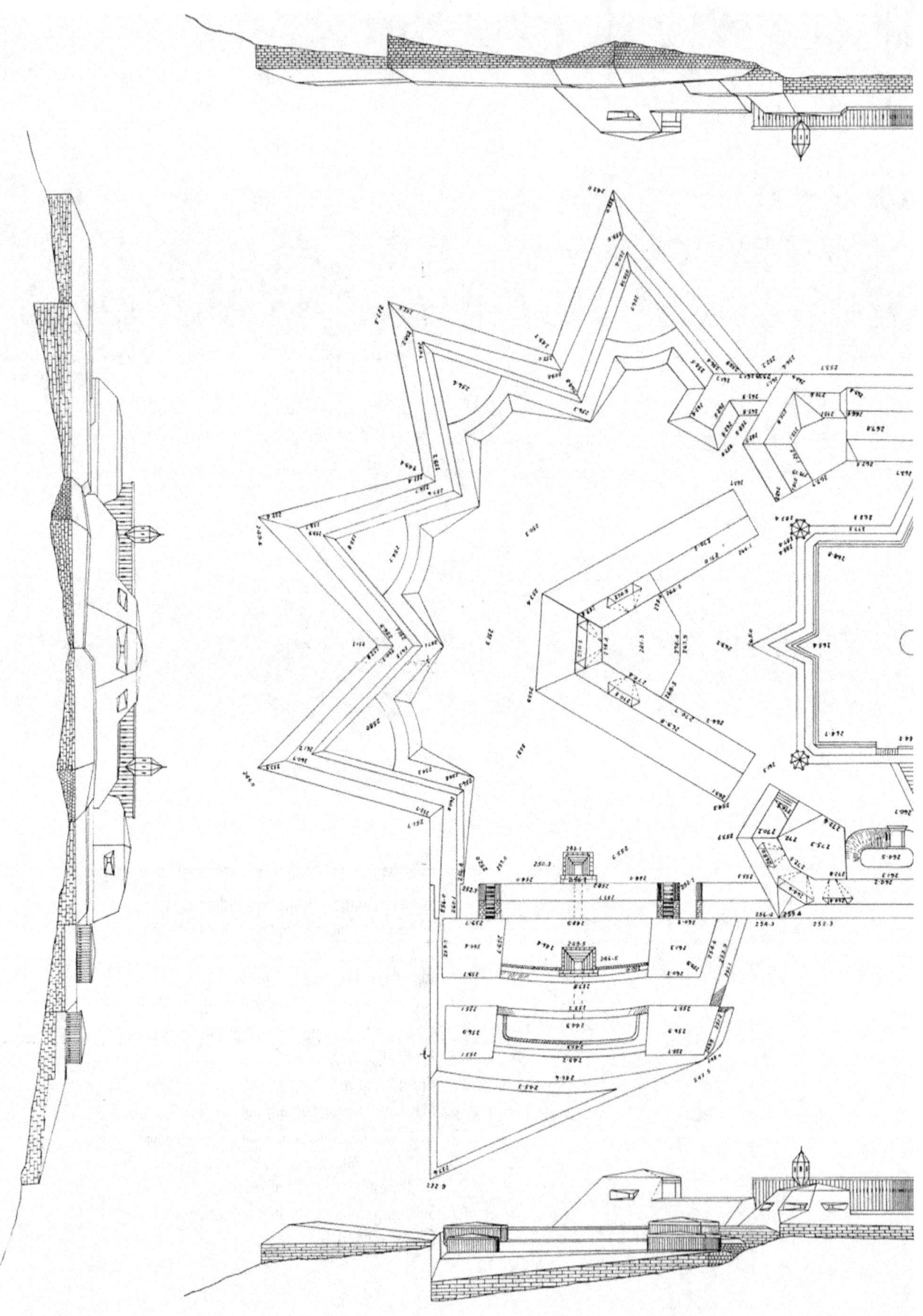

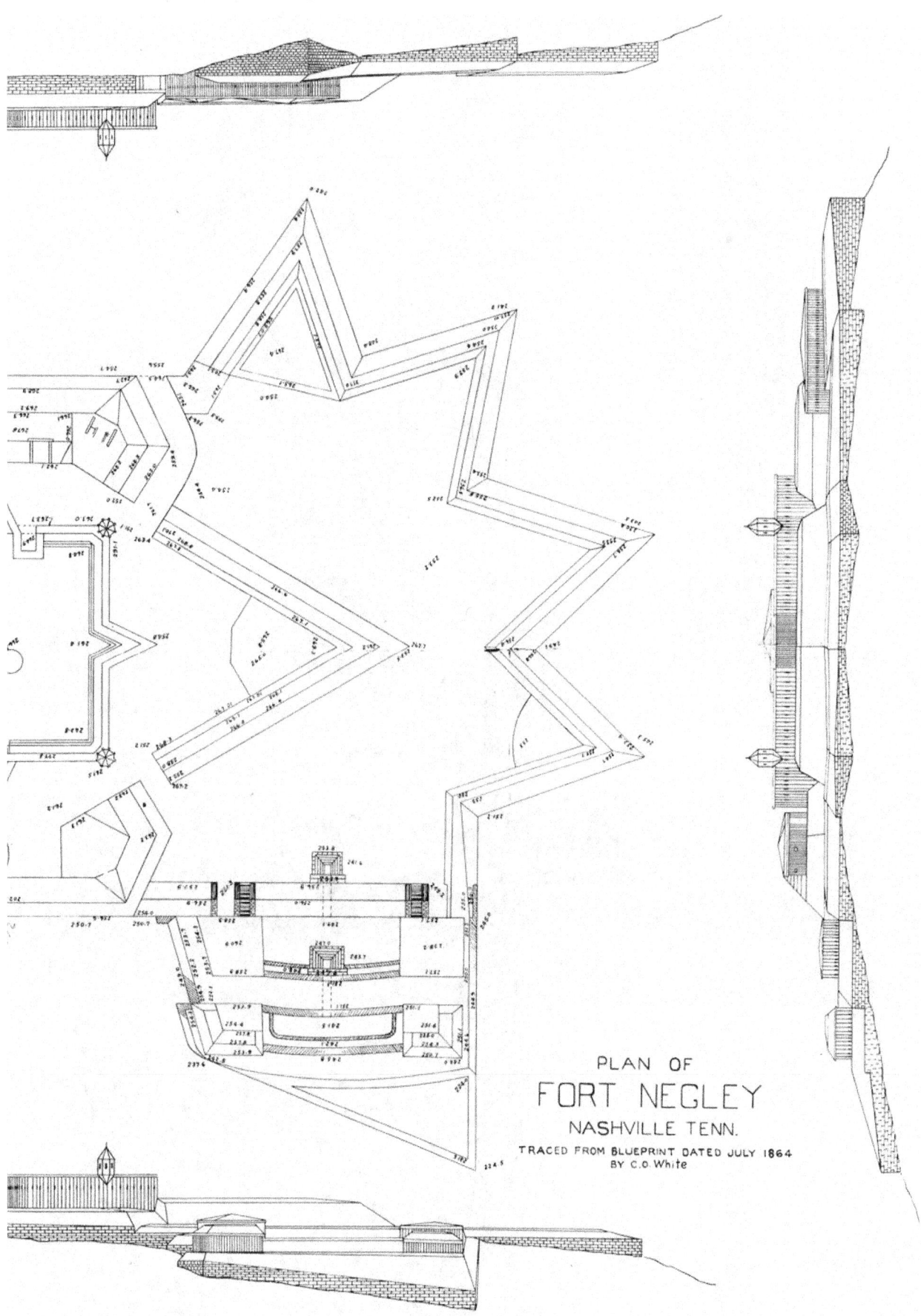

PLAN OF FORT NEGLEY NASHVILLE TENN.
TRACED FROM BLUEPRINT DATED JULY 1864
BY C.O. White

(Nashville Public Library Metro Nashville Archives)

92 Fortress Nashville: Pioneers, Engineers, Mechanics, Contrabands & U.S. Colored Troops

(Library of Congress)

(Library of Congress)

Fort Negley artillerymen at heavy gun platform next to sally port. Looking northeast, Nashville can be seen in the distance. Interestingly, the men are posing on a two-wheeled wooden naval-gun carriage. (Tennessee State Library and Archives)

(Continued from Page 84)
Franklin, Granny White, Hillsboro, Harding, and Charlotte turnpikes. The guns were placed so that they could be turned against the city itself in case of a civilian uprising.

When General Don Carlos Buell and his Army of the Ohio were forced to chase General Braxton Bragg's Army of Tennessee north into Kentucky, Gen. James Negley's division was left behind to guard Nashville. About this time, the South Tunnel on the L&N Railroad north of Nashville was disabled by Confederate cavalry. The Cumberland River hit low levels during a summer of drought, hindering the steamboat trade. This brief, anxious period was called the Siege of Nashville.

On Aug. 6, 1862, Gen. Buell assigned his chief engineer, Capt. James St. Clair Morton of the U.S. Army Corps of Engineers, to stay behind in Nashville and begin to build fortifications guarding all the major turnpikes and railroads into the city, which was already partially protected by a huge bend in the river to the east and north and the gunboats that patroled those waters. Buell told him to "go at once to Nashville and select sites and give plans and instructions for redoubts to protect the city…These works must all be practical and as simple as possible in the beginning, so that they can be constructed with the greatest promptness and occupied immediately by a small force. They can then be elaborated on and made more formidable. Start the works at once, the most important first. The commanding officer will call in slave labor on it."

Morton chose lofty St. Cloud Hill, south of town, for Fort Negley, his most elaborate fortification, since it commanded views of several pikes and the Nashville & Chattanooga Railroad. The fort would be the keystone of three other major forts in the vicinity — Fort Morton on Currey Hill and Blockhouse Casino on Kirkpatrick Hill, which would mutually support Fort Negley, and Fort Houston farther to the west. Contrary to Buell's directive, Morton produced complicated, elaborate designs for all of these fortifications (except Blockhouse Casino), which greatly hampered their construction and required major modifications. It is unlikely that Morton would name one of the forts for himself; Fort Morton was likely named by officials around May 1863 after Morton left the city.

Today, as it was during the war, Fort Negley overlooks the spacious City Cemetery, laid out in a grid of "streets" like a small town. Approximately 15,000 Federal soldiers were buried in areas adjacent to the Nashville City Cemetery, including the area between the railroad tracks and on the southern slope of St. Cloud Hill. An unknown number of civilians, black and white, may have been buried in the same areas. While the soldiers were moved to the Nashville National Cemetery, civilians would have been left behind.

A structure named Fort Confiscation appears in written accounts but not on any official map. It has been speculated that Fort Confiscation was actually the first name of Fort Morton. However, according to research

SOUTH WEST VIEW OF FORT NEGLEY.

conducted by Krista Castillo, museum director at Fort Negley Visitors Center, Fort Confiscation, composed of a stockade, earthworks, and approximately 2,000 cotton bales was constructed first (*New York Times*, Nov. 23, 1862). On May 9th, 1863, *The Ottawa (Ill.) Free Trader* reported…"Fort Confiscation, which is at present nothing more than a stockade, and serves as a mask to Forts Morton and Negley." Fort Confiscation became one of Nashville's 21 minor fortifications once Fort Morton was constructed. On Dec. 10th, 1864, *The Louisville Daily Journal* reported that a shell from Fort Confiscation hit Gen. John Bell Hood's headquarters. Fort Confiscation was still there in January 1868 when Dr. William P. Jones, the property owner, sold the wood (around 250 cords) to the city for the benefit of the poor (*The Republican Banner* and *The Nashville Daily Union*, Jan. 3, 1868). Unfortunately, all of the wood was consumed when the laborers tasked with removing and cutting up the stockade left a camp fire unattended.

Morton was outspoken in his views that, in the age of heavy rifled artillery, earthwork fortifications would prove more durable than stone or masonry fortifications. This view was substantiated in Savannah, Ga., where the brick Fort Pulaski was reduced to rubble while nearby Fort McAllister, made of sand, was basically indestructible. At St. Cloud Hill and other Nashville sites, however, Morton was forced by the nature of the terrain to work with limestone instead of earth (Nashville's nickname at the time was Rock City).

The design of Fort Negley is unlike that of any other major fort in the world. Morton would combine the star-fort configuration that every cadet at West Point learned by studying the works of French military architect Sébastien Le Prestre de Vauban with the sturdy tiers of two large bastions facing the probable approach of the enemy. It is likely that Morton had already drawn the designs for such a fort by the time he arrived at Nashville.

Fort Negley was named for Gen. Negley, provost marshal and commander of Federal forces in Nashville at the time. The fort was later renamed Fort Harker in honor of Brig. General Charles G. Harker, who was killed at the Battle of Kennesaw Mountain in June 1864. For some reason, the new name did not stick. Decades after the war, Negley would raise a grandson, Negley Farson, whose 1930s travels, writings, and stunts would rival those of Hemingway.

The fort envisioned by Morton was of three tiers, 600 feet long by 300 feet wide, covering four acres, and laid on an southwest-northeast axis. Two sides of the fort are of star design with four redan projections on each side. The ravelins inside are large enough for troops to encamp there. Binding the two star structures together are main works on each opposing side. On the south side of the fort are two tiered, bombproof bastions which could house and protect

(Continued on Page 100)

Letters Home From Fort Negley

Pvt. William Cammack, 30, of Deming, Indiana, served with the 12th Indiana Light Artillery at Fort Negley. Here are excerpts of letters he wrote home during the period surrounding the Battle of Nashville:

Nov. 28, 1864
Well we have had nothing new here yet. Are looking for Hood to take dinner with us one day this week. I propose setting at the head of the table. Well I guess he won't get here for a spell. There is lots of troops here and passing up to Gen Thomas who is only from fifteen to twenty-five miles from here. We are ordered to be ready at a moment's warning. I have had to Drill a few times on a small Frafr gun (64 lbses). Oh yes our old wash woman left us last night. I know where she went. They boxed her up. That was the last I saw of her.

Dec. 3, 1864
Cal, we have a pretty good prospect for a hot engagement. Hood has made an attach on this place. There has been fighting in sight for two days. Our fort threw Six Shots this evening which told the Jonnis that old fort Negley was here. I write as soon as its over and give a full description.

Dec. 9. 1864
We have been shooting som every day from this fort since last Friday. Rebs are in sight. Scrimishing is going on in Sight all the time. This evening several houses were burnt that stood between the contending forces. Our big guns belch forth their firey indignation every few minutes. It begins to look pretty hot. Cal, I can Stand and Smoke my old pipe just as contented as if sitting on the pegs at the end of the counter. They cant fool me out of a Smoke.

April 17, 1865
More than one man has already lost his life here by saying he was glad to hear of (Lincoln's) death. When the soldiers hear a man express himself as one of that class they just shoot him down like a dog and there is nothing said about it eather. Any man that would say any thing against it would share the same fate…
The news has just come in that Gen Johnson (Johnston) has surrendered his arms. Bully news ant it? We will have to fire another salute this afternoon. We have been at that kind of buiz for the past 2 weeks. I think the rebellion is about played out.

Sketches of Fort Negley—Drawn by W.H.H. Fletcher of the 12th Battery Ind. Vols. "This fort is garrisoned by the 12th Ind. Battery and Battery C, 1st Tenn. Light Artillery. The officers of the 12th Ind. Battery are James E. White (Capt. commanding the fort), and Lts. James A. Dunwoody, James W. Jacobs, Isaac Hamilton, James Robinson. Officers of Battery C, 1st Tenn. Lt. Artillery are Capt. Vincent Meyers and Lts. Joseph Grigsby and Caleb Goin. Lithographs printed by Gibson & Co., Cincinnati, Ohio. (This and all similar lithographs Library of Congress)

Impressing the Contrabands at Church in Nashville.

(Annals of the Army of the Cumberland)

Impressing Negroes to Work on the Nashville Fortifications.

A depiction of the central stockade with turrets at Fort Negley with Casemate No. 1 in the foreground. The fort is guarded by members of the U.S. Colored Troops. In the distance to the left is Fort Morton and Federal encampments. Original artwork by Philip Duer, used with permission.

(Continued from Page 96)
scores of infantrymen guarding the southern approaches. The bastions on the south side of the fort were constructed first, since these bastions protected the railroad. Each bastion had tunnels which protected men moving through the works. Off the northeast main work was the sally port (entrance), guarded by a small stockade with loopholes. On the uppermost level of the fort were two heavily fortified casemates housing 30-pounder Parrott rifles, capable of hurling a 29-pound shell 3.8 miles. These casemates were clad with railroad iron. There were 11 guns in the fort, manned by 75 artillerymen. The guns were enbarbette; no embrasures were prepared either in the upper or
(Continued on Page 106)

(National Park Service)

The 30-pounder **Parrott rifle** at Fort Negley could hurl a 29-lb. shell 6,700 yards (3.8 miles) in 27 seconds. The 4,200-lb. gun had a rifled bore 4.2 inches in diameter and a barrel ten-and-a-half feet long. The gun was invented in 1860 by Captain Robert P. Parrott, a West Point graduate. He resigned from the service in 1836 and became the superintendent of the West Point Foundry in Cold Spring, New York. Parrotts were manufactured with a combination of cast and wrought iron. The cast iron made for an accurate gun, but was brittle enough to suffer fractures. A large wrought-iron reinforcing band was overlaid on the breech to give it additional strength. Parrott rifles were manufactured in different sizes, from the 10-pounder up to the 300-pounder.

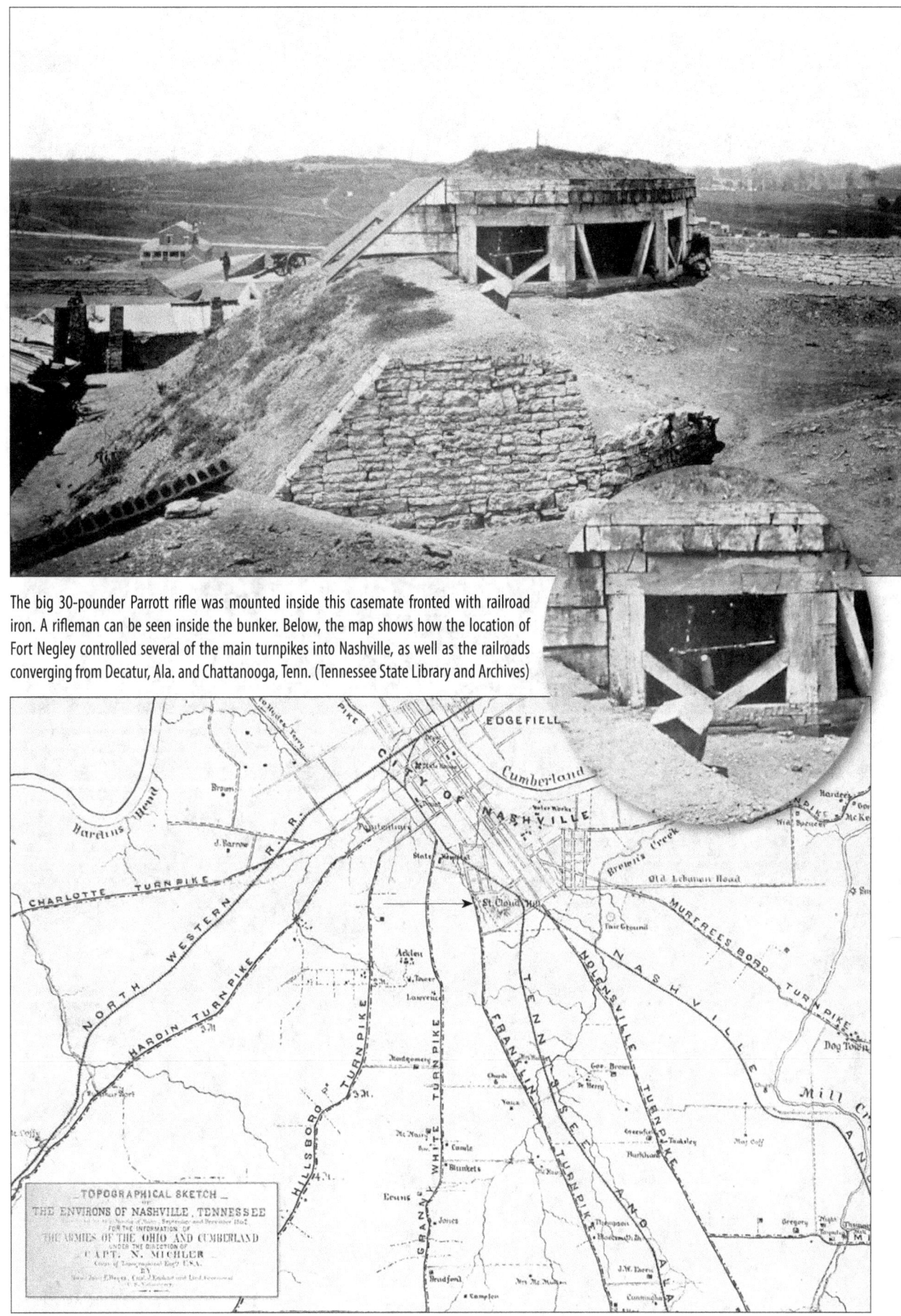

The big 30-pounder Parrott rifle was mounted inside this casemate fronted with railroad iron. A rifleman can be seen inside the bunker. Below, the map shows how the location of Fort Negley controlled several of the main turnpikes into Nashville, as well as the railroads converging from Decatur, Ala. and Chattanooga, Tenn. (Tennessee State Library and Archives)

1937

(Nashville Public Library Metro Nashville Archives)

1958

(Nashville Public Library Metro Nashville Archives)

Fortress Nashville: Pioneers, Engineers, Mechanics, Contrabands & U.S. Colored Troops

1960s
(Nashville Public Library Metro Nashville Archives)

1980s
(Nashville Public Library Metro Nashville Archives)

A Report on Fort Negley-1865

OFFICE OF INSPECTOR-GENERAL OF FORTIFICATIONS,
MILITARY DIVISION OF THE MISSISSIPPI,
Nashville, Tenn., May 15, 1865.

Maj. Gen. GEORGE H. THOMAS,
Comdg. Mil. Div. of the Miss. West of Alleghany Mountains:
I have the honor to submit the following inspection report
of the Defenses of Nashville:

DEFENSES OF NASHVILLE.
Nashville was first occupied by the U.S. army in March, 1862. General Morton, then captain, U.S. Corps of Engineers, commenced fortifying the position soon after its occupation. His plan was to hold Morton and Houston Hills and that on which Fort Negley stands by three large works controlling Casino Hill by a block-house and the fire of the two forts in rear. He also built defenses around the capitol, which is situated on a high hill within the city. It is presumed that these works were to be connected by an intrenched line when the necessity should arise.

Forts Morton and Houston were designed as very large works, the double bastions of Choumara with a demilune, and were to be built in a permanent manner, with detached stone scarps. I have been informed that he expected these works to hold out after the city had been taken, and therefore devised them with interior capacity for the defensive materials and provisions for resisting a siege in the event the lines around the city could not be maintained. The magnitude of these works prevented the carrying out of his views. They would have required more labor than building all the necessary redoubts to completely inclose the city.

Fort Negley (now called Fort Harker)
This large work was nearly completed by General Morton, assisted by Captain Burroughs, Corps of Engineers. It is a complex fort. Within stands a square stockade twelve feet high, with flanking projections on each face. It is surrounded by a redoubt essentially square, with redan projections on the east and west sides.

Its parapets are heavy, and the scarps were walled with dry stone, over which, however, the earth of the embankment falls, so as to give a continuous slope.

On the south are two bastions, the flanks of which join to the south face of the main work, as a curtain, thus forming a bastion front. Each bastion has two interior intrenchments rising in stages, which are themselves small bastion fronts, the bastions being small bombproofs loop-holed, flanking the interior ditch, and with infantry and artillery fire to the exterior. These small bomb-proofs are surrounded nearly to the height of the loop-holes by a parapet with low, dry stone scarps.

Immediately below the main parapet to the east and west, are outer parapets about nine feet thick, apparently for infantry, with sharp salients and dry stone scarps. They connect on the north side with the main work and on the south with the bastion front.

Near the entrance in one of the salients is a bomb-proof, loopholed, which flanks the gateway front, serves as a guard.house, and as a keep to the east star-shaped outwork. The main work connects with each of the outworks by two open passages without gates, wide enough for artillery. Within this work are two casemates of timber, covered on the slope toward the enemy with railroad iron and made bomb-proof with earth. The other guns, four in number, are en barbette. No embrasures were prepared either in the upper or lower parapets.

Brig. Gen. Zealous B. Tower

A strong work against assault, its power to resist siege is weakened by uncovered dry stone walls and exposed wood-work. In some measure it throws away the advantages of a simple earthen redoubt in an effort to gain security against coup de main. It is, however, a very imposing fort, and its appearance alone would keep an enemy at a good distance.

Its offensive power would be much increased by excavating the interior of the east outwork and placing guns there in embrasure. The terre-plein of the western outwork is sufficiently low; guns could be placed in embrasure there also, as well as in the main work.

If Casino Hill were strongly held, Fort Negley could only be attacked from much lower ground than its own site, and the emplacements for the attacking batteries would be distant. The hill slope is too rocky for the construction of trenches. Nothing has been done to this work under my direction further than the arrangement of the lower parapets on the western front for placing two guns in embrasure. The accompanying drawing explains this complex work.

Very respectfully, your obedient servant,
Z. B. TOWER,
Brig. Gen. and Insp. Gen. of Fortifications, Mil. Div. of the Miss.

From The War of the Rebellion: Official Records of the Union and Confederate Armies, Series I-Vol. XLIX, Part II-Correspondence, Etc., Page 775

(Continued from Page 100)
lower parapets. At the center of the fort was a 12-foot-tall timber stockade with protrusions, corner turrets, and loopholes. This is the citadel or fortification of last resort. Trees were left standing inside the stockade to serve as lookout posts and a tall pole with large flag signaled the nationality of the fort. Most of the structure of the fort consists of dry stacked limestone covered with dirt and turf. As designed and built, the fort lies low on the hill, seemingly built into its slopes. The cost of construction was roughly $130,000 ($3.5 million in today's dollars).

One Federal officer noted that Fort Negley was "a very imposing fort, and its appearance alone would keep an enemy at a good distance."

Another noted that "in some measure (the fort's design) throws away the advantages of a simple earthen redoubt in an effort to gain security against coup de main (sudden surprise attack)."

Morton would need a large work force to build this fort, not to mention the blasting and quarrying of stone required. As it turned out, Fort Negley required 62,500 cubic feet of stone and 18,000 cubic feet of earth to be moved. The limestone used to construct the fort was quarried from the top of St. Cloud Hill, with the quarry pit forming the building trench for the footers (the below-ground masonry courses) of the fortification walls. Nearby, the newly built Asylum for the Blind, costing $40,000, was blown up and destroyed to open fields of fire and eliminate shelter for any attacking forces. Hundreds of the popular oak trees enshrouding the hill were felled to facilitate construction and provide open fields of fire.

Morton called for the commandant of the city to supply a labor force of one thousand slaves who would begin work immediately. But by the morning of August 13th, the day construction was to commence, Morton had received only 150 enslaved workers and no tools or teams. "I lost forty-eight hours by the tardiness of the citizens in answering the requisitions of the commandant of the city for negroes, teams, tools, cooking utensils and provisions," Morton complained to Buell's staff. "Work was begun on the (railroad) bridge on morning of the 11th; on Saint Cloud's Hill this morning. I am not responsible for any delay the general may remark upon."

In 1860, Nashville had a population of almost 17,000, of which 3,226 were enslaved and 719 were free blacks. Upon Federal occupation, the forts became gathering spots for fugitive slaves who had run from their masters to the safety of the Union lines. These fugitives, whose legal status was in limbo, became known as contrabands of war. The Federal military was in a quandry as how to treat these contrabands, consisting of men, women, and children. At first, some were returned to their masters, particularly if the white owners were Union loyalists. Then in July 1862, the U.S. Congress passed legislation which provided for two new provisions:

The Second Confiscation Act freed the slaves of owners actively engaged in rebellion and authorized the military to appropriate such slaves "in any capacity to suppress the rebellion." The Militia Act authorized "persons of African descent" to be employed in military service "for which they may be found competent."

Since slaves were considered property, the legislation assumed, they could be seized as property from Southern sympathizers.

By 1863, one thousand fugitive slaves and their families were pouring into Nashville each day. According to historian Gary Shockley: "Runaway slaves — perhaps the most volatile 'property' issue confronted by the occupation — presented a special challenge. From Buell's initial policy of returning runaways to disloyal owners, the federal authorities moved

Ku Klux Klan Terrorized Freedmen

Originally the Ku Klux Klan (KKK) was organized in mid-1866 in Pulaski, Tenn. by six Confederate veterans as a social club incorporating oaths of secrecy, mystical initiations, outlandish titles for officers, and costumed ceremonies, according to historian Mark V. Wetherington.

He added, "In February 1867 Tennessee enfranchised freedmen, and Republicans established local chapters of the Union League, a political arm of the party, to mobilize the new black voters. During the spring of 1867 the KKK's innocent beginnings gave way to intimidation and violence as some of its members sought to keep freedmen in their traditional place. The official reorganization of the Klan into a political and terrorist movement began in April 1867, when the state's Democratic Party leadership met in Nashville."

"The violent tactics of the KKK soon spread to parts of Middle and West Tennessee, where bushwhacking and general lawlessness were already common, and throughout much of the South in 1868. Klan activity was especially strong in Giles, Humphreys, Lincoln, Marshall, and Maury Counties in Middle Tennessee..."

Governor William G. Brownlow encouraged the legislature to pass the Ku Klux Klan Act, which criminalized the terrorist organization, and he subsequently declared martial law in nine counties in Middle and West Tennessee. Nathan Bedford Forrest, the KKK's first Grand Wizard, believing that the Klan had served its purpose, called for the members to destroy their robes.

Fifty years later, in 1915, the Klan revived its activities at Stone Mountain, Ga. The Klan became popular in the South and Midwest. Membership declined during the Great Depression, and the Klan disbanded as a national organization in 1944.

Capt. James E. White, 12th Indiana Light Artillery, one of the last commanders of Union Fort Negley, his captain's shoulder knots, and ceremonial sword presented to him. The inscription on the sheath:
"Presented to
Capt. James E. White
Comdg, Ft Negly
By the officers of his command,
July 22nd 1864."

to a position of denying slaves to the rebellion and began impressing their labor for the Union, including extensive work on Fort Negley and the defenses of Nashville. From this interim step came active recruitment of blacks for military service, and thence, finally, de facto emancipation. Throughout, however, housing and other conditions for contrabands in Nashville were often appalling."

Military authorities in Nashville wasted little time in forcing or impressing any and all freedmen and contrabands to work on the fortifications. Federal patrols would seize barbershops and other places of business to round up workers. Cavalrymen surrounded black churches during Sunday services and grabbed any able-bodied celebrant they could muster. Soon, Morton had the labor force he needed to build the forts, especially Fort Negley, which was built in record time considering its size and complexity.

Later, he wrote: "It was my first idea to make use of enlisted men and of officers as assistants; but the continual movements of the regiments and changes to detail render such system unpracticable…I have over 1,300 Negroes, about 30 teams, and 30 carts constantly busy."

At times, there were 4,000 soldiers and 2,000 impressed freedmen (many of whom were skilled) and contrabands working the site. The contraband camp at the site, estimated by one witness to contain 2,000 persons, had few facilities to house or shelter the workers, who slept on the grounds. Sanitation was poor, meals were minimal, and disease made the rounds. Eventually, 600 to 800 black laborers would perish from the hard work, disease, and exposure to the elements. The dead were likely buried below the southern slope of St. Cloud Hill. There were 2,771 names on the labor rolls of free and impressed blacks working at Fort Negley from August 1862 to April 1863. Only about 310 men were ever compensated for their work.

According to historian Thomas Flagel, for blacks working on fortifications, the financial plight stayed very much the same; by April 1863, forts, blockhouses, and miles of trenchworks grew by the week in Nashville, but pay did not. Of nearly $86,000 owed to laborers, less than $14,000 had been received. In December 1863, War Secretary Edwin Stanton sent Brig. Gen. James S.

(Continued on Page 110)

Estimated cost of completing the defense of Nashville (October 1864).

Battery on Reservoir Hill	$5,000
Works on University Hill	10,000
Crest between University Hill and Fort Negley to sweep	5,000
Modifications of Fort Negley	20,000
Work on Casino Hill	15,000
Finishing Fort Morton	30,000
Finishing Fort Houston	30,000
Two batteries between Houston and Hill 210, defended by rifle pits and stockade blockhouse	20,000
Work on Hill (210 ref.)	40,000
Modifications of Fort Gillem	15,000
Battery on knoll to right	10,000
Work on second knoll	30,000
Contiguous battery	10,000
Work on riverbank	20,000
Total	260,000
Contingencies, rifle pits, etc.	40,000
Grand total	$300,000

Statement of expenditures by General Morton and Captain Burroughs

The following statement of expenditures by General Morton and Captain Burroughs, and the estimated allotment of these expenses to the different objects named and to the works, will enable you to form a more positive opinion in reference to the engineering operations here for the past two years and a half:

As nearly as can be ascertained, General Morton paid—

For material	$16,502.04
He left non-payment rolls amounting to	116,711.91
Total expenditures and obligations of General Morton	133,213.95

It is probable that there will be claims for trees cut, and houses demolished by his orders. The following is an approximate estimate of amounts applied to the different objects of expenditure while he was chief engineer at this place:

Preparing gun-boat	$5,000.00
Clearing ground of trees in vicinity of lines, and on the northeast bank of river	1,500.00
Work on temporary bridges, dismantling suspension bridge to obtain wire, and removing brick house from site of work	3,000.00
(Nearly all his force was engaged on temporary works while General Buell was in Kentucky, and Nashville was beleaguered.)	
A smaller estimate	10,000.00
Material received, stored, hauled, and forwarded to other points	2,000.00
Total	21,500.00
Cost of surveys	1,000.00
Cost of blockhouses for Louisville road	5,000.00
Work clearing tunnel	400.00
Temporary buildings as store-houses, stables, barracks, shops, etc.	3,000.00
Blockhouse on Casino Hill	1,000.00
Work on Capitol Hill	10,000.00
On Fort Negley	91,313.95
Total expended and debts incurred by General Morton, omitting claims for sites of forts and trees cut down	$133,213.95

Statement of expenditures by General Morton and Captain Burroughs (cont.)

Expenditures by Captain Burroughs:
Amount of pay-rolls from April 1863 to October 1, 1864......................................212,747.52
Paid for materials...8,913.57
Due for materials, mostly trees..11,000.00

   Total...232,661.09

Application, approximately estimated:
Surveys in Department of the Cumberland..2,000.00
Receiving, hauling, storing, and forwarding materials for Deparment
   of the Cumberland...5,000.00
Constructing 150 blockhouses, water tanks, and three reservoir tanks,
   Department of the Cumberland..3,500.00
Labor on pontoon bridge prepared at this place...1,000.00
Blockhouses built here for Department of the Cumberland 13,000.00
Temporary buildings and sawmills 2,500.00
Temporary defenses..2,337.54
Estimated cost of ordnance magazine, store-houses, and branch trestle railroad.....100,000.00
Expended on—
   Fort Negley..10,000.00
   Old Fort Morton..15,000.00
   New Fort Morton 40,000.00
   Fort Houston 38,323.53

   Total...232,661.09

Summary of expenditures

Expended on—
   Fort Negley...$101,313.95
   Old Fort Morton 15,000.00
   New Fort Morton..40,000.00
   Fort Houston...38,323.53
   Capitol Hill...10,000.00
   Casino Hill..1,000.00

    Expended on forts..205,637.48
Expended in Department of the Cumberland... 32,900.00
Temporary defenses 21,837.54
Temporary buildings, Nashville... 5,500.00

                                                              60,237.54
Expended on magazine 100,000.00

   Expended and due at Nashville..$365,875.02

In addition to the expenses incurred here as stated above, there has been expended:

By engineer agency, mostly for carts, drays, harness, barrows, etc.......................$148,246.27
Of which there has been turned over to other places
   and to the quartermaster's department..69,360.27

    Leaving a chargeable balance of..78,886.00

Of which amount 721 drays on hand cost..46,865.00
91 carts...3,640.00
1,118 sets harness..16,211.00
Saw-mill...2,500.00
   Total...$69,216.00

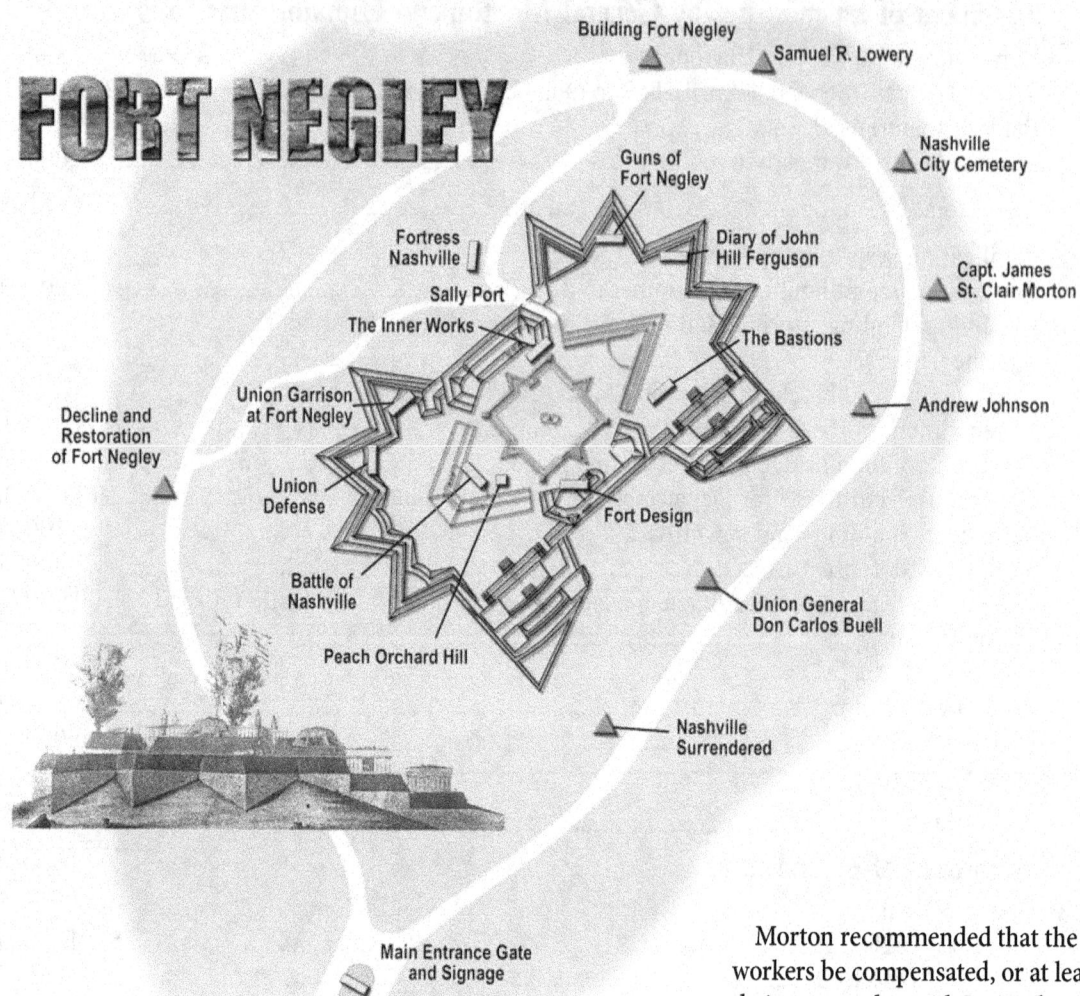

(Continued from Page 107)
Wadsworth westward to monitor the progress of contraband labor, where the inspector subsequently discovered conditions similar to involuntary servitude. Wadsworth expressed fears of an emerging type of serfdom, yet he still entertained hopes that the free marketing of said labor would someday "make the people of the South homogenous with those of the North."

At least 226 of those workers who survived were recruited into the U.S. Colored Troops infantry. As described by Castillo, museum coordinator at Fort Negley, they were captives of the Federal army, trading one master for another. One of those men was Jerry Jackson, who worked at Fort Negley and was compensated 42 dollars for six months of labor. Most workers received no pay at all. On Aug. 12th, 1863, Jackson was recruited at Fort Negley into Company F of the 12th USCT, which was used for labor on the Nashville & Northwestern Railroad and then later as garrison troops guarding the railroad trestles and bridges. Some USCT saw brief action at Johnsonville on Nov. 4th, 1864, before being recalled to Nashville to fight in the great battle there. Suffering from illness and injury while working on the railroad, Jackson died on March 1st, 1865 at the age of 40 at Nashville's smallpox hospital and was buried at Nashville's National Cemetery.

Morton recommended that the workers be compensated, or at least their masters be paid. In April 1863, Morton requested that the U.S. Corps of Engineers provide funds for the laborers. "The general commanding this department has expressed his wish that the negroes employed upon the fortifications at Nashville be paid wages, and so enabled to support their families. Their case being at present a very hard one, I respectfully ask your attention to the letter I wrote to the Department on this subject, I think in November last I will shortly prepare and forward estimate to put this matter in definite shape. At present the commanding general directs me to request you officially to give it your consideration. The chief difficulties are, of course, the obtaining the money and the doubt as to propriety of paying slave negroes of loyal and of rebel owners not present, or free negroes who cannot prove their being free."

The big controversy was whether to pay the worker, the fugitive slave, or the slave owner. It was agreed that the worker would be paid if his owner was pro-Confederate and that the slave owner would be paid if he was a Union loyalist. This distinction largely became null and void as few persons of any stripe ever got paid by the Federals.

One of the soldiers stationed at Fort Negley was Sgt. John Hill Ferguson, 34, of the 10th Illinois Volunteer Infantry. A native of Scotland, he came to America in 1840 and became a citizen in 1856. During the war, he kept a diary. His entry on Oct. 18th, 1862, described how four companies of his regiment erected tents inside the perimeter of the fort. Also, during the night a telegraph wire was strung from the fort to the headquarters of Generals Negley, Morgan, and Palmer.

An earlier entry described how the army hauled two 64-pound Confederate guns up from the Cumberland River and installed them at the fort. The Confederates had shipped the guns before Union forces occupied Nashville. When the guns arrived in Nashville, the Union forces commandeered them.

Ferguson described the first attempt to fill the cisterns. The "fiar (fire) company of Nashville under took to fill our Sisterns in the fort by forceing watter up the hill with 2 Ingines through gumalastic hoos [hose] but the presure was 2 great, the hoos would burst before the watter would reach the Sisterns…[they] worked the bigest part of the day then gave it up for a bad job."

On November 4th, Ferguson wrote that water-well drillers had reached a depth of 60 feet, and that the well was only five inches wide at the top. The well, eventually 200-foot-deep, was drilled "through solid rock by horse power."

The authorities in Nashville and Washington were fearful of Confederate cavalry attacks while Buell's army was off in Kentucky. The workers pleaded to be armed in case of direct attack but they were denied. They armed themselves with shovels and pick-axes, representing perhaps the first black soldiers of the Civil War, according to Castillo.

On Nov. 5th, 1862, General Nathan Bedford Forrest and 2,000 troopers rode on Murfreesboro Pike to within 1.5 miles of the fort. The booming of the guns, probably by the 12th Indiana Light Artillery, kept the troopers at bay. At that time, however, Confederate cavalry commander John Hunt Morgan was attacking Edgefield.

On Dec. 16th, 1862, the *Nashville Banner* reported that Fort Negley was finished. Firsthand accounts indicate that guns were in place on Saint Cloud Hill and were firing at Confederate forces scarcely two months after construction began.

WPA crews rebuild Fort Negley during the 1930s. (Tennessee State Library and Archives)

In late 1862, thousands of soldiers and civilians witnessed the Dress Parade of the 19th Illinois Volunteer Regiment on open ground near Fort Negley, the first of many such spectacles. Artillery crews camped within the ravelins of Fort Negley to be close to their guns. Federal infantry camped on the grounds surrounding St. Cloud Hill. Among the units stationed at or near Fort Negley:
Illinois – 13th, 19th, 73rd, 86th, 89th, 105th, 112th, 115th, 127th;
Indiana – 12th Light Artillery, 21st, 123th, 129th;
Kansas – 8th;
Ohio – 21st, 26th, 104th, 106th, 173rd, 175th, 188th;
Pennsylvania – 77th, 137th, and 15th Cavalry;
Tennessee – 10th Cavalry;
Wisconsin – 44th.

The 73rd and 112th Illinois, and the

Federal Army interpreters, Abe Lincoln, and city officials march up to Fort Negley during its grand 2004 reopening.

Visitors pose atop the stockade at Fort Negley during the 1940s. (Nashville Public Library-Metro Nashville Archives)

A marker at Fort Negley describing the Battle of Nashville, looking to the southwest. Visible in the background, left to right, are the Brentwood Hills, the City Reservoir and pumphouse (wartime site of Blockhouse Casino), and Rose Park, site of Fort Morton.

A smooth canteen — the largest artifact recovered from archaeological digs at Fort Negley, on display at the Visitors Center.

175th Ohio regiments were at or near Fort Negley during the Battle of Nashville.

In late 1862, the *Philadelphia Press* sent a reporter to Nashville and published a detailed description of Fort Negley and its armament in a "local boy does well" article on Morton, thereby educating the Confederates in regards to Nashville's defensive capabilities and possible weaknesses.

The firing of guns at Fort Negley were mostly in recognition or celebration, not combat. In fact, Fort Negley was so impressive and imposing, no enemy commander would entertain thoughts of attacking it (although there is speculation that Hood briefly considered sending suicide squads against the fort during his December 1864 campaign). The celebratory barrages produced an amusing incident in mid-1863.

In the pivotal month of July 1863, 1st Lt. James W. Jacobs of the 12th Indiana Battery was in charge of Fort Negley, which boasted 180 gunners and a bevy of heavy guns, ranging from a 64-pounder pivot gun to 24-pounder brass field pieces. As prearranged on a certain day all forts were to fire their guns in salute to the capture of Vicksburg, Mississippi. The guns at Fort Negley were prepared but no orders were received, so at noon all the guns in all the forts fired in unison except at Fort Negley. As expected, the commanding officer, General Robert Granger, came riding furiously to the fort and demanded to see the commanding officer, his face distorted with rage.

"What is it, General?" asked Lt. Jacobs.

"Why in hell didn't you fire a salute?" Granger demanded.

"I had no orders to do so, General."

"Do you stand there and tell me that you had no orders to fire your guns at 12 o'clock?"

"Yes, General, and I am telling you the truth."

The general reflected a moment, and then said, "Lieutenant, you are right. Never execute an order before you get it."

Smiling pleasantly, Granger grasped the lieutenant's hand and shook it heartily, announcing, "Somebody will catch hell for this."

As the war ground on and as the Federal armies of Rosecrans, Thomas, and Sherman moved inexorably toward Tullahoma, Chattanooga, Dalton, Atlanta, and then

(Continued on Page 116)

Fort Negley Visitors Center

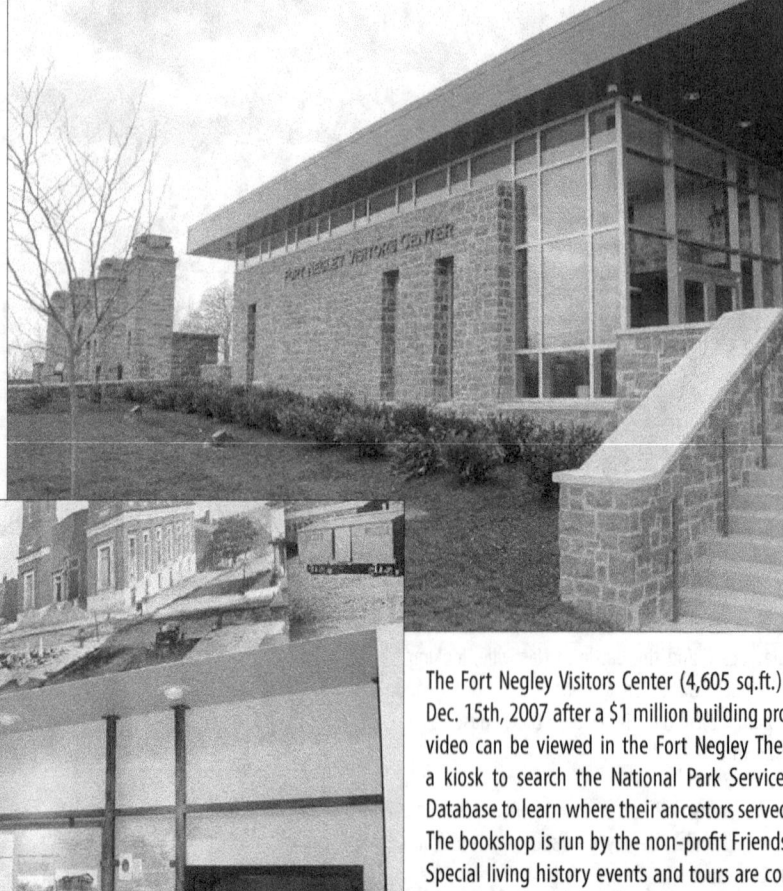

The Fort Negley Visitors Center (4,605 sq.ft.) opened its doors on Dec. 15th, 2007 after a $1 million building project. An introductory video can be viewed in the Fort Negley Theatre. Visitors can use a kiosk to search the National Park Service's Soldier and Sailor Database to learn where their ancestors served during the Civil War. The bookshop is run by the non-profit Friends of Fort Negley Park. Special living history events and tours are conducted on a regular basis. The center is also used for meetings by the Nashville Civil War Roundtable and the Ft. Donelson Camp 62, Sons of Union Veterans. The modern facility is open every day except Sunday-Monday, and the grounds are open every day from dawn to dusk. There is no admission fee. Restroom facilities are available during visitor center hours.

Nashville Chew Crew

The Nashville Chew Crew, short-hair sheep and herd dogs, at work reducing unwanted vegetation around Fort Negley via targeted grazing. (Photo by Krista Castillo)

Fort Negley Bastion Stoneworks

(Above) Southeast bastion looking from the south main work.

(Right) Bastion tunnel.

(Left) Southwest bastion looking toward south main work.

(Below) Southwest bastion looking toward Blockhouse Casino.

(Continued from Page 113)
through Georgia to the sea, Nashville and Fort Negley were relegated to the backwaters of the war.

Morton assumed more responsiblity and took command of Rosecrans' newly formed Pioneer Brigade. Much of the supervision over Nashville's fortifications fell to his assistant, 2nd Lt. George Burroughs, who had previously served as assistant engineer for Brig. Gen. G.W. Morgan's division in the Army of the Ohio. Only 20 years old, he was a graduate of the West Point Class of 1862 (3rd in class). He had drawn up fortifications at Cumberland Gap and then engineered defenses for the Louisville & Nashville Railroad. He served in the Tullahoma Campaign and at the battles of Chickamauga and Nashville. As a brevet major, he assisted in disposing of Federal property at Nashville until May 1867. He died at Charleston, S.C. in 1870 at age 28.

In October 1864, Bvt. Brigadier General Zealous B. Tower took charge of the Nashville fortifications. A West Point graduate and Mexican War veteran, Tower worked on Pacific Coast defenses, including Fort Point in San Francisco Bay, and after the onset of war served in defense of Fort Pickens in Pensacola and at the battle of Cedar Mountain. He was severely wounded at Second Bull Run and served in 1864 as superintendent of West Point. He served as chief engineer of Nashville defenses from Sept. 28, 1864 to July 1865. After the war, as a brevetted major general, he presided over the Board of Appraisers of the Tennessee Division. He served on many other engineering projects, mostly maritime, until his retirement in 1883. He died in 1900 at age 81.

The routine of garrison duty all changed in November 1864 as General John Bell Hood's Army of Tennessee invaded their namesake state, moving northward until it invested much of Nashville beginning December 2nd. Mauled at the battle at

View of Nashville from the Fort Negley stockade during the 1940s. Visible in the background are the State Capitol, Customs House, and Ryman Auditorium. (Nashville Public Library Metro Nashville Archives)

Franklin, Hood's depleted army hunkered down south of Nashville and waited for Federal General George Henry Thomas to attack them. Artillery shells were periodically exchanged, the Confederates hoarding their scarce ammo. On December 6th, the correspondent to the *New York Times* reported, "There has been heavy cannonading since 4 o'clock this afternoon. The guns of Fort Negley are shelling a rebel force in their front to prevent the enemy erecting batteries."

At that time, Capt. James E. White of the 12th Indiana Light Artillery was in charge of Fort Negley. Commanding the outer guard at the fort was Capt. W.S. Cain of the 8th Kansas Infantry. Fort Negley sported eleven heavy guns while nearby Fort Morton boasted eighteen.

Most Nashvillians were confident in the strength of the city's garrison and fortifications, as reflected in the purple prose of S.C. Mercer, editor of the *Nashville Daily Union:* "Nashville is powerfully fortified with forts and cannon, but we believe that the heroic hearts that throb within her line of bristling breastworks are of themselves a living and impregnable Gibraltar of freedom and Patriotism."

On the foggy morning of Thursday, December 15th, 1864, the guns of Fort Negley opened the ball on Hood's Army of Tennessee, more as a starting signal than to inflict much damage. The fort was not attacked during the two-day battle, which ended with the rout of Hood's army.

After the war, Fort Negley was decommissioned in September 1867 and the property rights reverted back to the heirs of Judge John Overton. All useable equipment and materials were auctioned off and the artillery pieces sent to the Western frontier forts. An urban legend has it that the limestone used to build Fort Negley was used to build the city reservoir (1887) nearby at the former site of Blockhouse Casino. The Metro Nashville Water Department denies this assertion, adding that the limestone was quarried at Rose Hill. "The outer walls were built of stone taken off the top of the hill above the present bottom of the reservoir," according to a 1913 engineering report. Regardless, in 1912 the city reservoir gave way in the middle of the night and flooded the neighborhood with 25 million gallons of water. Miraculously there were no fatalities. Another urban legend contends that there were tunnels between Fort Negley and the nearby City Cemetery. Archaeological studies have not confirmed that assertion.

In 1868, what remained of a magazine at old Fort Negley was used by the local den of the Ku Klux Klan for meetings led by Grand Wizard Nathan Bedford Forrest, according to newspaper accounts. The Klan was determined to terrorize and threaten the residents of Rocktown, a black neighborhood which had grown from the remnants of the old wartime contraband camp. Leander Woods, a veteran of the U.S. Colored Troops, formed a 50-man militia in the neighborhood, drilled on St. Cloud Hill, and worked to counteract the KKK. On or about Feb. 25, 1869, Forrest led Klan members to the fort, where they burned their robes and officially disbanded, believing that the Klan had served its purpose.

For decades following the war, St. Cloud Hill was bordered by neighborhoods of both white and black residents, neighborhoods that existed until they were broken up by the construction of interstate highways in the 1960s. The historic bond had been broken. Fort Negley fell into disrepair and obscurity, except for a brief restoration in the 1930s, and remained out of mind until the 1990s. Minor efforts were undertaken to preserve the fort and clear vegetation during the 1960s centennial of the Civil War and in the mid-1970s in association with the National Register of Historic Places nomination.

There is no national battlefield park at Nashville, despite the enormity and significance of the Battle of Nashville. In 1910 and again in 1928, Congressman Joseph W. Burns introduced legislation to designate a battlefield park at Fort Negley, but both bills died due to budget restraints. The oppressive Federal occupation of Nashville from 1862-72 and the perception of Fort Negley as a symbol of Yankee aggression may have led local officials to downplay any recognition of Civil War activity. In at least one instance, a local city councilman vowed never to spend one dime of taxpayer money for "that Yankee fort."

The iconic stone gate to Fort Negley built by the Works Progress Administration in the 1930s.

The Overton heirs, the Brinkleys of Memphis, planned to sell off the property into building lots, but the auction never took place. The heirs then offered to sell the land to the city for $100,000 to be used as a park. Likely due to the fact that taxes were delinquent on the land, the City of Nashville ended up issuing bonds to purchase the Overton property in 1928 for $20,000. Then the Great Depression hit hard the following year.

The Fort Negley that amazes tourists today is not the fortification built by the Federal military in 1862 (although some stonework has been identified as Civil War-era due to evidence of quarry marks). By the time of the Depression, the site was overgrown and only "traces of the breastworks" remained, according to the 2014 study of the site. For several reasons, the federal Works Progress Administration (WPA) selected Saint Cloud Hill as one of its employment projects. Fort Negley was chosen for its historic significance, tourism potential, urban greenspace, and proximity to a large group of unemployed workers.

In 1934, the authorities evicted a black squatter neighborhood on the hill and demolished their houses, also resulting in the closing of Auntie Cora's popular barbecue restaurant. Clearing the site in April and May of 1935 revealed what little remained of Fort Negley.

Newspaper accounts stated that two shifts of 575 workers each worked on the site. The project was completed in 1936 by a work force of 2,300 men under the direction of project engineer J.D. Tyner. The reproduction was remarkably similar to the original, based on plans found at the War Department in Washington, D.C. (those plans have since been lost).

In 1992, local stone mason Graham Reed was hired to examine Fort Negley; based on a brief visual inspection he speculated that at least half of the visible stonework dated back to the Civil War.

According to Zada Law, historical

Fort Negley Descendants Project

The Fort Negley Descendants Project (FNDP) is an oral history digital archive aimed at preserving the voices and stories of the descendants of the African-American laborers and soldiers who built and defended Fort Negley. Recent ground-penetrating radar reports have indicated a high likelihood that their remains still lie on the grounds of Fort Negley Park. After the war, those who survived settled the nearby historically black neighborhoods of Chestnut Hill, Wedgewood Houston, historic Edgefield, and Edgehill. At the turn of the century, several prominent families from these neighborhoods founded North Nashville and all of the prestigious black institutions residing there — the historically black colleges, businesses, and churches. In the 1950s, these same institutions trained and supported some of the sharpest minds of the civil rights movement. There is a long and unbroken connection between the builders and defenders of Fort Negley, and Nashville's current African-American population. Many members of this population see the fort as sacred, and they memorialize it with ceremonies, oral traditions, and historic reenactments. As gentrification pushes more black families from these neighborhoods out of the city limits, and this development robs the community of a place that has secured a spot in black oral tradition, stories will become less accessible. People will forget. In response, a handful of colleagues at Vanderbilt's Digital Humanities Center and the Robert Penn Warren Center for the Humanities have created a working group (Digital Initiatives in Public Engagement) to explore the ways in which digital solutions could contribute to a local public history concern — in this case, the loss of descendant voices. With generous help and support, the FNDP has been able to track down several descendants of the fort, both builders and defenders, and film their stories for the creation of the Fort Negley Oral History Archive to be housed in Vanderbilt University's repository in perpetuity. In partnership with the Nashville Chapter of the Afro-American Historical Genealogical Society, the project has also filmed a video to help individuals start their own genealogical research to discover if they are a descendant of Fort Negley themselves. The FNDP has also collaborated with various organizations and community leaders to organize events open to the public such as Shayne Davidson's lecture about her genealogical research and illustration techniques to create her art exhibit, "The Seventeen Men," which is discussed in further detail in her book *Civil War Soldiers: Discovering the Men of the 25th United States Colored Troops.*

[Information obtained from FNDP website]

---

archaeologist and the director of the Fullerton Laboratory for Spatial Technology at Middle Tennessee State University, the footers of the Civil War walls extended at least ten feet below the ground surface. The WPA workers likely left some Civil War masonry in place and used this coursework for the base of their reconstruction. Much of the footprint of Morton's Fort Negley may still exist below the surface.

The reproduction project took longer than the original Civil War fort. It was necessary to quarry 61,875 cubic feet of stone and add 18,000 cubic yards of dirt trucked to the site. A stone marker which still stands at the site (although barely legible) states: "Fort Negley. Built by Federal Forces 1862. Restored by WPA 1936."

According to the 2014 study: "Park board minutes indicate that additional work continued in stages through 1940 and included new access roads, water and lighting systems, ball diamonds and bleachers, a comfort station, garage, and storehouse. A small museum was located in a reconstructed subterranean munitions magazine inside the fort on the north side. The fort received two cannon from the United States Army Watervliet (New York) Arsenal in 1937, and two other cannon were also acquired. The park opened in the spring of 1941."

The new facilities included four small softball fields and one larger hardball field. Stone bleachers were built back into the hill, which can still be seen today. Also, the WPA built the elaborate stone entranceway.

Then World War II intervened and the park and the fort fell once again into disrepair from neglect. The park was closed for repairs in 1945, but only the baseball diamonds reopened in the spring of 1946. Most remarkably, the park was segregated for use only by whites. Two cannon were loaned indefinitely to Montgomery Bell Academy in 1947. The ballpark was used into the 1950s and through the early 1970s although the fort became an eyesore. Vagrants were often seen camping in the fort. Denuded during the Civil War and the WPA project, Saint Cloud Hill over the subsequent years became overgrown with vegetation, including evasive species, and dense stands of trees. The site became unrecognizable from a distance.

Little mention was made of Fort Negley during the centennial celebration of the Civil War in the early 1960s. In 1964, the city parks board voted to temporarily locate a zoo at the park but neglected to appropriate any funds. The zoo was never built. However, the Sixties brought major changes to the site. The construction

Fort Negley looking toward the northeast bastion. (Below) Hydraulic blow-out mars south main-work wall just below tourist overlook.

of Interstate Highway 65 and the downtown loop (and implementation of the Edgehill Urban Renewal Plan) displaced historic black neighborhoods and broke the site's link to those neighborhoods. In 1967, the city leased sections of the park north of the fort to relocate the children's museum from downtown in 1974. The museum later became known as the Cumberland Science Museum, now the Adventure Science Center. In 1978, the southeast portion of the site, where the ballparks had been located, was leased for the construction of Herschel Greer Stadium, home of the minor-league Nashville Sounds baseball team.

Fort Negley Park totals 64 acres (35 occupied by the fort, 18 by the old Greer Stadium site, and 11 by the Adventure Science Center).

In 1975, at the urging of the Metropolitan Historical Commission (MHC), Fort Negley Park (the eight acres of fort inside the "ring road") was listed on the National Register of Historic Places.

In 1980, with growing public awareness of Fort Negley as the only "attraction available now for interpretation of the role of the city during the Civil War," MHC secured a grant from the Tennessee Historical Commission and engaged the firm of Miller, Wihry & Lee, Inc., of Nashville, to prepare a study of Fort Negley Park. The study recommended that the historic site be excluded from the area leased by the science museum and that it be maintained instead by park management staff. The study also called for developing a plan for self-guided interpretation and "living history" presentations, but proposed "no major effort on the fortress" itself.

The 1980 study recommended that the remains of Fort Negley be "interpreted as ruins" and not rebuilt again.

Meanwhile, the decaying fortifications remained closed to the public, surrounded by a chainlink fence. The site was vandalized and compromised by unscrupulous relic hunters by day and occupied by vagrants, drug dealers, and various miscreants at night. In one instance, a body was found at the fort, propped up against some stonework, a victim of a carjacking. A local Civil War expert was leading a tour of the site for the governor and his state trooper escort when a homeless man popped his head up out from a tarpaulin and asked if they were touring the fort. "Did you know," he asked the governor, "that Fort Negley was the largest inland stone fortification built during the Civil War?" Another time, a tour leader had to quickly divert a group of ladies upon approaching a couple enjoying each other's company amidst the ruins.

The fort remained closed and untouched until the early 1990s when the parks department removed shrubs and trees from the fort's walls. The vegetation removal caused numerous "blow outs" of the fort's stone walls as the roots that had essentially held the dry-stacked walls together. The vegetation rotted and water infiltration loosened the stones from their settings. A pilot effort to rebuild and stabilize several sections of the fort's walls using modern building practices was unsuccessful.

An archaeological study was conducted and numerous artifacts from several different eras found. The largest Civil War artifact found was a soldier's canteen, which can be viewed at the Fort Negley Visitors Center.

In 1996, the city sanctioned a Fort Negley Master Plan, which determined that the fort's deterioration stemmed from several factors—the original construction techniques, the steep topography, unsuitable soil conditions, uncontrolled vegetation growth, inadequate site drainage, hydrostatic pressure, and vandalism. The study noted that the city could improve and enhance only the last four factors listed.

In 2002-03, the City of Nashville appropriated approximately $2 million to stablize the stoneworks, build walkways for tourists so as not to damage the ruins, erect interpretive signage, and eventually build a modern visitors center. This was one of the largest municipal appropriations ever to bolster Civil War tourism and preserve historic works.

On Dec. 15th, 2004 (the 140th anniversary of the beginning of the Battle of Nashville), Fort Negley Park and the historic fort opened to the public for the first time in 60 years. History interpreters, including U.S. Colored Troops, led a ceremonial procession up the walkway to the fort. Making presentations were Mayor Bill Purcell, Councilman Howard Gentry, who had roamed the site as a kid, President Abraham Lincoln (Dennis Boggs), and country music entertainer and Civil War enthusiast Kix Brooks. In December 2007, the Fort Negley Visitors Center opened to the public.

During the past 15 years, Fort Negley has witnessed more challenges and controversies.

STATION AT FORT NEGLEY, NASHVILLE.

Figure 79. Signal station in a tree (from Brown 1896:469).

In 2016, the Parks Department supervised the clearing of trees on 11 acres of the fort, creating a minor public outcry. Citing a "failure of communication," the department stated that the clearing had been recommended by long-term management strategies to return the landscape to its appearance during the Civil War, Reconstruction, and WPA eras; slow the deterioration of the fort due to uncontrolled growth and spread of exotic invasive vegetation (tree roots damage archaeological deposits); and expose the fort for better and safer public access.

One innovative approach to controlling unwanted vegetation has been targeted grazing by the Nashville Chew Crew, a flock of short-haired sheep that eat most anything in their way. The sheep, controlled by a portable electric fence and sheep-herding dogs, are also a major draw for visitors, especially children. Nashville Chew Crew owner Zach Richardson says he uses sheep instead of goats because the sheep are more docile and predictable.

In 2014, the Greer Stadium site was abandoned in favor of a new ballpark site downtown. Three years later the City of Nashville hired a commercial real estate firm to develop that site. The Cloud Hill Partnership proposed condominiums, retail space, and a commons area. Despite being favored by then-Mayor Megan Barry's office, the project was dropped in January 2018 following opposition from historical preservationists and neighborhood activists. Weighing heavily in the decision was the determination that the tract might still contain the remains of some of the workers who built Fort Negley in the first place. No actual graves have been found, but evidence exists to indicate the possibility.

Fort Negley Park is also home to a 250-square-foot fossil collection site donated by Vulcan Materials Company. The visitors center is surrounded by a pollinator garden maintained by the Friends of Fort Negley Park.

As of early 2022, a new master plan for the site is being formulated, in addition to $1 million in appropriations to continue stabilization work. Portions of the walkways have been closed in recent years due to deterioration. Historical preservationists hope that someday a major museum will be constructed near the visitors center that will serve as a hub not only for Nashville's "Civil War to Civil Rights" history but for that of all Middle Tennessee.

Each year, 2,771 small U.S. flags are planted in the grassy hillock near the fort, symbolizing the African-Americans (slaves, contrabands, impressed freemen) who worked on the fort.

In 2020, the foundation of a small building was discovered which served as Captain Morton's headquarters during the construction of Fort Negley. If alive today, the West Point engineer would probably be amazed that his showpiece fortress was still being appreciated today.

Modern aerial photography of Fort Negley-Fort Morton-Fort Houston layout along with Blockhouse Casino. All remnants of Greer Stadium have now been removed.

Researchers Use Hi-Tech Techniques to Interpret Nashville Battlefield

New technologies such as Ground Penetrating Radar (GPR), satellite imagery, story mapping, Geographic Information Systems (GIS) and Global Positioning System (GPS) are generating new interpretations of military activities at Nashville during the Civil War.

At Fort Negley, Brandon Hulette, Vanderbilt University military science professor, and his students have uncovered (without disturbing the terrain) anomalies along the wartime trench line northeast of the fort that indicate the existence of small, underground ammo storage structures. Noted landscaper Edward Law Olmstead produced a detailed survey of the site in 1891 in preparation for residential development. Anomalies detected by Hulette's crew aligned perfectly with an overlay of Olmstead's survey of the entrenchments which figuratively ran through Fort Negley. Hulette also noted that it is quite likely that wartime burials exist underneath the remains of the nearby parking lot of the former Greer baseball stadium—the rationale for further scientific research to determine whether this speculation is true.

A GPR survey of Shy's Hill, in coordination with the Battle of Nashville Trust (BONT), was conducted in 2021 by the "spatial specialists" to provide further information on the Confederate army earthworks there and subsequent "disturbances" caused by residential development. It was noted that at one point in the 1950s a water tank was sited at the summit, which had to be flattened by earthmovers. Shy's Hill is now eight to ten feet lower than during the Civil War.

Vanderbilt students have developed story maps on Nashville fortifications, military hospitals, and naval activity during the Battle of Nashville. In their study of Nashville fortifications, researchers used historical maps and documentation, satellite imagery, period photographs, and technical information to create overlays of new information, e.g., interlocking fields of fire and corresponding positions of enemy troops. To the east of the city, near the river, there was no interlocking fields of fire from the forts, leading to the conclusion that the Federals relied on the firepower of their river gunboats much more than acknowledged.

From left, Vanderbilt research analyst Natalie Robbins, students Jordan Rhym and Alyssa Bolster, geospatial librarian Stacy Curry-Johnson, and professor Brandon Hulette with a ground-penetrating radar machine at Shy's Hill. (Photo by John Banks)

"Up until the Civil War, the hospital was really the place you went to die…Federally occupied Nashville is just one of the sites of major change in hospital function…The hospital system we know today developed during this time, and Nashville was integral to this," noted James Atkinson, M.D., surgical pathologist at Vanderbilt Medical Center.

Doctors in military hospitals in Nashville used bromine to drastically reduce fatalities from infections, for example. Dr. Middleton Goldsmith used topical application and targeted injections of bromine to reduce the fatality rate of gangrene cases from 45 percent to three percent. The army physicians also tackled the problem of soldiers infected with venereal disease. The eventual solution was to legalize and regulate prostitution. The percentage of soldiers in Nashville with sexually transmitted diseases fell from 40 percent to four percent, according to Hulette.

In the near future, Vanderbilt researchers plan to post a Battle of Nashville story map, which can be updated as needed, on the BONT website.

In other developments, two civilian researchers—Jim Kay and Fowler Low—used metal-detecting technology and historical documentation to establish the true site of the Battle at the Barricade, Dec. 16th, 1864 on Granny White Pike—a mile south of the conventional established site.

At the Battle of Franklin site, historians used metal detectors (with permission) to uncover spent metallic cartridges from repeating rifles. They have been able to trace the movements of specific units and even specific riflemen during the battle 160 years ago (each rifle leaves specific identifying marks on every cartridge that it fires and expels onto the ground).

View of downtown Nashville from Fort Negley's sally port.

In this extraordinary photograph looking from the State Capitol, three fortifications can be seen on the horizon (magnified detail above, left to right) Fort Negley, Blockhouse Casino, and Fort Morton. Also visible just below Blockhouse Casino is Polk Place, the home of President James K. Polk (his widow during the war). The heavy guns at Fort Andrew Johnson have been covered due to a rainstorm. (Library of Congress)

Each fall, the Friends of Fort Negley plant 2,771 miniature United States flags on the grounds to commemorate the African-American laborers (slaves, impressed freedmen, and fugitive slaves) who built the fortifications under harsh conditions. Few were compensated for their hard labor; many died due to disease and exposure. Above, the flags can be seen in front of the stone gate, the massive U.S. flag, the visitors center, and the fort up on the hill.

Fort Morton

Fortification named for chief Federal military engineer

Fort Morton was an elaborate fortification built in 1862 on a bare limestone hill (Curry Hill, now site of Rose Park on Edgehill Avenue) as one of three anchors of the interior line guarding the southern approaches to Nashville (the Lebanon, Murfreesboro, Nolensville, Franklin, and Granny White pikes). It was designed by Capt. James St. Clair Morton to hold more heavy guns than Fort Negley itself (14 versus 11) and was literally too complicated to build according to its initial design. The fort was sited atop nearly solid limestone and the rocky nature of the terrain, the complex design, and the lack of adequate labor, in addition to other priorities, delayed work on this fort and required more funding than anticipated. The fort was named for Morton after he left Nashville in 1863. He was killed in combat on June 17, 1864 at Petersburg, Virginia, shot by a sharpshooter as he was scouting the front lines.

In September 1863, Charles A. Dana inspected the Nashville works and reported to Secretary of War Edwin Stanton:

"The central work, known as Fort Morton, is scarcely yet commenced. Simpler in design and more powerful when done than Negley. It is situated on a hill of hard limestone, and the very extensive excavations required must all be done by blasting. At the present rate of progress it will take two years to finish it. A part of it, namely the demi-lune in its front, is partly done, so far in fact that its parapet might be used as a rifle-pit and might afford some protection to field guns. This work will require a garrison of 1,500 to 2,000 men. The two redoubts and barracks connecting them, of which its main body consists, will be altogether 700 feet long."

At the beginning of 1864, Fort Morton enclosed one 30-pounder Parrott, one 32-pounder sea-coast, and one 24-pounder siege gun (the last two mounted on carriages like casemate carriages without the chassis).

When Brig. Gen. Zealous B. Tower was put in command of the Nashville defenses in October 1864, he noted that Fort Morton was not even half

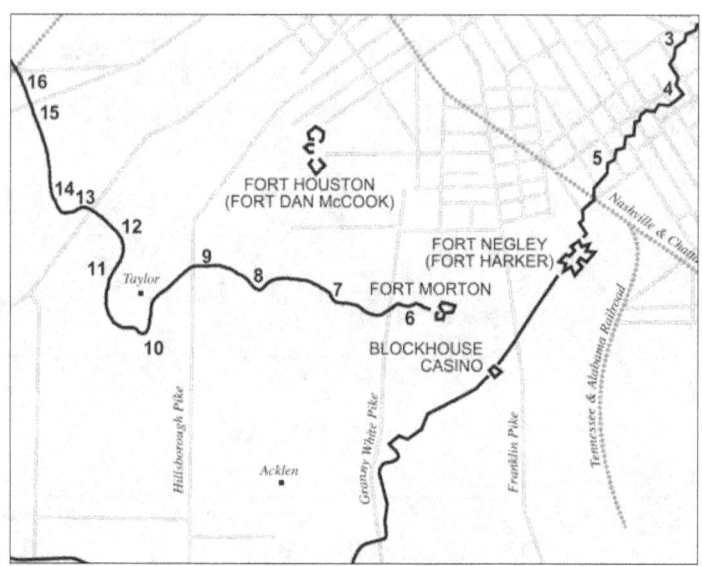

completed. He said that Fort Morton, after an expenditure of at least $15,000, was abandoned by direction of Col. Merrill, Engineers, when he took charge of the Department of the Cumberland Engineers. Fort Morton is a simple polygon, sufficient for the purpose intended. He recommended: "Fort Morton to be completed as now being built. The rear parapet will, however, be reduced to the minimum. It may be necessary to pile up rock and earth on the exterior for a glacis, and as some exterior obstacles, as the work is neither flanked nor has a ditch, and the ground near the fort is not seen from the parapet. The interior block-house covered by the parapet against direct fire, will serve as a keep and bomb-proof."

In May 1865, Tower issued another report. At that time he was Inspector General of Fortifications, Military Division of the Mississippi: "This work had made some progress, according to the original plans, when Col. Merrill (captain, Engineer Corps), foreseeing that it would never be finished, directed its abandonment and the substitution therefore of a polygonal redoubt, with guns en barbette and an interior block-house. When I assumed general direction of the Defenses of Nashville, this fort was not half finished. I modified it slightly by increasing the number of guns and placing them in embrasure, diminishing the parapets unnecessarily thick, introducing two service magazines, which would serve also as traverses, and reducing the block-house from 120 to 80 feet in length. It was my intention also to build a glacis around the work, revest the scarps with dry stone, and put flanks in the redan, so as to sweep the ditches of the fronts of attack;

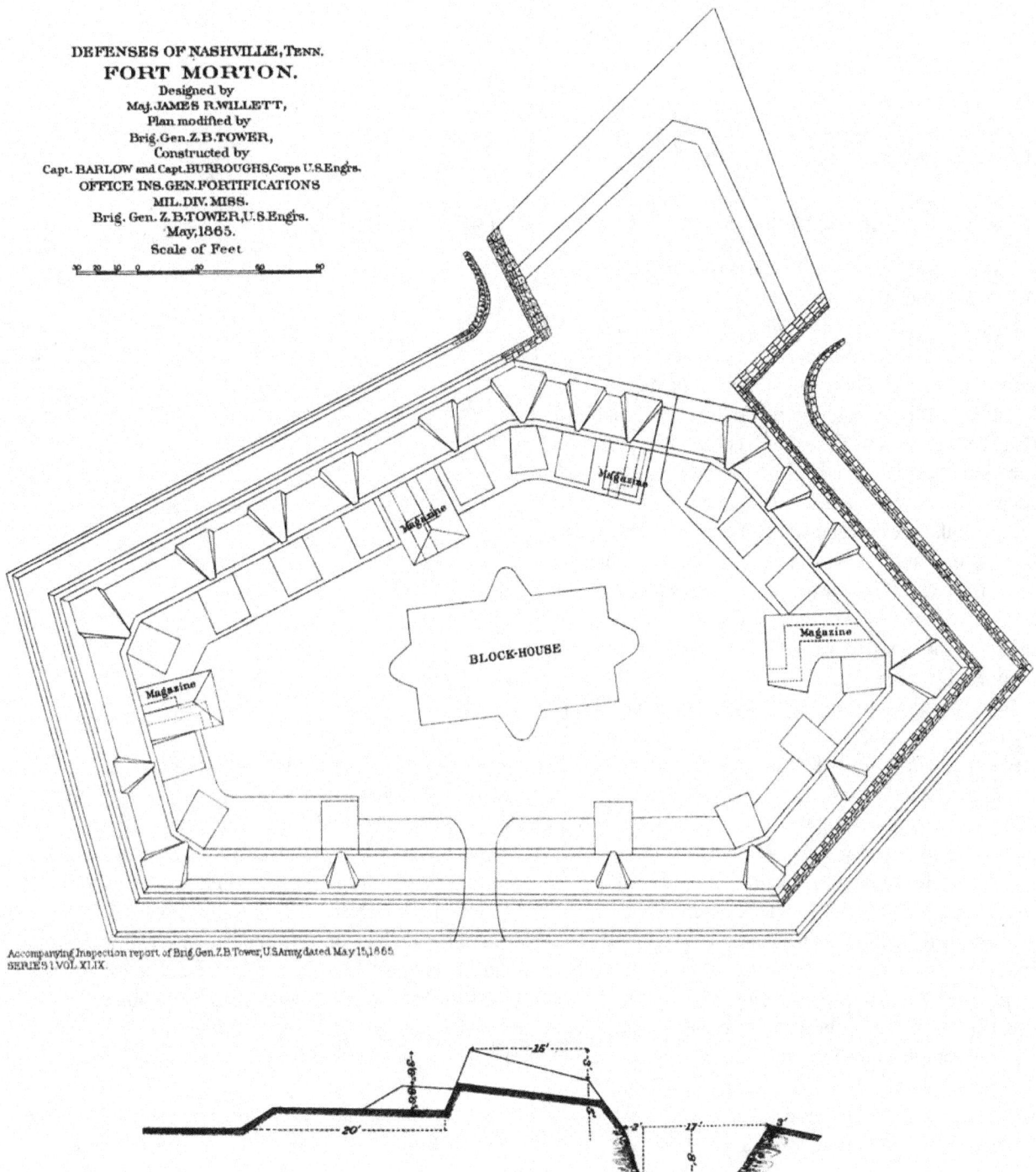

A difficult fortification to build due to the rocky nature of the terrain, this elaborate fort anchored the eastern end of the inner line of Federal defenses of Nashville. Shown above is a cross-section of the outer earthworks of Fort Morton.
(All Nashville fortification diagrams from Atlas to Accompany the Official Records of the Union and Confederate Armies)

this has in part been done. The accompanying sketch shows these arrangements. The rocky character of the site of Fort Morton, its position on a high hill, the necessity for blasting the terreplein and for the magazines, and for hauling earth from a much lower level, and the large keep have made this work expensive and retarded its progress. Fort Morton is nearly finished."

The fort was abandoned in 1867 after the end of the war and after Tennessee had returned to the Union. No traces of the fortification remain today.

Blockhouse Casino

Timber blockhouse in the form of a cross with earthen parapets

Blockhouse Casino was built as a timber-reinforced blockhouse on Kirkpatrick Hill. This was a small fort with earthen parapets, a crushed stone glacis and armed with light artillery pieces. The location was on a commanding hilltop along the Franklin Turnpike that lead into town. A more elaborate battery was envisioned by General Tower to surround the blockhouse but was never built. He suggested "a strong, double-cased block-house for Casino Hill, covered against direct fire from the high ground to the west by a parapet or battery for guns, the battery to be protected by external obstacles, and connected with the blockhouse by a palisade. The block-house to be a bomb-proof, surmounted by a parapet."

Gen. Tower's report of May 1865 included the following: "Casino Hill is half a mile distant from Fort Negley and one-third mile from Morton and is ten feet higher than this last fort. Gen. Morton placed on this hill a single-cased block-house in the form of a cross, relying upon the combined fire of Morton and Negley to drive an enemy from the position should he attempt to build batteries there. Had Fort Morton been finished of the magnitude originally intended, its powerful armament might have accomplished that object by deluging the hill by its fire. I designed for this position a simple battery, with a deep ditch and eight-foot rock scarps. The two faces were directed upon Morton and Negley, so as to expose the hill to the fire of these forts. The forge line, simply a stockade closing on the block-

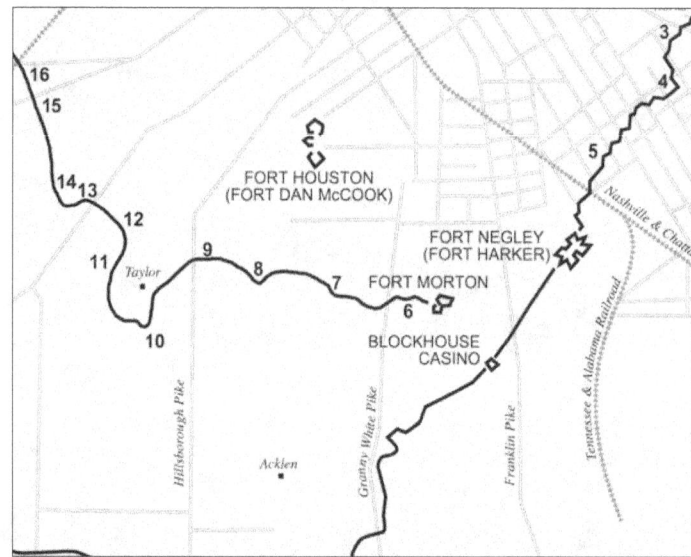

house, leaves the interior open to fire from the works in the rear, so that no enemy could hold the battery, should he succeed in carrying it. Lack of men and the urgent necessity for forwarding more exposed points on the defensive line prevented the commencement of this battery. The hill is limestone rock with scarcely any soil, and steep on the line of approach."

Fort Casino was abandoned in 1865 at the end of the war. Decades later, in August 1889, the city completed construction of a new reservoir that could hold more than 50 million gallons of water. Legend has it that the limestone blocks used to build the reservoir came from old Fort Negley; actually the stone was quarried nearby from what is now Rose Park. The structure, which cost $365,000 to build, stood nearly 34 feet high, shaped in an ellipsis 603 by 463 feet and divided into two basins by a center wall. A little after midnight on Nov. 5th, 1912, the southeast wall broke loose and sent 25 million gallons of water roaring into the adjacent neighborhood. Some houses were swept off their foundations, but no lives were lost. The damage totaled $2.5 million in today's dollars. The wall was fixed for about $100,000 and the reservoir is still in use today. A huge bladder in one half of the reservoir holds clean water; the other half is empty and reverting back to nature. The crack where the reservoir broke is quite noticeable. Today, however, the grounds are restricted due to security measures. The gatehouse on top of the reservoir is easily noticeable from the interior of Fort Negley.

(Metro Nashville Water Services)

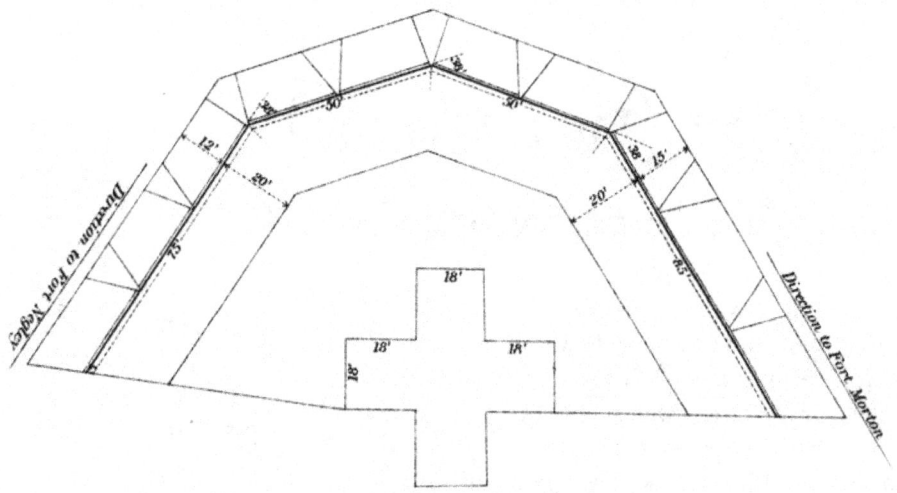

DEFENSES OF NASHVILLE, TENN.
Redoubt not yet constructed.,
May, 1865.

BATTERY FOR CASINO HILL.
Designed to cover Block-house from
ENEMY'S ARTILLERY
AND
DEFEND THE POSITION.
Devised by
Brigadier-General TOWER,
December, 1864.

Scale of Plan

Scale of Sections

Brig. Gen. Z. B. TOWER, U.S. Eng'rs.
INS. GEN. FORTIFICATIONS
MIL. DIV. MISS.
1865.

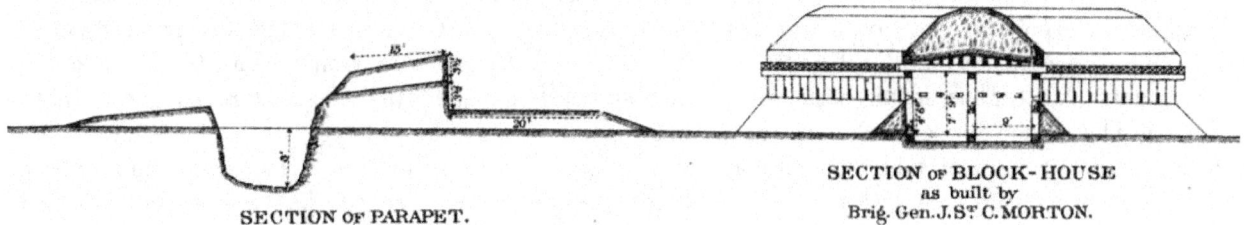

SECTION OF PARAPET.

SECTION OF BLOCK-HOUSE
as built by
Brig. Gen. J. ST C. MORTON.

Accompanying Inspection report of Brig. Gen. Z.B Tower, U.S Army, dated May 15, 1865.
SERIES 1 VOL XLIX.

Fort Houston (Dan McCook)

Fortification nearly finished in time for Battle of Nashville

Fort Houston was built in 1863 and named for Russell Houston, a Union sympathizer and owner of the land on which the fort was built. He probably donated the land for the fort with an understanding that it would later be purchased or returned. Houston allowed his home to be blown up to make way for the fort. The fort was located near Belmont & Broad Streets and was designed to hold more than 35 guns providing cover for the Charlotte Pike. The structure was renamed Fort Dan McCook in honor of Brigadier General Dan McCook, who was mortally wounded at Kennesaw Mountain and died July 17th, 1864. The fort was abandoned by Union troops in 1865 after the end of the war. None of the fortification exists today. The site is now the Music Row traffic roundabout.

In September 1863, Charles A. Dana reported to Secretary of War Edwin Stanton that Fort Houston was about one-quarter done, and could be completed with "comparative rapidity and cheapness."

In October 1864, Gen. Tower recommended "Fort Houston to be completed as cheaply as possible. Instead of the line of casemated bomb-proof connecting the two polygons, I propose a double caponiere. The parapets of these works will be made the minimum on the rear line, and the northern one left much lower than the plans indicated. The immense traverse bomb-proof will be omitted, and perhaps a small blockhouse bomb-proof put in their place. Of course, the independent scarp will not be constructed."

Tower's report of May 1865 states: "Fort Houston (now called Fort Dan McCook). More labor has been expended on this fort than would have been required to build a large bastion work. In November, 1864, it was in a very unfinished condition. It progressed very rapidly for the period of three weeks, by the hands of a large number of workmen, mostly from the quartermaster's department. It was made ready for twenty-six guns at the time of the battles of Nashville, though the polygons were not

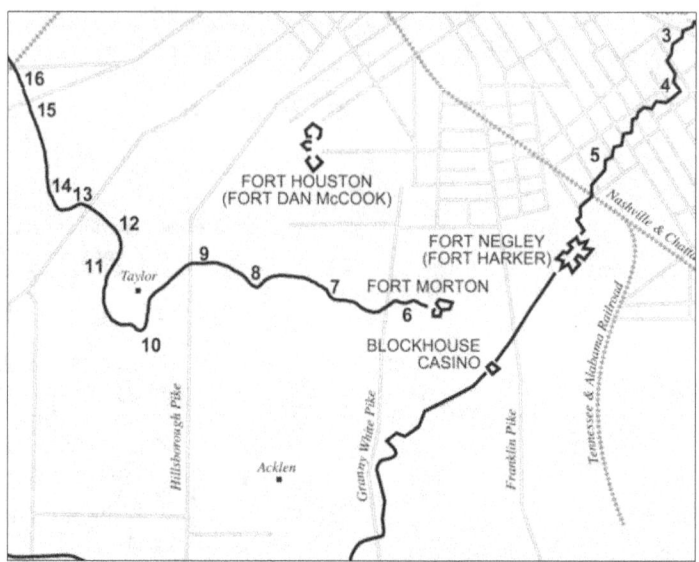

inclosed. A small force has been employed upon Fort Houston since December last. Nearly all the gabion embrasures have been constructed, and entrances walled, and the works inclosed. Much labor is required to finish it. Its dimensions are so great that a small number of workmen make slow progress upon it. When completed, it will mount thirty-five guns for direct fire and ten flanking guns. The original design was very costly, involving independent scarp walls, an immense traverse, and bomb-proof store-houses. All these structures have been omitted in the modified plans. The north polygon, not being inclosed, was reduced in size, to avoid heavy embankments, and the reference of the interior crest dropped. The accompanying sketch will show the magnitude and character of this fortification. The almost unprecedented rains of December, January, February, and March have greatly retarded progress upon all the forts about Nashville."

Three and one-quarter acres of the original site of the fort were purchased from Russell Houston after the war in August 1865 and the sale later ratified by Congress in 1874. In the same 1874 act the land was to be conveyed to Fisk University for educational purposes with no restriction on the resale of the land. The fort is referred to in all of the documentation for this transaction and in the actual act as Fort Russell Houston, not Fort Dan McCook.

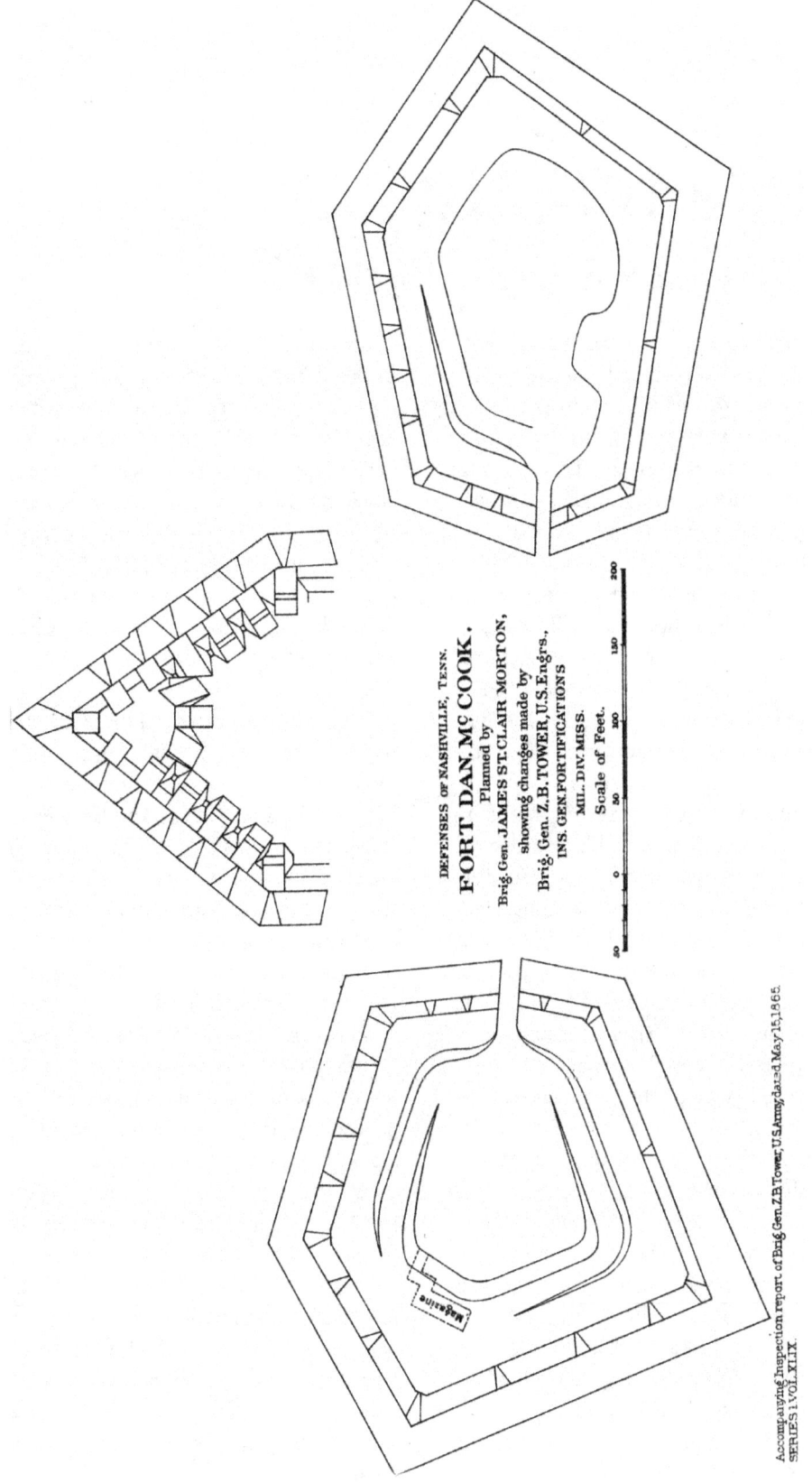

Redoubt for Hill 210
Lofty link in lengthy entrenchments encircling Nashville

Hill 210 is situated half a mile west and beyond Fort Gillem and is higher than that redoubt (near Merry Street & 25th Avenue North today). From its crest an enemy could fire at long range into the suburbs of the city and could make Cumberland Hospital and the large store-houses on the Northwestern Railroad untenable, according to General Tower. "I therefore planned a redoubt in October last to hold this hill. It was not commenced for the lack of means. When, however, Hood commenced his movement on Nashville, a large battery of two bastion fronts for fifteen guns, supported on either side by rifle-pits, was built, by the aid of employees from the quartermaster's department. The 30th of November, by my request, the commanding general directed large forces of the quartermaster and railroad departments to report to me for constructing an infantry line around the city. This line was built before the battles. It commenced at the reservoir and passed over Cemetery Hill to the railroad track, and was continued thence by Gen. Schofield to Casino Hill. From Fort Morton it passed around the Taylor barn, and thence north in rear of the Ellison house, to Hill 210. Most of the line from Hill 210 to the Cumberland River, touching at Gillem, Donaldson, and Hyde Ferry Forts, was a rifle-pit. This line was supported by twenty batteries, constructed with embrasures.

"The entrenchment is seven miles long; no shorter line, however, would inclose the store-houses and hospitals. The high range of hills, distant about three miles from the city, was entrenched by the army occupying them while Gen. Steedman threw up lines in front of the south suburbs of the city. Thus Nashville was doubly entrenched. The line of the hills was the best army line. It in part rested on Forts Negley, Morton, and Casino Hill, but received no support from Houston, Gillem, Donaldson, and Hyde Ferry Forts, and could not, therefore, be held except by an army. The interior line, while serving as a reserve to the exterior, would enable the usual garrison of Nashville, aided by the quartermaster employes, to hold the city against ordinary attacks from large raiding parties, under such generals as Forrest and Wheeler. Had the war continued it was my intention to put a redoubt on Hill 210 and support the two batteries to the left by block-houses. The battery at the Taylor barn would have been converted into a redoubt with a block-house keep. One small block-house between Morton and the Taylor house, and two between Negley and the reservoir would have completed the line of defense, and made it amply secure. These block-houses have all been prepared by a detachment of the 182nd Ohio Volunteers from timber cut down in the vicinity of Johnsonville. The spring floods destroyed the bridges on the Northwestern Road, and prevented the transportation of this material to the city.

"It is useless now to build these structures. As Nashville will probably have a garrison for one year at least, if not for a much longer period, I propose to complete Forts Morton, Houston, Gillem, and Hyde Ferry, almost finished, by the aid of soldiers. Negley and Donaldson are finished. Capt. Burroughs, U.S. Corps of Engineers, up to October, 1864, had charge of the works around Nashville, mostly under the direction of Gen. Morton. Maj. Willett, then lieutenant, also assisted Gen. Morton, and built the magazine. Col. Merrill gave little attention to the defenses of this depot, being principally occupied with those at Chattanooga. For so important a place, held so long by our troops, the Nashville defenses certainly were not pushed forward as much as they should have been. Little aid is given by commanding officers of posts when those posts are not in the front or constantly exposed. In such positions building redoubts is the first operation, while far back on the line of communication it is very difficult to get a detail to throw up lines. Every other labor takes precedence. Capt. Barlow took immediate charge of the works around Nashville the 13th of November, under my direction, and has performed his duties faithfully and intelligently. Capt. Jenney gave me much assistance, superintending at Forts Houston and Gillem and upon the lines.

Majors Dickson, Powell, and Willett assisted in the construction of the entrenchments around the city, which were mostly executed the first week in December, 1864, by the quartermaster and railroad employees."

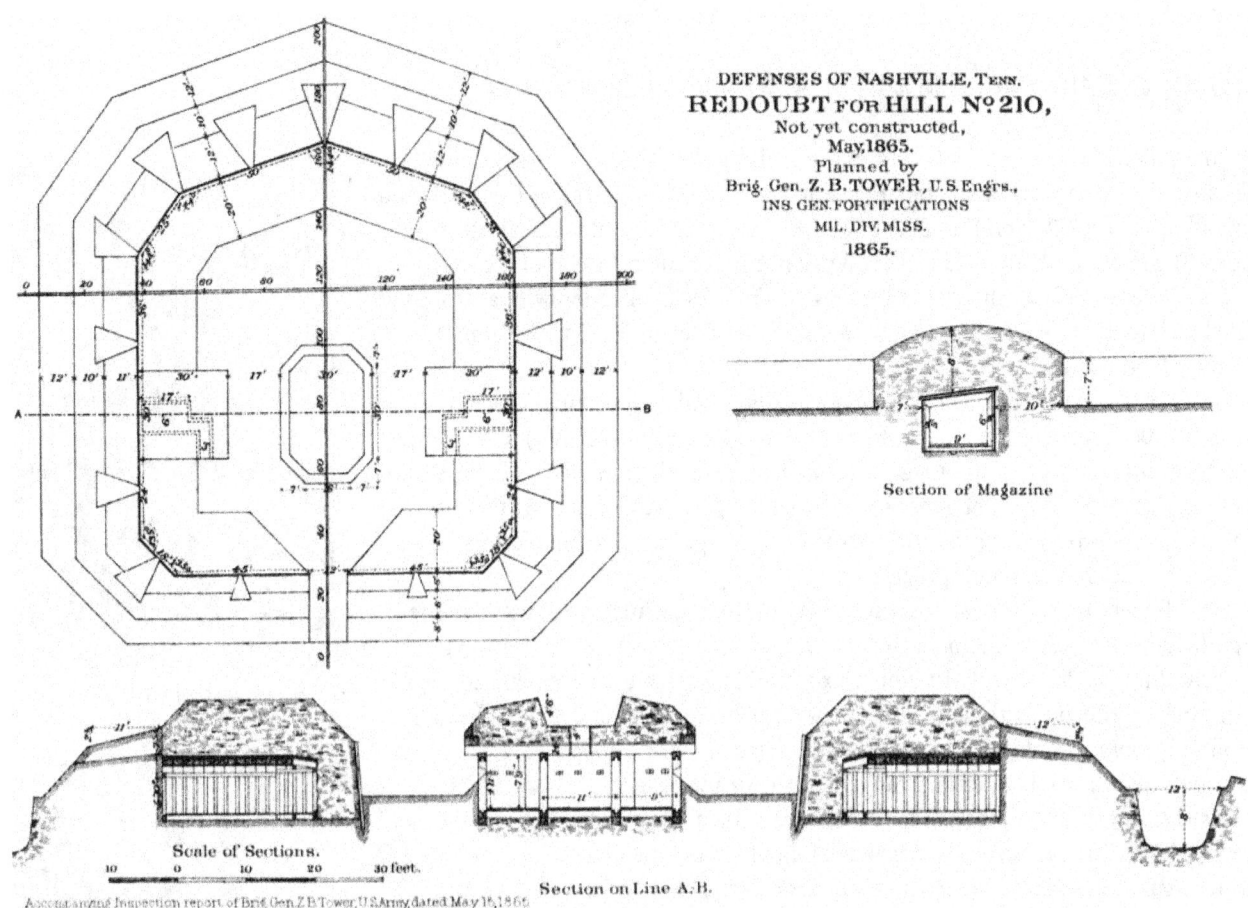

Fort Gillem (Fort Sill)

Strong earthworks fortification with central blockhouse

Fort Gillem was named for Brigadier General Alvan C. Gillem, who supervised construction of the fort. It was renamed Fort Sill in 1863 for Brigadier General Joshua W. Sill, who was killed on Dec. 31, 1862 at the Battle of Stones River near Murfreesboro. The fort was abandoned by Union troops in 1865 after the end of the war.

Fort Gillem was built in 1862 as a strong earthworks fortification with a strong central blockhouse. It was located on the interior line of defense 1.75 miles from the Cumberland River.

Gen. Tower's report of October 1864 noted: "Fort Gillem was built by Gen. Gillem; it is about 100 feet square; it is not defiladed from the near hill (210 ref.) to the southwest, has no bomb-proof nor magazine, but has a deep ditch, walled with dry stone. The emplacements for eight guns are in barbette. I propose to defilade this work, to throw up merlons for the protection of the guns and gunners, to build a small magazine and block-house, bomb-proof."

Gen. Alvan C. Gillem

Gen. Tower's report of May 1865 noted: "Gen. Gillem, while in command of the Tenth Tennessee Regiment, built this fort. It was a redoubt about 120 feet square, with narrow ditches, walled with dry stone, six feet high, having emplacements for eight guns in barbette, but without magazines or bombproof, and not defiladed from hill (ref. 210) looking into it. It was neatly constructed and was a good redoubt. I modified its interior arrangements with a view to increased strength and protection to its defenders. The parapet toward hill (210) has been raised two feet for defilement; two service magazines, which also serve as traverses, constructed on the faces, which would naturally be subject to ricochet from attacking batteries; thirteen embrasures, finished mostly with gabions, and a block-house keep set up. This structure has not been covered for lack of timber. Much blasting was required for the magazines, the drains, and of the terreplein to prepare the site of the block-house. It is proposed to finish this block-house, set up a gate at the entrance, and build a suitable bridge across the ditch. When thus completed the work will be ready for a small garrison and should be kept in repair."

Jubilee Hall at Fisk University

Gillem (1830-75) was a native of Gainesboro, one of only six native Tennesseans to serve as Union generals. With the outbreak of the war, Gillem became a captain on May 14, 1861, initially serving under George H. Thomas. Gillem was chief quartermaster of the Army of the Ohio and was brevetted as a major for gallantry at the Battle of Mill Springs. He was appointed colonel of the 10th Tennessee Infantry in May 1862 and served for a time as the provost marshal of Nashville during the Federal occupation.

Fort Gillem was abandoned in 1865. Eventually, the site became part of Fisk University (1866) and some of the fort buildings were used by the school for a time. Today, Jubilee Hall, a National Historic Landmark, sits on the site. Through the efforts of Fisk's Jubilee Singers (who sang Negro spirituals), funds were raised to relocate the school to 40 acres in North Nashville. Funds raised by the Jubilee Singers during an 1871-74 international concert tour were used to construct the school's first permanent building, Jubilee Hall. This imposing six-story building was designed by architect Steven D. Hatch of New York. The massive Victorian Gothic structure was built in 1873-76.

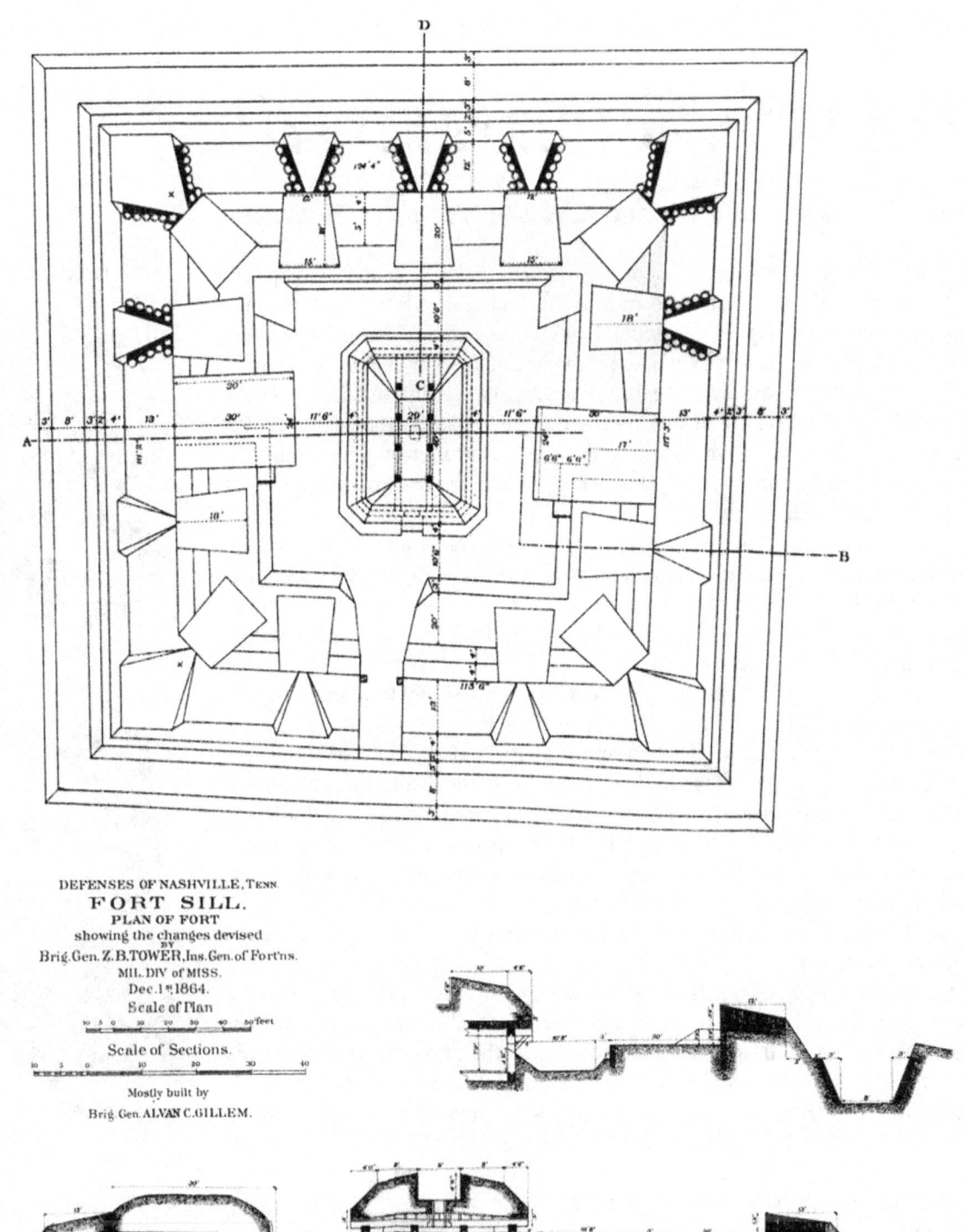

DEFENSES OF NASHVILLE, TENN.
FORT SILL.
PLAN OF FORT
showing the changes devised
BY
Brig. Gen. Z. B. TOWER, Ins. Gen. of Fort'ns.
MIL. DIV of MISS.
Dec. 1st 1864.
Scale of Plan

Scale of Sections.

Mostly built by
Brig. Gen. ALVAN C. GILLEM.

Fort Whipple (Redoubt Donaldson)
Built to reinforce interior line as Hood approached Nashville

Named Fort W.D. Whipple in honor of Brigadier General William D. Whipple, the redoubt, built in 1864, was also known as Redoubt Donaldson in honor of the city's quartermaster, Brig. Gen. James L. Donaldson. A native of Baltimore and West Point graduate, Donaldson served in the Second Seminole War in Florida and then in the 1st U.S. Artillery during the Mexican War, where he was distinguished at the Battle of Buena Vista. Serving under his former West Point classmate, Montgomery C. Meigs, Col. Donaldson was chief quartermaster of the Department of the Cumberland from Nov. 9, 1863 to June 21, 1865. He was appointed to the grade of brevet brigadier general in the Regular Army.

Donaldson organized the men of his quartermaster's organization into a combat unit and served in the Battle of Nashville. Donaldson efficiently and effectively managed the huge supply bases that served the armies of Grant, Sherman, and Thomas, and received their commendations in official reports. He is buried at Mount Auburn Cemetery, Cambridge, Mass.

Brig. Gen. William D. Whipple

Whipple (1826-1902), a native of New York, became Assistant Quartermaster General of the Army and Department of the Cumberland in November 1863 and the following month was appointed to be Thomas' chief of staff. In the latter capacity, he took part in all of the operations of Chattanooga and in the Atlanta Campaign as well as the Nashville campaign. He continued with Thomas after the war, until the latter's death in San Francisco in 1870, when Whipple was appointed aide-de-camp to Sherman, the General-in-Chief of the Army, a capacity which he served in for five years. He is buried in Arlington National Cemetery.

Fort Whipple was located near the modern intersection of D.B. Todd Jr. Boulevard and Formosa Street.

Brig. Gen. James L. Donaldson

Gen. Tower's report of May 1865 notes: "Redoubt Donaldson (now called W.D. Whipple) is situated midway between Hyde Ferry Fort and Gillem. It is a small battery with seven exterior and two interior embrasures. On the gorge, closed by a stockade, is a little octagonal block-house of ten feet sides, made bomb-proof. This small redoubt, intended for a six-gun field battery, covers the ground between Gillem and Hyde Ferry Fort, and is supported by infantry entrenchments on either side. I devised it for a model battery. The faces from angles of 144 degrees, while the embrasures open 40 degrees, so that the guns on each face can fire parallel to the contiguous capitals. By this arrangement there are no sectors without fire; in fact, the fire on the bisecting line of the angles is equal to that in any other direction. Such batteries, placed at intervals of 600 yards along infantry entrenchments, constitute a good defensive line for inclosing a city. Key points should be occupied by redoubts as large as Hyde Ferry Fort. Within this inclosing line should be built one or more strong redoubts to serve as citadels or keeps to the outer line, and arranged to fire into the gorges of the batteries, which, being simple stockades, would not shelter the enemy should he succeed in acquiring temporary possession. Battery Donaldson, commenced while Hood's army was approaching Nashville, is completed. For its preservation the exterior slopes have been sodded by the soldiers of the field battery stationed near."

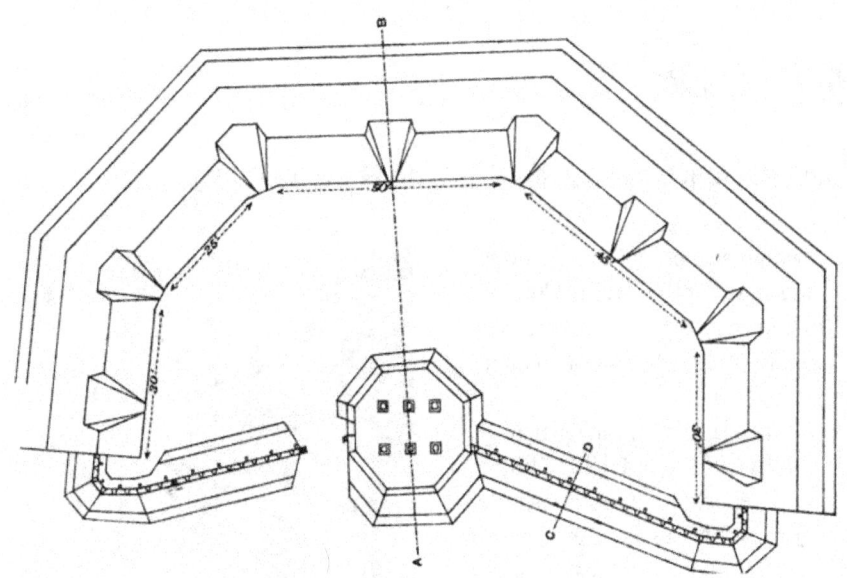

DEFENSES OF NASHVILLE, TENN.
FORT W. D. WHIPPLE.
Planned by
Brig. Gen. Z. B. TOWER, U.S. Engrs.,
INS. GEN. FORTIFICATIONS
MIL. DIV. MISS.
1865.
Constructed under the direction of
Capt. BARLOW, U.S. Engrs.

Scale of Plan.
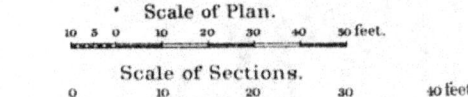

Scale of Sections.

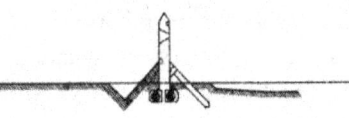

Section on Line C.-D.

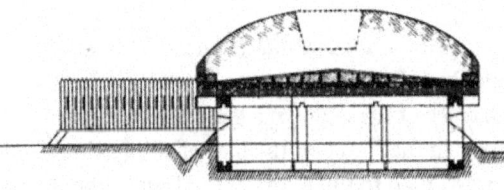

Section on Line A.-B.

Accompanying Inspection report of Brig. Gen. Z. B. Tower, U.S. Army dated May 15, 1865.
SERIES I VOL. XLIX.

Fort Garesche'
On far right flank; also known as the Fort at Hyde's Ferry

Named Fort Garesché in honor of Lt. Colonel Julius P. Garesché, who was killed at the Battle of Stones River, this fort was also known as the Fort at Hyde's Ferry. Today, the site is near the intersection of Buchanan Street and Buchanan Court.

Fort Garesché was built as a large polygonal earthworks that mounted 14 guns with a good field of fire in all directions. Gen. Tower's report of May 1865 included the following: "Hyde Ferry, Fort Garesché - As Fort Gillem is nearly one mile and three-quarters distant from the Cumberland River, it became necessary to close this space by one strong redoubt, at least. Having therefore obtained from the commanding general the aid of the 182nd Ohio Volunteers November last, they were set at work building a strong redoubt on the knoll crossed by the Hyde Ferry road about three-quarters of a mile distant from the ferry and one mile north of Fort Gillem. This position had a good command over the approaches in every direction. Rapid progress was made, so that the fort was prepared to mount a battery at the time of the battles of Nashville. The regiment was called upon to do military duty after the battles, resuming labor upon the work in strength about the middle of January. The ditches and parapet have been finished, and the latter mostly sodded; three magazines, serving as traverses, completed and also sodded. Gabion embrasures have been formed for fourteen guns and twelve platforms laid. The large block-house keep with flanking redans is set up and covered with timber. This covering, after being made water-proof, will be loaded with its parapet. The gateway has yet to be completed. This fort when finished will be very strong and a good specimen of polygonal redoubt. Its angles are made open so that the guns of the faces fire parallel to the capitals. It should be garrisoned and preserved. Were the scarps revetted it would be easily kept in order."

From the American Battlefield Trust (ABT): Lt. Col. Julius Peter Garesché, who served as Rosecrans' chief of staff, was the highest ranking Hispanic officer killed at the Battle of Stones River. Born near Havana, Cuba, in 1821, Garesché graduated from West Point in 1841 and fought in the Mexican War. Garesché later became a

Lt. Col. Julius P. Garesché

leading U.S.-based Catholic scholar, and in 1851 was decorated by Pope Pius IX.

In 1861 Garesché turned down a general's commission, preferring to earn that rank on the battlefield. In the fall of 1862, he was assigned to the Army of the Cumberland at Rosecrans' request.

Late on the afternoon of December 31st, Rosecrans, Garesché and staff watched Breckinridge's final charge from a knoll in full view of both sides. Rosecrans spurred his horse toward the Round Forest, his staff following. Suddenly, a Confederate shell whistled past Rosecrans' head and decapitated Garesché. The headless body continued on horseback for another 20 paces, before slumping to the ground, his blood spattering on Rosecrans. General Sheridan recalled this horrible public death "stunned us all, and a momentary expression of horror spread over Rosecrans' face; but at such time the importance of self-control was vital, and he pursued his course with an appearance of indifference."

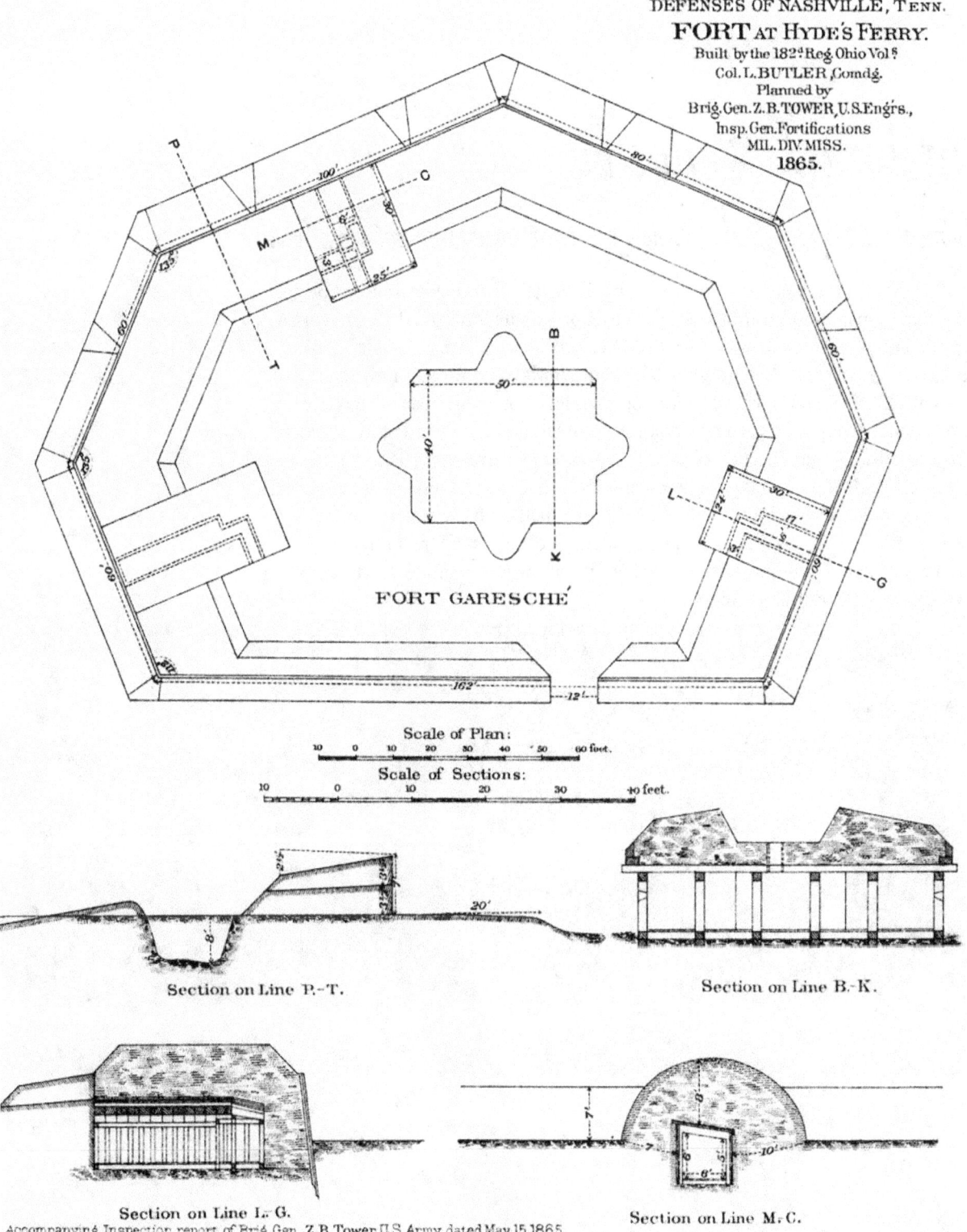

Railroad Redoubt

Elaborate octagonal fort planned for East Bank but not completed

In 1862, heavy guns were positioned near the Louisville & Nashville Railroad bridge over the Cumberland River to prevent capture or destruction from the east bank. A large redoubt was planned for the east bank but never completed. In his May 1865 report Gen. Tower writes: "Defenses north bank of Cumberland River.-At my request the Thirteenth U.S. Infantry, Capt. La Motte commanding, commenced an octagonal redoubt about three-quarters of a mile from the railroad bridge, at bend of track, where there is usually a large collection of cars. The work would cover approaches to the bridge. The ditch was excavated, parapet raised and revetted with openings left for embrasures. Little has been done to this work since the battles. It is not necessary to complete it." The redoubt was most likely located near the Foster Street intersection with the railroad. A two-gun battery was manned on the west bank of the river, just downstream of the railroad bridge.

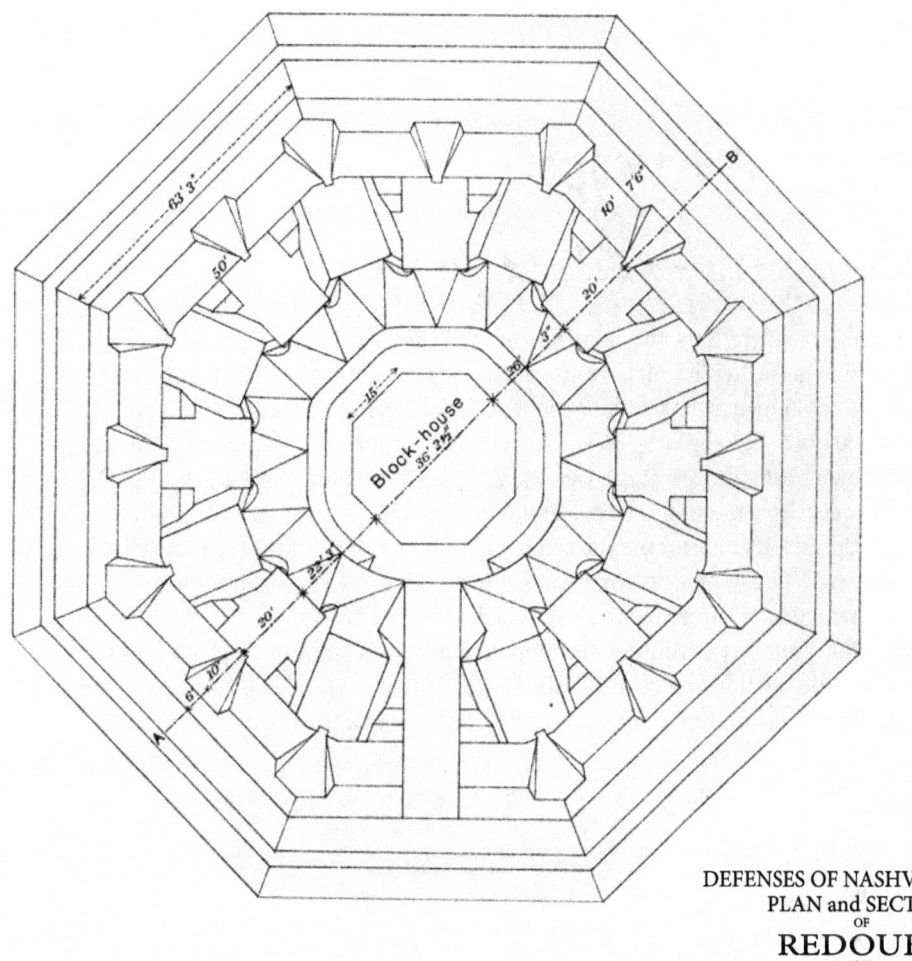

DEFENSES OF NASHVILLE, TENN.
PLAN and SECTION
OF
REDOUBT
ON
North Side of CUMBERLAND RIVER
3/4 Mile from the Railroad Bridge.
Designed and partly built under the direction of
Brig.Gen. Z.B. TOWER,
Insp. Gen. of Fortifications
MIL. DIV. MISS.
NASHVILLE, TENN.,
Dec. 1st, 1865
Interior Planes constructed by Capt.BARLOW, US Engrs.

Scale of Plan:
0   20   40   60   80   100

Scale of Section:
0   10   20   30   40   50

Partly built by the 13th Reg. Infantry.

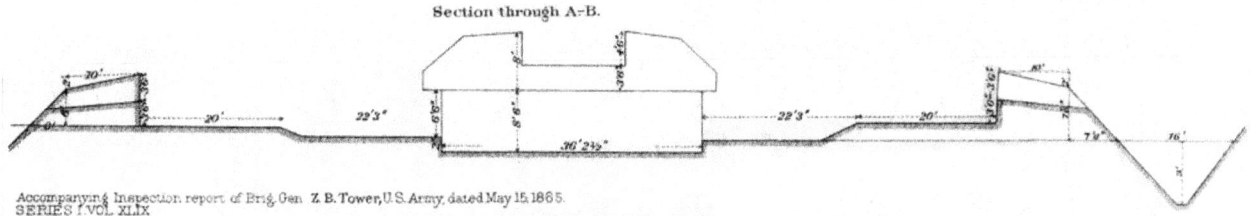

Section through A-B.

Accompanying Inspection report of Brig. Gen. Z. B. Tower, U.S. Army, dated May 15, 1865.
SERIES I VOL. XLIX

Magazine Granger

Huge gunpowder magazine provided safety for populace

For safety and supply considerations, the engineers of the Federal army built a massive black gunpowder magazine in Nashville which was completed by the summer of 1864. The exact location of the facility is not known; it was built on eight acres formerly occupied by the City Hospital, which burned down in February 1863. There still is a Magazine Street in the "Railroad Gulch" area of downtown Nashville (see map). General Z.B. Tower remarked in his report of October 1864: "The engineer department has built a grand depot magazine, the largest and best devised that I have ever seen. Its interior measurement is 150 feet by 60, high, airy, and well ventilated, solidly constructed, and lighted at either end by locomotive reflectors placed in small masonry rooms. The structure is covered with earth to a depth of eight feet. A covered roadway with stone masonry side-walls passes through the embankment and communicates with the magazine entrance. A solid trestle-work branch railroad from the main track has been built into the magazine yard, and a long building erected to receive the large quantities of fixed ammunition in transit."

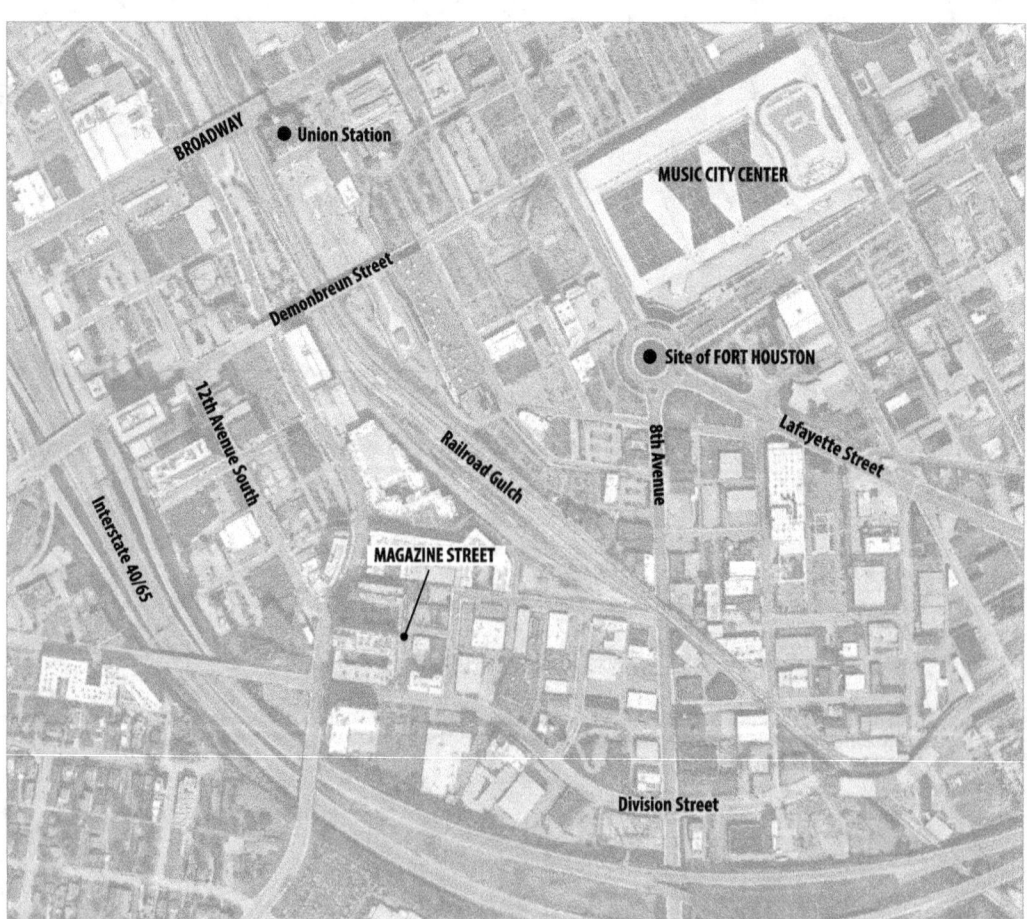

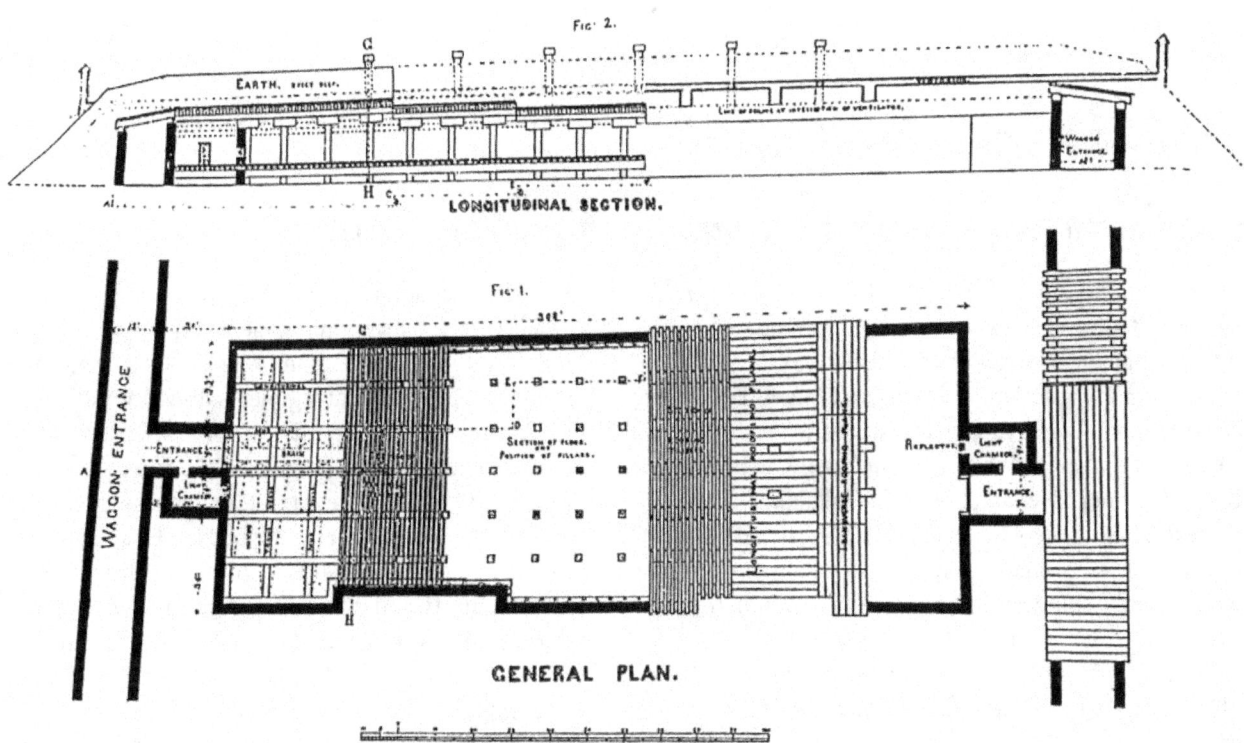

Portion of 1864 plan entitled "Magazine Granger, Nashville, Tenn." Prepared by Major James R. Willett under the direction of Colonel William E. Merrill, Chief Engineer, Department of the Cumberland (National Archives, Record Group 77, Z76-1.)

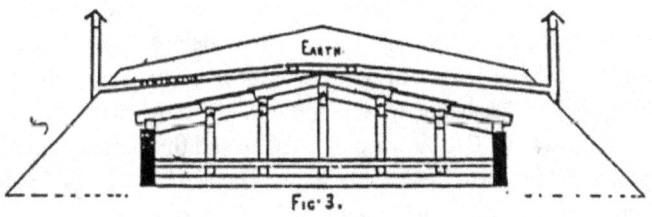

Brentwood Stockade

Fortification south of Nashville protected the railroad over the river

The main objective of Confederate Brigadier General Nathan Bedford Forrest and his cavalry (mounted infantry) during 1862-63 was to wreak havoc behind the Federal lines by attacking isolated garrisons and destroying communications and supply infrastructure. At this, the "wizard of the saddle" was highly successful. Following the Battle of Stones River and before the Tullahoma Campaign, Forrest served under General Earl Van Dorn and earned a victory at Thompson's Station (March 5, 1863), outflanking a Federal probe southward from Franklin and inflicting 1,800 casualties.

Further north, operatives and the so-called "40 Thieves" of irregulars under Forrest's brother, Capt. Bill Forrest, indicated that the stockade at Brentwood Station, between Nashville and Franklin, was vulnerable to attack. This stockade was located west of the railroad station on a high hill in the center of town. A second stockade of elaborate design was located nearly two miles south at the Nashville & Decatur Railroad bridge over the Little Harpeth River. Forrest received Van Dorn's approval to attack the fortifications.

On March 25th, 1863, N.B. Forrest positioned two companies of the 19th Tennessee cavalry on the Hillsboro Pike and two more companies north of Brentwood to protect his rear and block any Federals approaching from Nashville. Manning the Brentwood Station stockade was Lt. Col. Edward Bloodgood and 500 men of the 22nd Wisconsin Regiment. Forrest's first demand for surrender was ignored, but when Confederates advanced from the southwest and the south, white flags appeared over the fortification. Hundreds of Yankees along with wagons and supplies were captured.

Forrest moved south against the Little Harpeth River stockade with the 4th Mississippi and 10th Tennessee. He set up artillery and fired one shot against the fort, which had no artillery. The garrison immediately surrendered, with the loss of 275 men captured and 11 wagons confiscated. The troopers also burned down the railroad bridge.

Back at Brentwood, the 4th Federal Cavalry Brigade from Franklin caught up with the retreating Confederate cavalry and their captured wagons, driving them six miles westward from Granny White Pike and causing a stampede. Four Federal regiments drove the 10th Tennessee and then the 6th Mississippi westward, recapturing most of the wagons. At Hillsboro Pike, Forrest and his escort and the 11th Tennessee caught up with the whirlwind and stopped the Union advance. At this point, Col. James W. Starnes' Brigade arrived from the north and drove the Federals back to Brentwood Station, the Yankees compelled to burn and destroy the recaptured wagons.

During these raids, Forrest lost four killed, 16 wounded, and 39 captured. The Federal cavalry commmander, Brig. Gen. Green Clay Smith, reported four killed, 19 wounded, and four missing.

Forrest's men were able to arm themselves with captured Federal weapons, standard procedure for the troopers but a course of action challenged by his superior Van Dorn, who also contended that Forrest was a self-promoting glory hound. The two nearly came to blows before they reconciled. In May, at his Spring Hill headquarters, Van Dorn, a known womanizer, was shot and killed by a local doctor who had learned that his wife and the general had been keeping company.

The *Nashville Daily Union* of March 26th reported: "During yesterday, the city was in a considerable state of excitement in consequences to rumors of a battle raging at Brentwood, about nine miles from Nashville."

This was the first time in history that full companies of men went into battle fully armed with repeating rifles, according to Brentwood historian Preston Bain in a 2013 newspaper account. "Repeating rifles had been around and dispersed throughout the Federal Army but never before had whole companies of men been so armed. The Confederates were caught by surprise and were left stunned by the repeating rifle fire."

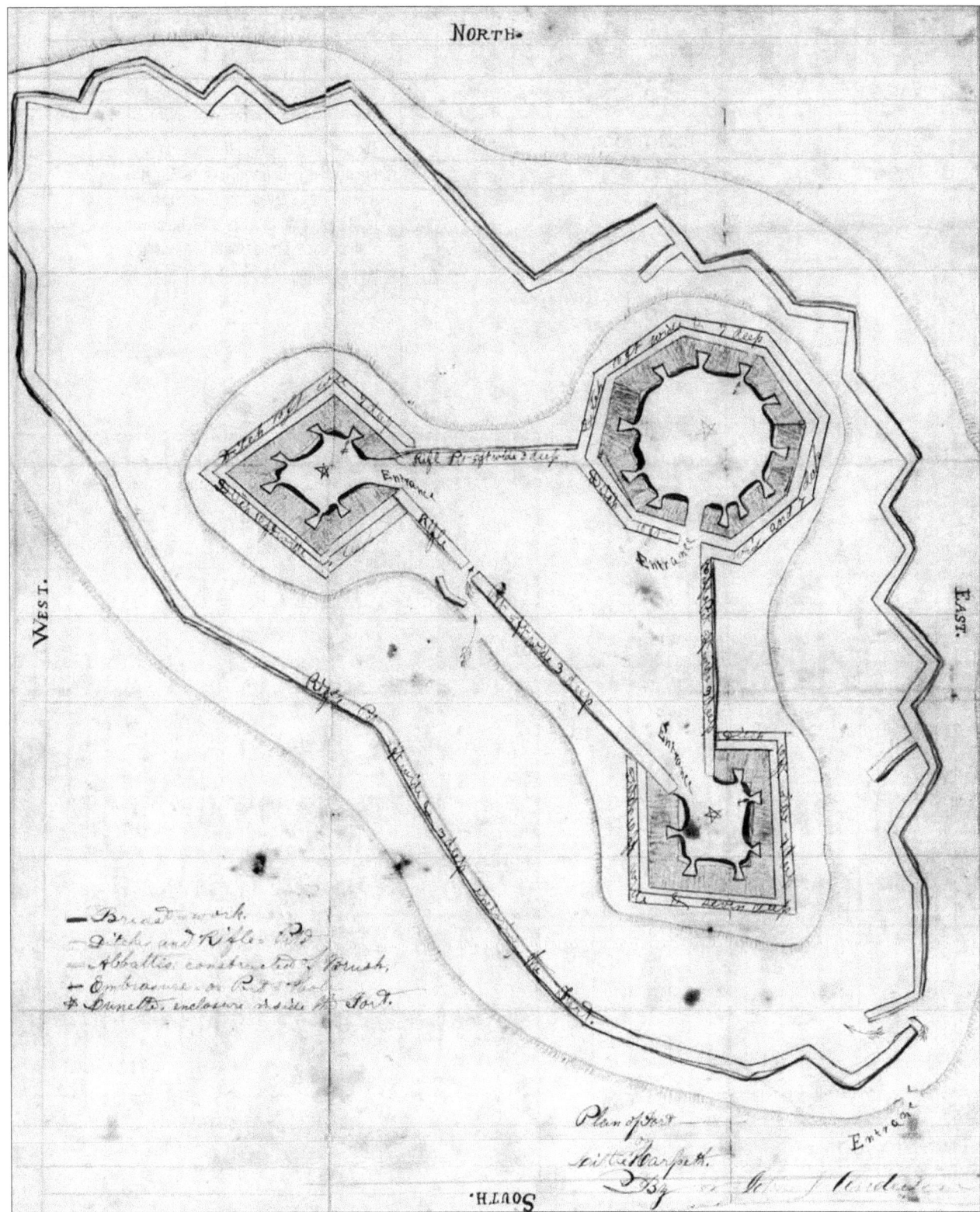

Brentwood Stockade

Copy of an original drawing by John L. Anderson, probably drawn in 1863. The only other known sketch of the stockade was completed by Anson Smith (see page 147) and is in the National Archives. Built during 1862 to guard the railroad bridge at the Little Harpeth River, Fort Brentwood housed troops from the 104th Illinois as well as the 19th Michigan. On March 25, 1863 the stockade was captured by the 4th Mississippi and the 10th Tennessee Cavalry units under the command of Major General Nathan Bedford Forrest. Captain Elijah Bassett surrendered the garrison of the 19th Michigan, which included about 200 men, all supplies, arms, and a dozen wagons. Forrest retreated west towards Hillsboro Road, where he was pursued by Brigadier General Green Clay Smith and 700 Union Cavalry. (Courtesy of the Jim Kay Collection.)

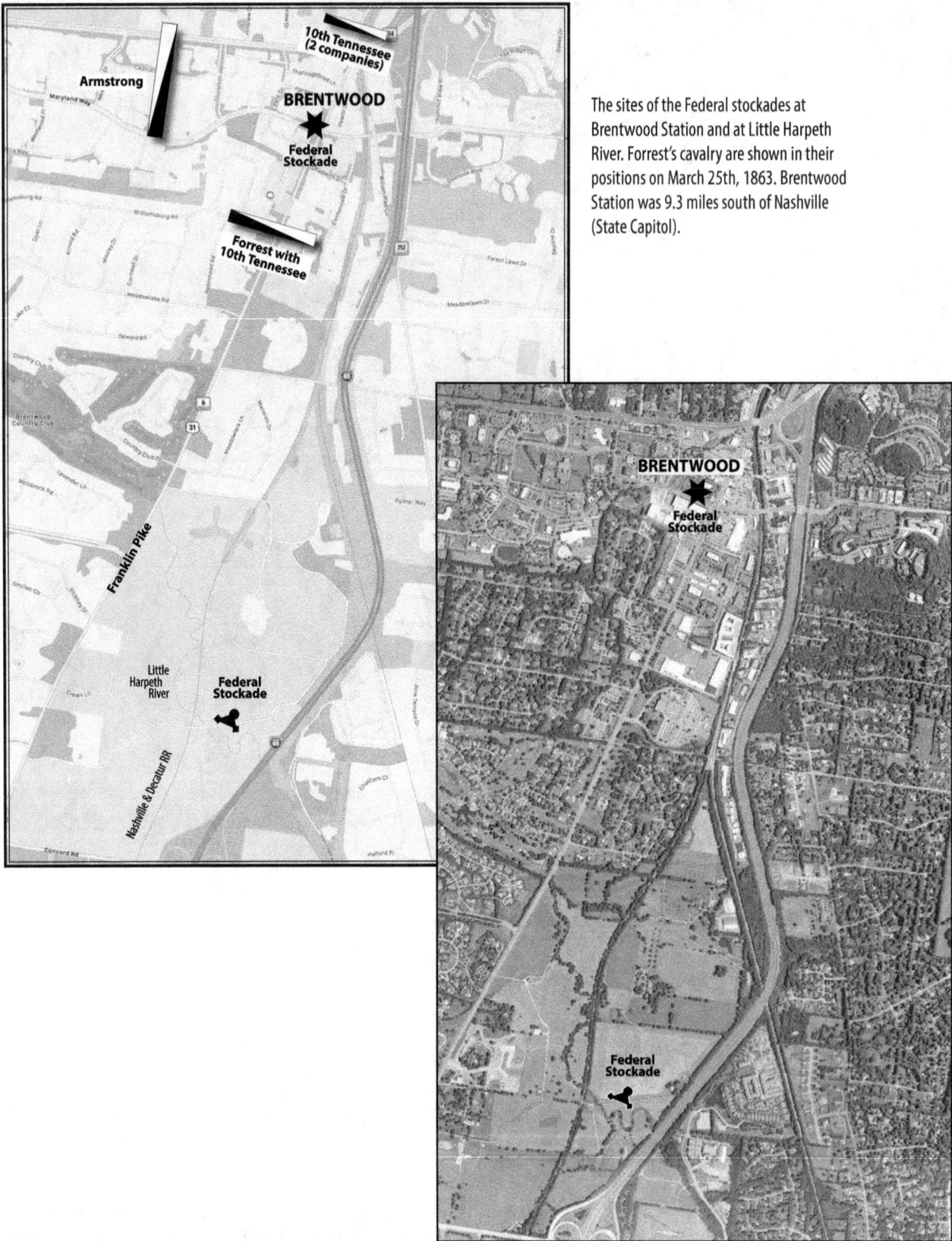

The sites of the Federal stockades at Brentwood Station and at Little Harpeth River. Forrest's cavalry are shown in their positions on March 25th, 1863. Brentwood Station was 9.3 miles south of Nashville (State Capitol).

Fortress Nashville: Pioneers, Engineers, Mechanics, Contrabands & U.S. Colored Troops

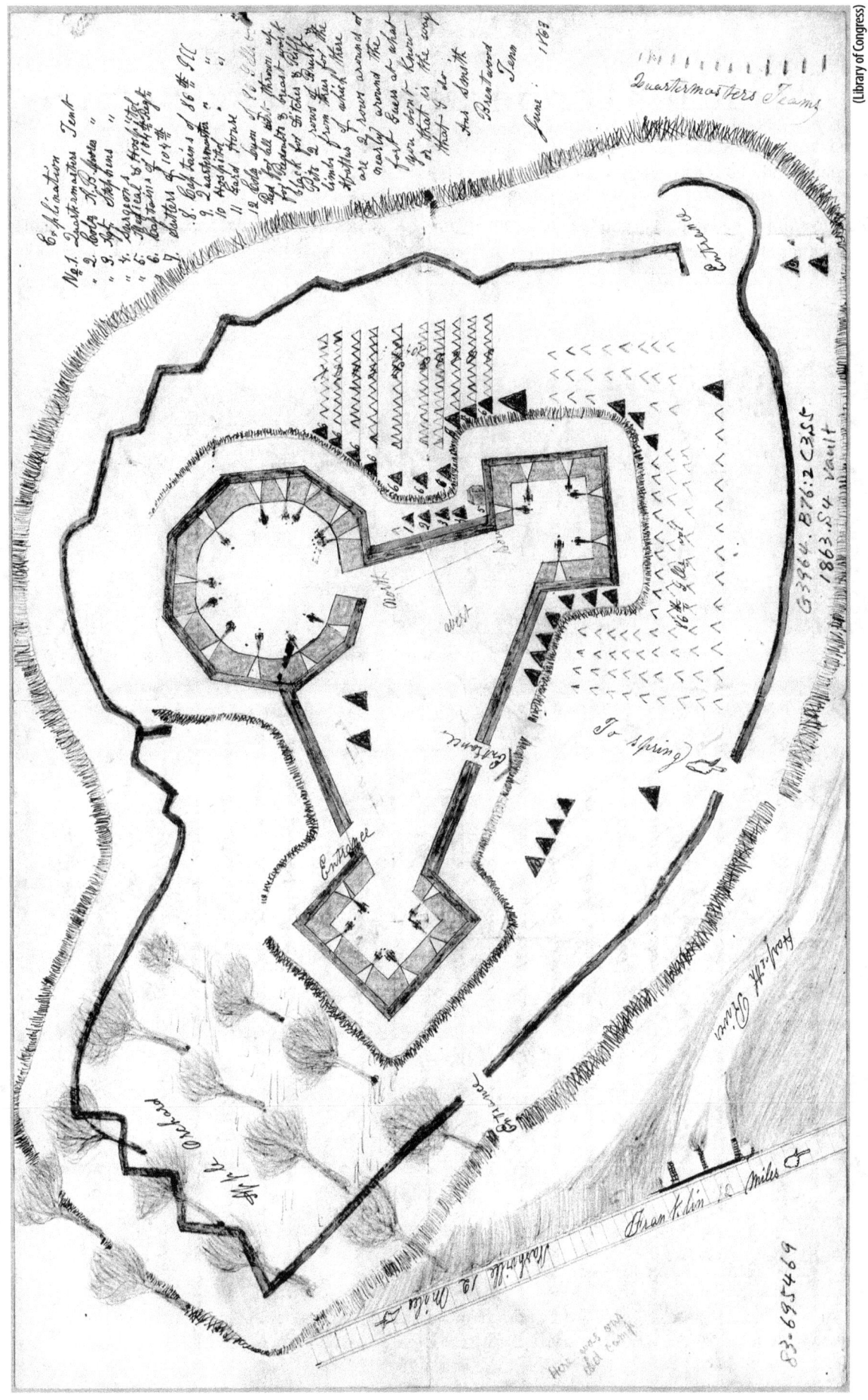

Middle Tennessee Infrastructure

From Louisville to Atlanta, Federal armies in the field relied on a steady flow of supplies to sustain them. Engineers built and rebuilt railroad tracks, trestles and bridges, blockhouses, warehouses and depots, fixed and field fortifications, in addition to maintaining locomotives and rolling stock, horses and mules, and steamboat freighters and river gunboats. Men and animals had to be fed, and fuel for steam engines had to be mined or felled. In addition, troops had to fend off roving bands of Confederate cavalrymen and opportunistic guerillas. No matter what, the supplies must get through!

The Eaton Depot in south Nashville stored war materiel. Buildings downtown can be seen in the background. (Tennessee State Library and Archives.)

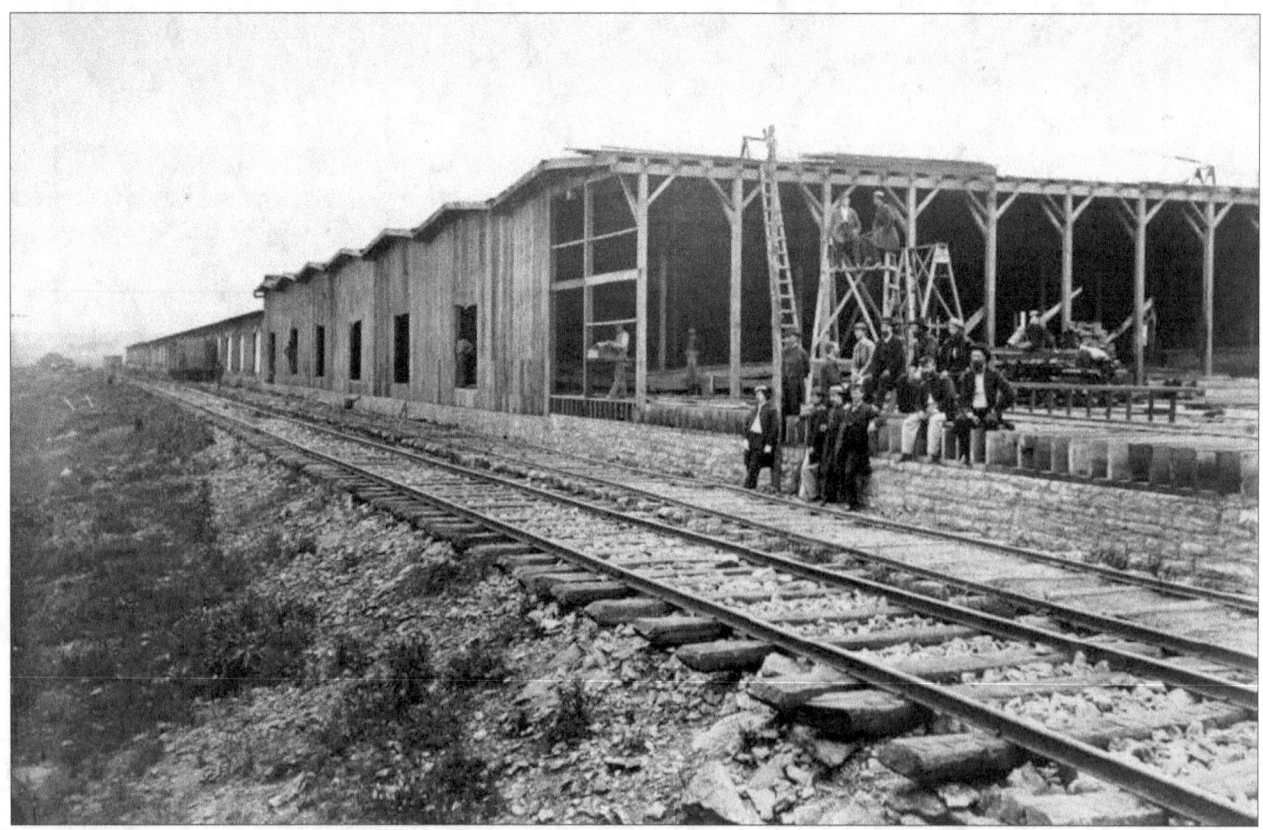

Unfinished forage house on N&NWRR in Nashville with a group of officers of the Quartermaster Department. The facility was to be built in two sections — 904x156 and 805x112. Only 201 feet of the second section had been completed. A forage house was used to store hay and oats for the Army's horses and mules. (Tennessee State Library and Archives.)

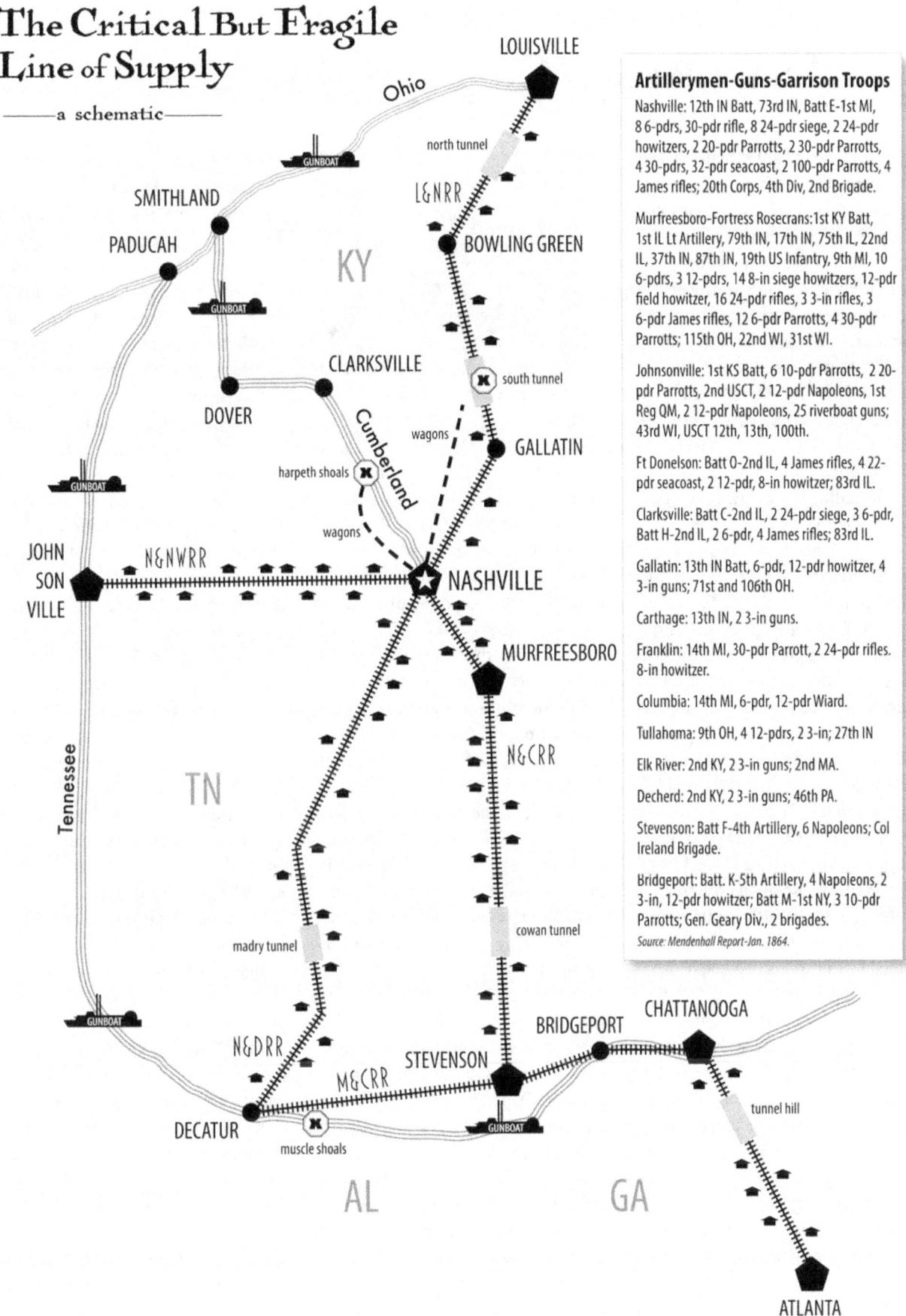

U.S. Army Corps of Engineers

The history of United States Army Corps of Engineers can be traced back to June 16, 1775, when the Continental Congress organized an army with a chief engineer and two assistants. Colonel Richard Gridley became General George Washington's first chief engineer; however, it was not until 1779 that Congress created a separate Corps of Engineers. Army engineers, including several French officers, were instrumental in some of the hard-fought battles of the Revolutionary War, including Bunker Hill, Saratoga, and the final victory at Yorktown.

In 1794, Congress organized a Corps of Artillerists and Engineers, but it was not until 1802 that it reestablished a separate Corps of Engineers. At the same time, Congress established a new military academy at West Point, New York. Until 1866, the superintendent of the academy was always an engineer officer. The first superintendent, Jonathan Williams, also became the chief engineer of the Corps. During the first half of the 19th Century, West Point was the major and for a while, the only engineering school in the country.

Throughout the 19th Century, the Corps supervised the construction of coastal fortifications and mapped much of the American West with the Corps of Topographical Engineers, which enjoyed a separate existence for 25 years (1838-1863). The Corps of Engineers also constructed lighthouses, helped develop jetties and piers for harbors, and carefully mapped the navigation channels. Throughout the 19th Century, the Corps built coastal fortifications, surveyed roads and canals, eliminated navigational hazards, explored and mapped the Western frontier, and constructed buildings and monuments in the nation's capital.

Once reestablished, the Corps of Engineers began constructing and repairing fortifications, first in Norfolk and then in New Orleans. The fortification assignments proliferated during the five years of diplomatic tensions that preceded the War of 1812.

While Congress reduced the size of the country's infantry and artillery forces after the war, it retained the increased number of officers that it had authorized for the Corps of Engineers in 1812. Pleas from several secretaries of war for more engineers to work on fortifications led Congress to double the size of the Corps again in 1838. The fortifications, which the Army engineers built on the Atlantic and Gulf coasts, and after 1848 on the Pacific coast, securely defended the nation until the second half of the 19th Century when the development of rifled artillery ended the earlier impregnability of the massive structures.

Much of the "civil engineering" work was done by the topographical engineers or "Topogs," who reported to a separate Topographical Bureau in the Engineer Department. In 1838, the topographical engineers—surveyors, explorers, cartographers, and construction managers—became a separate corps and remained that way until 1863 when they were reunited with the Corps of Engineers.

Engineer officers also superintended railroad work after 1824. They surveyed railroad routes and, once construction commenced, the War Department loaned engineers to various railroad companies. By 1830, many officers were being granted furloughs to work on railroads, in either construction or surveying activities. Finally, in 1838, Congress passed legislation that prohibited granting leave to Army officers to allow them temporary employment with private companies.

U.S. Army engineers played significant roles in the Mexican War, providing both mapping and construction services and troop leaders in theaters of operations while largely suspending work on navigational improvements. Engineers of all ranks gained renown for their military efforts during their service in Mexico in 1846-48. Chief Engineer Joseph Totten directed the successful siege of the port city of Veracruz, from which General Winfield Scott launched his decisive assault on the interior of the country. Captain William Williams, who had directed the Great Lakes survey, served as chief topographical engineer for General Zachary Taylor until his death at the battle of Monterey.

During the Civil War, Army engineers built pontoon and railroad bridges, constructed forts and batteries, demolished enemy supply lines, and conducted siege warfare. In December 1862 they laid six ponton bridges across the Rappahannock River, under devastating fire from Confederate sharpshooters, in support of the Union attack on Fredericksburg, Virginia. The 2,170-foot ponton bridge, which Union engineer troops laid across the James River in June 1864 as the Army of the Potomac approached Petersburg, Va., was the longest floating bridge erected before World War II. Drawn largely from the top of their

West Point classes, the engineers in the Corps before the Civil War included many excellent military strategists who rose to leadership roles during the war, including Union generals McClellan, Halleck, Meade, and Confederate generals Lee, Joseph Johnston, and Beauregard.

Until 1866, West Point's curriculum was modelled on that of the French Ecole Polytechnique, and designed to produce officers with skills in engineering and mathematics. Dennis Hart Mahan (1802-71) was a noted American military theorist, civil engineer and professor at West Point from 1824-71. An important influence on the military conduct of the Civil War, Mahan is best understood as an educator and technology transfer agent, not a theorist. Mahan almost singlehandedly compiled and transferred the best of European engineering to the United States. Virtually all 19th-Century American engineering schools were started with West Point-educated faculty, or adopted its texts. James St. Clair Morton, who engineered most of the fortifications in and around Nashville, was a disciple of Mahan.

On March 6, 1861, once the Southern states had seceded from the Union, the Confederate Congress passed an act to create a Confederate Corps of Engineers. The South was initially at a disadvantage in engineering expertise. Of the initial 65 cadets who resigned from West Point to accept positions with the Confederate Army, only seven were placed in the Corps of Engineers. The Confederate Congress passed legislation that authorized a company of engineers for every division in the field; by 1865, the CSA had more engineering officers serving in the field of action than the Union Army.

Between 1819 and 1860, a total of 87 graduates had earned the distinction of being placed immediately into the engineers, according to historian Thomas Army. Sixty-eight of them came from a free state and nineteen from a slave state. By the time of the Civil War 48 officers were serving in the corps of engineers. Of the 40 who remained in the Union army, 29 served as northern engineers during the war and 13 served as Union field commanders. Of the eight engineers who joined the Confederacy, five served as field commanders.

U.S. Signal Corps

The Wig Wag

In 1854, Albert J. Myer was commissioned as an assistant surgeon in the Regular Army. In 1856, he drafted a memo on a new system of signals and obtained patent letters on it. Two years later the War Dept. recognized the possibilities of the system and appointed a board to examine it. In 1860, the U.S. Army adopted Myer's system of signaling.

Because Myer's system of signaling used flags, and the flags to an untrained person seemed to swing around in no particular manner, the term wig wag was given to the flags.

Signaling as practiced by the Signal Corps using the wig wag flag was, for the most part, a method of conveying ideas by motions of a flag during the day or torch by night. Myer's system of wig wagging consisted of a four-element and two-element code. The two codes are essentially identical.

The two-element code was better known as the General Service Code. There were two basic wig wag flags, one white with a red center and the other a red flag with white center. Only one wig wag flag was used at any one time. The white flag with red center was used at dawn or dusk or when visibility was low. The red flag with white center was used during bright, sunlit days. There were different sizes of wig wag flags, the most common during the Civil War was four foot by four foot.

Myer's code for signaling was used until 1912 (except 1886-96 when the international Morse code was used), when the international Morse code was once again used.

`The wig wag flag was replaced by the Semaphore flag.

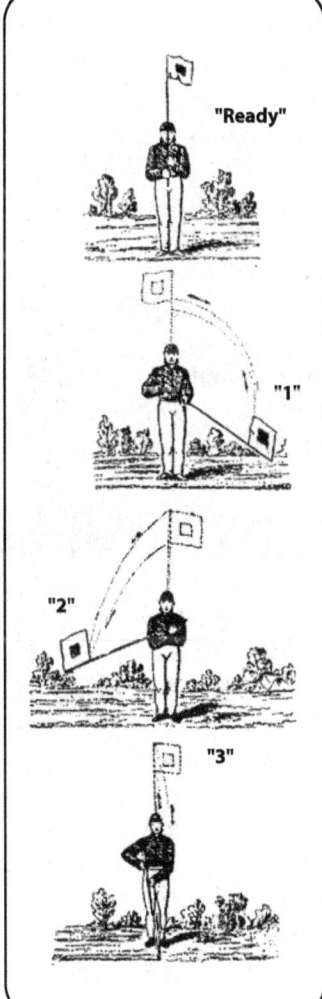

The alphabet (A-Z) General Service Code and example of the positions of the wig wag flag are as follows:

A	22	O	21
B	2112	P	1212
C	121	Q	1211
D	222	R	211
E	12	S	212
F	2221	T	2
G	2211	U	112
H	122	V	1222
I	1	W	1121
J	1122	X	111
K	2121	Y	111
L	221	Z	2222
M	1221	&	1111
N	11	ING	2212
		TION	1112

To speed delivery of messages, a code was worked out for common phrases, for example:

3	End of word
33	End of sentence
333	End of message
22.22.22.3	I understand or Message received
22.22.22.333	Cease signaling
121.121.121.3	Repeat
212121.3	Error
211.211.211.3	Move a little to the left
21112	Wait a moment
12221	Are you ready?
22122	I am ready
12222	Work faster
11222	Did you understand?
11112	Use white flag
22222	Use red flag

U.S. Military Hospitals

Contagious deadly diseases killed far more soldiers during the Civil War than combat. Nashville became a major medical center, with more than two dozen military hospitals tending to the wounded, especially following the battles at Shiloh, Stones River, Chickamauga, and Franklin-Nashville. Fourteen thousand wounded were brought to Nashville after Shiloh, and 60 to 100 died each day in the hospitals. The dead were buried at City Cemetery; after the war, a National Cemetery was sited north of the city.

Downtown Nashville housed 25 hospitals converted from schools, churches, mercantile, and other buildings. Many hospital buildings were white-washed with lime around the foundations to prevent "contagion." An inspection of Nashville hospitals by a local newspaper reporter in 1863 found them quiet, efficient, well-staffed, and extremely clean, under the circumstances. The *Nashville Dispatch* series of articles were so complimentary, in fact, they must have been intended to curry favor with the local Federal officials.

The officers and directors of Nashville hospitals were: Medical Director of Department, Surgeon Ebenezer Swift; Medical Inspector of Department, Lt. Col. Wyman; Medical Director of Post, Staff Surgeon A. Henry Thurston; and Roman Catholic Post Chaplain, Reverend J.A. Stevens. The Medical Purveyors Office was located in the Gardner Building on the Public Square.

Hospital No. 2 included one building with a room devoted entirely to "dead men's knapsacks."

Hospital No. 16 for "coloreds" had housed a furniture store, a brass band, painters, and two saloons. The former Planter's Hotel with two long balconies was converted to the U.S. Sanitary Commission Soldiers Home for officers.

The 26-hole latrine in the side yard of Hospital No. 8 was replaced for sanitary reasons by a two-story outhouse with drains that leaked, not a satisfactory solution to the problem.

Patients were directed to the prison hospital by the provost marshal. At the prison hospital, the sergeant-of-guard had 30 men detailed to guard the patients. Most hospitals had a chaplain. In many hospitals, Catholic nuns served as nurses. The size of the staffs varied from approximately 100 employees at the Morris & Stratton hospital to 2,580 at the H.S. French and Sons hospital.

Near the end of the war, the attendants and patients of Hospital No. 8 presented acting assistant surgeon George Duzan, 23, of the 52nd Indiana Regiment, a silver pocket watch valued at $75 ($1,000 today) in recognition of his "kind attention, skillfull treatment, and gentlemanly deportment."

Field hospitals sprawled beyond the city limits. The Cumberland Field Hospital, about a mile west of the Capitol, covered 30 acres and 384 floored tents, each with six patients (2,304 beds). The hospital also included 21 frame buildings. The army's smallpox field (tent) hospital was located west of town on the Charlotte Pike at the H.P. Bostick estate.

Hospital No. 2 included the former Western Military Institute (top) and Lindsley Hall (above) on College Hill. Formerly they were the University of Nashville's Literary Department and dormitory, respectively. (Below) The Ensley Building on the Public Square housed 200 beds. (Bottom) The Hynes High School served as Hospital No. 15. Located near notorious Smoky Row, it treated victims of venereal disease. (Opposite Page) The twin-towered First Presbyterian Church and the four-story Masonic Hall across Spring Street formed Hospital No. 8. The church sanctuary housed 206 beds. Next to the Masonic Hall is a stoneyard. The former military institute building, now a municipal office, and the church, now Downtown Presbyterian Church, still stand. (All photos Tennessee State Library and Archives)

Medical Facilities in Nashville During the War

By the middle of 1863 there were 25 military hospitals in Nashville, scattered throughout the city. The brick Cherry Street Baptist Church on Elm was used as the Post Hospital, with 125 beds. (The City Hospital was destroyed by fire on Feb. 20, 1863, with all 240 patients safely evacuated.) St. Mary's Catholic Church at Charlotte Ave. and Fifth St. also treated patients.

1. College Hill Armory, Third Presbyterian Church, and Primitive Baptist Church, South Cherry St. on College Hill (650 beds total).
2. University buildings, South Market St. on College Hill. Formerly the Western Military Institute (300 beds) and Lindsley Hall, 200 beds for officers.
3. Ensley buildings, SE corner of Public Square, 250 beds.
4. Howard High School, South College St. on College Hill.
5. Gun factory and state armory, upper end of Front Street.
6. Meredith Building, College Street near Broad.
7. Buildings, College Street between Church and Broad.
8. Masonic Hall, 368 beds, and First Presbyterian Church, 206 beds, Church St. near Summer. The Cumberland Presbyterian Church, corner of Cumberland Alley and Summer St., had 41 beds.
9. Carriage factory, North Market St. below the Public Square, 150 beds.
10. Medical College, South College Street. Also called City Hospital. Confederate soldiers were treated there.
11. Pest House on University Pike, 720 beds. Treated soldiers with venereal disease, and subsequently, infected prostitutes.
12. Broadway Hotel, Broad St. between Cherry and Summer, 500 to 600 beds.
13. Hume High School, South Spruce Street and corner of Broad.
14. Nashville Female Academy, Church St. near Nashville & Chattanooga R.R. Depot, 775 beds.
15. Hynes High School, Line St. at corner of Summer, 400 beds. Treated soldiers with venereal disease.
16. Gordon Block, corner of Broad and Front, served "colored" soldiers and contrabands, 375 beds.
17. Planters Hotel, North Summer corner of Deaderick, for officers, 120 beds.
18. Corner of Church and College.
19. Morris and Stratton (grocer's) Building, 14 North Market St, 300 beds.
20. First Baptist Church, North Summer St. between Deaderick and Union.
21. McKendree Methodist Church, Church St.
22. H.S. French & Son, corner of Clark and Market, 136 beds.
23. Corner of Vine and Broad.
24. Prison hospital, Second Baptist Church, South Cherry St. on the Hill.
25. Maxwell House, corner of Cherry and Church, convalescent barracks.

U.S. Army Legalized, Regulated Prostitution

During the war, hundreds of prostitutes came to Nashville to ply their trade. Venereal disease reached epidemic proportions among the U.S. Army soldiers stationed in or near the city, threatening military preparedness. Most of the business was conducted in a six-block-square of downtown known as Smoky Row.

On July 6, 1863, Gen. Robert Granger, the post commander, ordered all prostitutes out of the city. Three hundred to four hundred "soiled doves" (and a few victims of mistaken identity) were rounded up, forced onto steamboats at the wharf, and sent to Louisville. Black prostitutes were exempt from Granger's order, for unknown reasons. Louisville and Cincinnati refused to allow docking. By the end of July, the prostitutes were returning. By Aug. 4, one steamboat, shunned by every city on the river, returned to Nashville. More prostitutes than ever plied their trade in the city.

Granger rescinded his order and accepted the recommendation of the Provost Marshal that prostitution be legalized and regulated. "Public women" were ordered to be examined by U.S. Army surgeons for venereal disease. Those free of disease were granted a $30 license by the Provost Marshal to conduct business. By August 21st, at least 360 prostitutes had been licensed. By Sept. 1, 1864, the army had licensed 460 prostitutes in Nashville. Infected prostitutes were treated at a hospital building on the western city limits that had previously been used for smallpox patients. Later, Hospital No. 11 on University Pike, once a pest house, was converted into a second hospital for prostitutes. Nearly a thousand cases of venereal disease had been diagnosed among prostitutes the first six months of 1864. Hospital No. 15 treated soldiers with VD, cutting the infection rate from 40% to four percent. The legalization of prostitution was repealed after the war.

Fort Granger and Triune Works
Powerful artillery batteries pounded attackers at Battle of Franklin

Fort Granger was built by Federal forces in March-May 1863 to guard the Nashville & Decatur Railroad bridge over the Harpeth River near Franklin, 17 miles south of Nashville. Named for General Gordon Granger, commander of Federal forces in Franklin in 1863, this earthen fort was constructed by laborers working 16 hours a day under the supervision of Col. W.E. Merrill of the U.S. Topographical Engineers. Figuer's Bluff was chosen as the site for the fort because it held command over the southern and northern approaches to Franklin and held military control over the Harpeth River railroad bridge.

Granger (1821-76), a native New Yorker, was a West Pointer (Class of 1845), a Mexican War hero, and frontier soldier. He is best known for coming to the aid of Gen. George Thomas at the Battle of Chickamauga and quite possibly saving Rosecrans' army from annihilation. He is also known for making the Juneteenth proclamation in 1865 in Galveston, celebrating the end of slavery in Texas. Granger's outspokenness and bluntness with his superiors, including Grant, who disliked him, prevented him from gaining more prominent commands. He is buried in the Lexington (Ky.) Cemetery.

He is not to be confused with Gen. Robert Granger, who served as post commander in Nashville.

On Feb. 12th, 1863, the 125th Ohio Volunteer Infantry, led by Colonel Emerson Opdycke, arrived to begin construction of the fort under the supervision of Corps of Engineers Captain William Merrill. Silas N. Jones, Sergeant of Co. C, 125th Ohio, was placed in charge of contraband (fugitive slave) labor. A March 11th report stated, "He has now on his roll, able for duty with pick, axe, and shovel, over 250 names." The fort took 10 weeks to complete.

Fort Granger was 781 feet long and 346 feet wide, encompassing 11.76 acres (comparable to four football fields) and contained two fortified fronts, on the northern and eastern sides. The powder magazine was constructed in the basement of a house which stood on the site and burned down before the war. It was lined with bricks from nearby Harpeth Academy, which at that time stood west of Hillsboro Pike. The magazine was capable of holding 1,200 rounds of artillery ammunition. In 1864, a shed was built over the magazine to keep the ammunition dry. The ammunition was frequently taken out and aired because of dampness. A cistern from the house could hold 9,000 gallons of water. The storehouse could hold 70,000 rations.

The cavalier, or fort within a fort, was built in the southernmost area where the ground was the highest. This was the strongest area of Fort Granger and, in case of overwhelming attack, the place where defenders would make a final stand. This was an ideal location for artillery and provided the best view of the surrounding area. The cavalier

Bastioned Stockade with Tents Protected Railroad in 1862

In late 1862, Buell's engineers constructed stockades to protect the railroads, including this stockade at the river crossing at Franklin (prior to Fort Granger being built). A favorite form was a square redoubt with four circular bastions with diameters the same as a Sibley tent. The bastions were covered with tents and used as men's quarters. The stockade was one of the first defensive structures built by the Federals in Middle Tennessee, and was demolished after Fort Granger was built. This image was drawn by artist Henry Mosler and featured in James C. Kelley's 1989 *Civil War Drawings from the Tennessee State Museum*.

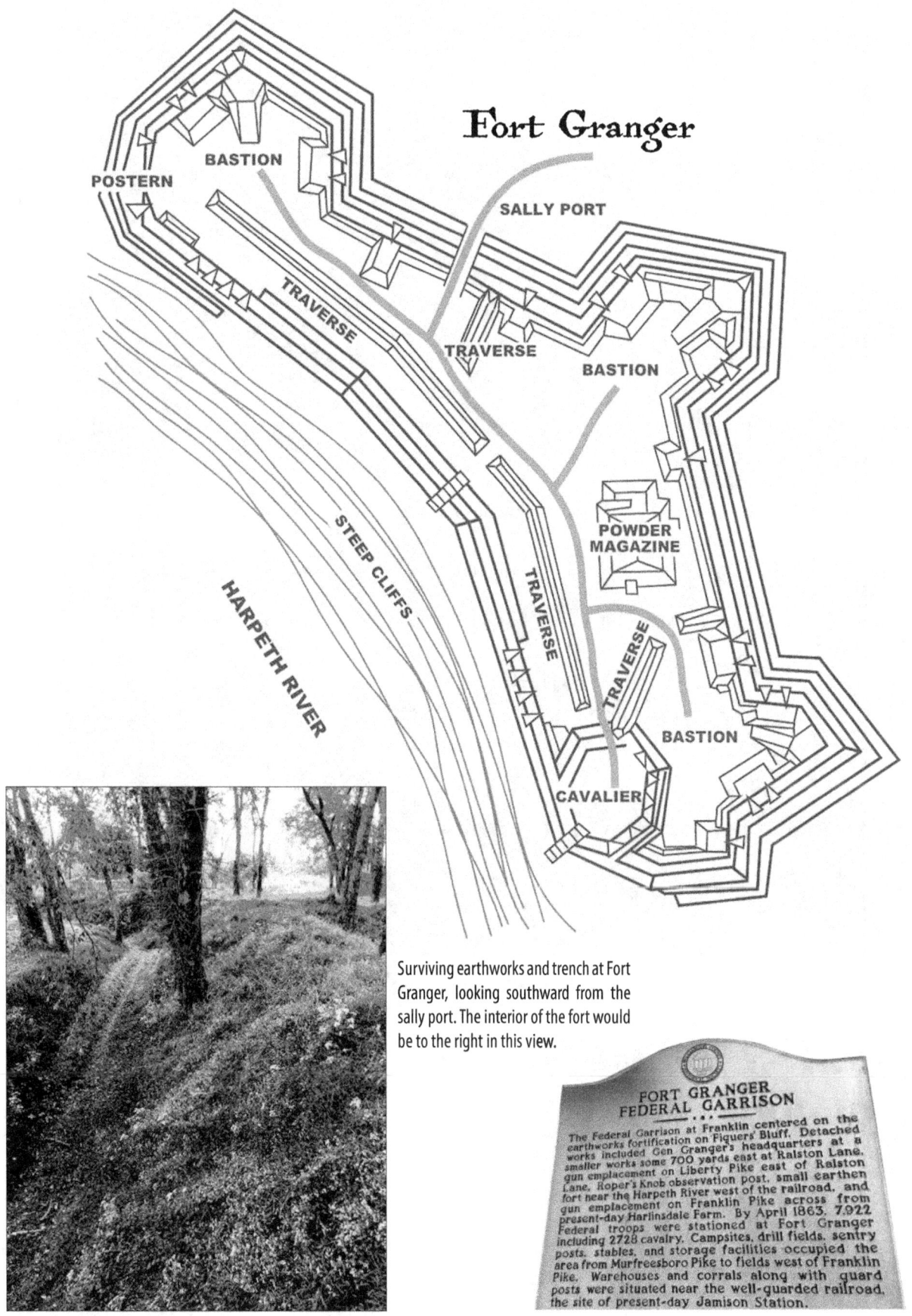

Surviving earthworks and trench at Fort Granger, looking southward from the sally port. The interior of the fort would be to the right in this view.

FORT GRANGER FEDERAL GARRISON

The Federal Garrison at Franklin centered on the earthworks fortification on Figuers' Bluff. Detached works included Gen. Granger's headquarters at a smaller works some 700 yards east at Ralston Lane, gun emplacement on Liberty Pike east of Ralston Lane, Roper's Knob observation post, small earthen fort near the Harpeth River west of the railroad, and gun emplacement on Franklin Pike across from present-day Harlinsdale Farm. By April 1863, 7,922 Federal troops were stationed at Fort Granger including 2728 cavalry. Campsites, drill fields, sentry posts, stables, and storage facilities occupied the area from Murfreesboro Pike to fields west of Franklin Pike. Warehouses and corrals along with guard posts were situated near the well-guarded railroad, the site of present-day Jamison Station.

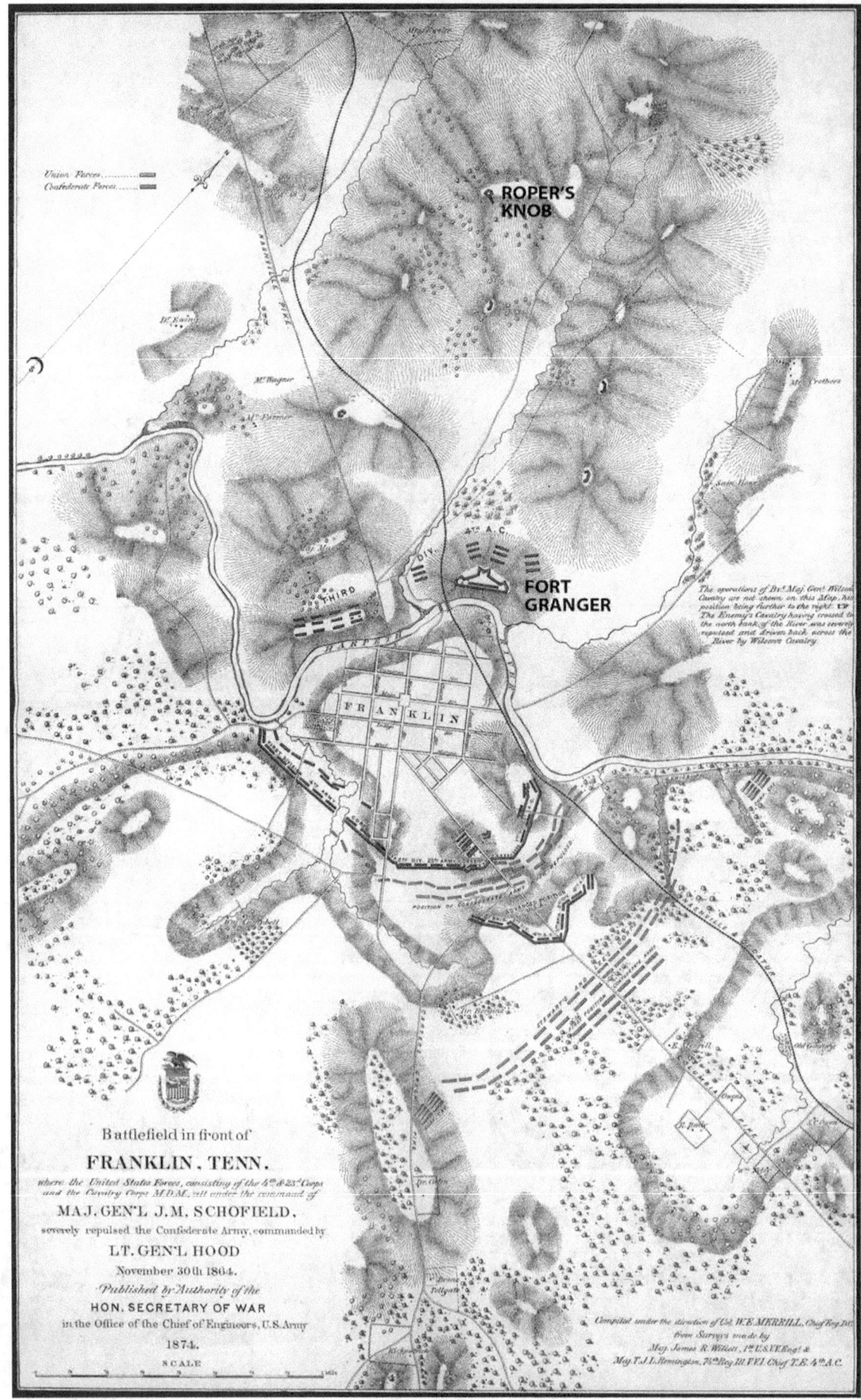

General Gordon Granger

The execution of two Confederate spies at Fort Granger. (Harper's Weekly, July 4, 1863)

was hexagonal with each side measuring 30 feet. Several traverses are located within the fort. Rifle pits placed the railroad bridge under intense infantry fire. The fort was surrounded by abatis (felled trees with sharpened branches), which formed a double row between the rifle pits at the southeastern corner of the fort. Along the rock cliffs between the river and the fort one Federal soldier on garrison duty made his mark by etching the image of an eagle into the limestone.

Gun batteries were stationed at four other hills nearby. Roper's Knob held a redoubt of four heavy guns and a blockhouse for 60 men. It contained a signaling station, two cisterns, and a "good size" magazine. At Roper's Knob, 50 men could hold off 5,000, according to Capt. Merrill. Roper's Knob was 250 feet above the surrounding plain and its garrison could see for six miles in all directions. The Federals constructed a sturdy tramway from the terrace on the side of the hill to the redoubt on the top. A steam engine and derrick were installed with ropes and a drum to pull heavy artillery to the top.

The old turnpike bridge at Franklin was burned by the Confederates and replaced with a pontoon bridge, which in turn was replaced by a trestle bridge. The work at Franklin was supervised by the 4th Battalion of the Pioneer Brigade, 11 detachments or about 220 men. Generally, 1,200 men (infantrymen) were worked in two shifts — 600 men for eight hours.

By April 1863, Fort Granger held 18 field guns and two 30-pound siege cannons. At full capacity, the fort could house 5,194 infantry troops, 2,728 cavalry, and 24 artillery pieces. Battery crews drilled at least once a week,

firing cannons for practice. During the war, Fort Granger was attacked several times by Confederate cavalry. On April 10th, 1863, the artillerymen used the cannons in combat against Forrest and Van Dorn. By May, Fort Granger contained six large siege guns (including two 24-pounder howitzers) and 24 smaller field cannon. On June 4-5th, the guns defended against Confederate cavalry roaming on the outskirts of town.

Shells from the fort landed in some Franklin houses during the Battle of Franklin on Nov. 30th, 1864, during which it served as the headquarters for Gen. John M. Schofield, who observed the battle from its heights. During that engagement the fort held 8,500 soldiers and 24 pieces of artillery. The fort's artillery inflicted serious damage on the right wing of the Confederate Army of Tennessee, Loring's Division, which attacked along the railroad cut.

Capt. Giles J. Cockerill's four 3-inch rifled guns of Battery D, 1st Ohio Light Artillery, fired 163 rounds upon the 35th Alabama and 12th Louisiana, Gen. Thomas Scott's Brigade, and the 43rd Mississippi, Gen. John Adams' Brigade. Overall, Stewart's Corps suffered 3,000 casualties, many inflicted by fire from Fort Granger.

Lt. Frederick W. Fout wrote, "From our post at Fort Granger, we could see every troop and every gun in our line, as long as it was day and the cloud of gun smoke allowed it. After sundown, the sparks of rifle fire and the lightning, thunder and groaning of the heavy cannons was splendid and awe-inspiring for the eye and the ear."

Fort Granger was also the site of a mysterious occurrence in the midst of the war.

On June 8th, 1863, just before dark, two Federal officers rode into Fort Granger and presented their credentials to Colonel John P. Baird. They identified themselves as Col. Lawrence W. Auton, an acting special inspector-general, and Major George Dunlop, assistant quartermaster, straight from the Army of the Potomac at Washington on a mission to inspect the departments of the Ohio and Cumberland. Their orders were signed by Adjutant General Edward D. Townsend. They also bore a letter of introduction and special pass signed by Chief of Staff James Garfield (the future U.S. President), who was misidentified as assistant adjutant general. The forms were written entirely in cursive, which was unusual. The two strange Federal officers told Baird they needed to get to Nashville that night and asked for a loan of 50 dollars. They said they had been attacked by rebels near Eagleville and lost their baggage and military overcoats.

Col. Baird accepted their story, loaned them the money, and bid them on their way. On their way out, Col. Louis D. Watkins of the 2nd Kentucky Cavalry noticed something strange and ordered the two men arrested. One of the men resembled Lt. William Orton Williams, whom Watkins had replaced in the 2nd U.S. Cavalry two years earlier (before the war). The two Federal officers were quickly exposed as frauds, and when consulted, Garfield told Baird to "call a drumhead court-martial to-night, and if they are found to be spies, hang them before morning."

The men confessed to be Confederates. Auton was actually Col. William Orton Williams of Bragg's staff, and a cousin of Robert E. Lee more or less engaged to Lee's daughter Agnes. Dunlop was actually Lt. Walter "Gip" Peter, Williams' cousin and a member of Gen. Joseph Wheeler's staff.

During the court-martial, both men claimed that they "were not spies in the ordinary sense." At 4:40 am Rosecrans ordered them to be hanged at once. By 10:30 am, Baird reported that the two men had been tried, found guilty, and executed, hung from trees just outside Fort Granger.

Baird said the men had not asked any information about the fort or Federal forces, that they confessed they were going to Canada or Europe or some such thing. Their papers listed Federal commanders in various Northern states. "Their conduct was very singular, indeed; I can make nothing of it," Baird stated. The bravery of the two men facing execution made an impression on the garrison; men were subdued for several days. "I have never saw anyone meet death with more courage than they did," wrote Major William Broaddus of the 78th Illinois.

Another Fort Granger mystery is the testimony of Park Marshall in 1935 that the fort was abandoned and vacant during the Battle of Franklin. Marshall lived 500 yards from the fort and witnessed the battle. Marshall also claimed that his observations were backed by Dr. Hugh Ewing.

The fort was manned by various detached regiments until May 1865. In 1909, Federal veteran J.T. King of Illinois visited Fort Granger in Franklin, where he had been stationed in 1863. "It was a beautiful fortification," he recalled, "with trenches sixteen feet deep, and with firing trenches or rifle pits…we felled the timber for a distance of three miles from the fort and drawing the trees with sharpened branches pointed outward to within a quarter mile." Not much was left of the old fortification. "The old fort stands neglected, a labyrinth of brush vines and timber growing on the parapet and from the sides of the trenches." He could still trace out the main embankments, the thirty-some embrasures from where the cannon muzzles peered, and the location of the magazine that held thousands of rounds of ammunition. To him, it was far from an instrument of devastation. To the contrary, King described the fort as a symbol of order in an otherwise chaotic war, yet he recognized it could have unleashed itself upon the surrounding secessionist population if its officers so desired. "Our artillery had a clear sweep of the country for several miles in every direction, and we could at any time have made kindling wood of Franklin in a very few minutes." But, King concluded, "there seemed however to be a tacit understanding between our commander, General Granger, and the citizens of Franklin, that if the Confederates would not use the buildings for shelter that we would spare the town." Long neglected after the Civil War, Fort Granger was purchased by the City of Franklin in the 1970s. Fort Granger Park (14.5 acres) can be accessed from the rear portion of Pinkerton City Park off Hwy. 96 between Franklin and I-65 South or from Eddy Lane. The fort is open daily and free of charge. There is interpretive signage and an overlook of the city but no other facilities. Only the earthworks remain today.

The Battle of Franklin

On Nov. 30th, 1864, a great battle was fought just south of Franklin, as Gen. John Bell Hood's Army of Tennessee assaulted the 23rd Corps of Gen. John Schofield, which was secured behind field fortifications centered on the Columbia Pike. Ground Zero was the farm of Fountain Branch Carter. The farmhouse sat atop a gentle hill; nearby on the other side of the pike was a large cotton gin. During the battle, 28 members of the Carter and Lotz families hid in the stone cellar of the farmhouse as the battle raged around them, well into the night. The next morning, the battle tally was horrendous, especially for five hours of fighting—2,000 dead, 6,500 wounded, and 1,000 missing. The Confederate dead included five briga-

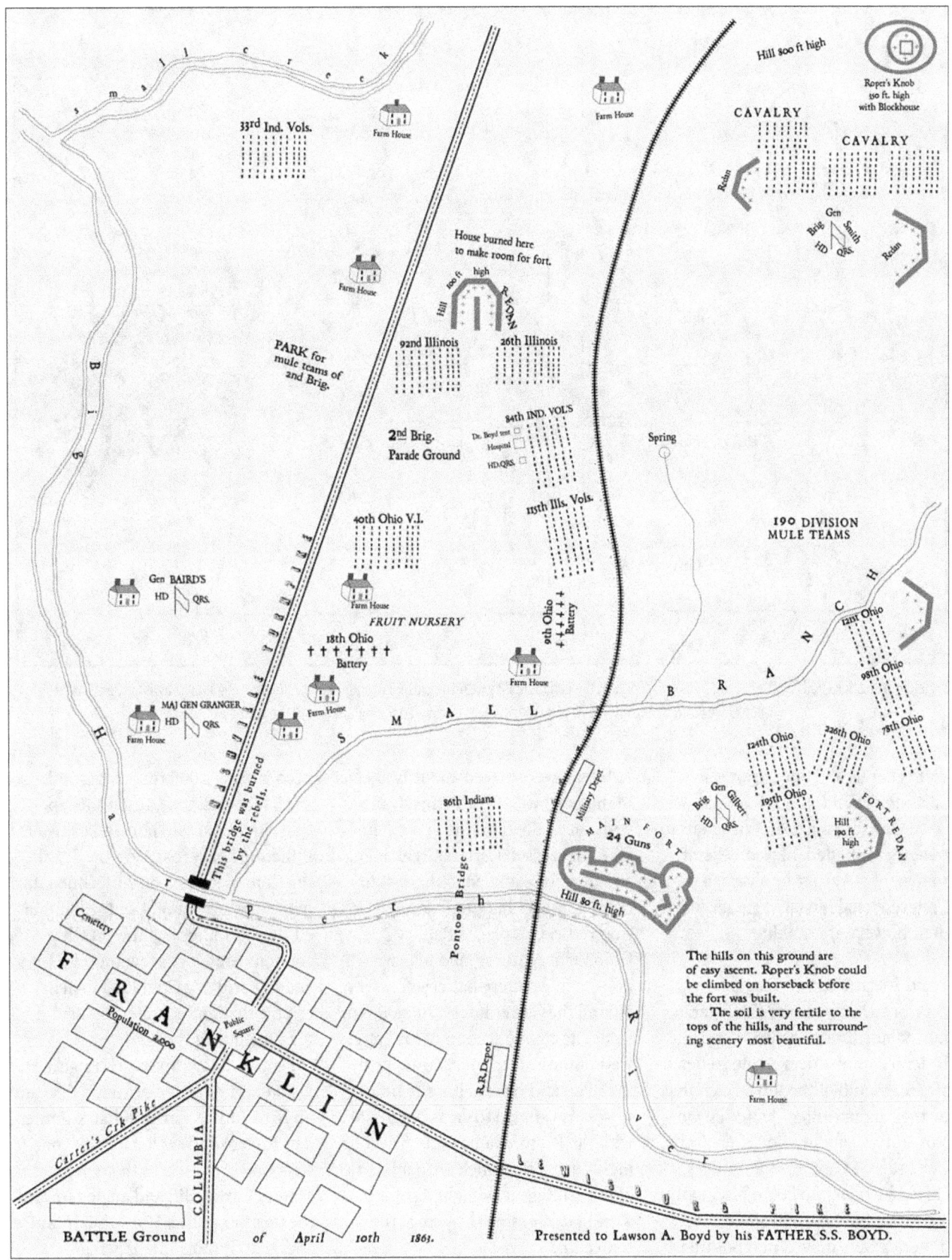

Like most Federal fortifications, Fort Granger near Franklin provided a base of operations for infantry regiments camped nearby, as depicted in this reproduction, by the author, of a wartime (April 1863) map drawn by Dr. Samuel Boyd, surgeon of the 84th Indiana Regiment. (Original at Boyd Family Papers, Bancroft Library, University of California at Berkeley.)

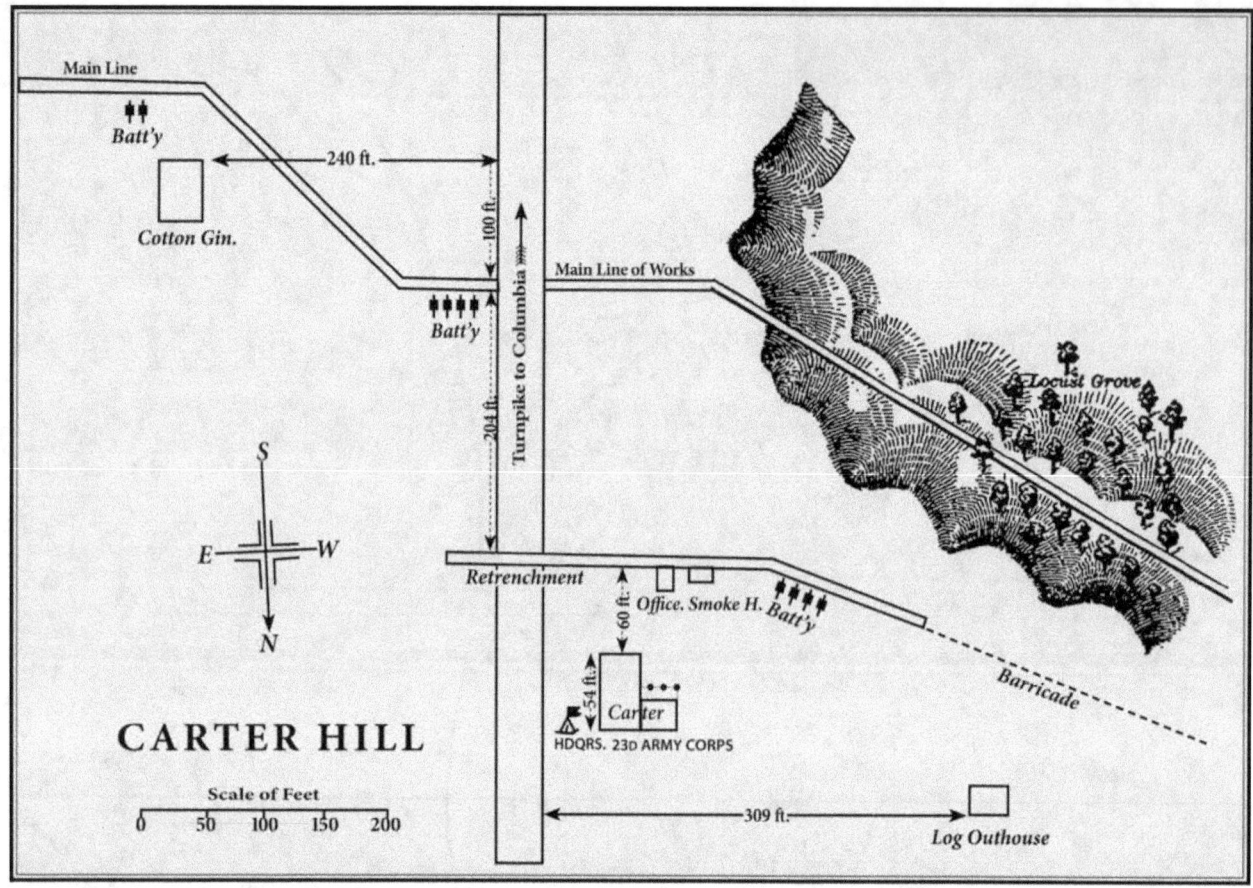

Field fortifications built by the Federal 23rd Army Corps at the Columbia Pike near the Carter farmhouse. Reproduction of a map drawn for Gen. Cox's 1897 book, *The Battle of Franklin*. In 2015, archaeologists uncovered the foundation of the cotton gin.

dier generals and one major general. Forty-four buildings in Franklin were converted into makeshift hospitals to treat the wounded. Eleven Federal soldiers would later be awarded the Congressional Medal of Honor for their bravery at Franklin.

Field fortifications erected

Prior to battle, as the Confederates moved northward from Spring Hill, Federal commanders created a defensive line south of the town, anchored by their light artillery batteries. On Nov. 30th, at about noon, Gen. Jacob Cox ordered Capt. Lyman Bridges, chief of artillery, to begin placing the guns. West of Columbia Pike, four batteries were moved into position. The 20th Ohio Light Artillery's four 12-pounder Napoleons were situated west of the Carter smokehouse. Four more 12-pounder Napoleons belonging to Battery B, Pennsylvania Light Artillery, were placed about 1,900 feet to the northwest. The Pennsylvania guns sat just behind the Federal line where it crossed Carters Creek Pike, and they were angled to the east to cover approaching Confederate troops. Bridges placed Battery A, 1st Ohio Light Artillery, and his own Illinois Light Artillery battery in reserve behind the Carter house. At 3:00 pm, he deployed one section of his battery west of the Carter house to support the main and reserve lines of infantry. It was also angled to create a crossfire with the Pennsylvania guns. The four highly accurate 3-inch ordnance rifles of the 1st Kentucky Light Artillery were placed beside the pike at the trench line.

During the battle, Bridges moved his other section into position alongside the first one. Over four hours, those guns, combined with Capt. Jacob Ziegler's Pennsylvania battery, fired almost 1,000 rounds of solid shot, spherical case, fused shell, percussion shell, case shot, and canister, inflicting heavy losses on the attacking Confederates. The 20th Ohio guns fired only 169 rounds, but still inflicted significant losses. The 1st Ohio cannons fired just a few rounds. The Federal artillery combined to help cripple the Confederate assault and turn the tide of the battle.

Federal Gen. Thomas H. Ruger's Division began to arrive around 6:30 am on Nov. 30th, a warm Indian summer day. Ruger soon set his men to constructing breastworks that wrapped around Carter Hill and angled toward the northwest. The 50th Ohio Infantry and 72nd Illinois Infantry regiments—approximately 600 men—held this section of the main line. The 44th Missouri Infantry, about 650 strong, was placed in a reserve position about 200 feet to the north and

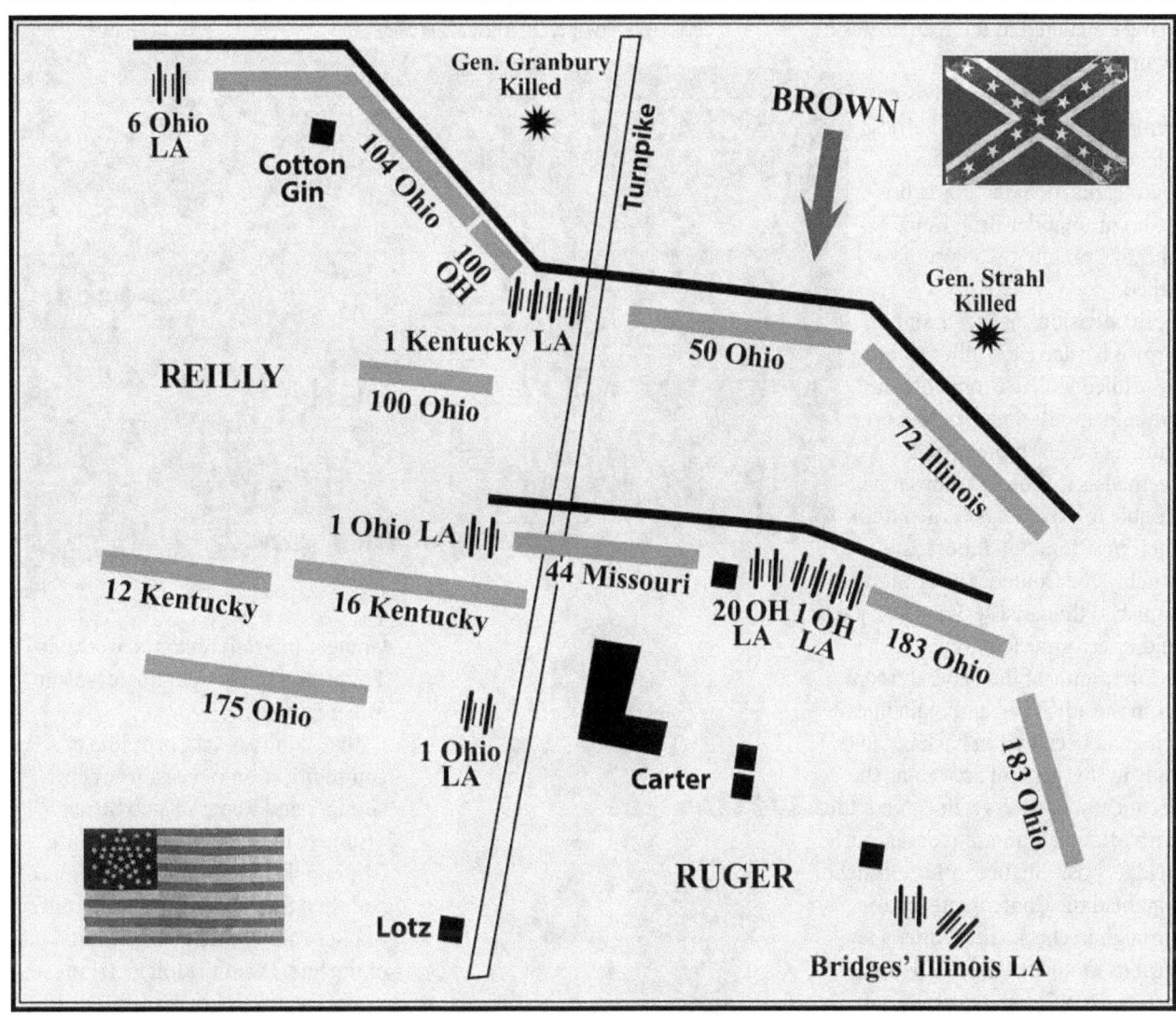

General Brown's Division attacked along the west side of the Columbia Turnpike. In addition to Strahl and Granbury, four other Confederate generals were killed in the assault and seven others wounded. Although technically a Confederate victory, Franklin was a very costly one indeed.

northwest. The 183rd Ohio Infantry took a reserve position to the right of the 44th Missouri. Other units, which included men from Illinois, Michigan, and Indiana, extended the main line west to Carter's Creek Pike. In this sector, the 20th Ohio Light Artillery and Bridges' Illinois Light Artillery supported Ruger's Division. Ziegler's Pennsylvania battery was posted behind the far right of Ruger's line. Altogether, about 5,000 troops were under Ruger's command. During the morning hours, Ruger's men cut down small locust trees and a row of fruit trees to create "a fair abatis" and slight obstruction in front of the works. They also grabbed whatever they could from the Carter farm, including fence rails and outbuilding timbers, to add to the earthworks. By 2:00 pm, the Federal fortifications were mostly completed. (In 2010 and 2017, archaeological digs determined the exact location of this portion of Ruger's defensive line.)

Confederate General John C. Brown's troops initially confronted Federal soldiers on an advanced line about half a mile south of the main line. When that position collapsed under the weight of Brown's attack, the Confederate troops charged forward. As they swept toward the main Federal line, the rebel yell began to reverberate across the landscape. Some of Brown's men went pell-mell into the tangled locust-tree abatis, and an officer later recalled that many men were "arrested by it." Nonetheless, just before 4:30 pm, Brown's men, who were aligned into two front-line brigades and two reserve brigades, hit and crushed a substantial portion of Ruger's main line. The 50th Ohio and 72nd Illinois infantry regiments suffered heavily, and Federal troops scrambled back to the secondary line. Brown's men surged across the Carter garden and up the slope out of the locust abatis. They pressed the attack but took heavy losses for their efforts. Brown was wounded, and Gen. Otho Strahl, one of his brigade commanders, was shot a short distance from here while loading rifles for his men. As Strahl was being evacuated to the rear, he was hit again and killed.

Across the turnpike, division commander Major Gen. Patrick Cleburne was shot and killed leading his men, including brigade commander Brig. Gen. Hiram Granbury, who was also killed.

The divisions of Ruger and Brown battled each other with unbridled violence long into the twilight and darkness of that evening. For a few moments, it seemed as if Brown's troops might be able to drive Ruger's men from their position. But almost as quickly, the Confederate advantage vanished though a series of fast-paced and interconnected events.

First, some of the Federal troops from the advanced and main lines who had been driven back began to rally in this general area along the secondary, or reserve, line. Next, the 44th Missouri Infantry courageously held its position, becoming an anchor that held the Confederate breakthrough in check. Then, after a few furious moments, at least four regiments from Colonel Emerson Opdycke's Brigade rushed into the breach and helped stabilize the line closer to Columbia Pike. All three actions doomed Brown's battered division.

Colonel Ellison Capers, 24th South Carolina Infantry, Brown's Division, wrote afterward, "Torn and exhausted, deprived of every general officer and nearly every field officer, the division had only strength enough left to hold his position." The human toll in this area was heavy. Ruger suffered about 700 casualties, and Brown's Division tallied almost 1,200 killed, wounded, and missing. One of the wounded was Fountain Carter's son, Capt. Tod Carter of the 20th Tennessee. Located on the battlefield the next day, he was found alive but with nine bullet wounds. He was taken to his childhood home, where he later died.

Bullet holes in wall of Carter House farm office.

Col. Moscow Branch Carter

The Carter House, along with the Carnton Plantation, the Lotz House, and other battle sites are accessible to the public today, although much of the battlefield has been claimed by modern development. Over the course of a decade, seven houses, two pizza restaurants, and a strip mall were purchased at the cost of $10 million and demolished to reclaim this important part of the battlefield.

Triune Earthworks

During 1863-64, the Federals established a defensive line of works and signaling stations ranging from Fort Granger in Franklin to the works at Triune and to Fortress Rosecrans in Murfreesboro.

The Union Signal Corps line of communication ran east from Fort Granger and Roper's Knob through Triune to downtown Murfreesboro (the cupola of the county courthouse) and then further east ten more miles to Pilot Knob, which was the left flank of the line. From the hill in Triune where Gen. Granger made his headquarters, the cupola in Murfreesboro and Pilot Knob can be seen, as well as Roper's Knob to the west.

In March 1863, men under Major General James Steedman, commander of the Third Division of the XIV Corps, began construction of a line of defensive earthworks near the village of Triune, halfway between Franklin and Murfreesboro, and 22 miles south of Nashville on the Nolensville Pike. The works consisted of two strong lofty redoubts, several artillery emplacements, a powder magazine, and miles of entrenchments. Stationed at the works were an infantry division and a cavalry division. Steedman made his headquarters at the Watt Jordan house a few hundred yards east of the pike. On April 6th, Gen. Gordon Granger took over command of the Third Division at Triune.

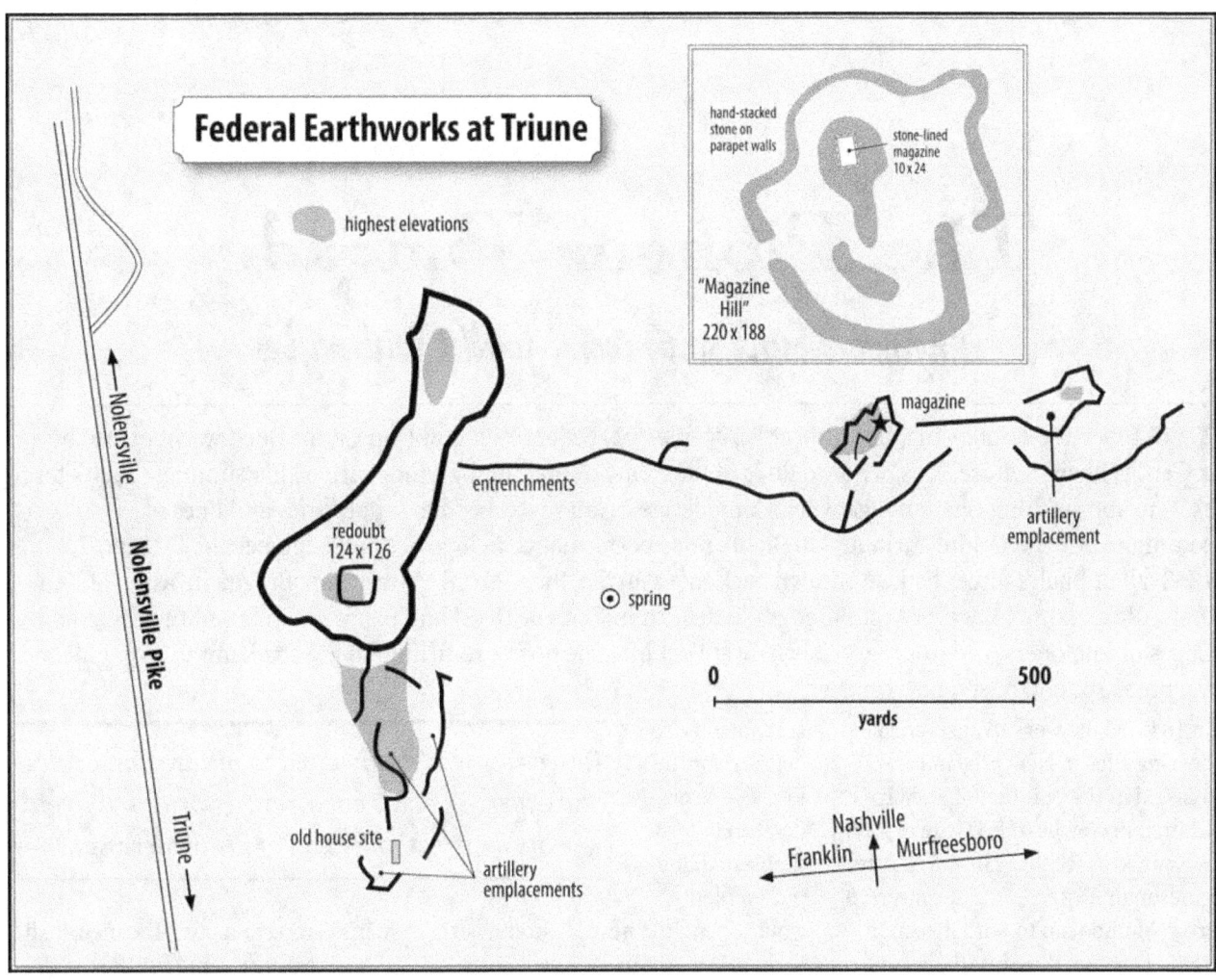

The house site, used by Federal officers as their headquarters, was destroyed by N.B. Forrest's artillery, the inhabitants "rudely awakened" one morning when a shell passed through the walls. Most of the Triune earthworks are located on private property. Original map drawn by F.M. Prouty for the Tennessee Wars Commission following the site survey for the Tennessee Division of Archaeology.

At least 15 engagements, mostly cavalry skirmishes, are known to have taken place at Triune, involving units of cavalry chieftains Forrest, Wheeler, and Van Dorn. During the Tullahoma Campaign in June 1863, Major Gen. Alexander McCook's four divisions of infantry and one of cavalry—a total of 30,000 men—moved through the area.

The village of Triune, home to 80 wealthy white homeowners, was considered "aristocratic" and a "thriving center of wealth, education, and culture." It boasted five general stores, a tailor shop, a weekly newspaper and printing shop, shoemaker, millinery, undertaker, two male boarding schools and two female boarding schools. The 18th District of Williamson County (the third richest county in the state) was quite affluent and included 1,200 slaves. The night of Dec. 26-27th, 1863, with the Federals nearby, a dance was held until 4:00 am at the Samuel Perkins house attended by local belles and officers of Confederate Lt. Gen. William Hardee, with the 33rd Alabama Infantry stationed as guards.

At the beginning of the war, the local Triune men first formed Company B and then Company D of the 20th Tennessee Infantry Regiment. Col. Joel A. Battle was elected captain of Co. B, and Moscow B. Carter (son of Fountain Branch Carter) lieutenant colonel. The unit fought in nearly every major battle of the Western Theater—its casualty rate at the Battle of Stones River was 55 percent. Of the 1,165 men who mustered in at the beginning of the war, 34 remained at the end. The residents of Triune were treated harshly during the war due to their heavily pro-Confederate stance. In April 1863, General David S. Stanley ordered the torching of all homes associated with men serving in the Confederate army. Before the order was rescinded, 20 fine homes and many buildings, including the Methodist and Baptist churches, the Masonic Hall, and the female academy were destroyed.

The Triune earthworks exist today but are located on private property.

The Pioneer Brigade
Engineers built with tools, fought with rifles

In October 1862, following the battle at Perryville, Ky., Federal General Don Carlos Buell was replaced by Gen. William S. Rosecrans, who would lead the Army of the Cumberland. Earlier that summer, Buell's forces were approaching Chattanooga when Confederate cavalry wrecked his supply lines and lines of communication, a failure attributed to Buell's poor performance in logistics and engineering. By Sept. 7, 1862, all of Buell's forces had withdrawn back to Nashville, then forced to turn back Bragg's invasion of Kentucky. Rosecrans, 44, was a West Point graduate (5th in Class of 1842) and former member of the US Army Corps of Engineers. Old Rosy served briefly at Fort Monroe before returning to the academy as an engineering professor and post quartermaster.

In 1847-53, he worked at several engineering sites in New England and assisted in the building of St. Mary's Church in Newport, R.I. (site of John F. Kennedy's wedding in 1953). A religious man, Rosecrans had converted from Methodism to Catholicism in 1845. Due to failing health, he quit the Army in 1854 and took several civilian positions. He took over a mining business in western Virginia and ran it successfully. He designed and installed one of the first complete lock and dam systems in wtestern Virginia on the Coal River. In Cincinnati, he and two partners built one of the first oil refineries west of the Allegheny Mountains. While Rosecrans was president of the Preston Coal Oil Company, in 1859, he was burned severely when an experimental "safety" oil lamp exploded, setting the refinery on fire. It took him 18 months to recover, and the resulting facial scars gave him the appearance of having a perpetual smirk.

Rosecrans was determined to nuture and exploit the expertise of his engineering corps and avoid the logistical mistakes of his predecessor. Instead of random selection of laborers, Rosecrans made sure that select craftsmen were detailed for the job. By his order, three thousand hand-picked men would spend the next 18 months of their enlistment in the newly formed Pioneer Brigade. Rosecrans issued General Order No. 3 on Nov. 3, 1862, which stated:

"There will be detailed immediately, from each company of every regiment of infantry in this army, two men, who shall be organized as a pioneer or engineer corps attached to its regiment. The twenty men will be selected with great care, half laborers and half mechanics. The most intelligent and energetic lieutenant in the regiment, with the best knowledge of civil engineering, will be detailed to command, assisted by two non-commissioned officers. This officer shall be responsible for all equipage, and shall receipt accordingly.

"Under certain circumstances it may be necessary to mass this force: when orders are given for such a movement, they must be promptly obeyed. The wagons attached to the corps will carry all the tools, and the

Rosecrans was determined to nuture and exploit the expertise of his engineering corps and avoid the logistical mistakes of his predecessor.

men's camp equipage. The men shall carry their arms, ammunition, and clothing.

"Division quartermasters will immediately make requisitions on chief quartermasters for the equipment, and shall issue to regimental quartermasters on proper requisition.

"Equipment For Twenty Men - Estimate For Regiment:
Six Felling Axes
Six Hatchets
Two Cross-Cut Saws
Two Cross-Cut Files
Two Hand-Saws
Two Hand-Saw Files
Six Spades
Two Shovels
Three Picks
Six Hammers
Two Half-Inch Augurs
Two Inch Augurs
Two Two-Inch Augurs
Twenty lbs. Nails, Assorted
Forty lbs. Spikes, Assorted
One Coil Rope

One Wagon, with four horses, or mules

"It is hoped that regimental commanders will see the obvious utility of this order, and do all in their power to render it as efficient as possible."

Although a generalization, basically the difference between US Army Corps of Engineers personnel and the volunteers such as the Pioneer Brigade is that the volunteers would move with the army, building any and all infrastructure required, and would fight as infantry if necessary.

Three thousand men were needed for the brigade, which was divided into three battalions (one from each wing of the whole army). The First Battalion of the Pioneer Brigade came from the 14th Corps or center wing; the Second Battalion from the right wing or 20th Corps; the Third Battalion from the left or 21st Corps. The actual numbers rose and fell due to circumstances. At one point, the brigade numbered 5,000 men. Each battalion was subdivided into ten to twelve companies of 80 to 100 men aggregated into four or five regiments. These battalions were organized to work independently of each other.

At first, many men resented being separated from their regiments and having to do what was considered menial labor. But the pioneers became a unit unto itself. The men ate better and they didn't have to drill incessantly like before or perform picket duty, and they were supposed to receive additional pay.

The men wore an insignia of cloth consisting of two crossed hatchets on the left sleeve above the elbow.

One of the consistent problems with the Pioneer Brigade was that the regular army commanders were led to believe that the engineering duty was temporary and that the pioneers would rejoin their regiments during movements of the army. This may have been a ruse on the part of Rosecrans and his officers "intended to ensure that the regiments detailed their best men and not their dregs"

Troops cavorting on a corduroy road, actually a walkway of logs to stabilize the muddy ground.

for this detail. This created some bitterness among the regular regimental officers toward the brigade.

The biggest problem was the pay issue. The pioneers weren't considered special by the Washington authorities and rarely received the pay they thought they were due. According to historian Geoffrey L. Blankenmeyer: "Further, as the men of the Pioneers were still considered to be a part of their original regiment, payday often came and went with no wage at all. Although these men remained to be listed on their regimental muster roll, they were also listed as being officially 'detached' and thus not due any pay through those channels. These organizational flaws created dissension within the ranks of the Pioneers and later developed into a breakdown of internal discipline and a rash of external sniping which became most evident during the Tullahoma campaign, the Pioneers 'unfinest' hour... By creating the Pioneer Brigade, Rosecrans now had an elite, cohesive unit of tradesmen who could quickly erect field fortifications, corduroy roads, build and repair bridges and move and fortify his army with skill and alacrity. What Buell had once attempted to do with the 800 men of the First Michigan Engineers, Rosecrans would improve on by adding the three thousand men of the Pioneers."

In essence, the Pioneer Brigade was the only military unit in the Civil War that used both shovel and musket.

The pioneers were commanded by Captain James St. Clair Morton, 33, of Philadelphia, a brilliant graduate of the University of Pennsylvania and West Point (2nd in Class of 1851). Sporting long, curly blond hair, he was described as "equal parts brilliant and brash," a protégé of professor Dennis Hart Mahan, who had rejected the contemporary military strategy of the time by preaching defensive tactics including entrenchment. His father, Samuel Morton, was a distinguished physician who founded craniometry and supported scientific racism, which designated the races of mankind as different species, with varying intelligence based on cranial capacity. James taught engineering at West Point. His first military assignment was in Charleston, South Carolina, as assistant engineer in the completion of Fort Sumter. In 1860, he led the Chiriqui Expedition to Panama in search of a route for a railroad or canal crossing. He was supervising engineer for the construction of Fort Jefferson on Dry Tortuga, Florida. He authored several books on engineering prior to the war, including a paper on the flaws in New York's existing harbor defenses.

Morton has been compared to the young George Armstrong Custer, but Morton identified more with fellow

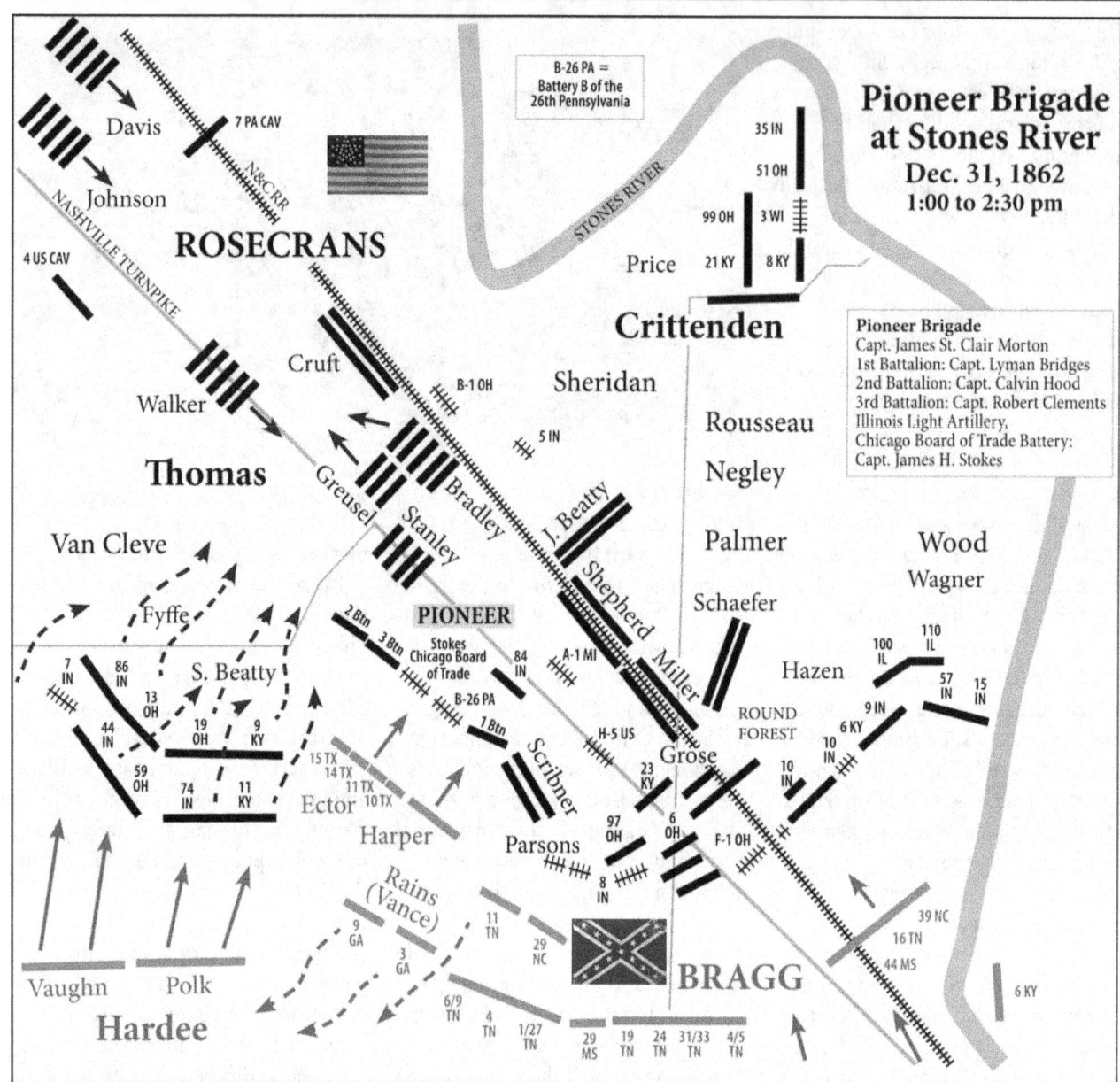

engineer Rosecrans. Morton would become well-known for his design of the fortifications at Nashville and many other sites in Middle Tennessee.

Historian David Powell: "James St. Clair Morton had a profound impact on the Army of the Cumberland, for he both constructed much of the defensive and support infrastructure the army came to rely on, as well as trained and organized the men who built it. His work continued to prove its worth long after he left. Sherman's advance on Atlanta rested on the bedrock of support he built, and might not have been possible without it."

Morton's war experience began in May 1862 when he was assigned to General Buell as chief engineer of the Army of the Ohio, and in April 1863 he became chief engineer of Gen. Rosecrans' Army of the Cumberland. It was written that his "iconoclasm, outspoken individualism and willingness to challenge engineer dogma endeared him to Rosecrans, who shared the same traits."

About 1,600 men of the 2,600-man Pioneer Brigade joined the army's march to Murfreesboro in late December 1862 in a drenching rain. The pioneers cut roads for ammunition trains and ambulances and repaired the bridge at Stewart's Creek. On the fateful day of Dec. 31st, 1862, the Confederate army under Gen. Braxton Bragg hit the Army of the Cumberland hard just northwest of Murfreesboro. Fearing the collapse of his right wing, Rosecrans placed the Chicago Board of Trade (Stokes) artillery battery at a clearing to protect the vital Nashville Pike and the railroad, and placed the Pioneer Brigade with them in reserve. Sergeant Henry Freeman stated: "The sounds of battle were unmistakably coming nearer and nearer with omens of disaster. Rumors of trouble began to run through the lines. A few minutes later the woods were suddenly filled with stragglers, riderless horses, and ambulances driven with frantic speed. Our time for action had come." At-

Standard pioneer chevron pattern
H: 3.5 inches
W: 5 inches

Variant pioneer chevron pattern
H: 3.2 inches
W: 4.75 inches

tacked vigorously five times, the battery inflicted many casualties with torrents of canister, and the pioneers held firm. The Confederate attacks upon the pike — the Federal army's only route for escape or reinforcements — were repulsed. For the next few days, the battery and brigade held firm, not seeing any more fierce action although on January 2nd the pioneers took part in General Negley's assault of the Confederate lines across McFadden's Ford of Stones River following Mendenhall's artillery barrage. On Jan. 4th, Bragg's army retreated from the field, the bloody battle a tactical draw but a huge strategic Federal victory. The brigade lost 12 men killed and 23 wounded. The artillery battery lost three killed and ten wounded.

Morton was promoted to brigadier general for "distinguished service in the fortification of Nashville…and behaving like a hero during the whole battle of Stones River." The victorious Gen. Rosecrans noted, "Among the commands which deserve special mention for distinguished services in the battle is the Pioneer Corps, a body of 1,700 men composed of details from the companies of each infantry regiment…(They) marched as an infantry brigade with the left wing, making bridges at Stewart's Creek; prepared and guarded the ford at Stones River on the night of the 29th and 30th; supported Stokes' battery, and fought with valor and determination on the 31st, holding its position 'til relieved on the morning of the 2d; advancing with the greatest promptitude and gallantry to support Van Cleve's division against the attack on our left on the evening of the same day, constructing a bridge and batteries between that time and Saturday evening. The efficiency and esprit du corps suddenly developed in this command, its gallant behavior in action, and the eminent services it continually rendered the army, entitle both officers and men to special public notice and thanks."

After the battle, the troops bivouacked for the winter while the Pioneer Brigade went to work building the largest fortification and depot constructed during the war — Fortress Rosecrans on the northwest corner of the town. The fort spanned 200 acres and 14,000 feet of earthworks and enclosed portions of the Nashville & Chattanooga railway, the Nashville turnpike, and Stones River. The First Michigan Engineers also worked on building the fortress. Huge amounts of supplies would be stockpiled at the fortress depot for shipment south during the campaigns to take Chattanooga and Atlanta. The fortress also served as a fallback position in case of disaster. The pioneers also worked on the fortifications at Nashville and repair work on the Nashville & Chattanooga Railroad.

In late June, Rosecrans finally moved against the Confederate forces scattered throughout Middle Tennessee. Four companies totaling 200 men from the First Battalion of pioneers marched at the head of Gen.

Uniform of Pioneer Brigade soldier. Note the crossed hatchet insignia on sleeve just above the elbow, red in color. (US Army Quartermaster Museum, Fort Lee, Va.)

In essence, the Pioneer Brigade was the only military unit in the Civil War that used both shovel and musket.

Thomas' 14th Corps, with the remainder stationed back at Murfreesboro. Although the Tullahoma Campaign was a resounding success, driving Bragg's forces from Middle Tennessee, the performance of the Pioneer Brigade was dismal and a blemish on their record. Historian Philip Shiman noted: "While the Pioneers performed well enough when concentrated for engineer work under Morton's watchful eye, they did particularly badly when scattered on pioneer duty with the army. During the march to Tullahoma during the summer of 1863 their behavior was a minor scandal. They did little drilling and at times there was much drunkenness. One Union general noted that '…They were straggling along, no one having particular charge of them, their tools never being unpacked, and when there was work to be done, a detail was always made from the regiment to do it'."

On June 28th, General Alexander McCook's wing fell behind the pioneers and was "severely delayed in (their) march." McCook complained to Rosecrans, who called for Morton and "abused him in a rough and violent tirade" in front of other officers. The Pioneer Brigade drew no more duties for the final two days of the campaign, remaining encamped at Manchester. For the remainder of July, however, the pioneers were extremely busy repairing roads and bridges, under the watchful eye of Morton, who went to great lengths to instill discipline and issue detailed orders to his officers. In August, they continued their advance upon Chattanooga, building large bridges over the Elk River and the Tennessee River (a pontoon bridge).

Assisting Morton were Lt. Cornelius Lamberson, assistant adjutant-general; Lt. Abram Pelham, quartermaster; Lt. Kilburn W. Mansfield, commissary of subsistence; Lt. Thomas J. Kirkman, inspector; Lt. John B. Reeve, aide-de-camp; and Francis Pearsall, assistant engineer and volunteer aide-de-camp.

As of the summer of 1863, the Pioneer Brigade consisted of four battalions totaling 3,800 men, 50 wagonloads of tools, implements, and construction equipment, and a pontoon train of 80 boats (40 of which can move at one time in line of march).

In mid-August, the Pioneer Brigade had a temporary change in command as General Morton was on leave due to disability. Command of the brigade was handed down to Captain Patrick O'Connell. The First Battalion was put under Capt. Charles J. Stewart.

On Sept. 8th, Rosecrans forced Bragg out of Chattanooga. The pioneers entered the city on Sept. 15th and busied themselves building bridges, repairing roads, operating sawmills, and constructing fortifications.

During the Battle of Chickamauga, Sept. 18-20, 1863, a mistake in maneuvering Federal troops led to the rout of Rosecrans' army by Gen. James Longstreet. The Federals fled back to Chattanooga; only the heroic efforts of Gen. Thomas and Gen. Granger saved the bluecoats from total annihilation. During that day, Morton became separated from Rosecrans while surveying the front and was caught up in the rout and fled back to the city. Bragg's Confederates then laid siege to Chattanooga, nearly starving the Federals. The Pioneer Brigade was instrumental in establishing the so-called "Cracker Line," which opened the route of supplies into the city. Rosecrans was relieved of duty and Thomas placed in command of the Army of the Cumberland. U.S. Grant arrived and took command of all U.S. forces at Chattanooga, which included the newly arrived army of Gen. William T. Sherman.

On October 10th, Morton was officially relieved of his duties and replaced by General William F. Smith as chief engineer of the Army of the Cumberland. Morton requested to be demoted to rank of major and was transferred to General George Meade's army in Virginia. Morton was subsequently killed in action at Petersburg on June 17, 1864 while conducting a survey at the front.

Pioneer Pvt. William Perkins-3rd Battalion

Gen. Smith, 39, known as Baldy, was a contentious character. He was praised for his gallantry in the Seven Days Battles and the Battle of Antietam, but was demoted for insubordination after the disastrous defeat at Fredericksburg. He was lauded for his engineering work on the Cracker Line at Chattanooga. Later, leading the first operation against Petersburg, Virginia, Smith's hesitation cost the Federals an opportunity to quickly end the war, and he was relieved of command.

In October 1863, General Smith moved quickly in establishing a position in Lookout Valley five miles upriver from the Confederate guns on Lookout Mountain. He then consolidated the Pioneer Brigade with the First Michigan Engineers, putting the Pioneers under the command of Colonel George P. Buell. A first cousin of D.C. Buell, George Buell was a Hoosier, the city engineer of Leavenworth, Kansas, and a gold miner and civil engineer. He led the 58th Indiana Infantry at Perryville, Stones River, and Chickamauga. He was promoted to colonel in the regular army in 1879, in command of the 15th U.S. Infantry Regiment. He died in 1883 while on duty in Nashville, and is buried at Mount Olivet Cemetery.

The pioneers' first duty under Buell was to establish their part of the Cracker Line by connecting Chattanooga to Stevenson and Bridgeport.

Then the pioneers were ordered to construct a pontoon bridge over the West Chickamauga on which Sherman's men would advance, but the pontoon train was not available. The engineers managed to construct a trestle bridge over the river with the materials at hand.

Lieutenant Henry C. Wharton reported: "I have the honor to submit the following brief report of the building of the trestle bridge across the West Chickamauga, in the advance of your troops upon Ringgold. The bridge was built by the First, Second, and Third Battalions, Pioneer

> *"I have never beheld any work done so quietly, so well, and I doubt if the history of war can show a bridge of that extent laid down so noiselessly and well in so short a time."*
>
> — *Gen. William T. Sherman*

Corps, under the command of Col. Buell, assisted by a small detail of men from the Fifteenth Missouri Infantry. Col. Buell arrived with his command between 8 and 9 p.m., and his wagons, with chess-plank and balks, about one hour afterward. The timber for the remaining portions of the bridge was cut down, and had to be carried to where the bridge was to be constructed. Orders were given to commence work at about 9 p.m. and the bridge was completely finished by half past 6 the next morning. Owing to the cold, the pioneers were divided into three reliefs, each taking one-third of the night. Taking this fact into consideration, and also considering the depth of the stream (often over five feet), the building of this bridge reflects credit on the Pioneer Corps, officers and men. I would state that it was originally intended to throw a pontoon bridge over the Chickamauga, but (for the lack of pontoons) this object was defeated."

General Sherman noted, "I have never beheld any work done so quietly, so well, and I doubt if the history of war can show a bridge of that extent (viz., 1,350 feet) laid down so noiselessly and well in so short a time."

Historan Blankenmeyer: "The Pioneers improvised their bridge; a span of over four football fields, in nine hours of work in chest high, icy winter waters through a cold November night. An amazing feat by any standard."

The Pioneer Brigade remained under the direction of Colonel Buell in Chattanooga and went to work on the Union depot there. The last documented work performed by the pioneers in the field of operations was that of repairing the road over Lookout Mountain from Nov. 29th to Dec. 13th, 1863. The brigade was subsequently moved to Nashville for the remainder of its existence.

On Feb. 15th, 1864, fourteen months after its creation, the Pioneer Brigade was recognized by Congress as an official engineering unit, a regiment to be called the First United States Veteran Volunteer Engineers (USVVE). Half of the men in the Pioneer Brigade re-enlisted in the USVVE, the others were placed under command of Captain Patrick O'Connell, who replaced Buell. A report on work at Chattanooga stated: "The command has been engaged during the past week grading the streets, building culverts, blasting rock, building levees, constructing water works on Cameron's Hill, and magazines in the different fortifications, doing picket duty, repairing roads between Ringgold and this place, building bridges over the creeks and erecting at Hiawasse River near the RR crossing."

Cities such as Nashville and Chattanooga may have been transformed into ugly, crowded, muddy, smoke-choked industrial centers, but local economies were boosted by U.S. military engineering. Historian Thomas Flagel noted: "Rail and road repair, barracks construction, material purchases, and river spans were almost exclusively for military purposes, but the effect upon adjacent populations sometimes resembled a kind of nineteenth-century New Deal. Federal work projects took over where local economies had collapsed."

The Pioneer Brigade, now numbering only 200 men,

> *The campaign qualifies as one of the most remarkable feats for any unit of military engineers.*

was officially disbanded on Sept. 10, 1864 with Special Order No. 129 after the fall of Atlanta. Most of the pioneers who did not enlist in the USVVE ended the war serving as infantrymen in northern Georgia.

The First Regiment of the USVVE remained in Tennessee, repairing bridges, fortifications, and railway infrastructure. Eight other regiments were formed out of Washington, D.C., with most mustered out in 1866.

Historian Blankenmeyer summed it up: "What the Army of the Ohio could not achieve in August, 1862 under General Don Carlos Buell, they were able to achieve as the Army of the Cumberland under General William Starke Rosecrans the following year. That campaign qualifies as one of the most remarkable feats for any unit of military engineers. It is unfortunate, but history tends to shed its brightest lights on the fighting units."

Fortress Rosecrans

Morton's other masterpiece ~ Largest earthen fortification built during the war

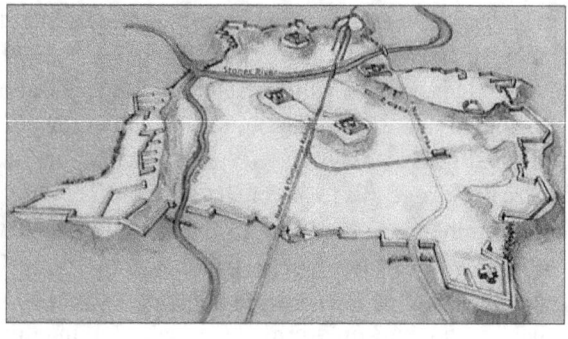

In the months following the Battle of Stones River near Murfreesboro, engineers and thousands of Union soldiers and black laborers toiled around the clock to build a huge fortified supply depot on the Tennessee & Chattanooga Railroad, named Fortress Rosecrans, which would supply future Federal campaigns against Chattanooga and Atlanta.

The fort was designed by Brig. Gen. James St. Clair Morton, Chief Engineer of the Army of the Cumberland, and named after the commanding general, William Rosecrans. It was built by the newly created Pioneer Brigade and the First Michigan Engineers, and gangs formed from the 40,000 infantrymen stationed there.

The construction of the fort, the largest earthen fortification built during the Civil War, began on Jan. 23, 1863 and was completed in June, although functional before then. The resulting 200-acre fort was large enough to protect an army of 50,000 troops and could stockpile enough supplies to feed that army for up to 90 days. It was sited 1.5 miles northwest of the antebellum Rutherford County courthouse in Murfreesboro. Following the fort's completion, a heavy artillery piece targeted the courthouse in case the local population rose up in rebellion. Other artillery pieces were sited at points 900 to 1,450 yards distant.

The project was massive, and along with Fort Negley, one of Morton's masterpieces. Due to the terrain, Morton designed Fortress Rosecrans from earthworks rather than stone.

On Feb. 15, 1863, John C. Spence of Murfreesboro, owner of the Red Cedar Bucket Factory, wrote: "Preparation is being made for building fortifications and rifle pits near this place. Large quantities of timber trees are cut and hauled to the grounds. The work is commenced and pushed on rigerously-digging and blasting rocks. A great number of negroes are employed at this kind of work, under pay, of course. Field hospitals are being built by tearing down and hauling away houses in town without even posting notice to the owner. Such is power — it can tyranize over weakness. Fortifications — They will scarcely ever have an opportunity of firing a gun at an enemy at this place, but military men probably know the best."

Work was hard, but supplies were abundant. In the period of one month, the First Michigan's 800 men built a magazine 140x30 feet and 12 feet high, an ordnance building 100x30 and 14 feet high, and a storehouse large enough to hold five million rations.

"The Army of the Cumberland was well protected and housed during the winter of 1863," according to historian Geoffrey L. Blankenmeyer. "It was also the best fed during any other time spent in this war."

A vast complex of weapons, ammunition, and food storehouses sprawled across the open ground within the fort. The fort also included four sawmills and a 50-acre vegetable garden used by the large local hospital. In addition to 14,600 feet of exterior defenses — ten lunettes linked by curtain walls and other obstructions — Fortress Rosecrans included four interior fortifications, called redoubts.

Blankenmeyer described the layout of the fortress depot:

"Two lunettes, Negley and Stanley and battery Cruft were located north of Stones River. Redoubt Schofield was located behind and in support of these three strong points. Lunettes Palmer, Thomas, McCook, Crittenden, Granger, Rousseau and Reynolds and battery Mitchell were located in a wide arc along the south bank of the river. Within this perimeter were the three other redoubts Brannan, Wood and Johnson. Demi-lunettes Davis and Garfield were located on the high ground. The magazine, ordnance, quartermaster and engineer's depots were south of the river and the commissary depots were on the north side. In all, the fortress measured 1,250 yards by 1,070 yards, spanned

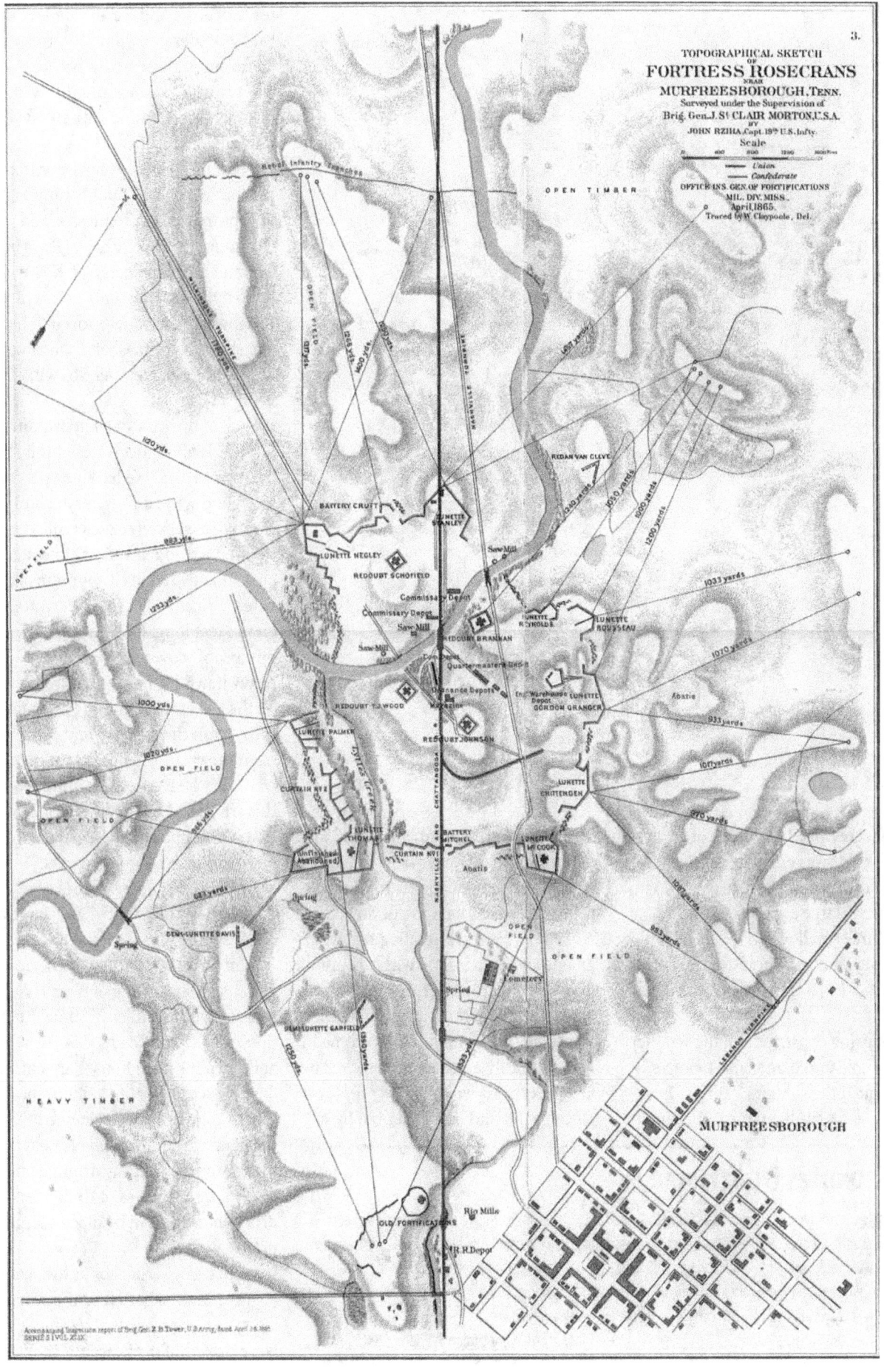

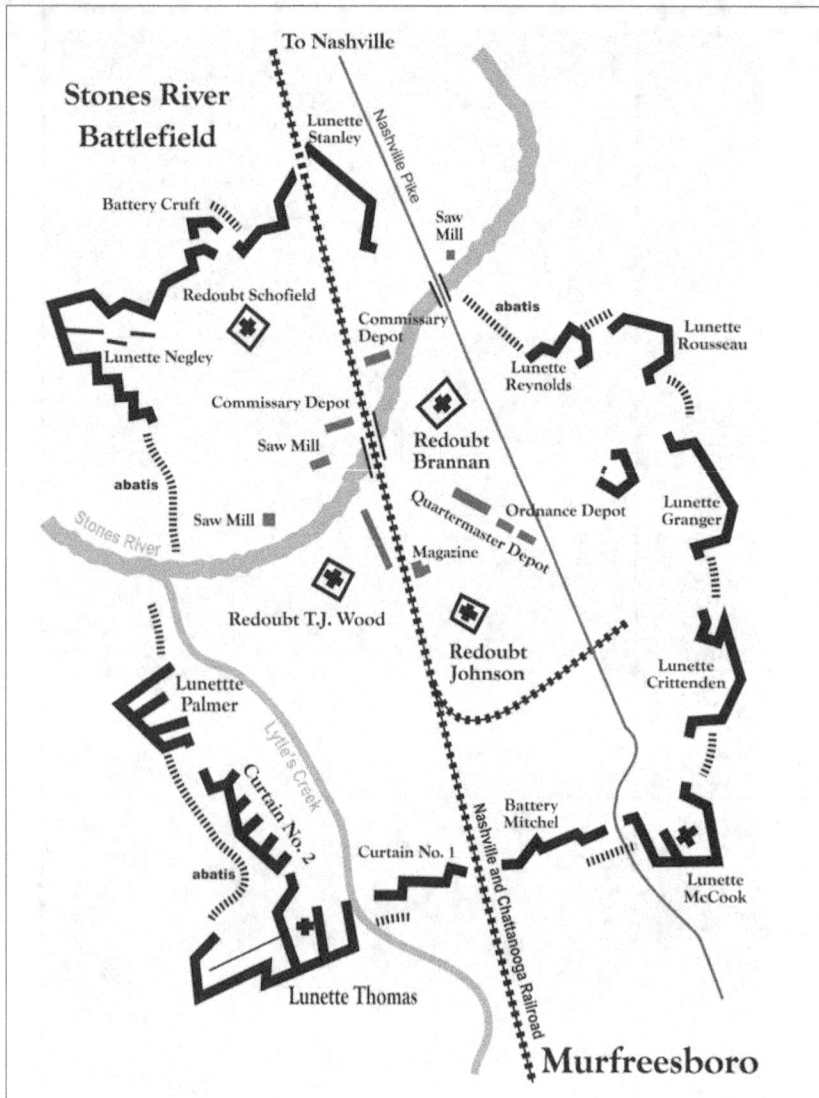

200 acres and 14,000 feet of earthworks. It was estimated that the fort could withstand a siege of an enemy force as large as 60,000 men."

Three small works — redan Van Cleve, and demi-lunettes Davis and Garfield — were built outside the fortress.

During construction, the weather was cold and damp, and sickness plaqued the soldiers' camps. Desertion was a problem for some units.

On May 12th, infantryman and laborer Lyman Widney wrote in his diary: "It does not appear to us that we are helping much to put down the Rebellion. Here we are digging away in one spot apparently with the purpose of remaining here in security until the end of the war as though we expected the Confederates to come to us voluntarily and surrender. We have the idea that it will be necessary for us to go out and bring them in by force of arms and we also believe that the Confederates will wait for us to come out of these works before they will venture to attack us. This war business is becoming entirely too deliberate for us."

In June 1863, Rosecrans garrisoned the fortress and began his campaign to capture Middle Tennessee and Chattanooga. He left Brig. Gen. Horatio van Cleve in charge of 2,000 light-duty convalescents. Rosecrans' Tullahoma Campaign proved to be a smashing success. Chattanooga was occupied on September 9th without firing a shot.

Back at the fort, in a letter home on July 18, 1863, James Jones of the 57th Indiana wrote: "We have powder in sacks and we just ram in a sackful at a time with wet coffee sacks rolled up tight on a wooden rod. It almost deafens a man when fired. Our guns are rifle parrots (64-lb. rifled Parrotts) eleven feet long. Our stockade is 'bumproof' covered over the top with heavy hewn timbers and gravel, overall three feet thick. We are but a few yards from the railroad and 30 rods from the river." He noted that there was "very heavy freight running on this (rail)road all the time." He added that his company had been feasting on blackburries for two weeks. "The surrounding country here is covered with buries, they are twice as large as I ever saw."

John Spence wrote about the aggressive clear-cutting of trees. "We can now see for miles in some direction from town. Ready, Bell, Murfree and Carney's farm houses are entirely destroyed and portions of numbers of others…Things are so changed that in the course of time it will be a hard matter to trace out the original landmarks. A wilderness of timber has disappeared and in its place a large prairie waste."

Spence also wrote about the slaughtering of livestock to feed the garrison troops: "The army were receiving large droves of beef cattle…They were

Luxuries from Home

Locomotive and train arrived from Nashville today…The shrill whistle evoked hearted cheers from all quarters of our camps. It conveyed to us…that we are linked again with home and friends, by an iron roadway over which may come plentiful rations, letters, newspapers, tobacco, whiskey and numerous so-called luxuries which soldiers love so well and miss so much when the wagons are overtaxed in carrying only the necessities of life."

— Diary of Sgt Major Lyman S. Widney, 34th Illinois

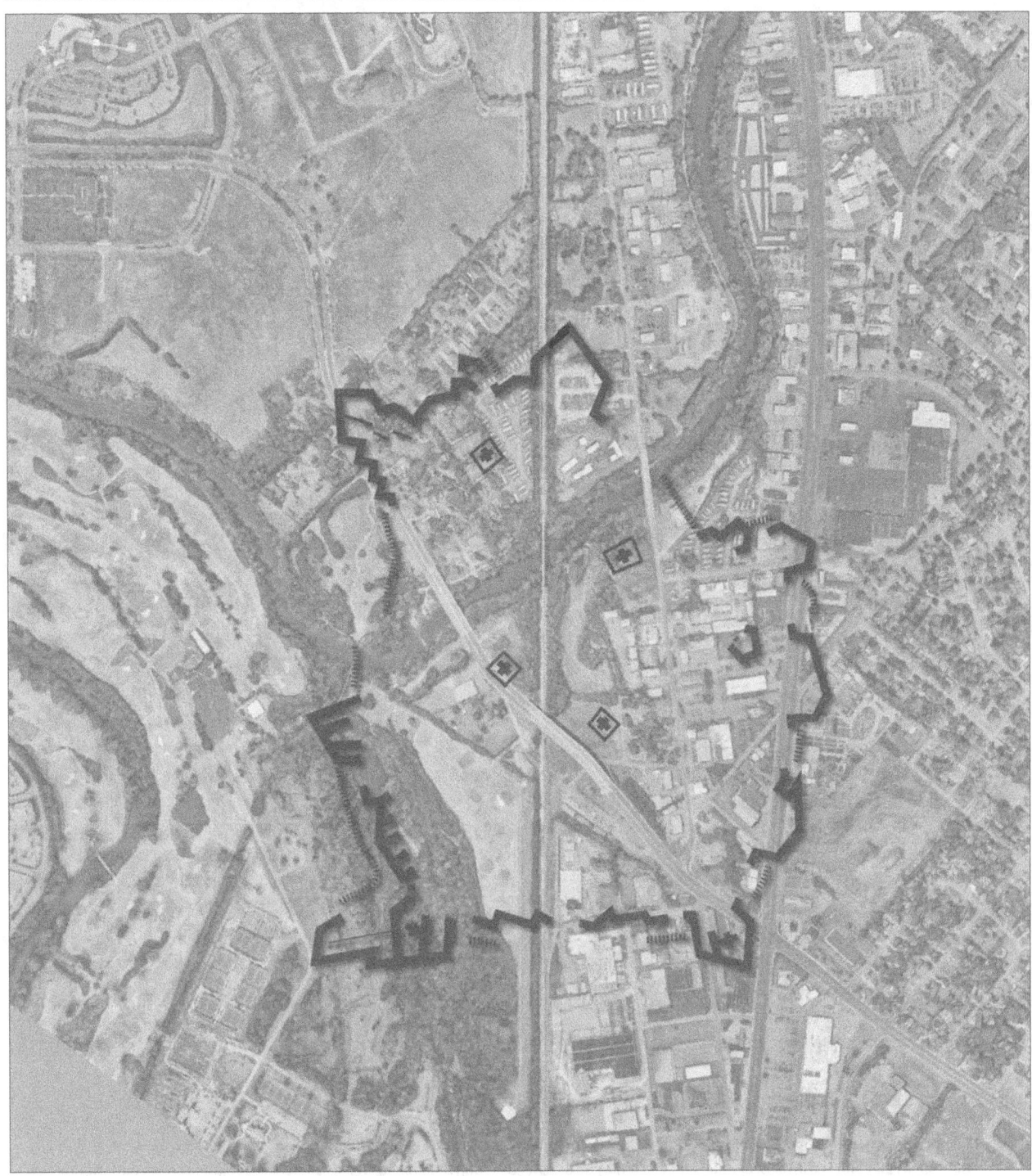

The outline of historic Fortress Rosecrans overlayed on a modern aerial view of Murfreesboro, with the Nashville & Chattanooga Railroad as the vertical axis, similar to Morton's 1865 wartime map. Visible also are Stones River and the old Nashville Turnpike and the municipal Old Fort Golf Course. Extant today are Redoubt Brannan — the northeasternmost interior fort — and the earthworks in the southwest or lower left portion of the layout. The nearest boundary of Stones River battlefield would be 1.52 miles to the northwest; the visitors center 2.36 miles.

generally kept in lots in and about town. It took about fifty or sixty every day to supply the demand of the army and hospitals. They would drive out that number shoot them down. When butchered, it generally covered over a half acre ground, the entrils [sic], heads and feet, left lying there – so in the course of time several acres was covered in this way, and it began to get warm weather. The smell became very offensive."

One of the most frequent and intense expressions of a garrison's power came from its booming artillery, noted historian Thomas Flagel. Volleys from artillery practice or celebrations could be heard from a dozen miles and more. He described the sounds. Rifled shells produced a piercing screech which ceased only

Reboubt Brannan was one of four earthen redoubts (forts of last resort) inside Fortress Rosecrans and the only one to survive. Featuring four guns at its corners and a bombproof stockade at its center, the redoubt overlooked the turnpike, the river, and the railroad. The earthworks above are the easternmost corner of the structure. The stockade no longer exists.

Engineering Inspector Brig. Gen. Z.B. Tower described Fortress Rosecrans:

"This large work is composed of a series of bastion fronts, with small, irregular bastions and broken curtains; or more properly it may be described as consisting of lunettes connected by indented lines, having in the interior four rectangular redoubts, and one lunette as keeps to the position.

"In large permanent works, with high scarps, the ditches are swept by guns in the flanks, because the depression of the guns prevent the canister-balls from rising above the parapet. In field forts, with ditches only six feet deep and long curtains, opposite flanks cannot fire in the same manner as in permanent works without risk to the defenders; but by breaking the curtain line the ditches are swept by close musketry. This is the manner of flanking the ditches of Fortress Rosecrans. Its lines give powerful cross fires and direct fires, both of artillery and infantry, on all the approaches. Placed on the crests of the elevations, they not only command the distant country, but effectually sweep the gentle slopes within canister-range.

"This fortress could not be taken except by siege, if properly garrisoned and well defended. The parapets have high commands and when built were well revetted with fascines. The work has many traverses, covering against ricochet fire. Most of the guns are in embrasures, made with gabions. Lunettes Thomas and McCook and the four interior redoubts have large block-houses in the form of a cross. The magazines, except in Fort Brannan, are small. That in Lunette Mitchell is subject to being flooded, and is consequently useless in the wet season. The ditches of the redoubts are not so well preserved as those of the main lines. In fact the exterior slopes of the parapets and the scarps have taken the natural slopes, about 45 degrees. These redoubts, however, are strong against attack, being defended by large keeps, which deliver their fire upon every part of the interior."

after impact. Spherical shots, when spinning, had the eerie ability to sound like buzz saws cutting into thick lumber. Shells from large Parrot siege guns – common entities in defensive structures – produced a deep hum, described by one witness as something "between a buzz and a groan."

By September, the fort was garrisoned by Col. John Coburn's brigade, along with the 5th Iowa Cavalry and the Ohio Light Artillery 9th Battery. In October 1863, Confederate cavalry leader Joseph Wheeler raided in the area, thought twice about attacking Fortress Rosecrans (quickly reinforced by Brig. Gen. William T. Ward), then burned the bridge at the Middle Fork of Stones River three miles south of town, torn up some railroad tracks, captured the blockhouse, and then retreated to Shelbyville.

On November 10th, Major Gen. Lovell H. Rousseau was put in charge of the Nashville District and its railroads.

In December, while Confederate Gen. John Bell Hood waited for General George Henry Thomas to attack him at Nashville, Confederate Gen. Nathan Bedford Forrest raided several railroad blockhouses between Nashville and Murfreesboro. He wisely declined to attack the fortress. On Dec. 7th, two brigades sent out under Brig. Gen. Robert H. Milroy defeated Forrest. The fortress was never threatened again; in March 1865 it mounted 57 guns.

In April 1866 the Federal army abandoned Fortress Rosecrans and sold all the structures at auction. Since then commercial development has destroyed most of the fort, along with erosion over time. General Z.B. Tower reported in April 1865: "It requires much labor to keep so large a work in repair; small portions of the parapets have sloughed off, due to frost and heavy rains. Some thirty feet of the parapet revetment of Lunette Thomas had fallen down...Parts of the revetted traverses in Lunette Negley are badly broken down..." Only a few sections of the fortress complex exist today—Curtain Wall No. 2 and Lunettes Palmer and Thomas in Old Fort Park, and Redoubt Brannan along the Old Nashville Pike. Signage interprets these works, which can accessed from the Stones River Greenway or from Old Fort Park. Redoubt Brannan is accessible from the greenway and the turnpike.

Stones River National Battlefield was established in 1927, covering about 10 percent of the actual battlefield. The fortress site is connected to the battlefield park by the Stones River Greenway.

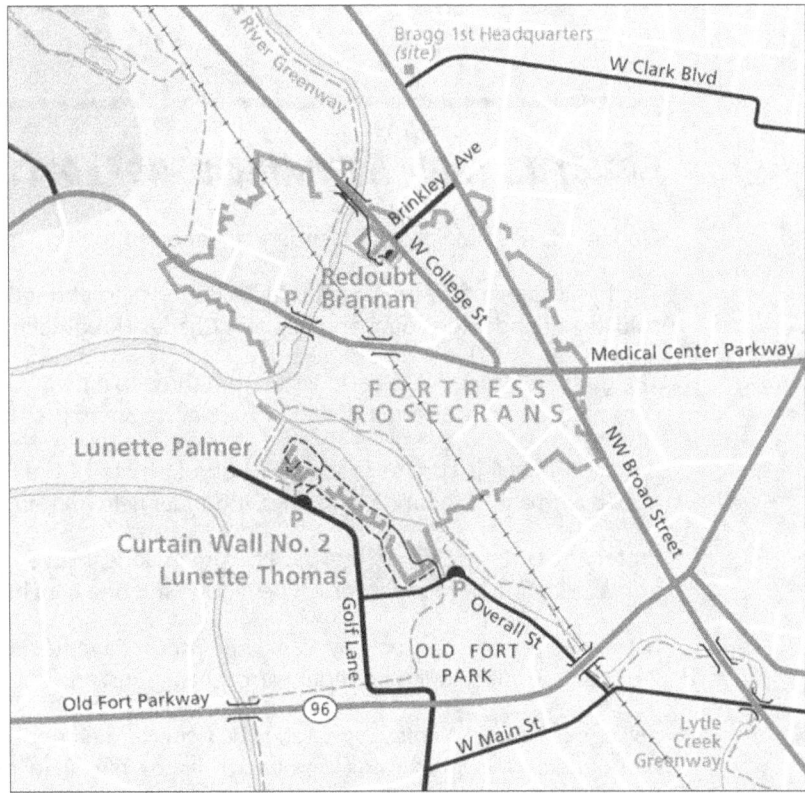

Stones River National Battlefield park map shows the outline of Fortress Rosecrans and what remains today —Redoubt Brannan, Lunette Palmer, Curtain Wall No. 2, and Lunette Thomas with dotted lines indicating walking trails.

The Hazen Brigade Monument at Stones River National Battlefield, built of native limestone, is the oldest standing Civil War monument on its original site.

Batteries and Armament at Fortress Rosecrans

On Jan. 14, 1864, Major John Mendenhall reported to Brig. Gen. John M. Brannan:

The fort is manned by the First Kentucky Battery and about 800 convalescent officers and soldiers, all under the command of Major Charles Houghtaling, First Illinois Light Artillery.

The guns are divided into batteries of from three to nine guns each, under the charge of a commissioned officer, and from 60 to 108 enlisted men present.

Battery Mitchell is commanded by Lt. John D. Irwin of the First Kentucky Battery and is armed with one 12-pounder and one 6-pounder field gun, and two 8-inch siege howitzers.

Battery at Lunette Palmer, by First Lt. Luman Jones, Seventy-Ninth Indiana Volunteers, armed with four 6-pounder Parrott field guns and one 8-inch siege howitzer.

Battery at Lunette McCook by Capt. J.R. Fiscus, Seventeenth Indiana, armed with one 24-pounder, rifled, four 6-pounder Parrott field guns, and two 8-inch siege howitzers.

Battery at Lunette Negley, by Capt. D.M. Roberts, Seventy-Fifth Illinois, armed with two 6-pounders, one 3-inch, one 6-pounder James rifle field guns, and one 8-inch siege howitzer.

Battery at Lunettes Rousseau, Sheridan, Thomas, and Reynolds, by Capt. W.A. Gregory, Twenty-Second Illinois, armed with three 6-pounder field guns, one 24-pounder, rifled, and one 8-inch siege howitzer.

Battery at Lunettes Granger and Crittenden, by Capt. W.N. Doughty, Thirty-Seventh Indiana, armed with one 6-pounder, and one 3-inch gun, and one 12-pounder field howitzer.

Battery at Redoubt Johnson, by Lieut. William Pool, Eighty-Seventh Indiana, armed with four 24-pounders, rifled.

Battery at Redoubt Schofield, by First Lt. William H. Leamy, Nineteenth U.S. Infantry, armed with one 30-pounder Parrott, four 24-pounders, siege, and one 6-pounder, field guns.

Battery at Redoubt Wood, armed with four 24-pounders, rifled.

Battery at Redoubt Brannan, by Second Lt. J.D. Williams, Ninth Michigan, armed with three 30-pounder Parrotts, two 12-pounder field guns, and one 8-inch siege howitzer.

The First Kentucky Battery (Capt. Theodore S. Thomasson), besides its own guns (two 6-pounders, one 3-inch and two 6-pounder James rifled field guns) has charge of one 24-pounder, rifled, and three 8-inch siege howitzers. Each battery, except the one at Lunettes Granger and Crittenden, has a magazine, all of which are in good condition.

Garrison, 115th Ohio and 22nd and 31st Wisconsin.

On March 23, 1864, General Brannan ordered three batteries assigned to Fortress Rosecrans—Battery D, 1st Michigan Light Artillery; 8th Battery, Wisconsin Light Artillery; and the 12th Battery, Ohio Light Artillery.

On Dec. 25th, 1864, General Van Cleve resumed command of Fortress Rosecrans, to include the 12th Indiana dismounted cavalry, the 61st Illinois, the 3rd, 4th, and 29th Michigan, and the aforementioned three artillery batteries.

Behind enemy lines, Confederate cavalry captain John Hunt Morgan struck against the Louisville & Nashville railroad in the Green River country of Kentucky the winter of 1861-62. Morgan led many raids against infrastructure in Nashville and Middle Tennessee. (Original art by John Paul Strain. Used with permission.)

U.S. Military Railroads and River Freighters

On September 2nd, 1864, General William T. Sherman's armies marched victoriously into Atlanta, an event that helped ensure Lincoln's re-election and the successful prosecution of the war. This following a four-month-long plodding campaign through northwestern Georgia against dug-in Confederate veterans. And it all hinged on a single railroad track back to Chattanooga and a vulnerable supply chain winding back to Nashville and Louisville. Sherman's victory simply would not have happened without that railroad line of supply.

During the campaign and ten-week-long occupation of the city, Sherman noted that his armies required the daily delivery of 1,600 tons of supplies on 160 rail cars over a rail system 473 miles long for a period of more than six months.

This logistics chain necessitated the use, at least to some extent, of at least seven railroad lines, including the completion of one line, the Nashville & Northwestern, from Nashville to the new supply depot at Lucas Landing on the Tennessee River. Confederate cavalry operating behind the lines were keen on disrupting this flow of vital goods, especially at choke points such as bridges, trestles, tunnels, and the depots themselves. Chief Engineer Capt. William E. Merrill reported that his men helped build 160 blockhouses for Federal garrisons to protect this infrastructure, 47 on the line between Nashville and Chattanooga alone.

During the 1864 campaign, Sherman's armies comprised 98,000 men and 35,000 horses and mules. Supplies were needed 20 days ahead of requirements. From November 1863 through August 1864, the following supplies passed through Nashville headed to the Atlanta campaign:
- 41,122 horses and 38,724 mules
- 3,795 wagons
- 445,355 pairs of shoes
- 542,693 pairs of pants
- 177,842 coats and overcoats
- 342,590 shirts
- 975,201 socks
- 253,136 hats and 75,436 caps
- 116,106 knapsacks
- 290,000 blankets
- 529,000 tents
- Millions of bushels of corn and oats, and tens of thousands of tons of hay.

Back at Louisville, the depot there built up a stock of 10 million rations and forwarded 300,000 per day. Five hundred and fifty barrels of flour per day were used by the cracker and bread bakeries. One thousand hogs were processed each day.

Brig. Gen. L.C. Easton was Sherman's quartermaster, while Col. Lewis B. Parsons was in charge of river and rail transportation in the district.

Gen. Robert Allen, chief quartermaster in the Western Theater, said, "No army in the world was ever better provided than Sherman's."

If the initial invasion of the South had been led by gunboats plying the rivers, it was the railroad that sustained the territorial gains of the U.S. Army and provided the main mode of transportation. At the time of occupation, Nashville was served by five railroad companies:

- **Louisville & Nashville R.R.** — The main line of supply between Nashville and the North. The station in Nashville was the city's largest and busiest. Two vital sections were the bridge over the Cumberland at Nashville and the South Tunnel near Gallatin in Sumner County. The L&N road was officially opened on Oct. 31, 1859. It took about nine hours for a passenger train and about 18 hours for a freight train to make the entire run, using the official timetable. The first train had gone through on Aug. 10th, 1859 and was celebrated by a barbecue in Nashville attended by 10,000 people.
- **Edgefield & Kentucky R.R.** — This short line linked Nashville at the Edgefield Junction with Clarksville via a connection to the Memphis, Clarksville & Louisville RR at the Kentucky state line. Extending from Nashville to Clarksville, and 61 miles long, it was composed of three links: first, the Louisville & Nashville Railroad from Nashville to Edgefield Junction, ten miles; second, the Edgefield & Kentucky Railroad to the State line, 37 miles; and third, the Memphis, Clarksville & Louisville Railroad to Clarksville, 14 miles.
- **Nashville & Chattanooga R.R.** — The main line of supply for the Union advances against Chattanooga and Atlanta. The railway ran southeasterly to Stevenson, Ala., where it joined the Memphis & Charleston and ran east to Chattanooga, a total of 151 miles. The tunnel at Cowan was 2,228 feet long, bored through Cumberland Mountain in 1849-53 by crews of English and Irish immigrants and black slaves using only black-powder blasting and hand tools.
- **Nashville & Decatur R.R.** — Also known as the Tennessee & Alabama, this line ran due south 119 miles to Decatur, Ala., where it connected with the Memphis & Charleston.
- **Nashville & Northwestern R.R.** — Incomplete and running only 28 miles to Kingston Springs at the beginning of the war, this railway in 1863-64 was extended the full 78 miles westward to the Tennessee River, where Union forces built Johnsonville, a huge supply depot.

Federal forces were authorized by an Act of Congress on Jan. 31st, 1862 to seize any railroad necessary to support military operations. Eventually, all railroads in and out of Nashville were commandeered by the U.S. military for their exclusive use.

Supplying the armies

Every soldier each day needed 12 ounces of pork/bacon and 4 oz. of beef, 1 lb. 6 oz. of bread or flour, 1 lb. 4 oz. of corn meal, plus vegetables, sugar, vinegar, salt and coffee. For horses, quartermaster regulations called for 26 pounds of food (14 pounds of hay and 12 pounds of grain, usually oats, corn or barley) per day per horse (23 per mule). One steamboat could haul up to 500 tons of freight, the equivalent of 250 wagonloads or 125 rail boxcars.

Massive quantities of war materiel, foodstuffs, and forage passed through Nashville on the railroads as the Union armies moved farther south. Supplies came from Louisville over the rails or on steam transports and barges moving downstream on the Ohio River and then upstream on the Cumberland to the Nashville wharf or upstream on the Tennessee River to the depot at Johnsonville (see separate chapter for details). Often the rails became doubly important as seasonal low water on the Cumberland River would temporarily prevent steam transports from reaching Nashville.

One mile of antebellum Southern railroad required a large investment in iron, local products and labor. The labor came mostly from slaves hired from local plantations. Frequently, plantation owners along the proposed line of a railroad would pay for shares of ownership in the new railroad with the labor of their slaves on the line near the plantation.

The local products were ballast material — sand, gravel, crushed rock, and wood. The wood was required for bridge, water tank, and station construction and for ties. The new railroad made do with whichever ballast material and wood it could obtain in the area.

The iron was the difficult part of building the railroad. Rail was usually purchased from England and produced in Wales. The cost, including shipping, was about the same as Northern produced rail, but English rail had a reputation for lasting longer. Rail produced in the South was almost non-existent. Chairs and spikes were frequently Southern.

Here are the numbers for one mile of track (both rails), made as cheaply as possible, consistent with accepted engineering practices:
A total of 117 tons of iron, to include:
T-rail
– 3,500 yards of 62#/yard rail for the main line
– 60 yards of 50#/yard rail for 1/5th of a 480-ft. station siding (1 siding per 5 miles)
Chairs – 422 18# cast iron (1 per length of rail)
Spikes – 4,700 1# iron (2 per tie)
Ties – 2,350 (1 per 2.25 ft.)
Culverts – 1 per 5 miles
Bridges – 1 per 10 miles, very rough average
Ballast – varied, but always as little as possible in the initial construction
Station – 1 per 5 to 10 miles
Water tank – 1 per 15 miles, or more

Source: Confederate Railroads, David L. Bright

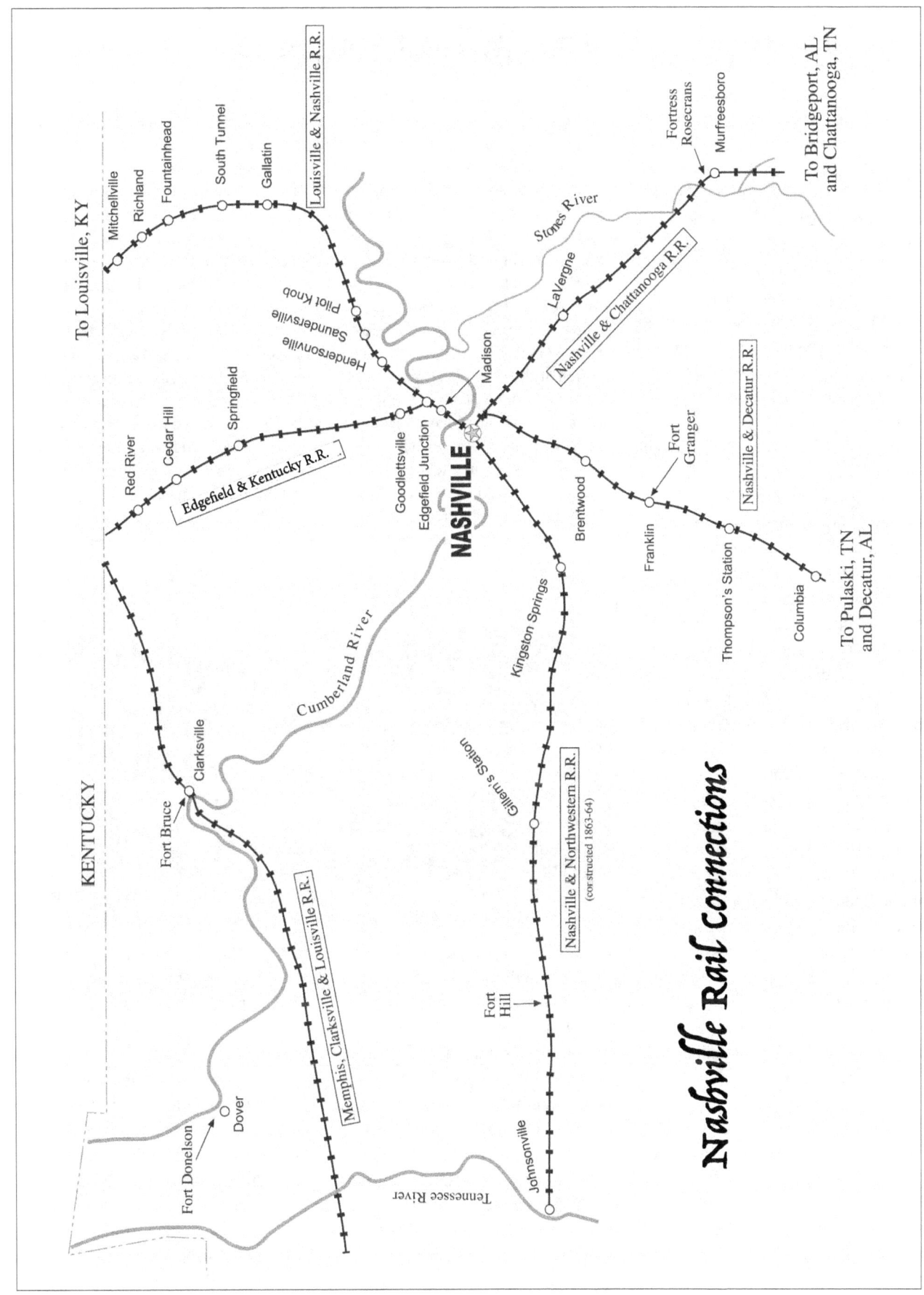

During 1863, 70 percent of the supplies reaching Nashville came by steamboat. Each boat, which could hold 100 to 500 tons of freight, made the 192-mile trip between Smithland on the Ohio and Nashville in 39 hours upstream and 20 hours downstream. Convoys of several dozen freighters usually required an escort of at least four tinclad gunboats. A typical steamboat would burn 36 cords of wood, at five miles per cord, and refueled four times (a cord in wood 12-18 inches long stacked four feet high and eight feet long). There were 20 woodyards between Smithland and Nashville.

By July 1862, more than 250 tons of supplies were reaching Nashville each day by steamboat. From February through May, 1864, 614 steamboats averaging 260 tons each delivered 158,616 tons of material to Nashville. More than 200 steamboats were chartered to supply Nashville during the war, first on a per diem basis, then by weight load delivered. From February to March 1864, the average tonnage per boat increased from 201 to 294.

Colonel James L. Donaldson was the quartermaster at Nashville. "For weeks, my levee thronged with transports of all sorts, and a force at least 3,000 men, and 4 to 500 teams were kept constantly at work, day and night, Sundays, and week days, in transferring the supplies to my various depots and store houses. My estimate is that, for three months, or more, I received and handled daily, an average of from 2,000 to 3,000 tons of freight, exclusive of the amount arriving here by railroad."

A typical freight train of 16 cars, each holding 10 tons, would take 19 hours to make the 192-mile trip from Louisville to Nashville.

City becomes logistics center

In many respects, the Federal quartermasters were the unsung heroes of the war — "indispensable cogs in the nation's war machine," according to

General William T. Sherman:

THE ATLANTA CAMPAIGN would simply have been impossible without the use of the railroads from Louisville to Nashville (185 miles), from Nashville to Chattanooga (151 miles), and from Chattanooga to Atlanta (137 miles). Every mile of this "single track" was so delicate, that one man could in a minute have broken or moved a rail…we had, however, to maintain strong guards and garrisons at each important bridge or trestle—the destruction of which would have necessitated time for rebuilding…Our trains from Nashville forward were operating under military rules, and ran about ten miles per hour in gangs of four trains of ten cars each. Four such groups of trains daily made 160 cars, of 10 tons each, caring 1,600 tons, which exceeded the absolute necessity of the army, and allowed for the accidents that were common and inevitable…that single stem of railroad, 473 miles long, supplied an army of 100,000 men and 35,000 animals for the period of 196 days, viz., from May 1 to November 12, 1864. To have delivered regularly that amount of food and forage by ordinary wagons would have required 36,800 wagons of six mules each, allowing each wagon to have hauled two tons twenty miles each day, a simple impossibility in roads such as then existed in that region of country. Therefore, I reiterate that the Atlanta Campaign was an impossibility without these railroads; and only then, because we had the men and means to defend them, in addition to what were necessary to overcome the enemy.

The locomotive roundhouse and turntable in downtown Nashville likely looked much like the one pictured here in Chattanooga, although it was only partially round. (OR Atlas)

historian Lenette Taylor, the biographer of Captain Simon Perkins Jr., an assistant quartermaster at Nashville. Long hours, multiple jobs, immense responsibilities, rare favorable recognition, and stagnation in rank were the realities of the quartermaster's position. Everything they needed to know had to be learned on the job.

Perkins served two and a half years as a quartermaster — officer in charge of forage on the Tennessee River; depot quartermaster at Stevenson, Alabama; headquarters quartermaster for Major General William S. Rosecrans; in charge of railroad transportation and quarters at Nashville; depot quartermaster and disbursing officer at Nashville; and depot quartermaster at Gallipolis, Ohio.

The quartermaster's job was stressful and precarious. Taylor: "Operating in territory occupied by large numbers of civilians of divided loyalties and different races, quartermasters had to determine the loyalty of citizens, procure civilian property for government use, and employ thousands of civilian workers, including huge numbers of free blacks and for-

(Continued on Page 184)

The Taylor Depot was a commissary warehouse for food supplies at the south end of the Nashville & Decatur Railroad station at Fifth Avenue and Broadway. This is the west side of the building fronting the railroad siding. In the background can be seen the solitary spire of the McKendree Methodist Church and the twin spires of the First Presbyterian Church. (Tennessee State Library and Archives)

The main depot of the Louisville & Nashville Railroad showing a row of steam locomotives and tenders. A shop building can be seen in the left background.

The same depot seen from the west showing switches and siderails. The tall structure in the background is roundhouse construction. (Library of Congress)

A composite photograph showing the south side of the Taylor Depot (on the left) and the rear of the building on the right. The structure measured 517 feet long by 190 feet wide, almost 100,000 square feet. (Tennessee State Library and Archives)

Guerrillas destroying a Railroad-Train near Nashville

(The Annuals of the Army of the Cumberland)

(Continued from Page 180)
mer slaves, as well as women."

Quartermasters often served many masters and satisfied few, despite their efforts. For example, during most of 1863, Perkins was accountable simultaneously to the commander of the Army of the Cumberland, the chief quartermaster of that army, the depot quartermaster at Nashville, the commander of the post of Nashville, the military governor of Tennessee, and the assistant quartermaster general in Cincinnati, as well, of course, as to Quartermaster General Montgomery Meigs.

By December 1863, Col. Donaldson had more than a dozen quartermasters, 12,000 civilian workers, more than 600 miles of railroad to manage, and more than 100,000 men to support. His office disbursed more than $5 million worth of business every month. "This is the biggest army depot to-day on the face of the earth,"

Asst. Quartermaster James Rusling proudly reported. The U.S. Military Railroad sent out of Nashville more than 29,000 cars of stores from July 1864 to June 1865, plus 5,673 cars loaded with troops (283,716 men). Almost a quarter of a million tons of supplies left Nashville for various posts by rail during that time period.

Although no prisoners were held at Nashville except for brief periods of time, thousands passed through the city, as did thousands of wounded soldiers from the battlefields. From November 1863 through August 1864, 10,000 Confederate prisoners of war were transported from the front to Northern POW camps. On July 10-11th, 1864, a total of 1,333 prisoners arrived from northern Georgia. It was not uncommon for citizens to witness Confederate prisoners of war marching through the streets of the city.

In the first ten months of 1864, 140,000 Federal troops were transported south from Nashville, and 40,000 sick and wounded and 50,000 volunteers were returned from the front.

Sometimes the heavy use of the railroads by the military resulted in deprivations on the part of civilians. In February 1863, 60 carloads of non-military freight accumulated at Louisville waiting for clearance. The supply of coal was exhausted by the Nashville Gas Light Co. and the lights went out, including those used by the military. Consequently, river barges of coal were brought in through the assistance of the local military.

Federal troops on furlough were forced to march back to Nashville, along with cattle droves. Cavalry troopers, of course, rode back.

The Federals greatly expanded the rail facilities in Nashville, building a new roundhouse and maintenance shops near the main N&CRR terminal

(Continued on Page 191)

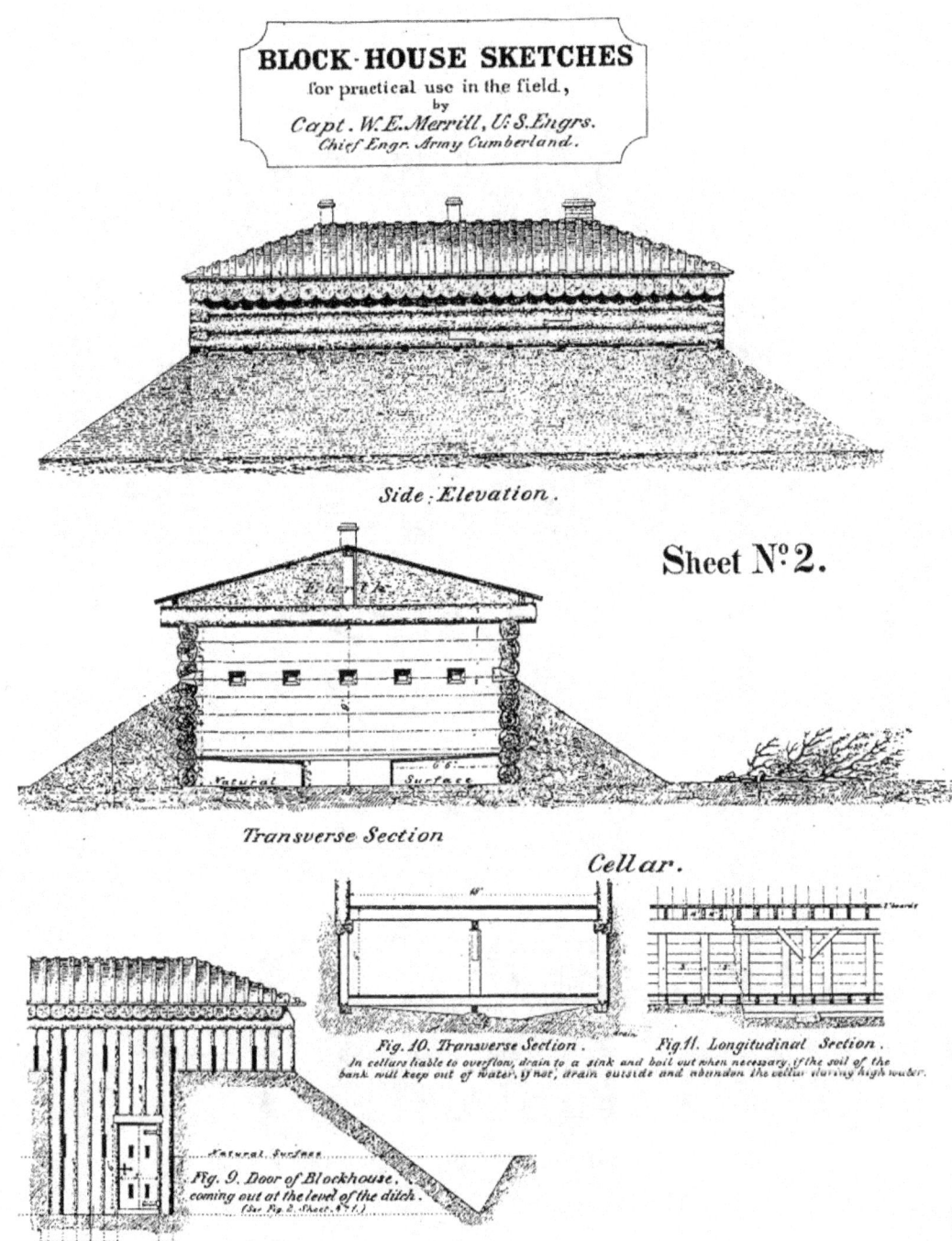

Adapted from Blockhouse Sketchs (3/1864) by Col. William E. Merrill (Buell-Brien Papers, Tn. State Library and Archives.)

Capt. William E. Merrill was meticulous in the prepartion and modification of plans for railroad blockhouses. Instructions and designs were provided to the officers and men working on their construction. Many of the design elements resulted from March 1864 experiments carried out by Capt. Merrill, with the assistance of Lt. Col. Kinsman Hunton of the First Michigan Engineers. Most of the more frivilous elements, such as the turrets, were never built in the rush to complete the blockhouses.
(Buell-Brien Papers, Tennessee State Library and Archives.)

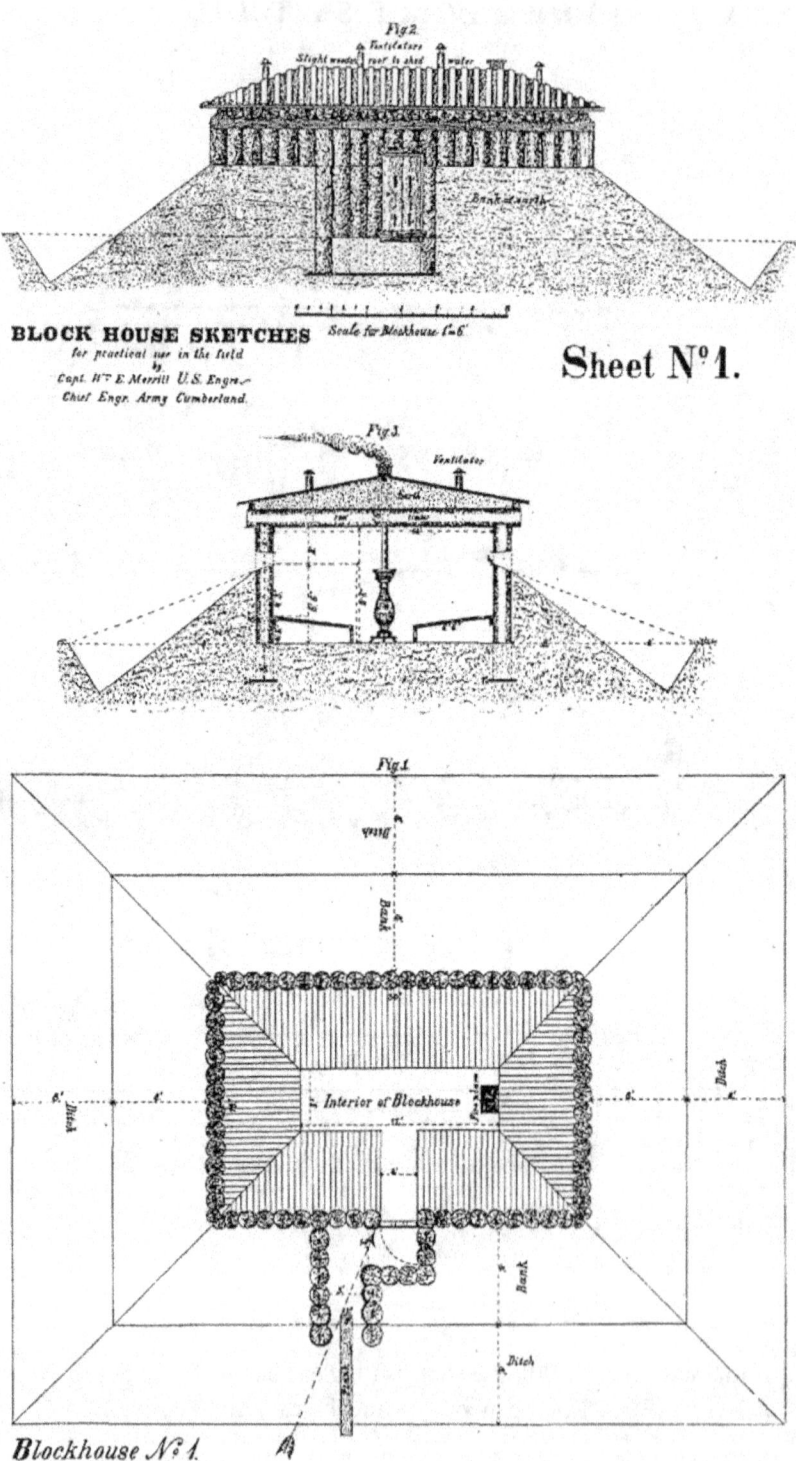

Adapted from Blockhouse Sketchs (3/1864) by Col. William E. Merrill (Buell-Brien Papers, Tn. State Library and Archives.)

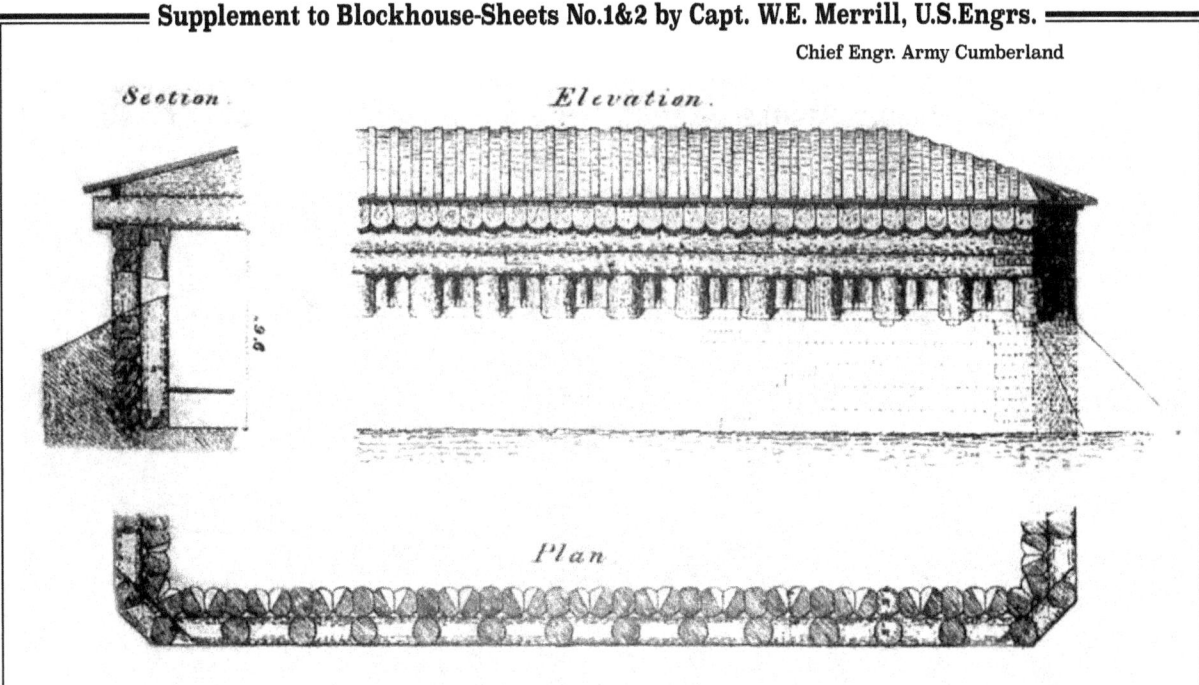

Supplement to Blockhouse-Sheets No.1&2 by Capt. W.E. Merrill, U.S.Engrs.

Chief Engr. Army Cumberland

Experiments made at Lavergne, Tenn., March 4th & 7th 1864, having demonstrated that the Blockhouses previously planned are not sufficiently strong, to resist rifled field artillery, the following plan has been devised, to remedy this defect.

Case the outside with one layer of horizontal logs piled on each other up to the lower edges of the loopholes. Place a similar casing above the loopholes to the roof. Support the logs above the loopholes upon the top log of those below, by short & thick pillars, placing them between the loopholes, as as not to interfere with their fire.

Bank up to the loopholes as previously ordered.

The average thickness of the main logs should be two feet, of the outside casing from 1'6" to 2'.

Adapted from Blockhouse Sketchs (3/1864) by Col. William E. Merrill (Buell-Brien Papers, Tn. State Library and Archives.)

Capt. William Merrill

Adapted from Blockhouse Sketchs (3/1864) by Col. William E. Merrill (Buell-Brien Papers, Tn. State Library and Archives.)

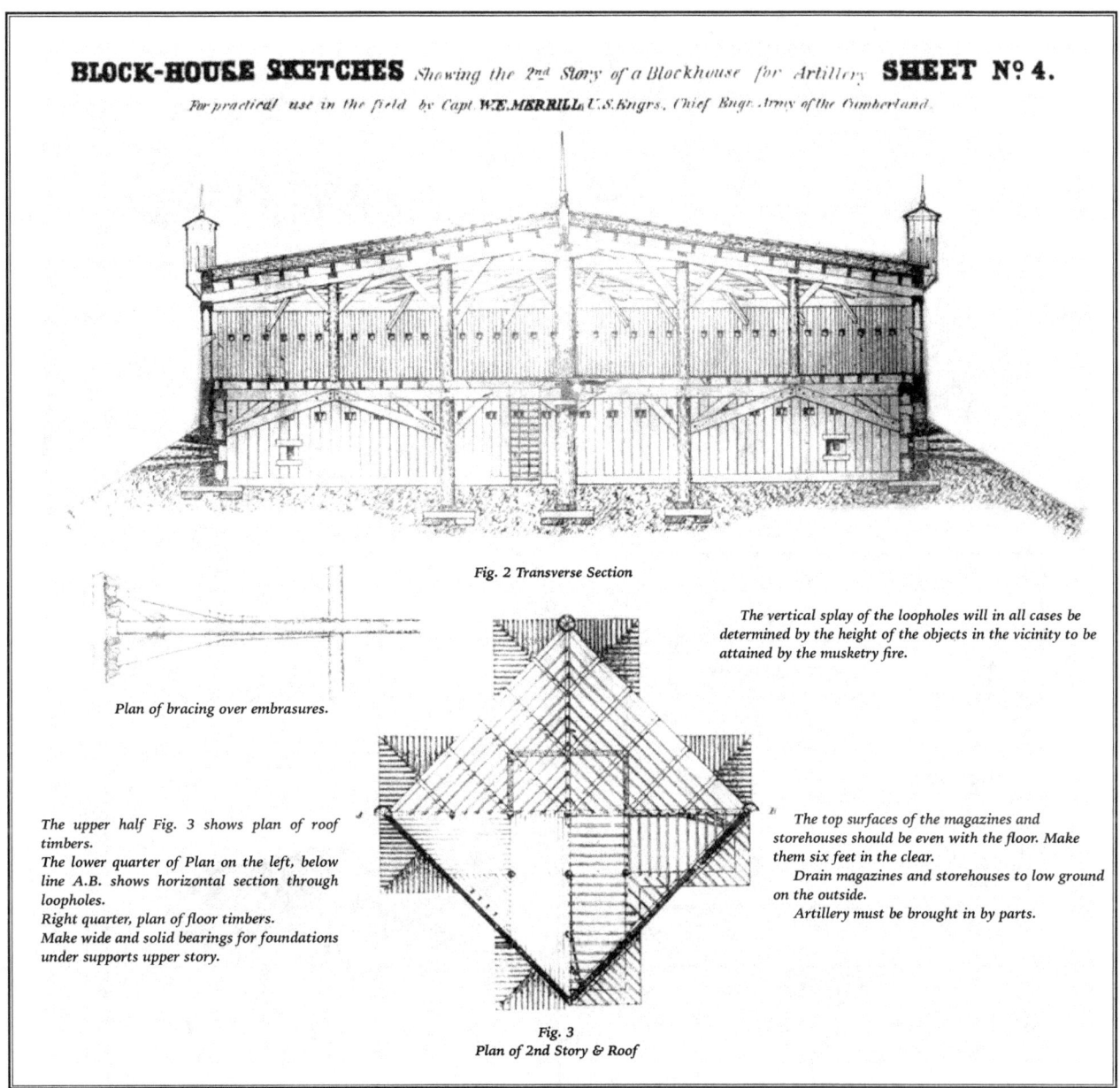

Adapted from Blockhouse Sketchs (3/1864) by Col. William E. Merrill (Buell-Brien Papers, Tn. State Library and Archives.)

Original Blockhouse Sketchs (3/1864) by Col. William E. Merrill (Buell-Brien Papers, Tn. State Library and Archives.)

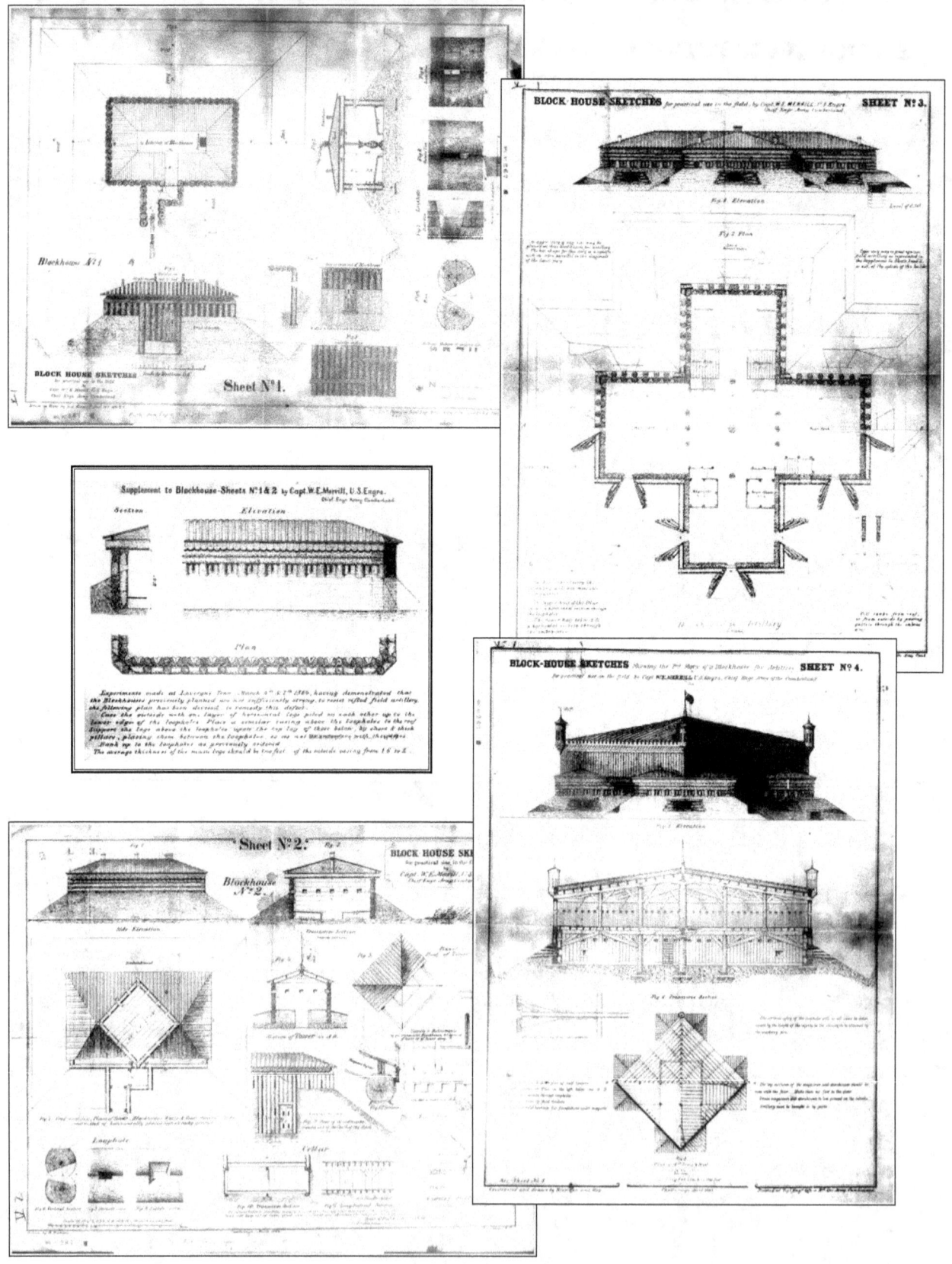

A track gang at work on the Nashville & Chattanooga RR tracks near Murfreesboro following the Battle of Stones River. The gandy dancers are spiking the rails to the ties without benefit of tie plates. They are working dirt instead of crushed rock or gravel for ballast, which tended to wear down during rainy seasons. In bad cases, wooden slabs or wedges would be inserted to stabilize the track. (Library of Congress)

(Continued from Page 184)

in the "gulch" west of the State Capitol. The coal yard could hold four million bushels.

Quartermaster Donaldson directed the construction of three huge warehouses:
• The Taylor Depot at the Broad Street terminus of the Nashville & Decatur R.R. covered two-thirds of the area from Broad to Demonbreun and extended west to the railroad. Fifty wagons could be loaded or unloaded at one time.
• The Eaton Depot, for subsistence stores, located just south of the Taylor Depot on the same railroad.
• A grain storage facility at the Nashville & Northwestern R.R. depot.

All in all, Nashville's "depot" sector covered 378 acres.

In February 1864 a few hundred yards of track were laid from the upper wharf on the Cumberland River to the Nashville & Chattanooga depot, to facilitate shipments from steamboats to southbound rail cars. Coach service between the wharves and rail depots was supplied by J. Lee Able.

By the spring of 1864, the principal shops of the United States Military Railroads, Division of the Mississippi, had been consolidated at Nashville. Blacksmith shops made use of 40 forges. Maintenance was performed on all 221 locomotives of the division as well as rail cars. A giant roundhouse with turntable was constructed for the mechanics. Thirty thousand U.S. government civilian employees in Nashville worked for the quartermaster or in the railroad shops. Most

The Master of Supplies

One of the unsung heroes of the Federal success in the Western Theater was the quartermaster responsible for keeping the beans and bullets moving along from Northern manufacturers to the fighting men at the front. That man was Major General James L. Donaldson, the quartermaster stationed at Nashville.

A native of Baltimore, he was the son of a lawyer, legislator, and officer who died in the War of 1812 while his son was an infant. Donaldson graduated from West Point in

1836 as a 2nd lieutenant in the artillery. He was assigned to the topographical service and saw duty in the Seminole War. Later, he built Fort Kent in Maine and drew the map upon which the "Disputed Territory" controversy was settled. During the Mexican War, he fought with General Taylor at Monterey and Buena Vista. He became a captain quartermaster in Mexico, then the Department of New Mexico. In 1861 he was promoted to major and brevet lieutenant colonel. He successfully completed a valiant, lengthy journey to protect much-needed supplies at his Santa Fe outpost.

On Nov. 10th, 1863, he became the Chief Quartermaster of the Department of the Cumberland, assigned the immense task of providing the besieged army at Chattanooga with provisions, forage, ammunition, and all other supplies to maintain that position, the troops being on less than half rations, and the animals in a starving condition. Subsequently and until the end of the war, Donaldson was assigned the task of forwarding from Nashville nearly all of the matériel for Sherman's Atlanta Campaign and March to the Sea. After the capture of Atlanta, Donaldson was called upon to provide for the Army of the Cumberland, falling back before Hood on Nashville. With the permission of General Thomas, Donaldson organized, drilled, and disciplined his quartermaster and commissary workforce as soldiers reporting for duty at the city's trenchlines during the Battle of Nashville. For this, he was breveted to major general. His chief assistant was Capt. J.F. Rusling. Donaldson served as quartermaster for the Department of Tennessee until 1866 and retired from the service on disability in 1869. Donaldson died in 1885 at the age of 72.

In praising Donaldson's services, General Thomas wrote to him, in part: "Joining me at Chattanooga, at the period when all looked gloomy and foreboding, you unraveled the intricate meshes then surrounding the Quartermaster's Department within my command, and restored system and order where confusion had triumphantly held sway. By the marked ability with which you administered the department from that time until the close of the late war, you greatly contributed to the success which crowned the efforts of the armies in the field ..."

A timber blockhouse along the Tennessee River showing the upper story set diagonally atop the first story, which is covered with earth. (Library of Congress)

of them had been recruited up North. One fourth brought their families with them, and most moved back North after the war, but not all.

"The shops at Nashville particularly were on a large scale," reported General David C. McCallum, general manager of the U.S. Military Railroads, "as at times one hundred engines and more than one thousand cars were there at once, it being the main terminal station of five hundred miles of road running from it east, south, and west."

Railroad engineering

The construction, maintenance, and repair of railroad beds and tracks, bridges and trestles, and locomotives and rolling stock required a substantial number of engineers, pioneers, immigrant and civilian workers, impressed fugitive slaves and freedmen, and soldiers from the ranks.

McCallum explained to Secretary of War Edwin Stanton: "The difference between civil and military railroad service is marked and decided. Not only were the men continually exposed to great danger from the regular forces of the enemy, guerrillas, scouting parties, etc., but, owing to the circumstances under which military railroads must be constructed and operated, what are considered the ordinary risks upon civil railroads are vastly increased on military lines. The hardships, exposures, and perils to which trainmen especially were subjected during the movements incident to an active campaign were much greater than that endured by any other class of civil employees of the Government—equalled only by that of a soldier engaged in a raid into the enemy's country. It was by no means unusual for men to be out with their trains for five to ten days, without sleep, except what could be snatched upon their engines and cars while the same were standing to be loaded or unloaded, with but scanty food, or perhaps no food at all, for days together, while continually occupied in a manner to keep every faculty strained to its utmost."

By July 1864, McCallum's responsibilities included 896 miles of track, 165 engines, 1,500 cars, and more than 10,000 employees in the Military Division of the Mississippi alone, according to author/historian Earl J. Hess. By the end of the war the western railroad crews constructed more than 18 miles of railroad bridges and 433 miles of track. Counting all military railroads in the Western Theater, McCallum's men operated 1,201 miles of track by war's end.

McCallum found it necessary to organize the Construction Corps in the Military Division of the Mississippi into seven divisions, with an engineer in charge of each and supervisors in control of each subdivision within the seven divisions. Men responsible for laying track constituted one subdivision, while those who constructed bridges made up another. Each subdivision was further divided into gangs, and the gangs into squads.

Rails attacked by guerrillas

When Gen. D.C. Buell captured Nashville in February 1862, the Confederates destroyed most of the Louisville & Nashville Railroad between Bowling Green and the Tennessee capital.

The railroads were attacked often by Confederate cavalry under the commands of Gen. John Hunt Morgan, Gen. Nathan Bedford Forrest, and Gen. Joseph Wheeler. In addition, smaller bands of partisans and guerrillas played havoc with the smooth operation of the railways. The most notable of these were Major Dick McCann and Capt. Duval McNairy.

In August 1862, Morgan successfully attacked the stockades at Pilot Knob, Saundersville, Drakes Creek, and Manskers Creek along the Louisville & Nashville R.R. and took 163 prisoners. The 20 Union infantrymen stationed at the Edgefield Junction stockade, however, repulsed three attacks over three hours and inflicted 26 casualties.

On Nov. 4th, 1862, the guns at Fort Negley opened fire on Forrest one-and-a-half miles down the Murfreesboro Pike. Due to weather conditions, the artillery could be heard in Bowling Green, Ky., according to historian/author Larry Daniel. That same day, Gen. William S. Rosecrans ordered McCook's Corps in Kentucky to relieve the city. His three divisions marched 72 miles in three days. Although low on rations, Negley's garrison in Nashville was in no real danger, newspaper hysteria to the contrary. The L&NRR tunnel south of Mitchellville remained badly damaged, so much so that the 35 miles to Edgefield Junction had to be covered by wagon. McCook hoped to repair the five bridges between Nashville and Gallatin by November 15th and then keep the 22 locomotives and 300 cars in Nashville constantly running to the terminus. Repairs to the entire road were not completed until November 26th.

A timber blockhouse on the Nashville & Chattanooga Railroad.

How to derail a train

In the year ending June 30, 1862, Confederates had destroyed or stolen 161 locomotives and 142 boxcars on the Louisville & Nashville RR alone.

Hess wrote about ruining the rails: "Dismantling the rails was systematic, when time allowed. The men were divided into parties, and the men of the first party distributed along the track, one man to each tie. At a signal the whole section of track was raised on edge and tipped over, ties on top. The ties were pried loose from the rails and the first party moved on to another section, while the second party stacked the ties and laid the rails over them. The ties were then set alight and when the rails were red-hot the third party, using pinchers or "railroad hooks," bent them around trees and also twisted them. The twist was important, for both the Southern and Northern repair crews became as adept at straightening rails as the soldiers were at bending them. Rails which were not bent in too small a "U" could be straightened, but a scientifically twisted rail had to go back to the rolling mill.

"The procedure for stopping a train, whether by guerrillas or detailed cavalrymen, was a minor science," wrote Hess. "The best way to do so was to choose a spot where the track made a sharp curve and remove the outside rail. That was guaranteed to produce a wreck."

Five days after the battle at Stones River, Dick McCann led irregulars on a raid that destroyed a locomotive and construction train on the Nashville & Chattanooga and captured the crew and guards. In retribution, the Nashville post commander, Gen. Ormsby M. Mitchell, ordered the 85th Illinois Infantry Regiment to burn all the houses and barns owned by McCann and his associate, Thomas Chilcut.

On April 10th, 1863, Gen. Wheeler attacked a Louisville & Nashville R.R. train nine miles northeast of Nashville with an artillery ambush from across the Cumberland River at 500-yard range. Twenty-one rounds destroyed the locomotive and killed most of the cargo in the 18 rail cars — horses and livestock. Wheeler also ambushed another train at Antioch south of Nashville

The Cowan Tunnel, or Cumberland Mountain Tunnel, looking at the south end. The tunnel was built by the Nashville & Chattanooga Railroad Company in 1852, a major engineering feat at the time. The tunnel was listed on the National Register of Historic Places in 1977. (Photo by Bryan MacKinnon.)

and captured 20 Federal officers and a large quantity of mail.

On Sept. 1st, 1863, Capt. McNairy, a Nashville Confederate cavalryman who organized and commanded partisan units behind the lines, led a raid on the Nashville & Northwestern R.R., burning wood cut for fuel and capturing black laborers.

In the late summer of 1864, Wheeler attacked Sherman's supply lines, cutting telegraph lines, burning crossties, and heating and bending rails on the Nashville & Chattanooga R.R.

On Sept. 1st, 1864, Gen. Lovell Rousseau moved out of Nashville and confronted Wheeler at Lavergne. Rousseau's men were driven two miles back toward Nashville, and three days later again engaged Wheeler south of Franklin on the Nashville & Decatur R.R., driving the Confederates southward.

Hazardous mode of travel

Travel by rail was dangerous, and not just because of Confederate raiders. On Sept. 28th, 1863, a steam locomotive exploded after leaving the Nashville station bound for Stevenson, Ala., while passing through the Broad Street cut. Fortunately, there were no reported fatalities.

On April 6th, 1864, a train enroute to Chattanooga exploded near Murfreesboro. The next day, a train bound for Louisville left the track in Kentucky after encountering a piece of equipment left by a construction crew. Three were killed and 60 injured on a siding near Gallatin in a massive pile-up. A southbound locomotive hit a stray horse on a bridge over the Duck River. One man was killed and several injured in an accident on the N&NWRR near Waverly.

Bridges over rivers, trestles over streams and depressions, and tunnels were attractive targets for the Confederates. The main railroad bridge over the Cumberland at Nashville was an impressive feat of engineering, built in the 1850s in a swing-span design with massive stone piers (which still stand today). The center span could pivot to allow passage of steamboats with tall chimneys. It had the longest draw of any railroad bridge at that time. On Feb. 19th, 1862, the retreating Confederates burned the wooden platform of the bridge. It was rebuilt by the occupying Federals and opened again to trains on June 11th. Buell ordered a stockade built on the Edgefield side of the river and that the bridge be fortified. Guard turrets were built at both ends of the bridge and the trestlework was planked, with loopholes allowing infantry on the bridge to fire at attackers.

There were five bridges between Nashville and Gallatin that had to be protected. Seven blockhouses were built between Nashville and Murfreesboro to protect bridges across the streams. In all, the U.S. military built 47 blockhouses on the Nashville & Chattanooga between those two cities to protect bridges and trestles. Such structures were built of wood and were vulnerable to heavy artillery. Overall, more than 160 blockhouses would be built along the railroads in Middle Tennessee and northwest Georgia.

A hospital train crossing a trestle on a run from Chattanooga to Nashville. (Theodore R. Davis, Harper's Weekly, Feb. 27, 1864.)

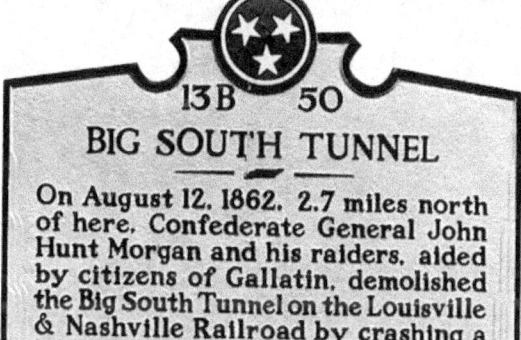

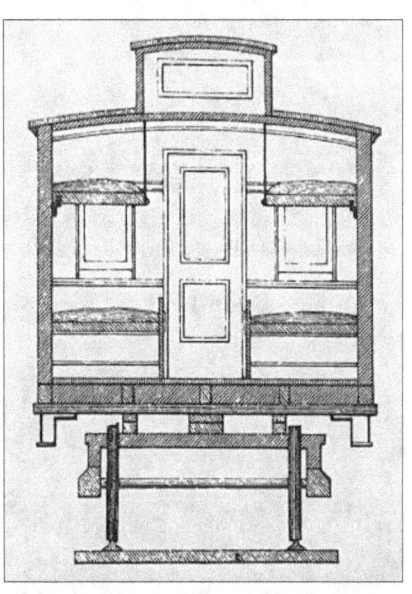

Transverse section of an Army of the Cumberland hospital car. Pattern of Drs. Cooper and Herrick. (OR Atlas)

Railroad tunnels vulnerable

One vital railroad structure was the South Tunnel located six miles from Gallatin, north of Nashville on the Louisville & Nashville R.R. between Fountain Head and Gallatin running through the Highland Ridge. (The North Tunnel was located at Muldraugh's Hill near Louisville. To confuse matters further, the South Tunnel at Gallatin had two tunnels, North and South, separated by 400 feet.) At Gallatin, the northern tunnel is 945 feet long, a long curve almost 100 feet below the summit of the hill. There is a space between the two tunnels about 390 feet long, then the southernmost tunnel is 600 feet long and some 165 feet below the summit. The two tunnels and their approaches through solid rock are more than three miles long. In those days of cheap labor, the tunnels cost the rail company more than $200,000. All the work was performed with star hand drills, blasting powder, and horse and mule teams to haul away the rock.

On Aug. 12th, 1862, General Morgan captured Gallatin, the Union force there, and destroyed a 29-car train along with the water tank and two bridges. Thinking the raiders gone, workmen were sent back to Gallatin. The cavalry returned in force and the workmen and their guards were driven almost to Nashville. Morgan captured the Union guard at South Tunnel hill, which left 46 miles of railroad north of Nashville unguarded, with all bridges out and the telegraph wires destroyed.

By this time, Gen. Braxton Bragg's Confederate forces had invaded Kentucky and were almost to Louisville. All the railroad except for 26 miles near Louisville were in rebel hands. Every bridge and trestle all the way to Nashville was closed. Morgan fired several freight cars and rolled them deep into the southernmost tunnel, where the supporting timbers burned and 800 feet of tunnel was filled to a depth of 12 feet with wreckage, rock, earth, and debris. It took months to clear the tunnel and rebuild the track;

(Continued on Page 198)

The Nashville & Decatur Railroad bridge over the Duck River, showing stone piers, box-type construction, and trestle approaches. (Library of Congress)

The "cornstalk and beanpole" railroad bridge at Whiteside, Tenn., showing four tiers of trestles, two blockhouses (one with a partially completed roof), a garrison encampment, and stockade – on opposite side of gorge. (National Archives)

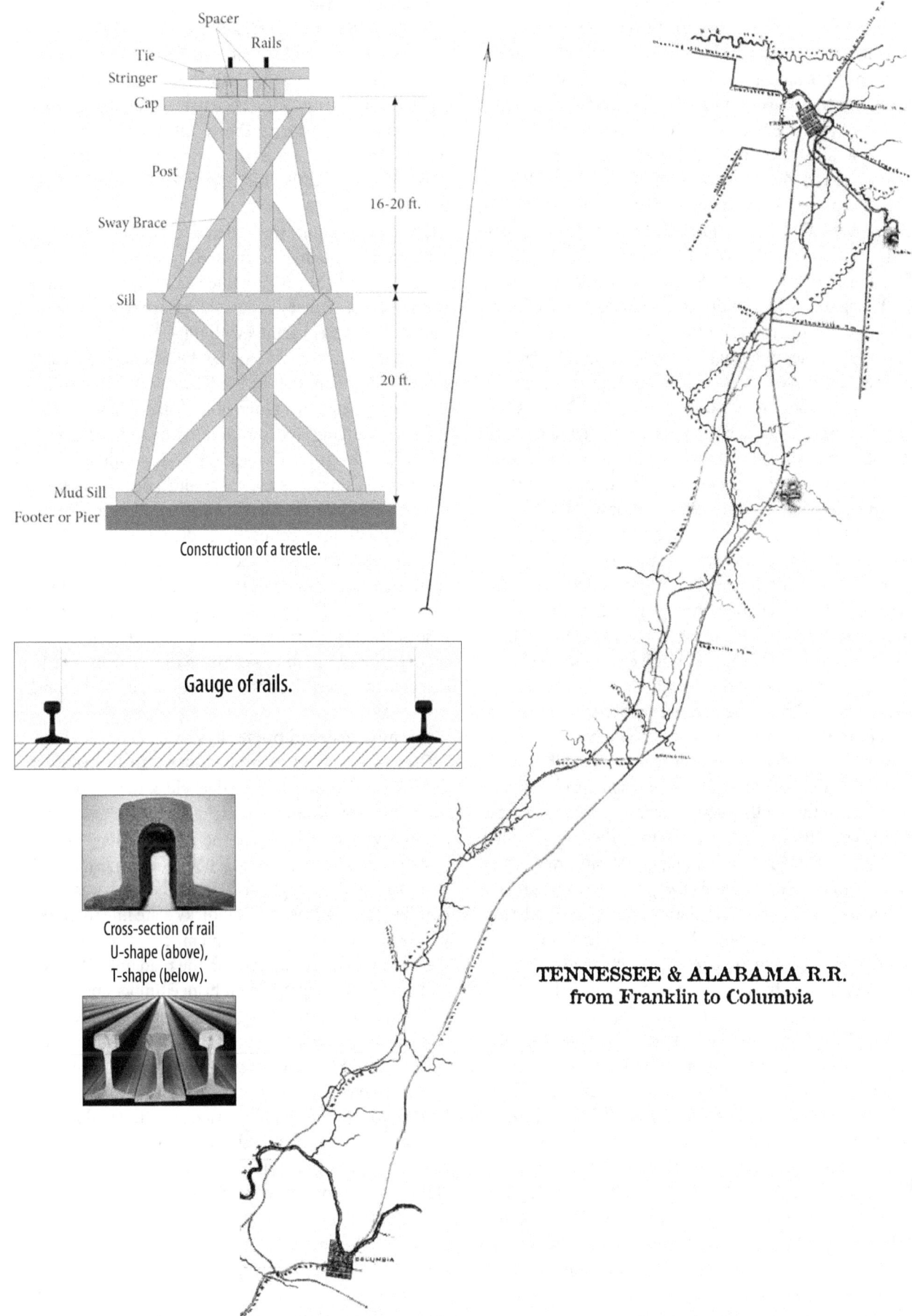

Construction of a trestle.

Gauge of rails.

Cross-section of rail U-shape (above), T-shape (below).

TENNESSEE & ALABAMA R.R. from Franklin to Columbia

(Continued from Page 195)

the tunnel was closed for 98 days. The quartermaster at Nashville was forced to build a railroad siding and unloading ramp at Mitchellville, 10 miles north of the disabled tunnel, and transport supplies 35 miles the remainder of the way to Nashville in guarded convoys of up to 500 wagons. Eventually the tunnel was cleared, and passage reopened to rail traffic on Nov. 26th. By mid-May 1863, infantry was placed at Muldraugh's Hill (328), Elizabethtown (236), Lebanon (985), Munfordville (2,372), Bowling Green (1,843), Gallatin (3,474), and Nashville (1,325).

Around Christmas 1862, Morgan's men burned the two 500-ft-long trestles at Muldraugh's Hill near Elizabethtown, Ky., and closed rail traffic for five weeks.

Blockage of the tunnel and low water on Cumberland River restricted the flow of supplies during the autumn of 1862, forcing troops to be put on half rations. Soldiers patroled the streets of Nashville to prevent theft and looting.

On March 4-5th, 1863, skirmishes occurred at and near Chapel Hill, Unionville, Spring Hill, and Thompson's Station. At Unionville, the 17th Pennsylvania and the 4th Michigan Cavalry attacked Col. Robert Russell's 20th Tennessee cavalry. The Confederates lost 50 killed and 180 wounded, many by saber strokes; 58 were taken prisoner. At Spring Hill, Confederate cavalry under Earl Van Dorn and Forrest drove off Federal cavalry on the 4th, then at Thompson's Station surrounded and engaged the remaining infantry. After heavy fighting on March 5th, the Federal garrison surrendered. The Confederate victory disrupted Gen. Philip Sheridan's move against Columbia.

On Oct. 10th, 1864, the guerilla leader Capt. Ellis Harper of Portland (then called Richland) and a hundred of his men attacked the South Tunnel, tearing up the tracks, and burning crossties and wood cut for fuel. They captured four of the U.S. Colored Troops walking their beat through the tunnel, with two others getting away and alerting the Gallatin garrison. Harper had no means of caring for prisoners and feared that gunfire would bring the Federals even quicker. He ordered them to be killed with an axe.

Another impressive structure was the Cumberland Mountain Tunnel near Cowan, Tenn. The Cowan Tunnel was built in 1849-52 by the Nashville & Chattanooga Railroad Company. At 2,200 feet long, it was a major engineering feat. The tunnel was listed on the National Register of Historic Places in 1977. It is still operational.

Work was performed by black slaves, Irish immigrants, and local workers, with Swiss engineers. Three ventilation shafts approximately 170 feet deep were bored during the construction to facilitate air circulation, provide additional work areas, and to enable evacuation of steam and smoke from the locomotives during use. The grade at mountainous Cowan is so steep that "pusher" engines or locomotives had to be used to push trains up the track, a practice still performed today.

In April 1864, Chief of Staff Brig. Gen. William D. Whipple recommended that 1,460 troops be stationed at 45 blockhouses along the Nashville & Chattanooga Railroad, running from Mill Creek to Chattanooga Creek, with garrisons at Murfreesboro, Tullahoma, Stevenson, and Bridgeport. The troops to be stationed included the 23rd Missouri, 115th Ohio, 33rd and 85th Indiana, and 31st Wisconsin.

The 23rd Missouri was ordered to McMinnville to relieve the 18th Michigan, which then rejoined its brigade.

Col. Coburn's brigade rejoined its division, with three companies of artillery assigned to Murfreesboro. The convalescents were armed with muskets.

Hess wrote about two small but quick raids — "Captain T. Henry Hines of the 9th Kentucky Cavalry (C.S.) led 14 of his men in a raid behind Union lines in February 1863, ranging across middle Tennessee and into Kentucky. The group traveled for 21 days and destroyed a train of 21 cars, plus a steamer and a depot. Hines estimated his men burned half a million dollars' worth of property without meeting any Federal troops.

"A guerrilla gang of 75 men out for plunder rather than patriotism stopped a train near Franklin, Kentucky, in March 1863. They robbed the passengers and stole the case belonging to express agents on board. A detachment of the 129th Illinois came on the scene in time to disperse the guerrillas and recovered much of the loot."

New route ordered opened

On Aug. 4th, 1864, Sherman ordered that Chief Engineer W.W. Wright open the Edgefield & Kentucky Railroad as an additional route of supply to Nashville. Seven days later, Wright and the First Division of the Construction Corps, under L.H. Eicholtz, division engineer, arrived at Springfield. Most of the work was building bridges from the Kentucky state line to Clarksville, which was completed September 16th. The construction force remained on the road until October 16th, employed in getting out bridge timber and crossties, and grading and laying a track with sidings 6,765 feet long from the main line to the levee at Clarksville. On October 25th, W.R. Kingsley, division engineer, was appointed engineer of construction and repairs. On March 4th, 1865, a freshet carried away the Red River bridge, which was rebuilt by March 25th, only to be carried away again on April 7th and not rebuilt. On May 20th, a freshet took away the Sulphur Fork bridge, which was quickly repaired.

Inspection finds railway deficiencies

In the spring of 1863, Secretary of War Edwin Stanton directed General Hermann Haupt to inspect the Western

Secretary of War Edwin Stanton

Gen. John Milroy

General David C. McCallum

Theater lines and report back. Haupt assigned F.H. Forbes, a Massachusetts newspaper reporter, to carry out the inspection. Forbes made a lengthy investigation and reported that waste, inefficiency, and speculation were hampering the railroads. Stanton then sent Federal quartermaster Montgomery Meigs to inspect the supply lines. In December 1863 General D.C. McCallum was directed to move the construction corps into Tennessee and make needed repairs. In February 1864, U.S. Grant relieved John B. Anderson of his duties and appointed McCallum as general manager of U.S. Military Railroads in the Western Theater. Adna Anderson was appointed general superintendent of transportation and maintenance of roads in use, and W.W. Wright, chief engineer of construction in the military division of the Mississippi. By 1865 the Construction Corps had built facilities for 90 engines, plus their own barracks and mess facilities.

McCallum reported, "The road between Nashville and Chattanooga is still in bad condition, and, in my opinion, no energetic means have been taken to put it in repair…The track was laid originally on an unballasted mud-road bed in a very imperfect manner, with a light U-rail on wooden stringers, which were badly decayed and caused almost daily accidents by spreading apart and letting the engines and cars drop between them."

At the time Federal engineers were constructing Fort Negley, the railroad companies were replacing the old "U" rail with the more efficient "T" rail. The old "U" rails were to be melted down for reuse, but many were used for cladding at artillery positions and blockhouses and on gunboats.

The "T" rail was held in place by spikes driven into each tie close to the rail. The ends of rails were joined by chairs, in which the rails sat, or by bolting straps of iron (fish plates) to the sides of the joining rails and bolting them together. T-rail would flake in spots as it wore, according to historian David L. Bright. This flaking (lamination) could be fixed one of four ways: on the spot with a blacksmith forge, by replacing the rail and treating the flaking in a company shop, by removing the flaking section and welding good rail in its place, or by replacing the rail and re-rolling the rail at a rolling mill.

More locomotion needed

McCallum reported that the 47 locomotives and 437 freight cars on hand at the beginning of 1864 were totally inadequate. There should be one locomotive and 12 railcars for every two miles of rail in use. He estimated that there should be 200 locomotives and 3,000 boxcars made available. In response, Sherman ordered Anderson to impress rolling stock from the L&NRR, Louisville & Lexington RR, and Kentucky Central RR, yeilding 21 engines and 195 cars. Stanton authorized McCallum to buy more stock from Northern manufacturers. The first three months of 1864, the U.S. military railroad took delivery of 140 new locomotives and 2,159 cars.

All locomotives and rolling stock had to be manufactured for five-foot (wide) gauge of tracks (Northern rails were standard gauge–4 ft.-8.5 in.).

M.W. Baldwin & Co. of Philadelphia built four steam locomotives (2-6-0 wheel arrangement) in 1860 and delivered them to the Louisville & Nashville Railroad in 1861. They weighed more than 32 tons each and featured 18-inch cylinders and 52-inch drivers. The shop numbers 982, 983, 986, 988 became L&N's road nos. 35-38. M.W. Baldwin produced more than 100 locomotives during the war and remained in business until 1956. In 1862, during the Confederate retreat, the L&N lost five new 4-4-0 locomotives built by Moore & Richardson (all were returned in 1865-66). More locomotives were desperately needed in Tennessee. In 1863 Stanton confiscated three locomotives being built by Baldwin for the old NY & Harlem line owned by Cornelius Vanderbilt. In early 1864, Stanton

Guard at stockade overlooking railroad bridge; pass to ride the US Military Railroad.

appealed to the builders of locomotives to supply the need in Nashville.

Trains running from Chattanooga to Nashville carried sick and wounded soldiers, discharged vets, refugees, captured Confederates, freed Negro slaves, and materiel sent to the rear. During 1863-64, the construction crews of the U.S. Military Railroad replaced 130 miles of the 151 miles of the Nashville & Chattanooga Railroad. The U-shaped rails were replaced with T-shaped rails, and 45 water tanks were erected. All at a cost of more than $4 million.

On Oct. 12th, 1864, the newly formed 11th Minnesota Regiment was assigned to guard the L&N Railroad from Edgefield Junction north to the Kentucky line. The regimental headquarters and three companies were based in Gallatin, described as "a lively little city." Companies were placed at Edgefield Junction (modern Amqui), Saundersville (between Goodlettsville and Gallatin), Richland (modern Portland), Buck Lodge (north of Portland), and Mitchellsville (on the state line). Two companies were stationed at South Tunnel. The regiment served on the railroad line until June 1865, when they were relieved by a regiment of U.S. Colored Troops. The 11th Minnesota returned to St. Paul and mustered out the next month.

Hood moves north to Nashville

Providing protection from sporadic cavalry raids was nerve-racking enough but all that changed in the fall of 1864 when General John Bell Hood led the Army of Tennessee north into Middle Tennessee in the hopes of recapturing Nashville.

Gen. Rousseau moved his headquarters from Nashville to Murfreesboro, bringing along the 61st Illinois. The 4th Michigan arrived from Tullahoma and the 21st Indiana battery from Columbia. The 8th Minnesota was routed to Murfreesboro after fighting the tribal Sioux in Dakota Territory. Other units ordered to Fortress Rosecrans were the 3rd, and 29th Michigan, the 174th and 181st Ohio, and the new 140th Indiana.

In November 1864, General Van Cleve was in charge of defending the N&CRR between Nashville and Murfreesboro. His brigade consisted only of the 115th Ohio and three batteries. The Buckeyes were used to garrison the seven blockhouses protecting the railroad bridges. His three batteries were stationed in Fortress Rosecrans, which was commanded by Major Frederick Schultz, who relieved Lt. Col. Walker E. Lawrence on Nov. 16th.

Major General Robert H. Milroy at Tullahoma defended the railroad from Murfreesboro to Decherd with the 177th and 178th Ohio, 12th Indiana dismounted cavalry, and the 13th Battery, New York Light Artillery, whose ten guns ranged from 24-pounder rifles to 3-inch ordnance rifles.

Col. Willard Warner garrisoned Decherd with the 180th

Chief Railway Engineer W.W. Wright. (Library of Congress)

The Railroad Regiment

The 89th Illinois was known as the Railroad Regiment, the men recruited from railroads around Chicago in August 1862. Led by Col. Charles T. Hotchkiss of the Galena & Chicago Union Railroad, the regiment fought at Stones River, Franklin, and Nashville, losing up to 50 percent in casualties. The 89th Illinois fought at Peach Orchard Hill under Abel Streight's 1st Brigade of the 3rd Division (Beatty), IV Corps (Wood).

Ohio. Major John C. Hamilton defended Shelbyville with the 178th Ohio.

Posted at Stevenson, Ala., was Colonel Vladimir Krzyzanowski, who commanded five infantry regiments (6th Kentucky, 58th and 68th New York, and 106th and 180th Ohio) and two batteries (Battery K, 1st Ohio Light Artillery and 9th Battery, Ohio Light Artillery) and defended the railroad from Decherd to Bridgeport. A Polish noble, Krzyzanowski took part in the 1848 uprising against Prussia and left Poland after its suppression. He worked as a civil engineer and surveyor in Virginia, concentrating on railroad work. During the Civil War he enlisted in the U.S. Army, recruited a company of Polish immigrants, and became colonel of the 58th New York Volunteer Infantry Regiment, listed in the official Army Register as the "Polish Legion." He was a cousin of Polish composer Frederic Chopin.

The Battle for Blockhouse No. 7

Hood sent two divisions of Gen. Nathan Bedford Forrest's cavalry corps to attack and/or isolate the garrison at Fortress Rosecrans in Murfreesboro in hopes of drawing the Federal army out of Nashville and into open battle. Part of the plan was to attack the blockhouse garrisons protecting the bridges of the Nashville & Chattanooga Railroad, of which there were seven between Nashville and Murfreesboro.

The division of Brig. Gen. Abe Buford was given the task of ripping up the rails and attacking the three blockhouses closest to Nashville, those guarding the Mill Creek bridges.

On Dec. 2nd, 1864, a train carrying portions of Major Gen. James B. Steedman's command from Chattanooga to Nashville, namely the 44th USCT and Cos. A and D of the 14th USCT, left Murfreesboro following a delay due to a train derailment. In charge of the train was the 44th's colonel, Lewis Johnson.

At about 11:00 am, Buford's troopers, some clad in dusty blue uniforms, approached Blockhouse No. 2 on Mill Creek, occupied by a detachment of the 115th Ohio under Lt. George D. Harter. While the Confederate horsemen surrounded the blockhouse, the train carrying Johnson's men rolled into view. The train approached the blockhouse slowly. Just when it pulled onto the bridge, a hidden Confederate battery opened fire, hitting the locomotive and first car and causing several casualties. Johnson ordered

Company Man

James Guthrie, a wealthy Kentucky politician, served as president of the L&N RR from Oct. 1860 to June 1868. He had served as Secretary of the Treasury in the Pierce administration. The rail line ran 269 miles, and consisted of 30 locomotives and 300 freight and passenger cars, all of 5-foot gauge. Although a Unionist, Guthrie was more loyal to his business and investors. He quarreled with William P. Innes, superintendent of railroad transport under Rosecrans, and he was accused of giving more priority to lucrative commercial contracts than government military contracts. In turn, Guthrie accused the Federal officials of violating contracts. Guthrie did not get along with John B. Anderson, an L&N transportation department man who was placed in charge of the military railroads in the Ohio, Tennessee, and Cumberland departments in 1863-64.

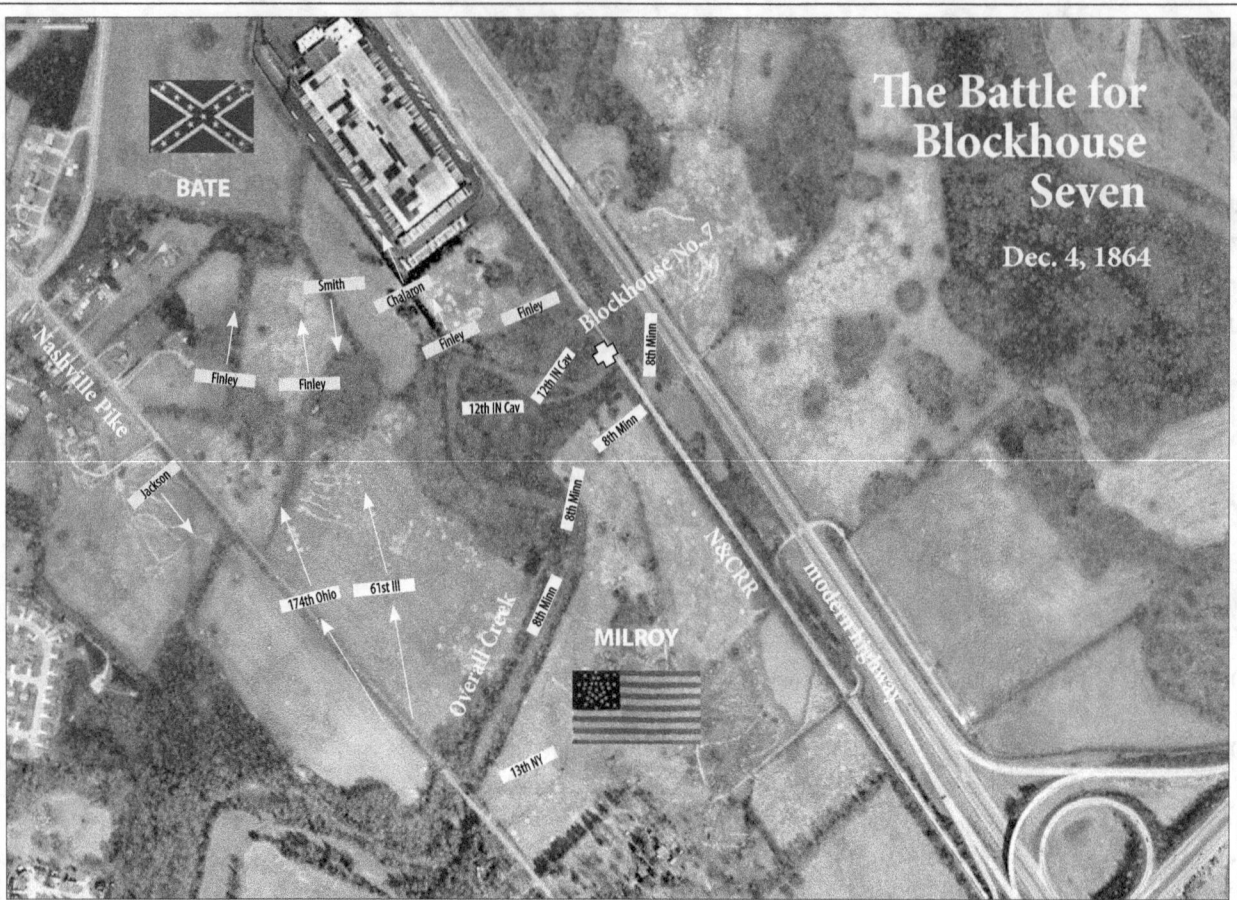

Blockhouses on Nashville & Chattanooga Railroad between Nashville and Murfreesboro:

No. 1—NE of Mill Creek crossing about one mile south of Dogtown.

No. 2—Half mile SE of No. 1 at Mill Creek crossing.

No. 3—1.5 miles SE of No. 2 near Mill Creek Pike, the creek and the railroad, two miles north of Antioch.

No. 4—Bridge at Hurricane Creek.

No. 5—Bridge across Hart's Branch, just north of downtown Smyrna.

No. 6—Bridge across Stewart's Creek.

No. 7—Bridge over Overall Creek, four miles NW of Fortress Rosecrans.

In April 1864, Chief of Staff Brig. Gen. William D. Whipple recommended that 1,460 troops be stationed at 45 blockhouses along the Nashville & Chattanooga Railroad, running from Mill Creek to Chattanooga Creek, with garrisons at Murfreesboro, Tullahoma, Stevenson, Ala., and Bridgeport, Ala. Troops to be stationed included the 23rd Missouri, 115th Ohio, 33rd and 85th Indiana, and 31st Wisconsin.

He recommended the following arrangement of troops along the line from Nashville southward:

Three batteries in forts at Nashville, already in position. This in addition to the infantry.

The 23rd Missouri is to be ordered to McMinnville to relieve the 18th Michigan, which regiment will then join its brigade.

Col. Coburn's Brigade to join its division. Three companies of artillery to be assigned to Murfreesboro. The convalescents to be armed with muskets.

Gen. Rousseau to man the blockhouses from Nashville to Murfreesboro.

Two regiments at Murfreesboro, and in blockhouses as far as Tullahoma.

One regiment at Tullahoma.

One regiment at Stevenson.

Bridgeport, two regiments proposed, although it requires 3,000 men on both sides of river, and three batteries.

his 350 men off the train and into the blockhouse, which he thought had been abandoned.

When he realized there wasn't enough room inside the blockhouse for his men and the Buckeyes already inside, he positioned his black troops around the blockhouse, which then came under artillery fire. Col. Johnson noticed that his men needed to capture a high hill 500 yards to the east occupied by Buford's men, who were sniping at them. Unsuccessful in their attempt to storm the hill, the USCT dug in at the bottom of the slope.

After several hours of skirmishing, Harter was able to supply the black troops 2,000 more rounds of ammunition, which they used to hold off the Confederates. At 5:00 pm, the rebels managed to haul a field artillery piece atop the hill and fire at the blockhouse, knocking the lookout tower to pieces. Each time they fired, the recoil would knock the gun back out of sight. One shell exploded inside the blockhouse, killing the railroad conductor and wounding several of the Ohioans. The structure was reduced to a "ruinous condition." Under the cover of darkness, the USCT fell back to the blockhouse. "The rebels had enough men to just eat us up," Pvt. John Milton recalled.

Aware that his men were low on ammo and worried that surrender might produce "butchery" similar to Fort Pillow, Johnson decided to fight his way back to Nashville. If the Federals were going to send reinforcements, they would have already arrived, he reckoned. Of the 332 men engaged, a dozen had been killed, 96 wounded, and 57 were missing. He left the surgeon and chaplain of the 44th USCT in charge of the wounded. Leaving the fort at 3:30 am the Federals caught the Confederates resting and made their way to Nashville, arriving at daybreak. That morning, Buford's men torched the blockhouse, the Mill Creek railroad bridge, the abandoned train and prepared to move against Blockhouse No. 1, a mile closer to Nashville. Incredibly, Johnson did not warn the occupants of Blockhouse No. 1 — the 110 men of the 115th Ohio commanded by Lt. Jacob N. Shaffer — of the debacle at Blockhouse No. 2.

At Buford's headquarters at the Nashville Insane Asylum, Forrest ordered Buford to send one of his brigades against Blockhouse No. 1 and the other against Blockhouse No. 3. Artillerymen pounded Shaffer's blockhouse with shells while those inside returned small-arms fire. The troopers repeatedly demanded the surrender of the blockhouse, but Shaffer refused. By late afternoon, the blockhouse had been pounded to pieces. Ten were dead and 20 wounded. The Federals feared they had been abandoned by their comrades. Under the fifth flag of truce, Shaffer surrendered his command. After burning the blockhouse and bridge, the Confederates returned to camp.

At Blockhouse No. 3 near Antioch, Capt. Denning N. Lowrey held out for 36 hours under an artillery barrage (90 shells hit the log structure) before surrendering his com-

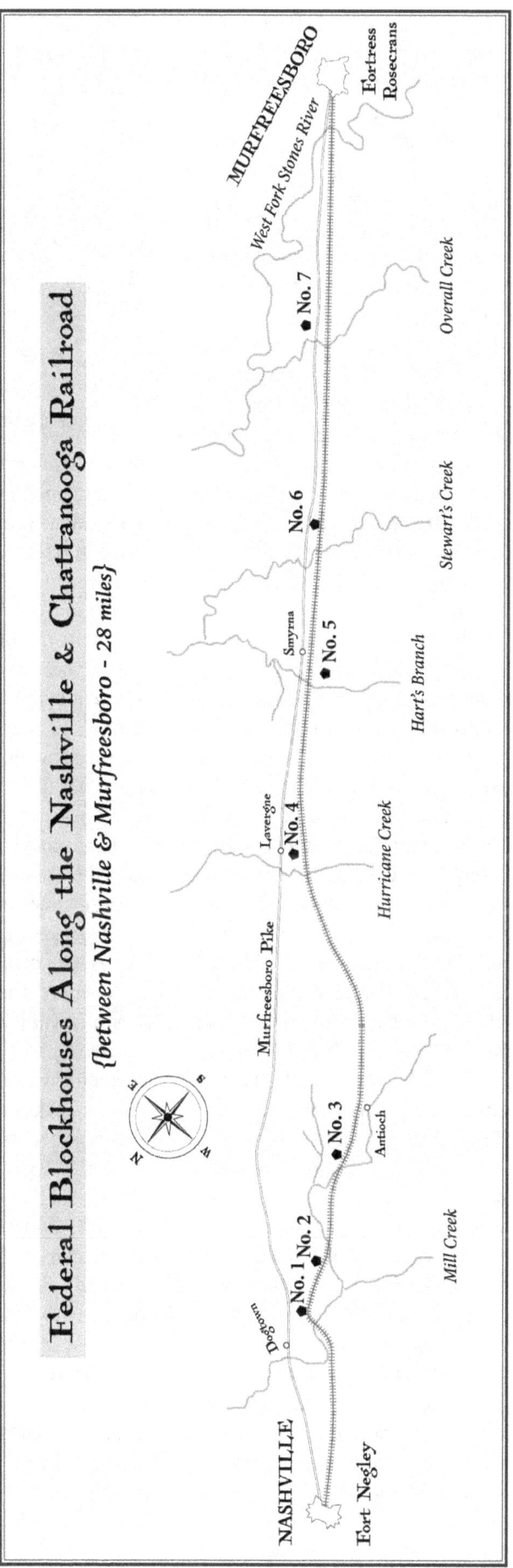

Federal Blockhouses Along the Nashville & Chattanooga Railroad
{between Nashville & Murfreesboro – 28 miles}

mand, 32 men of the 115th Ohio.

On December 4th, Hood ordered Forrest to move on Murfreesboro the next day, after a day of rest. Moving out the Murfreesboro Pike and crossing Hurricane Creek, Forrest ordered Brig. Gen. William H. "Red" Jackson to attack a Federal redoubt on a hill west of Lavergne. Buford would move against Blockhouse No. 4, guarded by the 115th Ohio under Sgt. William McKinney. On the evening of December 4th, Capt. Lewis F. Hake, garrison commander at Lavergne, received a telegram from General Thomas ordering the abandonment of all blockhouses. The telegram had been sent on December 1st, taking three days to arrive! Hake sent couriers down the tracks in both directions to relay the news.

Buford's men surrounded Blockhouse No. 4 before it could be abandoned. Under a white flag, the Confederates accepted the surrender of Sgt. McKinney. The blockhouse and Hurricane Creek bridge were burned.

Meanwhile, Jackson's men had surrounded the hilltop redoubt near Lavergne and demanded its surrender. Hake was loading wagons and preparing to leave at the time. Surrounded by troopers and artillery, Hake chose capitulation over annihilation. He surrendered his 72 men, two field pieces, 25 horses, wagons, and "a considerable supply of commissary and quartermaster stores."

Under Hood's orders, the small infantry division of Major Gen. William B. Bate was directed toward Murfreesboro to act in coordination with Forrest's cavalry. On December 4th, three regiments of Col. Robert Bullock's Florida brigade moved against Blockhouse No. 7 near Overall Creek, about five miles northwest of Murfreesboro. Pickets belonging to the Fifth U.S. Tennessee cavalry were driven in. The Washington Artillery (CSA) of Louisiana unlimbered their three 12-pounder Napoleons and began shelling the blockhouse at 11:00

Members of the railroad's construction corps at Chattanooga. (Library of Congress)

am. The artillery was commanded by Lt. J. Adolph Charlaron. Bate's brigade of Brig. Gen. Thomas B. Smith was placed in reserve, while the remaining brigade, that of Brig. Gen. Henry R. Jackson of Georgia, was ordered to tear up the railroad tracks.

Blockhouse 7 was occupied by a detachment of the 115th Ohio under Lt. Henry A. Glosser. The blockhouse was reinforced by the 13th Indiana Cavalry of Col. Gilbert M.L. Johnson, which had been ordered out of Fortress Rosecrans to reconnoiter the Nashville Pike. Soon the Federal cavalrymen were engaged with Bullock's Florida infantrymen. Johnson had his men dismount and seek cover in the cedars. The Louisianans stopped shelling the blockhouse and began firing at the Federal cavalrymen. Johnson sent a message to his superior back at the fortress. Rousseau subsequently sent a "flying column" to reinforce Johnson which consisted of three regiments (61st Illinois, 8th Minnesota, and 174th Ohio) and the guns of the 13th Battery, New York Light Artillery, all under the command of Major Gen. Robert H. Milroy. At 2:00 pm the flying column reached Overall Creek, and the artillery quickly threw their pieces into action. They engaged in an artillery duel with the rebel guns 900 yards away on the other side of the creek. Milroy determined to take the initiative. The 8th Minnesota under Col.

Minor T. Thomas was to advance to the blockhouse and send his men over the railroad bridge, if possible. They were covered by skirmishers of the 61st Illinois. Thomas advanced his men to the blockhouse, possible only because the skills of the rebel sharpshooters were "deplorable." They then crossed over the bridge, followed by the 174th Ohio, commanded by Col. John S. Jones, under a galling fire. Under the cover of gunsmoke and falling darkness, Milroy decided to use the 174th Ohio to rush the Confederate artillery battery. The 13th Indiana cavalry under Johnson thundered across the bridge, swung to the right, and encircled the hill which harbored the rebel guns. Bate called up the reserve brigade under Smith. The Washington Artillery turned their guns and blasted double-canister into the ranks of Indiana cavalry, which recoiled under the pressure. Retreating back to the creek, the Federal cavalrymen fell under friendly fire from riflemen on the east bank. Johnson reformed his regiment and prepared to attack the enemy artillery again. The 174th Ohio went on the attack, throwing several volleys into Bullock's Floridians, who were "rocked back on their heels." Col. Bullock was severely wounded.

Returning from their demolition duties, Jackson's men were ordered by Bate to congregate on the Nashville Pike. First they were met by the re-

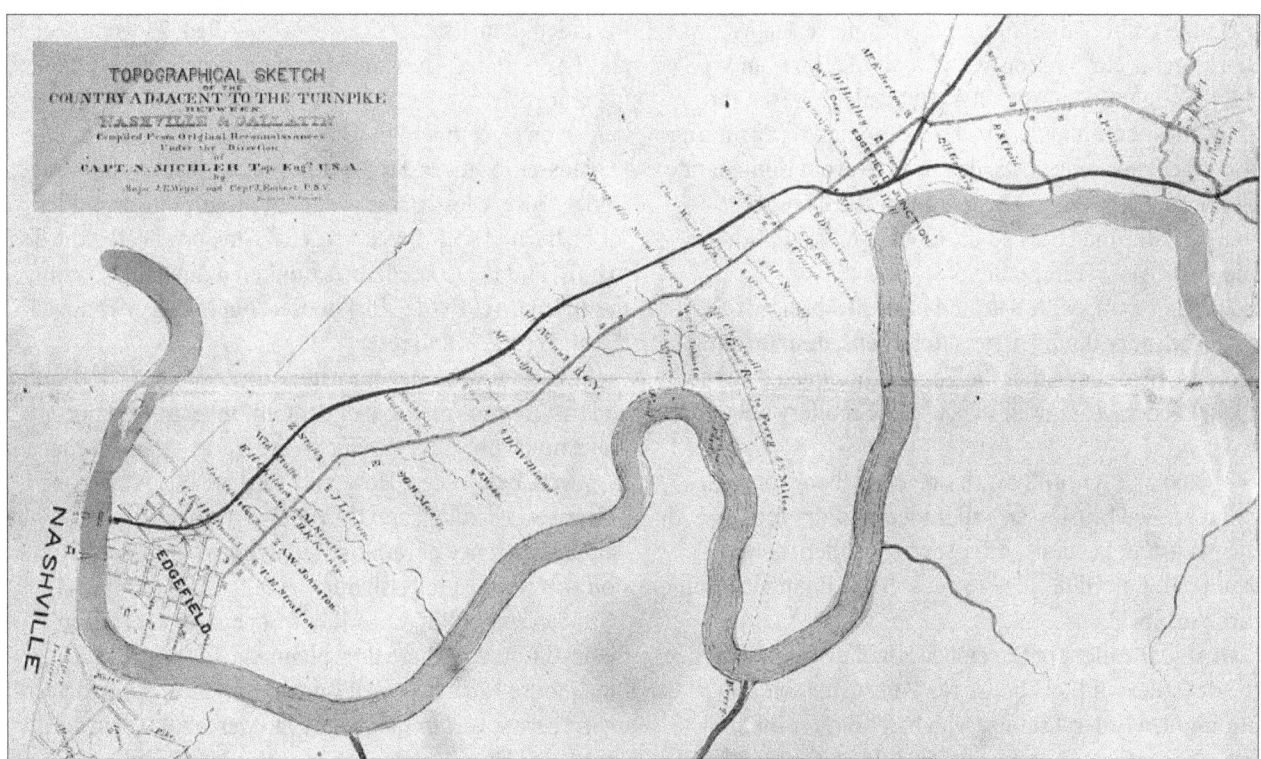

Sections of one continuous map drawn by J.E. Weyss of the Cumberland River, the turnpike, and the Louisville & Nashville Railroad (darker line) between Nashville/Edgefield and Gallatin. The map shows where the railroad splits at the Edgefield Junction (now known as Amqui). (Library of Congress)

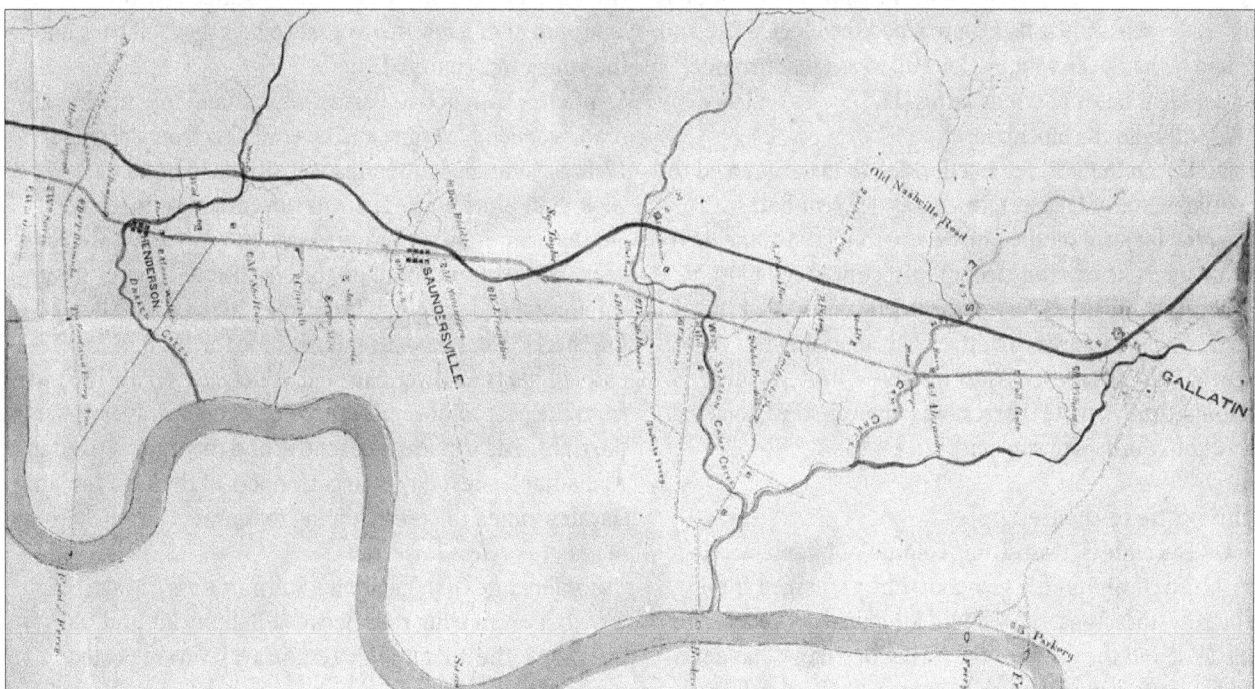

treating Floridians, then they sent a massive volley into the attacking Buckeyes. At this point, Milroy decided to call back the 174th Ohio due to the darkness and Bate's superiority (relayed to Milroy by several Confederate prisoners). The 8th Minnesota, plastered by the artillery, never did find their way across the bridge.

Johnson's troopers advanced upon the artillery position but the gunners had already retreated. The Federals held the bridge for several more hours, then retreated back to Fortress Rosecrans at 1:00 am. Glosser's men remained at Blockhouse No. 7 even though it had been hit by artillery shells 31 times. Bate failed to notice the Federal retreat. He was afraid they might maneuver behind him during the night so he pulled his command back to Stewart's Creek.

During the fight for Blockhouse No. 7, the Confederates lost 15 killed, 59 wounded, and 13 missing. Milroy reported capturing 20 prisoners, and losing 12 killed and 95 wounded.

On December 5th, Bate prepared to move against Blockhouses No. 5 (Capt. William M. McClure) and No. 6 (Lt. John S. Orr). Shortly after daybreak, however, the blockhouse commanders received Thomas' tardy message and evacuated their posts, detoured to avoid Confederate patrols, and arrived at Murfreesboro at 2:00 pm. Of course, Bate's men burned the blockhouses and bridges and destroyed several miles of track.

Forrest met up with Bate and informed him that they would advance upon Fortress Rosecrans, despite Bate's protests. Bate stated that the fort was occupied by up to 10,000 Federals and well-defended by artillery but to no avail.

At 2:00 pm, Buford's cavalry reached Blockhouse No. 7, still occupied by Glosser, and traded artillery rounds with the persistent Federals. After several artillerists were wounded, Buford decided to allow Bate's infantry to capture the blockhouse.

During the night of December 5th, Bate tried to induce the defenders of Blockhouse No. 7 to surrender. About midnight, a Federal soldier who had been captured by the rebels appeared at the fort. He told Lt. Glosser that if he would surrender General Bate would parole the entire garrison. Nothing happened until 12 hours later — a Confederate with a white flag approached the blockhouse and stated he had been sent by Col. Hill to ask for surrender under Gen. Bate's previous terms. Lt. Glosser replied, "We will hold the blockhouse."

On December 6th, Forrest decided to make a forced reconnaissance of Fortress Rosecrans. He sent Bate's infantry forward on the Murfreesboro Pike, advancing to three-quarters of a mile from the northwestern most lunette of the fortress. Forrest then determined that the fortress was too strong to attack. His new plan was to prevent Rousseau's forces at Murfreesboro from joining Thomas in Nashville (Forrest was unaware that Rousseau was under orders to stay put).

The Battle of the Cedars

On December 7th, a strong column of infantry, artillery, and cavalry under General Milroy emerged from Fortress Rosecrans, moved into Murfreesboro, veered to the west and then toward the Nashville Pike. What developed was a heavy skirmish known as the Battle of the Cedars, won by the Federals as portions of Bate's infantry fled in the face of a Federal charge (the details of which are beyond the purview of this publication). Eventually Milroy withdrew back into Fortress Rosecrans as Abe Buford's Confederate troopers attacked the town of Murfreesboro from the north. Rousseau launched a vigorous counterattack and drove the enemy cavalry from the town square.

On December 7th, the Federals lost 22 killed and 190 wounded. Bate listed his losses as 19 killed, 73 wounded, and 122 missing. The Confederate cavalry leaders did not file reports.

Also on December 7th, Col. B.H. Hill tried to persuade Lt. Glosser to surrender Blockhouse No. 7 at Overall Creek. Under truce, Glosser was given ten minutes to surrender "or be burned with Greek fire, etc." Glosser rejected the demands, and the Confederates hunkered down, blockading the approaches to the blockhouse, "hoping to wear down the bluecoats' will to resist."

The next day the warm weather turned to sleet and snow and the ground froze. Bate's demolition teams did not make much progress tearing up the tracks. Many of the men were barefooted. However, Murfreesboro's storehouses were overflowing, and the rebels drew a "superabundance of rations." Soon, however, Bate's division was recalled to Nashville.

On December 9th, Col. Hill's troopers advanced into Bedford County, where they planned to attack N&CRR Blockhouse No. 9 near Bell Buckle. The 115th Ohio detachment there was commanded by Lt. Merchant S. Hurd. Hill demanded Hurd's surrender, to which Hurd replied "if they wanted the blockhouse, they must come and get it." Hill sent his dismounted troopers to attack (without artillery) and after a few ineffectual volleys they fell back and the attack was canceled.

Infantry assigned to Forrest's command tore up the railroad between Lavergne and Overall Creek on December 12th. Several more attempts were made to induce Lt. Glosser at Blockhouse No. 7 to surrender but to no avail. On that same day at dusk, Rousseau at Fortress Rosecrans sent a 15-car train south on the N&CRR headed to Stevenson for needed supplies. Thirty men of the First Michigan Engineers (Co. L under Lt. Earl) rode the train in case repairs to the tracks were needed. A detachment of 150 men from the 61st Illinois under Lt. Col. Daniel Grass guarded the train. The trip was uneventful and the train arrived in Alabama the next day. Thirty troopers of the 12th Indiana Cavalry joined the train, hoping to reunite with their command at Fortress Rosecrans.

Confederate Gen. Lawrence Sullivan Ross' cavalrymen were busy removing rails between Bell Buckle and Murfreesboro. The 3rd and 6th Texas Cavalry were posted nearby.

On December 13th, Rousseu sent another train southward with Col. Thomas Saylor of the 4th Michigan in charge. Three miles south of Murfreesboro, Saylor's men discovered a culvert on fire. The train stopped and the fire was put out. About six miles later, the train conductor stopped again, stating it was time to gather firewood. By 4:00 pm the train was loaded with wood. After pulling out, the train came under fire from Sul Ross' Texans. Saylor ordered two companies of Wolverines to detrain and force

Gen. Nathan Bedford Forrest

the rebels to fall back. The Federals then had to re-lay about 50 feet of track. By this time, the Confederates had brought up John Morton's artillery battery. The bluecoats got back on the train but the engineer and brakemen had disappeared. They soon showed up and the train began moving. A projectile pierced the boiler of the locomotive, and about five miles from Murfreesboro all the steam had escaped. The Federals pushed the train the rest of the way back. Saylor reported losing 17 men — one killed, 10 wounded, and six missing.

About 2:00 am on December 15th, the supply train from Stevenson approached Christiana and slammed on the brakes to avoid the broken rails. Then the rebels tore up track behind the train. Col. Grass told his Illinois men to disembark and the Michigan engineers began to fix the track. Daybreak found the train about six miles south of Murfreesboro. The 3rd and 6th Texas Cavalry hemmed in the train and Morton brought up his guns, "bull-pups" as he called them. The bluecoats left the train and tried to cut their way back to Fortress Rosecrans. They were charged by the Texans. Half of the Federal regiment, along with 31 of the Michigan engineers and the dismounted Indiana

Officials of the U.S. Military Railroads, Division of the Mississippi:

1864.
A. Anderson, general superintendent, to November 1.
E.L. Wentz, general superintendent, after November 1.
W.J. Stevens, superintendent railroads running from Nasliville.
Colonel L.P. Wright, superintendent railroads from Chattanooga, to July 1.
W.0. Taylor, superintendent railroads from Chattanooga, after July 1.
A.F. Goodhue, engineer and superintendent railroads at Memphis, Tenn. and Columbus, Ky.
W.W. Wright, chief engineer.
John Trenbath, Auditor.
Colonel John C. Crane, assistant quartermaster, disbursing officer.

1865.
W.J. Stevens, general superintendent.
R.B. McPherson, assistant superintendent.
J.B. Van Dyne, chief master of transportation.
A.W. Dickinson, superintendent Nashville railroads, to July 25.
George H. Hudson, superintendent Nashville railroads, after July 25.
W.R. Griffin, superintendent Nashville, Decatur, and Stevenson railroad.
A.A. Talmadge, superintendent Chattanooga railroads.
A.J. Cheney, superintendent Knoxville and Bristol railroad, to Sept. 1.
A.J. Cheney, superintendent Chattanooga and Atlanta railroad, after Sept. 1.
A. Hebard, chief engineer repairs, Nashville railroads.
A. F. Goodhue, engineer and superintendent of railroads West Tennessee, Kentucky, and Arkansas.
Colonel L. P. Wright, superintendent Memphis railroads.
W.W. Wright, chief engineer.
L.H. Eicholtz, acting chief engineer, January 1 to July 1.
John Trenbath, auditor.
Captain F.J. Crilly, chief quartermaster and disbursing officer.

Important Sale, Ruins of Taylor Depot.

Will be sold at Public Auction, on Friday, December 8th, 1865, at 10 o'clock a.m., at Taylor Depot, corner of Summer and Demonbreun streets, Nashville, Tennessee, the entire lot of Copper, Brass, steel, Wrought, Cast and Sheet Iron materials, saved from the debris of the great conflagration of June 9th, 1865.

This sale will include an immense quantity and variety of Carpenter's, Ship-carpenter's, Wheelwrights', Blacksmiths', and Miscellaneous Tools. Also the machinery of (2) two Portable Circular Saw Mills, a large lot of Circular, Crosscut and Mill blades; a large lot of Shovel and Spade Blades, etc.

A complete list of these articles can be seen, and any information concerning same received, by applying to S.R. Butler, at Q.M.'s office, at the Taylor Depot. As the articles must be weighed after they are sold, persons purchasing a part or entire lot, will be required to leave a deposit in Government funds. This will be an attractive Sale, and well worth the attention of Foundrymen and Speculators.

By order of Col. A.J. Mackay, Chief Q.M. Dep't of Tennessee.
W.A. Wainwright
Capt & AQM, and CAA MD of T.

cavalrymen were either killed or taken prisoner, including Col. Grass. One wing of the 61st Illinois led by Major Jerome B. Nulton forced its way to the safety of Blockhouse No. 8.

Afterwards, Major Jerome Nulton of the 61st Illinois wrote, "In justice to the First Michigan Engineers, allow me to say that they behaved themselves with firmness and during the engagement they fought like veterans."

Ross' Texans had taken 200 prisoners and captured a supply train full of sugar, coffee, hardtack, and bacon — a full 200,000 rations. The looting was interrupted by a "flying relief column" under Milroy sent from the fort. The Confederates took all they could carry and then burned the train.

News arrived of the great battle at Nashville on December 15-16th, in which Hood's army was routed and now in full flight. Forrest ordered all of his men to meet the main army at Columbia, where he took control of the rear guard.

Back at Fortress Rosecrans, Rousseau ordered work parties to see that the Nashville & Chattanooga Railroad was restored to service in the shortest possible time. The right-of-way was opened by the morning of December 20th. Before nightfall, ten trains had reached Murfreesboro from Chattanooga, assigned to take Steedman's men back to Bridgeport.

The next day Rousseau moved his headquarters back to Nashville and sent Milroy to Tullahoma. Van Cleve resumed command at Fortress Rosecrans. By Christmas Eve, the railway between Nashville and Murfreesboro was back open. Nearly eight miles of track had to be re-laid and 530 feet of bridges rebuilt. Detachments from the battered 115th Ohio along with the 1st Michigan Engineers set out to rebuild all the damaged blockhouses.

The garrison at Fortress Rosecrans was reduced to only five infantry regiments — the 12th Indiana dismounted cavalry, the 61st Illinois, and the 3rd, 4th, and 29th Michigan. Artillery was manned by Battery D, 1st Michigan Light Artillery (Capt. H.B. Corbin), the 12th Ohio Battery (Capt. F. Jackson), and the 8th Wisconsin Battery (Capt. H.E. Stiles).

Massive clean-up needed

During the first six months of 1865, the military railroad wrecking train (developed by George Herrick, superintendent of rail cars at Nashville) picked up and carried into Nashville 16 wrecked locomotives and 294 carloads of wheels, bridge irons, and other materials salvaged from wrecks caused by Confederate guerrillas.

In 1866, Chief Engineer W.W. Wright provided a listing of all the work done on the railroads in Middle Tennessee and elsewhere.

He reported that 148 miles of track, including 15 side-tracks, had been rebuilt on the Nashville & Chattanooga Railroad (151 miles long). In essence, the road was "made new." Forty-two bridges totaling four miles in length had been built and/or rebuilt. Mill Creek Bridge No. 1 at 260 feet long had been rebuilt five times, two other Mill Creek bridges four times, and five others in the Nashville area three times each. Sixteen water stations with 35 tanks had been constructed. Sidings were put in at intervals, not more than eight miles apart, each capable of holding five to eight long freight trains, and telegraph stations were established at most of them. In all, 19 miles of new sidings were added to this road, and 45 new watertanks erected.

• On the Louisville & Nashville Railroad, in 1864 alone, 2,934 tons of new rail and 471 tons of rerolled rail had been laid, along with 96,709 crossties.

• On the Nashville & Decatur Railroad, more than 34 miles of track were rebuilt, along with 41 bridges, including the Big Harpeth River Bridge (187 feet long and 38 feet tall) rebuilt twice and partially rebuilt twice; Rutherford Creek No. 2 (265x27), rebuilt twice and partially rebuilt three times; Duck River (627x72), rebuilt twice; and the Kalioka Trestle at 1,130 feet long, rebuilt once. The work of reconstruction was commenced Dec. 19th, three days after the Battle of Nashville, and completed to Pulaski on Feb. 10th, 1865.

• Nashville & Northwestern Railroad: Crews laid 50.75 miles of track (5,161 tons of iron) and 107,000 crossties; built 44 bridges and trestles totaling more than four miles and consuming 4.1 million feet of lumber, including a 2,151-foot trestle at Nashville and a 1,326-foot trestle at Sullivan's Branch; built 34 structures, including 14 at the Johnsonville depot, consuming 1.8 million feet of lumber and 742,200 shingles; and built 14 water stations. On Nov. 30th, the road was entirely abandoned, and the movable property on it taken to Nashville. During General Hood's occupation, from December 1st to 16th, all the bridges were destroyed. Repairs were commenced January 2nd, and the road completed by February 13th. Two thousand two hundred lineal feet of bridges were rebuilt. On Sept. 1st, 1865, the road was turned over to the railroad company.

• Memphis & Charleston: Rebuilt 4.5 miles of track, 11 of 12 bridges totaling 4,943 feet, and 13 water tanks.

• Nashville & Clarksville Railroad: Built 3,433 ft. of bridges and trestles consuming 890,000 ft. of lumber; 1.75 miles of track using 15,000 crossties.

Director McCallum estimated that the cost of railroad building and maintenance for the first nine months of 1864 totaled almost $30 million (half a billion in today's dollars).

First Michigan Engineers & Mechanics
Volunteer Engineers Matched West Pointers in Building Skills

Civil engineers played a significant role in the Civil War, many volunteering for military duty as able mechanics and others performing as official engineers. Such was the case of the esteemed First Michigan Regiment of Engineers and Mechanics, which served in Tennessee and the mid-South as part of the Army of the Cumberland. During the war, the regiment was involved in the battles of Mill Springs, Perryville, Stones River, Chattanooga, and Sherman's March to the Sea. They were also heavily involved in building Nashville's fortified infrastructure and its rail lines.

The volunteer regiment was formed in Marshall, Mich. (Camp Owen) in 1861 with the approval of Secretary of War Simon Cameron and Governor Austin Blair. Their founder and leader throughout the war was William P. Innes, 35, a prominent railroad surveyor and civil engineer from Grand Rapids. Innes seemed fated for railroad work, leaving home at age 13 to work the rails. His only military service was six weeks as a private in the Mexican War. During the 1850s, he served as chief engineer for the Grand Rapids & Northern Railroad Company.

The regiment comprised about 1,000 men of various backgrounds, including mechanics, craftsmen, farmers, laborers, carpenters, and wagonmakers. About 15 percent of the men were officially too old or too young to serve. Many were related. One father-son team (there were 25 pairs) were 54 and 13 years old respectively. The officers ranged in age from 26 to 54, and were engineers, physicians, ministers, merchants, farmers, and a master railroad mechanic.

The unit first served under General

Colonel William P. Innes

Don Carlos Buell and arrived in Nashville in November 1862 following the Battle of Perryville, Ky. They were put to work fixing the Louisville & Nashville Railroad to Gallatin and building three bridges over Mill Creek on the Nashville & Chattanooga Railroad, all in the first two weeks.

About that time, General William S. Rosecrans replaced Buell as commander of the Army of the Cumberland. Morale was a big problem at the time. Rosecrans noted that "many soldiers have sought and allowed themselves to be captured and paroled by the enemy to escape from further military duty, and in order to be sent home." The First Michigan was not being paid the $17 per month as engineers, rather the $13 per month as infantrymen. Congress corrected that inequity in July 1862, but the men still hadn't been paid for nearly a year. About one hundred of the engineers actually mutinied and refused to work. These men were sentenced to 30 days of hard labor, three days of only bread and water, and confined to the Nashville Workhouse, which was so drafty and sooty it was known as the "smokehouse."

The pay issue was resolved, and other than turf wars and petty jealousies among the officers, the men worked throughout the war with high morale and justified pride in their accomplishments. Plus, they realized that they enjoyed more creature comforts than regular infantrymen.

On Jan. 1st, 1863, in the midst of the Battle of Stones River (Murfreesboro), the regiment found itself in Lavergne, a small village on the main pike between the battle and Nashville. Much of the Federal army had been routed the previous day. The village was smouldering, along with hundreds of supply wagons torched by Confederate cavalry. The First Michigan cleared the way for large trains of wounded headed back to Nashville and stemmed the tide of stragglers and deserters. They camped east of the pike, about a mile south of the village and half a mile from the railroad tracks. Using abandoned wagons, timber, and any other debris they could find, they built crude breastworks to crouch behind — 400 men armed mostly with old 1842 rifled muskets although 100 had newer Springfields and Enfields. Soon they were beset by thousands of troopers in Brig. Gen. John A. Wharton's command, which had just destroyed a train of 30 wagons. A full volley from Companies A and H of the engineers drove the horsemen away. Next came a full regimental attack, but that also was blunted by a massive volley. Attacks were made during the next three hours, backed by the two guns of Capt. B.F. White's Tennessee battery. On one occasion, the breastworks were attacked from the east, north, and south by the 14th Alabama Battalion and the Fourth Tennessee. All were repelled except for one brave trooper who charged right up to the cedar brush and aimed directly at Colonel Innes. The assailant was quickly shot down by a lieutenant in Company E. During the firefight, Innes was described as cool and reassuring, walking along the line and directing fire. Then, rebels dressed in captured blue overcoats tried to sneak into the field fortification, but their gray pants gave them away.

At 5:00 pm, an exasperated Wharton sent a white flag of truce to the engineers and demanded surrender or annihilation. Innes told the flag bearer, "We don't surrender much." Wharton sent a second flag for a truce to gather the wounded and bury the dead, but Innes responded that his men would take care of those duties and that any more flags of truce would be fired upon. Wharton finally retreated, stating later that his losses had been "very considerable," probably about 50 casualties. The First Michigan lost three killed, four seriously wounded, and six captured (pickets who had been cut off from the main body). Forty of the engineers' horses and mules had been killed.

The engineers celebrated their stand, carrying Innes around on their shoulders. About a week later, Innes wrote to his men: "You have all been tried and not found wanting." Writing after the war, New York newspaper tycoon Horace Greely said of Stones River, "The silver lining to this cloud

1st Michigan Engineers Regimental flag

"We don't surrender much."

is a most gallant defense made on the 1st by Colonel Innes' 1st Michigan Engineers and Mechanics."

A poem was written by an anonymous wag — "War Song: Battle of Lavergne." For many years, veterans at reunions would raise their glasses to the toast, "The Battle of Lavergne — a new year's ball at which we gave our visitors the best we had."

On Feb. 12th, 1863, the regiment camped on Rains' Hill, two miles south of Nashville, where the uncle of General Rains, slain at Stones River, still resided. Work details were sent out daily to make repairs and build bridges on the Nashville & Decatur Railroad. Much of the timber work and carpentry were performed outside camp, then transported with the men on flatbed railcars to the worksite. Timber was cut around their campsite; the first to go were the big 200-year-old oak trees on Rains' Hill. During the rail trips, so many men were using dogs and rabbits for target practice that they were threatened with arrest.

Many of the railroad bridges were wooden trestles. As described in Mark Hoffman's regimental history, trestle bridges consisted of a deck supported by a series of closely spaced, framed supports called trestles, each consisting of one or more "bents." The bents were driven into the riverbed or rested on platforms. The legs measured five or six inches square and were connected, in varying numbers, to a perpendicular cap. The caps were placed perdendicular to the direction of the bridge to a width of 12 to 16 feet. Transoms and braces supported the legs horizontally. Each arrangement of cap, legs, and supporting timbers was called a bent. The bents were usually cut from nearby standing timber, preferrable pine, spruce, or ash because they were light and stiff. Framed bents rested upon masonry piers or wooden cribs filled with ballast and were secured on their lower end with horizontal timbers called sills. They were often first framed on level ground and hoisted into position, one trestle at a time. Pile bents were secured directly to the riverbank by driving or anchoring their vertical legs. The men worked from boats or standing in chest-deep water for hours at a time during the setting of the trestle. Trestles were placed 12 to 15 feet apart. Transverse planking connected the caps and served as bridge flooring or the bed for railroad iron. Side rails were installed for infantry or artillery crossings. Bridges higher than 25 or 30 feet required two or more tiers of trestles. The trestle bridge at Whiteside, Tenn. required four tiers of trestles. President Lincoln himself had remarked that such bridges seemed to consist "only of beanpoles and cornstalks."

The railroad bridge over the Little Harpeth, 12 miles south of camp, was completed in two days. The men began work on the Big Harpeth River bridge at Franklin on February 19th but did not complete it until March 6th, delayed by high water, swift currents, and frequent heavy rains. The men worked long days standing in swift, cold water. One man almost drowned. Several times, work was destroyed by the raging river. Finally, the bridge, tested by the passage of a train, was planked for use by infantry

U.S. Engineer, Quartermaster Uniforms and Buttons

Engineer Private Uniform

US Engineer Officer's Button

US Topographical Engineer Officer's Button

Engineers First Sergeant Uniform

Infantry Quartermaster Sergeant Uniform

Light Artillery Quartermaster Sergeant Uniform

and artillery trains.

By March 18th, the regiment was back in Murfreesboro to work on the massive depot there called Fortress Rosecrans, 200 acres enclosed by fortifications protecting the railroad and the pike. The First Michigan built a warehouse capable of holding five million rations, a 4,500-sq-ft. powder magazine 12 feet high, and a 3,000-sq.-ft. ordnance building, all within 30 days. Morton's Pioneer Brigade also worked at Fortress Rosecrans and sometimes the units clashed, the West Pointers versus the volunteers. At one point, Morton attempted to incorporate the First Michigan into the Pioneer Brigade while Col. Innes was on leave; Gen. Rosecrans intervened on behalf of the Wolverines. Later, Innes complained to Rosecrans about the allegedly inferior quality of work performed by the brigade and threatened to resign, but Rosecrans would not accept his resignation.

Another point of contention between the First Michigan Engineers and Morton's Pioneer Brigade was the building of the Nashville & Decatur Railroad bridge over the Elk River down near the Alabama line. The bridge had to be 470 feet long and 60 feet tall. The Corps of Engineers stated they could erect the structure in three weeks; Innes countered by figuring ten days. Rosecrans gave the job to the Michiganers, who proceeded to complete the bridge in seven days. The train rolled over it as a band played "Hail Columbia." One of the engineers who kept a diary wrote, "For a large job it was the best and quickest done of any we ever undertook."

On Aug. 10th, 1863, Colonel Innes was named military superintendent of all railroads in the district. On or about Sept. 1, 1863, Tennessee Military Governor Andrew Johnson directed Innes to supervise the completion of the Nashville & Northwestern Railroad from Nashville west to the giant Johnsonville depot and fortress on the Tennessee River. This additional route of supply was going to be needed for campaigns against Chattanooga and Atlanta (due to the low seasonal water levels on the Cumberland River).

Twenty-three miles of the N&NWRR had been laid from Nashville to Kingston Springs before the war and a small section ran eastward from Johnsonville. Forty-four miles of track

An unidentified member of the First Michigan Engineers. (Library of Congress.)

needed to be laid to complete the route, requiring 115,000 cubic yards of grading, 5,200 feet of trestle, 107,000 railroad ties, 3,200 tons of rail, and 230,000 railroad spikes. Four companies of the First Michigan under Major John B. Yates were allotted for the work, along with protection provided by the Tenth (First Middle) Tennessee and the 13th U.S. Colored Troops (six companies) under Lt. Col. Theodore Trauernicht. General Alvan Gillem was in overall command of the armed forces along the rail line.

Seven bridges were required to be built over the Harpeth River within ten miles; most of these were box bridges built on stone piers. By the first week of October, the work had nearly been completed. However, three companies were pulled off to work on sections of the Nashville & Chattanooga Railroad destroyed by Wheeler's cavalrymen. By November 3rd, General George H. Thomas ordered the Michigan engineers back to work on the N&NWRR, along with the 12th USCT, the 8th Iowa Cavalry, and the First Kansas Battery.

The men lived fairly well in their camps. They foraged and stole livestock from nearby farms to feed themselves. At one point, they lucked into a cache of whisky. "Altogether one of the most pleasant times we had during the war," one engineer recalled.

By the end of January 1864, the First Michigan had built 2,300 feet of trestle and 30,000 railroad ties. Thirty-four miles of track were completed west from Nashville, with 20 more miles ready for grading and track. Eighteen miles extending from Johnsonville were ready for track with only six more miles to grade. Innes asked for more civilian workers and estimated that the railroad could be completed in 60 days. However, Federal authorities were not satisfied with the progress on the railroad, with most criticism falling on Gov. Johnson and military rail superintendent John Anderson. On March 5th, the four companies of the First Michigan left the N&NWRR, now being built by Missouri engineers, and joined the rest of the regiment down around Bridgeport and Chattanooga.

At that time, General William T. Sherman was preparing to march his armies of more than 100,000 men into north Georgia and on to Atlanta. His sole source of supply would be the Nashville & Chattanooga Railroad. His men would require 130 railcars, each loaded with 10 tons of supplies, to reach Chattanooga each day. Captain William E. Merrill, chief engineer of the Army of the Cumberland, was placed in charge of building and repairing bridges and blockhouses along the route. The eight companies of the First Michigan not at Chatta-

This Chattanooga Creek trestle under construction shows the foundation of stone piers, the added support of the underlying trestles, and the main framework of the box bridge, which could be constructed in camp and hauled by train to the bridge site. (Library of Congress.)

nooga were charged with assisting Merrill, divided into two battalions under Major John B. Yates and Lt. Col. Kinsman Hunton. The blockhouses at the railroad bridges would house a 20-man garrison, usually USCT soldiers, who could hold out inside long enough for reinforcements to arrive in case of Confederate cavalry attacks. That is, if the garrison did not surrender when threatened with annihilation.

The optimum shape for a blockhouse was octagonal, but square or rectangular plans were used because they could be constructed faster. Plans called for square or rectangular blockhouses with the second story or tower set diagonal to the first to allow increased fields of fire. The walls of the first floor were double thickness of timber, at least 40 inches combined, with 20-in. timbers erected vertically as the inner wall and another row set horizontally for maximum strength. The second-story walls were single thickness. The roof was built of logs with dirt thrown on top. Shingles of boards or battens provided waterproofing. The blockhouses also included ventilators, bunks, cellars, and water tanks. Engineers often erected only the inner vertical wall and left the garrison to complete the rest of the structure. Construction of the blockhouses was hampered by the frequent changing of garrisons.

The First Michigan finished out the war marching with Sherman to the sea and into the Carolinas. The regiment lost 13 men killed and 247 died of disease.

In 1912, a monument to the First Regiment Michigan Engineers was erected at the state capitol. (Michigan State Senate)

Cavalry, Irregulars Try to Cut Supply Lines

Guerrillas, Gunboats & Convoys

During a church service in Shelbyville, a pro-Southern preacher prayed: *"O, Lord, let the rain descend to fructify the earth and to swell the rivers, but O Lord, do not raise the Cumberland sufficient to bring upon us those damn Yankee gunboats."*

Because the water level of the Cumberland River greatly fluctuated with the seasons, the low water preventing the passage of large freighters, the Nashville & Northwestern Railroad was extended west of Nashville to the Tennessee River, where a huge fort and depot named Johnsonville was built, offering an alternative route of supply.

During most of the war, Forrest and Wheeler and a host of partisan rangers and guerrillas operated against the Union supply lines, causing many headaches for Federal officials, general officers, and quartermasters. "To the west of the city, Confederate units seemed to move at their own pleasure," stated historian Walter Durham.

Getting ready to launch the Atlanta campaign in 1864, Gen. William Tecumseh Sherman stated: "The great question of the campaign was one of supplies. Nashville, our chief depot, was itself partially in hostile country, and even the routes of supply from Louisville to Nashville, by rail and by way of the Cumberland River, had to be guarded." He admitted, "I am never easy with a railroad which takes an whole army to guard, each foot of rail being essential to the whole; whereas they can't stop the Tennessee (River)…"

Steamboats carried a lot of freight. One steamer transport could carry 500 tons of cargo, enough rations and forage for a 40,000-man army and 18,000 animals. The same amount of cargo would require 250 wagons or 125 rail flatcars. As the war progressed, Sherman became frustrated and angered by the rebel resistance: In December 1863, he wrote to a military friend: "For every bullet shot at a steamboat, I would shoot a thousand 30-pounder Parrotts into even helpless towns."

Brigadier General Philip Sheridan stated: "The feeding of our army from the base at Louisville was attended with many difficulties, as the enemy's cavalry was constantly breaking the railroad and intercepting our communications on the Cumberland River at different points that were easily accessible to his then superior force. The accumulation of reserve stores was therefore not an easy task."

Southern civilians thrilled to the exploits of their partisan warriors, but one Federal soldier spoke for many of his comrades when he called Southern guerrillas "thieves and murderers by occupation, rebels by pretense, soldiers only in name, and cowards by nature."

Lawlessness and atrocities were committed by both sides. U.S. General George Crook recalled that when his command landed in Carthage, Tennessee, in 1863, the Unionists disembarked the riverboats and immediately began looting the town. The general sighed: "I had my hands full, what with looking out for the enemy and restraining the lawlessness of our own people."

"Immediately after the fall of Nashville in 1862, a strict embargo was placed on all shipments of freight and high charges collected for its transportation," according to steamboat historian Byrd Douglas. "Before commercial freight could be shipped during the war, it was necessary to obtain a priority permit, which carried a high tax. The surveyor of Customs Office in Nashville collected these taxes and issued the permits. It soon became one of the busiest places in the city. Old records at this office prove that the merchants of Nashville, despite the intervention of war, continued to do a relatively good business during the occupation of the city by the Federal government."

Rear Admiral David Dixon Porter, commander of the Mississippi Squadron at Cairo, declared: "You can never go wrong in doing a rebel all the harm you can. I am no advocate for the milk and water policy."

On Oct. 18, 1862, Porter issued General Order No. 4 from his Cairo headquarters directing that Federal gun-

---

Much more information on this subject can be found in *Iron Maidens and the Devil's Daughters: Federal Gunboats versus Confederate Cavalry and Gunners on the Tennessee and Cumberland Rivers, 1861-65* by Mark Zimmerman, Zimco Publications LLC, 2012.

boats never tie up at riverbanks, keep deck guns loaded and aimed at the banks, and keep small arms loaded and ready to repel boarders. The gunboats were ordered to shell and destroy any houses near sharpshooter attacks regardless of civilian casualties.

"When any of our vessels are fired on, it will be the duty of the commander to fire back with spirit, and to destroy everything in that neighborhood within reach of his guns. There is no impropriety in destroying houses supposed to be affording shelter to rebels, and it is the only way to stop guerrilla warfare. Should innocent persons suffer it will be their own fault, and teach others that it will be to their advantage to inform the Government authorities when guerrillas are about certain localities."

Naval officer and historian Alfred T. Mahan stated: "The feeling in the country favored the Confederate cause, so that every hamlet and farmhouse gave refuge to these marauders, while at the same time the known existence of some Union feeling made it hard for officers to judge, in all cases, whether punishment should fall on the places where the attacks were made."

As noted previously, immediately following the Northern victory at Stones River (Murfreesboro), on Jan. 3, 1863, the commanding general, William Rosecrans, began a campaign of cajoling Federal officials in Cairo and Washington for more gunboats on the Cumberland River. The hearty cooperation promised him was compromised by a lack of crews and armament for new gunboats, periods of low water, lack of coordination between the army and navy, and faulty lines of communication, all tempered by constant rumors of Confederate armies on the move. It reached the point where Rosecrans implored the Washington authorities to allow him to purchase steam transports, convert them into army gunboats, and crew them with infantry. Fortunately, the powers in charge did not agree to this proposal.

The U.S. Army operated a small fleet nevertheless. "The U.S. Army Quartermaster Department operated a fleet of

Steamers lined up at the Nashville wharf in December 1862 with goods unloaded on the river bank.

"I am never easy with a railroad which takes an whole army to guard, each foot of rail being essential to the whole; whereas they can't stop the Tennessee (River)…"
— Gen. William T. Sherman

ersatz gunboats on the Western rivers," wrote brown-water navy historian Jack Smith. "The majority of these were very small, were manned by civilian crews with military gun crews for the usual two or three cannon, and were protected by cotton bales or planking. The most famous of these were the *Silver Lake No. 2* and *Newsboy,* which operated on the Cumberland River, and the Stone River, which steamed on the Tennessee. In March 1864, the *Newsboy* rescued three transports under attack on the upper Cumberland, while in October, the Stone River participated in the defense of Decatur, Alabama, from the army of General John Bell Hood."

Confederate Major J. R. "Dick" McCann was known as the Guerrilla Chieftain. He was captured in August 1863 at Weems Springs, Tennessee, and imprisoned at Johnson's Island in Ohio. A veteran of the Mexican War, he formed the Cheatham Rifles, later Company B of the 11th Tennessee. After the war he was court clerk of Nashville until his death in 1880. One of his captors wrote of McCann: "His name has been an epitome of the seven deadly sins, and if a dastardly act were committed by a Hottentot between Knoxville and Nashville, 'Dick McCann did it.' We assure our readers that he is not the fiend he has been written and painted. He is about five feet eight inches high, 140 pounds weight, fair complexion, bright blue eyes, brilliant, polished and humorous in conversation. He was born of Irish parents at Petersburgh, Va., and came to Tennessee with his family when a mere child. He is married, having four children now in Nashville, his wife being in East Tennessee. He is not now, nor ever was for a disruption of the American Republic. He never has had but two disunionists in his command." Against secession before the war, McCann nonetheless went with his adopted state. He claimed that if Lincoln offered general amnesty to the Southerners the war could be ended quickly.

Despite their assignment to the 9th Tennessee Cavalry, McCann's Squadron continued to do independent partisan service for much of the fall and winter of 1862-63. After the war, a writer for the *Confederate Veteran* magazine stated: "It would be impossible to relate all of his numerous adventures. He was busy prowling around night and day, and rarely permitted the enemy to venture beyond the fortifications of Nashville without some evidence of his thoughtful attention."

On Jan. 13, 1863, the transport *Charter* was attacked by McCann's guerrillas at Ashland, five miles upstream from Harpeth Shoals on the Cumberland. The cargo of hay, corn, and commissary stores was destroyed and the boat burned. The crew was paroled except for six captured contraband deckhands (former slaves), who were led ashore and each executed with a shot to the head. In September 1862, McCann's unit vandalised the Louisville & Nashville Railroad between those two cities. In January 1863, in retribution, General

Maj. Gen. Joseph Wheeler

Mitchell, commander of the Nashville garrison, ordered the houses and barns owned by McCann and an associate burned to the ground by the 85th Illinois Infantry. Up to a dozen other houses in that neighborhood were also torched that night.

On Jan. 13, 1863, the same morning as McCann's attack, three boats carrying wounded Federal soldiers, *Hastings, Parthenia,* and *Clio,* left Nashville flying hospital flags. Waiting for them at Harpeth Shoals was the brigade of Wheeler's troopers under Colonel William B. Wade. The lead boat, *Hastings,* was ordered to halt by Wade's men, but the civilian pilot turned the boat over to Chaplain Maxwell P. Gaddis of the 2nd Ohio Infantry, the military's senior representative aboard. Gaddis responded that the boat could not stop, and was fired upon by muskets. Gaddis decided to pull over to the bank, but his maneuvers were viewed by the Rebels as an attempt to escape. Two artillery rounds hit the hospital boat. When the boat finally grounded, Wade's men, many of them drunk, boarded and began to loot the passengers. The captains of the other two boats went to shore, believing the flotilla had stopped to refuel. Unknown

Col. Thomas M. Woodward

to Gaddis and Wade, the boats also contained cotton cargo. Wade ordered all of the wounded soldiers ashore so the boats could be fired. Gaddis demanded an official order from Wade's superior, Wheeler, and a courier was sent off to retrieve it. Wheeler ordered all of the 260 Federal soldiers paroled but only if Gaddis agreed to burn the cotton upon reaching Louisville. In the meantime, the U.S. Army steamer *W.H. Slidell,* armed with two or three field pieces and commanded by Lt. William Van Dorn, arrived upon the scene. The army gunboat ran aground on the opposite bank and surrendered. The Rebels managed to cross the river with the boat, and its cannon were thrown into the water. All of the Union men were put aboard the *Hastings* except for three contrabands, who were summarily executed. The three other boats were torched and soon the full blast of the heat was so intense it set the grass aflame on the opposite shore.

News of the capture and destruction of the boats spread quickly, the civilian population rejoicing and the Union authorities reacting with great

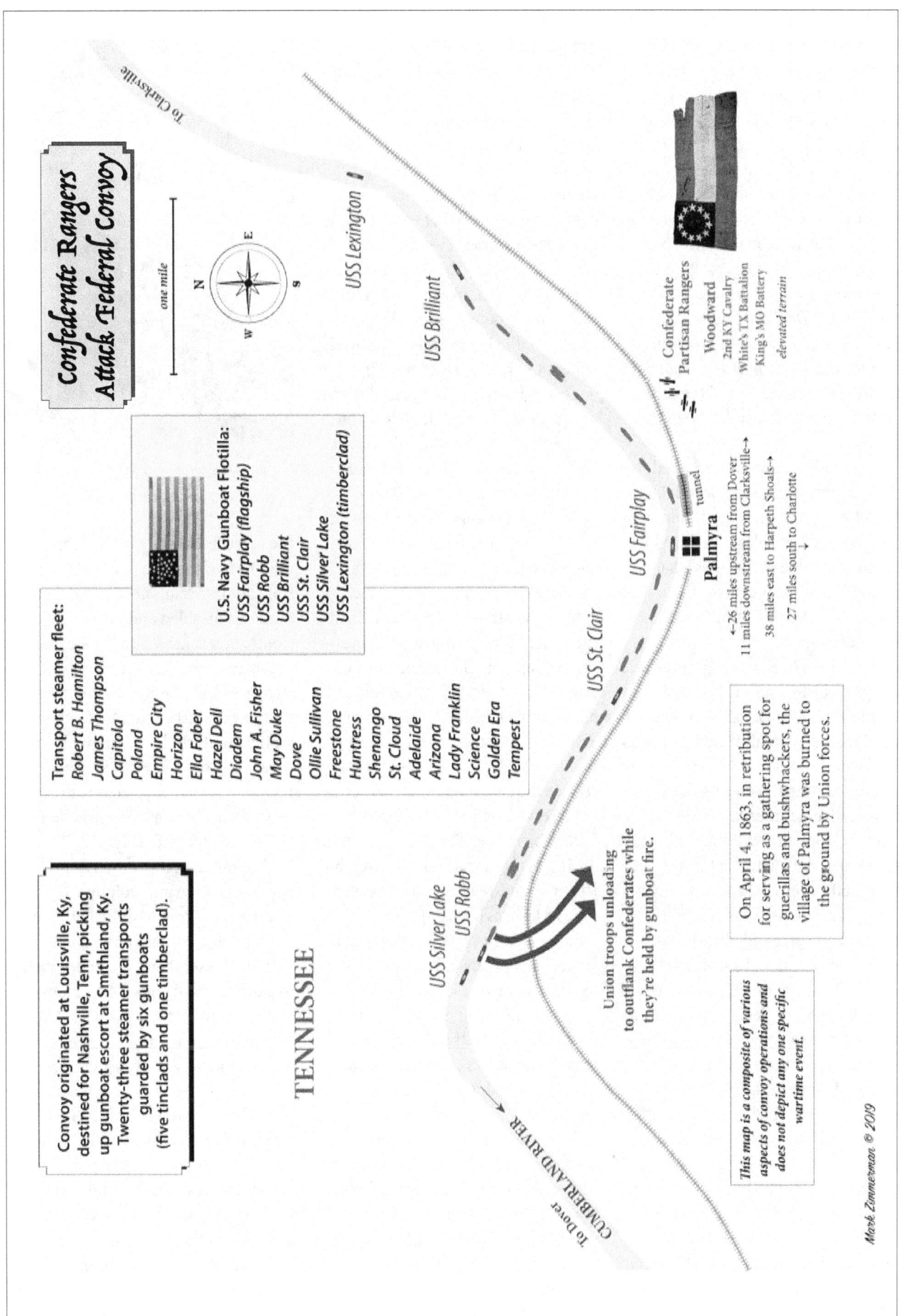

trepidation. The gruesome murder of the contrabands was further sensationalized by the Northern press.

On Jan. 18, 1863, a partisan ranger battalion led by Capt. D.W. Holman captured the transport *Mary Crane* as she was loading firewood at Betsy's Landing near Harpeth Shoals. She had been en route from Evansville, Indiana to Nashville loaded with beans, sugar, flour, corn, soap, and candles. The transport had been escorted by the tinclads *Brilliant* and *St. Clair,* but they had dropped out of formation to refuel. The partisans shot and killed the pilot, destroyed the cargo, and burned the *Mary Crane* to the waterline.

General Rosecrans complained that the partisan attacks at Harpeth Shoals and elsewhere represented "inhuman violations of the rules of civilized warfare by the rebel authorities" that revealed the "barbarism of these rebel leaders." Meanwhile, Confederate General Bragg recommended a promotion for Wheeler as a "just reward to distinguished merit."

On Feb. 20, 1863, following the battle at Dover, Federal tinclad gunboats led by Lt. Commander LeRoy Fitch began a reconnaissance up the Tennessee River all the way to Muscle Shoals. Along the way, the gunboats ran into a small battle at Clifton, where 60 troopers of the 3rd Michigan Cavalry under Capt. Cicero Newell had surrounded and attacked the infamous guerrilla leader Col. John F. Newsom and his men at dawn. Newell was wounded and succeeded by Capt. Frederick C. Adamson. Newsom was shot in the left arm and captured along with 53 of his men, who were described as a "ragamuffin collection." The remainder of the guerrillas fled and the town was torched. The gunboats saw the smoke and responded. Fitch agreed to ferry the Wolverines and their captured prisoners and horses across the river. The surgeon aboard the *USS Brilliant* remarked about the Confederates, "They are good marksmen and if they fire from planted guns, we stand a chance to be well pounded as a ball from a 12-pounder would go through us like so much paper."

More than anyone else, Lt. Commander LeRoy Fitch devised the counterinsurgency techniques and convoy maneuvers that ensured that the needed supplies would reach Nashville and the armies in the Western Theater. A native of Logansport, Ind., and a graduate of the U.S. Naval Academy, Fitch was appointed by Flag Officer Charles Davis to head up a fleet of light-draught tinclads known as the Mosquito Fleet to escort convoys and patrol the Ohio, Tennessee, and Cumberland rivers. His biographer, Jack Smith, stated that Fitch "gained a reputation for his prosecution of counter insurgency warfare and as an innovator of protective measures for the massive number of contract steamboats which churned the Ohio, Cumberland, and Tennessee rivers supplying the advancing Union armies." Duty-bound and indomitable, Fitch was not a flamboyant man and he has not received the recognition that he deserves, Smith said. In service the entire war, the energetic Fitch served well at Island No. 10, Fort Pillow, Memphis, Dover, on the Ohio during Morgan's famous raid, on the Upper Tennessee River, and during the 1864 Nashville campaign.

Steamboat historian Douglas said of Fitch: "He was the officer who did more than anyone else to keep the Cumberland open." Porter said of Fitch after his death: "The gallant Fitch never shrunk from the performance of any duty however hazardous."

Fitch learned about counterinsurgency from his half-brother Graham Newell Fitch, who was 25 years older than him and outlived him by 17 years. A former U.S. Senator, the elder Fitch raised the 46th Indiana Volunteers and led it into battle at Fort Pillow and Memphis. In June 1862,

Lt. Commander LeRoy Fitch

the elder Fitch was witness to the destruction of the *USS Mound City* ironclad in Arkansas. While the gunboat was disabled and under tow, Confederates fired into the steam drum and then shot survivors as they struggled in the water. Eighty-two sailors were either shot to death or scalded to death by steam, with another 43 drowned. Fitch vowed to retaliate against any persons firing on boats from shore and destroy their residences and property.

The younger Fitch devised the strategy of engaging and holding Confederates on the shore with gunboats while another gunboat or transport disembarked Union cavalry to ride hard and engage the enemy from behind. Although a sound tactic, the Confederates always posted pickets or lookouts and they were never really surprised or routed. Besides, several rounds of canister from a 32-pounder was usually enough to persuade cavalry, especially units without field artillery, to disperse.

Fitch was much more successful in devising methods of escorting convoys with gunboats. By January 1863 Fitch's flotilla numbered six gunboats, and each one of them carried powerful eight-inch naval guns that outclassed any Confederate field artillery in the region. Borrowing from army doctrine, Fitch placed the slowest steamers in the front of the convoy, with the most valuable cargos in the middle, and the fastest boats in the rear. The gunboats were dispersed along the line, with one always in the lead and one trailing. In strong currents, a light-draught might be lashed to one of the larger transports. Following the *Hastings* affair, naval headquarters ordered that at least two gunboats escort every convoy. Each convoy to Nashville would take a week, with boats leaving Smithland or Dover each Monday. While the transports were unloading at the Nashville wharf, the gunboats would conduct patrols on the upper Cumberland River all the way to Carthage.

On Jan. 23, 1863, a 22-boat return convoy left Nashville with the escorts *Brilliant* and *St. Clair*. At Betsy's Landing, raiders used three field artillery pieces to fire down on a transport, the *R.B. Hamilton*, and placed a shot into the vessel. The *St. Clair*, under Acting Volunteer Lt. Charles Perkins, steamed up and engaged the guerrillas, scattering them with several well-placed shells.

From January 24 to March 15, 1863, seven escorted convoys consisting of 180 steamers and 30 barges successfully made the trip between Smithland/Dover and Nashville. Soon, however, such success promoted complacency. On April 3, a convoy under the command of Acting Volunteer Lt. Jacob Hurd steamed toward Nashville with the most valuable boats up front and the slowest in the rear (contrary to orders). The convoy included the steamers *Eclipse* and *Lizzie Martin* lashed together, the gunboat *St. Clair*, the transport *Luminary*, the towboats *Charles Miller* and *J.N. Kellogg*, each drawing barges, and the slow gunboat *Fairplay*. The convoy reached Palmyra (between Dover and Clarksville) at 10:30 pm. From a bluff, Confederate gunners under Lt. Col. Thomas G. Woodward opened up with a 10-pounder Parrott rifle and a 12-pounder smoothbore. At 400 yards, the *Luminary* was hit with balls from 60 musketmen and the *St. Clair* was hit by small arms, canister, and at least six shells, one of which severed the steam supply pipe and disabled the gunboat. The executive officer's right knee was shattered by a six-pounder shell. The *Luminary* took the gunboat under tow. The *Kellogg* ended up towing the disabled gunboat back to Smithland, where Fitch was informed of the attack. He telegraphed headquarters: "I leave in 10 minutes for Palmyra with all the boats. Will whip them out."

The timberclad *Lexington* and the four tinclads in the avenging flotilla reached Palmyra the afternoon of April 4. While Fitch generally maintained a lenient attitude toward occupants of the surrounding countryside, in this case he agreed with the other naval officers that Palmyra should pay the price. The village had the reputation as "one of the worst secession places on the river" and as a notorious gathering place for guerrillas, bushwhackers, and other scoundrels. Guerrilla attacks were organized at meetings held in the saloons on the court square in Charlotte, seat of Dickson County. The town was equidistant from Palmyra to the north and Harpeth Shoals to the northeast. A landing party was sent ashore under Acting Master James

32-Pounder Naval Gun

Guerrillas used Charlotte as base of operations

In 1860, 300 people lived in Charlotte, the Dickson County seat. During the war, the residents witnessed considerable military activity, beginning February 17, 1862, when Confederate Col. Nathan Bedford Forrest arrived here to re-equip his men and horses after escaping the surrender of Fort Donelson. Late in 1862 and early in 1863, local guerrillas used Charlotte as a base. Col. Thomas G. Woodward's band of partisans and Gen. Joseph Wheeler's Confederate cavalry raided the Union transportation center at Harpeth River Shoals on the Cumberland River, 6 miles northeast. In 1863 and 1864, Federal forces built the Nashville and Northwestern Railroad from Nashville to the Tennessee River.

Historical Marker located at the Charlotte Co. Courthouse, 22 Court Square, Charlotte, TN 37036

Fitzpatrick with orders to burn down Palmyra but not to loot. Residents of the village were given notice to evacuate immediately. One Federal officer wrote later: "It was clean work — every building was in flames and falling." Fitch reported that they had "burned the town; not a house left; a very bad hole; best to get rid of it and teach the rebels a lesson."

An amphibious force of gunboats and infantry set out from Clarksville to pursue the partisans at Betsy's Landing upstream, but the rebels were forewarned and disappeared. Fitch had hoped to employ his counterinsurgency technique to bag the bushwhackers. He intended to engage the rebels on the shore and hold them long enough so that the Union cavalry and infantry could sneak up behind and attack them against the river.

On April 9, two small steamers, *Saxonia* and *R.M.C. Lovell,* left Dover without waiting for gunboat escorts and were captured at nightfall 15 miles downstream from Clarksville by Woodward's 2nd Kentucky Cavalry and local guerrillas. The boats and their cargoes were burned and eight blacks from the crew were shot and killed. When the captain of the *Lovell* protested, he also was shot and killed. Rear Admiral Porter noted that the transport captains "paid the penalty of disobedience."

In six weeks time, Woodward, commanding the 2nd Kentucky Cavalry, White's Texas Battalion, and King's Missouri Battery, sank two gunboats and four armed transports, crippled six other transports, killed 157 men, and destroyed $250,000 worth of property, according to Confederate records.

Despite the relative success of the convoy system and the constant presence of the Union gunboats, the pro-Southern local population remained resistant. Nannie Haskins, 16, belonged to a prominent family in Clarksville whose house overlooked the river. At dusk on May 12, 1863, Nannie wrote in her diary, "Those hateful gun boats! They look like they are from the lower regions. Now this the second night that four of them have been anchored in the river opposite our house. I know they are frightened, they have placed their gunboats so that if an attack is made, they can shell the town. Poor cowards, I can just turn my head now and see them crawling about on the boats like so many snakes." The teenager added: "The whole county is alive with robbers, every night we hear of a new robbery and perhaps a murder. Every minute of the day we hear of something startling, which four years ago would have made us 'shake in our shoes,' now merely give them a passing thought. War has hardened us."

Brevet Brigadier General Lewis B. Parsons served as the efficient quartermaster stationed at St. Louis. For the fiscal year ending July 30, 1863, Parsons estimated that steamboats had brought 136,000 troops and 338 million pounds of goods through Nashville, while the rails handled 193,000 troops and 153 million pounds of goods. By March of 1864, there were 50 steamer transports tied at the wharf at Nashville, the largest number ever, according to the *Nashville Times* newspaper.

How effective were the guerrillas against river shipping? According to Parsons, direct guerrilla action against river transports on all the Western rivers (Mississippi, Tennessee, Cumberland) during the entire war resulted in the sinking of 28 vessels weighing 7,065 tons and costing $355,000. Regular Confederate forces sank another 19 vessels weighing 7,925 tons and valued at $518,500. Most effective of all, Confederate secret agents using incendiaries sank double the tonnage (18,500) and 29 boats worth $891,000.

Confederate cavalry raids diminished after the spring of 1863. "Confederate overconfidence, combined with the inventiveness and flexibility of Federal commanders in confronting the raiding threat, negated raiding as a strategy in 1863," according to historian John Mackey. "After that summer, the Confederacy tried other raids, but when facing a veteran Union cavalry, fortified rail lines, the overwhelming material might of the north, and the slow attrition of their own mounted units, they

USS *Fairplay* (No. 17) tinclad gunboat

"You can never go wrong in doing a rebel all the harm you can."
— Rear Admiral David Dixon Porter

> *No one person exemplified the deadly impact of the partisan guerrilla than Jack Hinson, a lone sharpshooter whose vengeance against the occupying Federal forces resulted in dozens of kills.*

had little chance of changing the course of the war. The summer of 1863 was the last real opportunity for organized raiding by conventional cavalry forces…"

No one person exemplified the deadly impact of the partisan guerrilla than Jack Hinson, a lone sharpshooter (or sniper) whose vengeance against the occupying Federal forces resulted in dozens of kills, according to Hinson's biographer, Tom C. McKenney, who noted, "Elements of nine regiments, both cavalry and infantry, and an amphibious task force of specially built navy boats with a special-operations Marine brigade targeted the elderly man with a growing price on his head. They never got him."

Captain Jack Hinson, as he was known, was in his late 50s when he ordered a specially made .50-caliber rifle (the gunsmith was William E. Goodman of Lewis County) and began targeting Federal officers for death. One of his favorite shooting dens was at Towhead Chute on the Tennessee River near Hurricane Creek in Benton County, Tenn. As the river flows northward through the narrow channel, the current increases significantly, and the paddlewheelers headed upstream worked hard against the current, barely moving. A Union officer might stand on the deck, admiring the view in his navy blue uniform. Hinson aimed between the shiny brass buttons of the victim's jacket. For each kill, he would stamp a round notch onto the octagonal barrel of his heavy rifle. By war's end, the plain-looking rifle was adorned with 36 notches (McKenney estimated that Hinson's actual kill total was close to 100).

What had motivated this lone sharpshooter to seek such vengence? Hinson was a wealthy landowner and farmer in the area between the rivers, now known as the Land Between the Lakes. At first, he was a Unionist, opposed to secession, even though he owned slaves. The invasion and occupation of his homeland, and the increasingly harsh treatment of locals by the Yankees, forced him to change his mind. His estate was called Bubbling Springs, located three miles southwest of Dover. General U.S. Grant at one time was a guest in the house of Hinson and his wife, Elisabeth. The couple had eight sons and two daughters. One son was a Confederate soldier stationed in Virginia, who was wounded and died in 1865. Two other sons were briefly imprisoned as spies during the Battle of Fort Donelson. The eldest son, Robert, was involved with a local guerrilla group. At one point, Robert was imprisoned at the county courthouse but escaped by jumping out a second-story window. His horsemen became known as Hinson's Raiders. He was captured a second time and questioned by authorities in Nashville who never realized exactly who he was. He managed to either escape or he was paroled. On Aug. 6, 1863, Hinson's men attacked Fort Henry and seized the telegraph station.

One morning in the fall of 1862, two of Jack Hinson's younger sons were out on horseback, armed and hunting game, when set upon by a roving Federal cavalry patrol. The Hinsons, ages 17 and 22, were assumed to be spies. They were tied to trees and summarily executed by firing squad. Their lifeless bodies were put on display in the courthouse square in Dover. Then the lieutenant commander of the patrol beheaded the bodies with his saber. As the patrol made its way to Bubbling Springs to confront Jack Hinson, they placed the severed heads of his sons on the gateposts at the Hinson home. A doctor and friend of the family was able to subdue the enraged Jack Hinson and save him from being killed or imprisoned by the patrol, which eventually left the scene. At this point, Hinson, of Scots-Irish descent, determined to save his family and vowed vengeance, however long it might take. He needed time to secure his family and estate, order his custom-made sharpshooter rifle, obtain supplies, and scout shooting locations.

Hinson's first victim was the lieutenant of the patrol that killed his sons. Hiding and waiting in the woods, Hinson shot him in the saddle while leading a patrol. A short time later, Hinson was out and about and sighted a Yankee patrol which included the sergeant who had placed the boys' heads on the gateposts. On the patrol's return to base, Hinson was waiting. With cool efficiency, he shot and killed the sergeant and made his escape.

Eventually the Hinson family was forced to abandon Bubbling Springs, which was burned to the ground by soldiers, and relocate. Hinson took his position at Towhead Chute and began picking off Federal naval officers on the gunboats. At one point, Hinson's shots were so devastating that the captain of one gunboat "surrendered" his vessel before realizing that there were no Confederate troops to accept his surrender.

Jack Hinson died peacefully nine years after the war at the age of 67. Hinson's rifle is now owned by a judge in Murfreesboro, Tennessee.

Johnsonville and the Nashville & Northwestern Railroad
River Depot and Rail System Vital to Union Conquest

On May 19th, 1864, a large party of celebratory Federal officials, including Military Governor Andrew Johnson, rode a passenger train 78 miles west from Nashville on the recently constructed Nashville & Northwestern Railroad to dedicate the new supply depot at Lucas Landing on the Tennessee River. Following a brief celebratory speech upon reaching their destination, Johnson named the new military facility after himself.

Barely six months later, after Johnsonville had proven vital to Sherman's campaign against Atlanta, Confederate cavalry attacked the depot and destroyed much of the riverfront facilities, plus four gunboats. Several days later, Johnson was elected Vice-President of the United States on the Union Party ticket with President Lincoln.

The building of the depot and the Nashville & Northwestern Railroad "remains as one of the greatest engineering feats of the Civil War," according to historian Jerry Wooten, author of *Johnsonville*.

In 1863, the necessity for a terminal on the Tennessee River became apparent when low levels on the Cumberland River precluded steamboat deliveries at the Nashville wharf and the disabling of the Louisville & Nashville Railroad tunnel at Gallatin complicated shipments from Louisville.

[For more details on steamboat shipping, see author's *Iron Maidens and the Devil's Daughters*.]

The Nashville & Northwestern Railroad had been built from Nashville to

Andrew Johnson

Kingston Springs when the war halted construction. On Oct. 22nd, 1863, Secretary of War Stanton ordered the N&NWRR to be completed for military purposes and placed Johnson in charge of the project. Col. W.P. Innes, the commander of the 1st Michigan Engineers, was placed in charge of the construction. At the last moment, the planners switched the terminus of the route from the riverport of Reynoldsburg to Lucas Landing, where the river channel was 380 yards wide and 60 feet deep.

Dissatisfied with the pace of construction, U.S. Grant replaced Innes with a personal friend of his, General D.C. McCallum, a native of Scotland, in February 1864. McCallum then appointed W.W. Wright to supervise the railroad work. Wright chose Lt. Col. John Clark as his chief engineer. Wright then had Grant order 2,000 mechanics and laborers sent from the North to work on the road. J.B. Anderson, general manager of military railroads, provided the locomotives and cars.

The project involved "a rather formidable amount of grading, bridging, track laying, and other work incident..." reported Wright. More than 50 miles of track were laid using seven different pattern of rails and 107,000 crossties, most manufactured on-site. The road required the building of 45 trestles, more than four miles in combined length and requiring four million feet of lumber. The giant trestles at Nashville, Sullivan's Branch, and Johnsonville were 2,151 feet long, 1,326 feet long, and 1,525 feet long, respectively. The crews built 34 buildings, including tool houses, houses for trackmen and switchmen, telegraph offices, blacksmith and wheelwright shops, sawmill and freighthouses. These required 1.8 million feet of lumber and 742,200 shingles. Many of these buildings were later destroyed. Fourteen water stations were built, most destroyed.

On Sept. 1st, 1863, work commenced on the new section of track. On Sept. 27th, Johnson advertised for 1,000 men, white or black, to work on the railroad. Slave owners would be paid $300 for each worker, and freedmen would be paid $10 a month. By October 3rd, however, only 230 blacks had responded. Missouri engineers came from Corinth, Miss. to help

JOHNSONVILLE ~ OCTOBER 1864

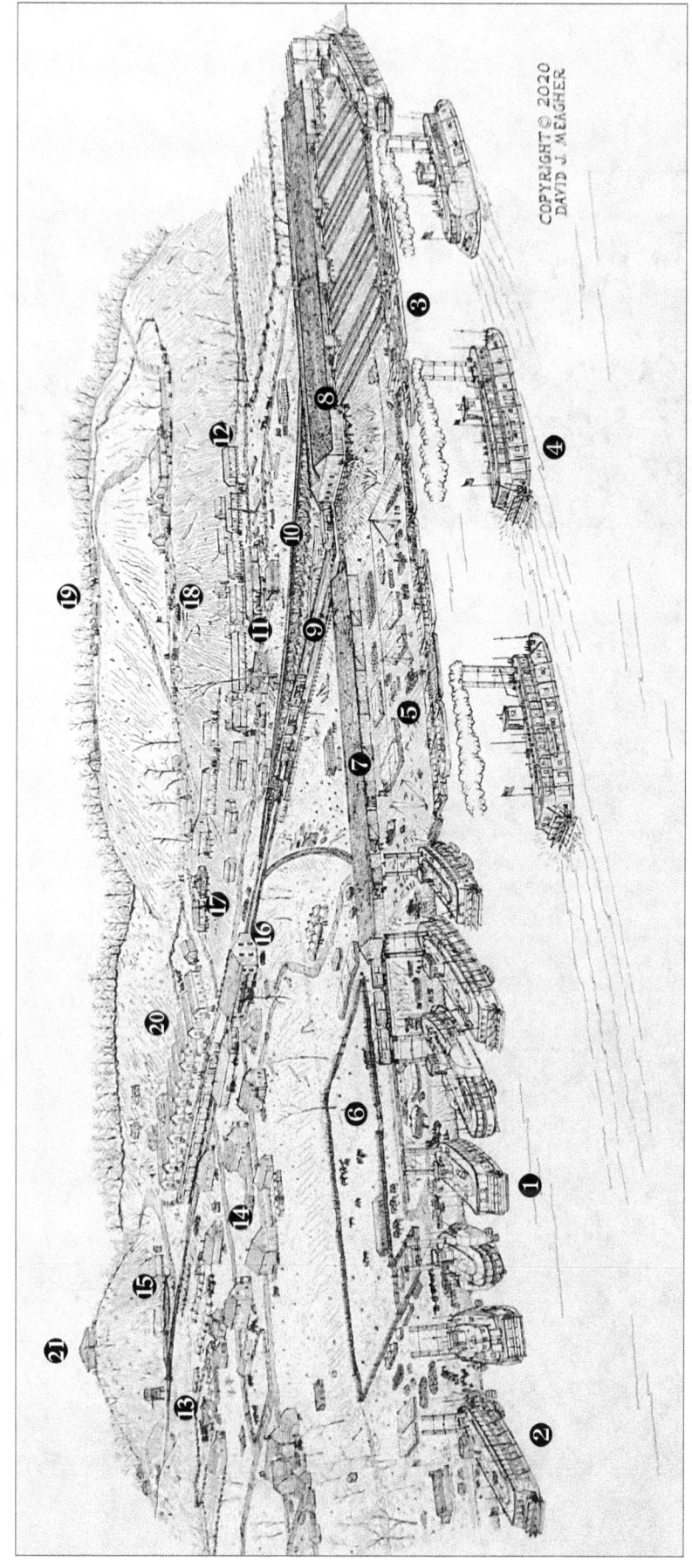

The Johnsonville depot looking southward, upriver on the Tennessee, showing steamboat transports (left) waiting to unload their cargoes at either the small transfer building (middle) or the larger mechanized transfer building (right). Tinclad gunboats on the river are the *USS Elfin* (52), *USS Key West* (32), *USS Tawah* (29), and *USS Undine* (55).

1. Transports
2. Transport *Venus*
3. Barges
4. Union gunboats
5. Open storage
6. Large corral
7. Small warehouse
8. Large warehouse, fitting machinery
9. Transfer sidings
10. Trestle to large warehouse
11. Sawmill complex
12. Worker housing
13. Soldier huts and tents
14. Civilian houses
15. Railroad turntable
16. Administration building
17. Small corral
18. Lower redoubt
19. Upper redoubt
20. Soldier barracks
21. Blockhouse

In 1864, the Nashville Quartermaster directed photographer Jacob Frank Coonley to document all bridges, trestles, buildings and railroad facilities used by the department. Coonley used a locomotive (No. 56), tender, and boxcar fitted out with darkroom, stove, cooking gear, bunks for five men, and a barrel of water. Many meals consisted of tough mule meat and boiled potatoes. Soldiers were detailed to protect the train, which rebels never managed to capture. Coonley photographed military railroads from Louisville to Atlanta. He was a protégé of George N. Barnard and also worked with Matthew Brady. He continued his photography career after the war.

Here, he has photographed his train at the beginning of the Sullivan Branch trestle on the N&NWRR ten miles west of Bellevue. The small redoubt from which he took the photo was surveyed and sketched in 1988 by Fred Prouty of the State Dept. of Archaeology.

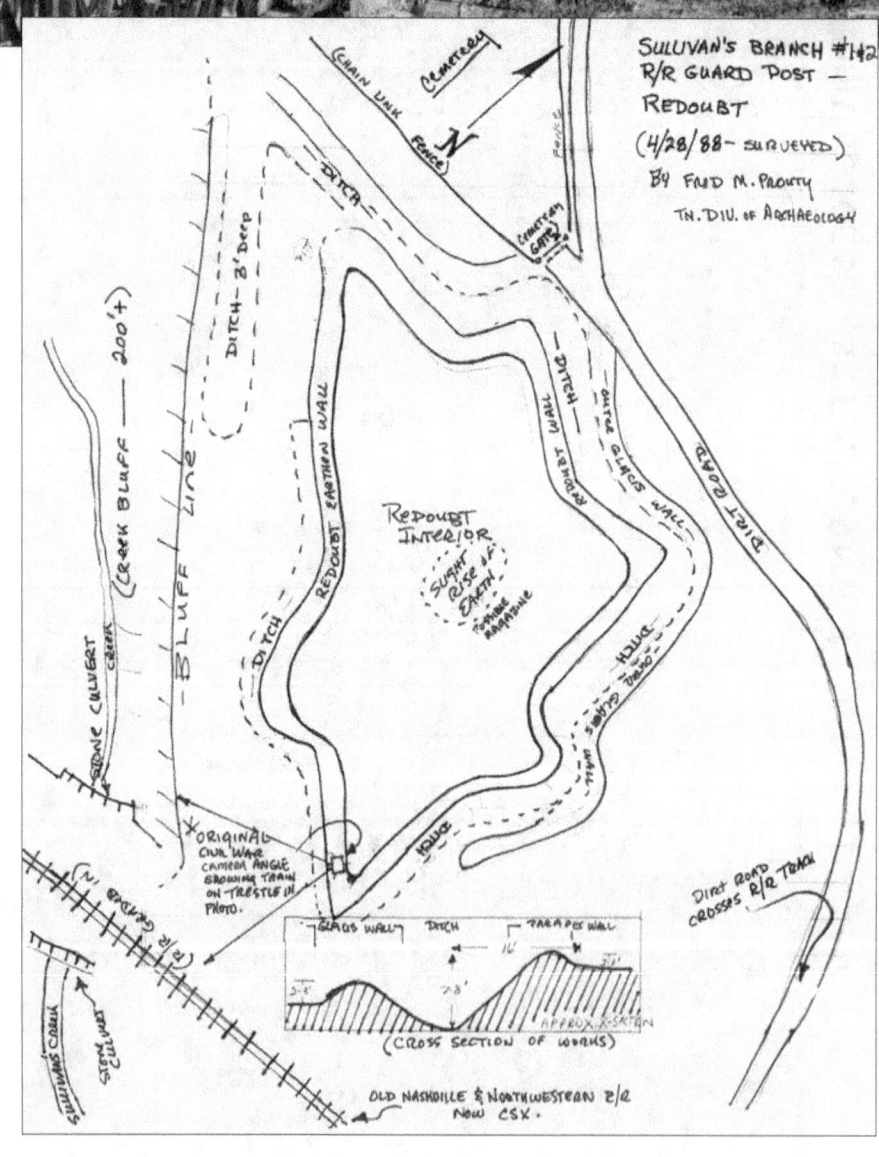

This is the first of five photos taken of Johnsonville by J.F. Coonley following Forrest's raid but prior to the battle at Nashville. This is the only wartime photograph of the civilian town at Johnsonville in the background. In the foreground is a burned-out railroad siding covered with protective tarpaulins. The small objects discarded along the railroad tracks are shoes fallen off from shipments. The damaged depot is muddy and wet from recent rains, with barren burned trees dotting the landscape littered with debris. (Library of Congress.)

build the new railroad. The First Regiment of Missouri Engineers began work on the railroad in February 1864, building the bed and then the blockhouses. They were organized by consolidation of Bissell's Engineer Regiment of the West (originally formed July 1861) and the 25th Regiment Missouri Infantry. In August, they were transferred to Sherman's army in northern Georgia and participated in the March to the Sea and the Carolinas Campaign. Officers included the brilliant engineer Col. Henry Flad, a native of Germany. The regiment lost 163 men during the war, 16 men killed or mortally wounded, and 147 men (including one officer) killed by disease.

While at Johnsonville, the engineers were ordered to disarm some government workers in the quartermaster department who had mutineered for a wage increase. The engineers then escorted the malcontents to the state prison in Nashville. On another occasion, a dispatch rider ran into a squad of guerillas between Johnson-

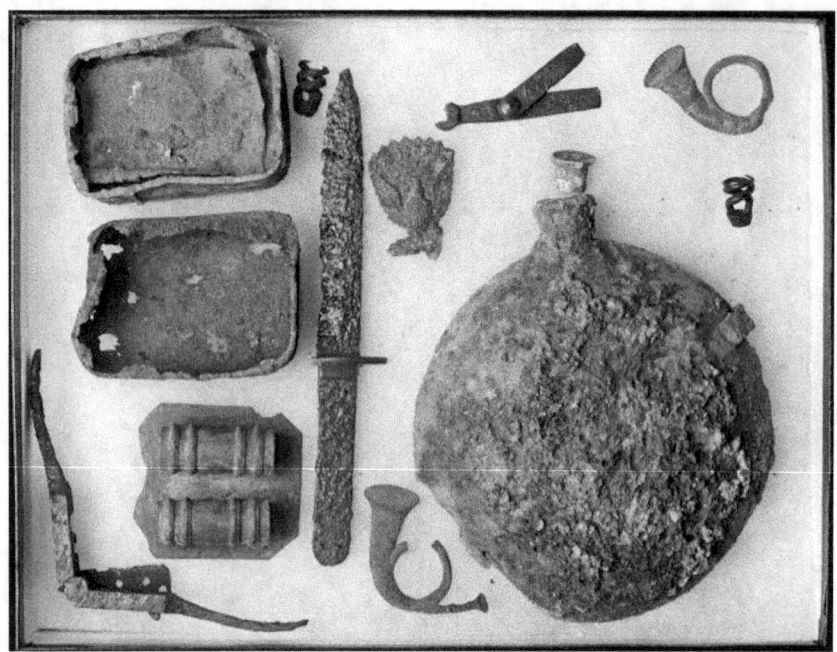

ARTIFACTS found in a trash pit at the campsite of the USCT 12th and 13th Infantry Regiments along the Nashville & Northwestern Railroad line near Johnsonville include sardine cans, a drumstick holder, a bullet mold, a side knife, a U.S. eagle hat device for a Hardee hat, two hunting horn insignia infantry hat badges, a Federal canteen with a homemade lead stopper, a gun tool, and two worms (wipers) used to clean out a .58 caliber musket. (Tennessee State Library and Archives)

ville and Waverly and was shot from his horse. One rebel then emptied his revolver into the victim. The rider was brought back to camp alive, with a bullet in his head, one in the shoulder, another in the left leg, and survived.

Free blacks and contrabands (runaway slaves) were impressed by the army to build the railroad. Many of the impressed blacks were recruited into the 12th and 13th Regiments of the U.S. Colored Troops and began work on the railroad in November 1863, one month after the 100th USCT, mostly from Kentucky, began work along with two companies of the 40th USCT. Between 5,000 and 7,300 USCT soldiers are estimated to have worked on building the N&NWRR.

The 12th USCT (200 men) was relieved on April 23, 1864 and the 13th (500 men) on May 10th. Many were then assigned to garrison blockhouses

Cut 29 on the Nashville & Northwestern Railroad east of White Bluff. Confederate cavalry attacked Iowa, Michigan and USCT positions here defending the railway. (Tennessee State Library and Archives)

One page of the labor roll of black slaves impressed in October 1863 to work on the Nashville & Northwestern Railroad shows that some actually worked at Fort Gillem in Nashville. Presumedly the owners of these slaves were Unionist. (Library of Congress)

and guard the bridges and trestles of the N&NWRR. They earned $11 per month for guard duty (freedmen earned $20 a month to work on the railroad). The blockhouses were built under the supervision of Capt. W.E. Merrill, chief engineer of the Army of the Cumberland.

General Alvan C. Gillem, a native of Jackson County, Tenn., was used as a utility man by Federal officials. He was charged with defending the railroad while under construction. His forces included three companies of the 10th Tennessee Cavalry, the 10th Tennessee Infantry, 1st Kansas Light Artillery under Capt. Marcus D. Tenny, 8th Iowa Cavalry of Col. John B. Door, Co. A of 14th Tennessee Cavalry under Lt. William Cleary, and the 43rd Wisconsin under Col. Amasa Cobb.

"This gang have their headquarters near Waverly, and they are supported and sustained by the whole community in that vicinity. Waverly is the nest of the vilest and most pestilential set of traitors that live, and the place ought to be destroyed."

— Col. William P. Lyon, 13th Wisconsin, July 29, 1863

By February 1864, crews were laying one-third to three-fourths of a mile of track each day. River barges brought two locomotives and several rail cars for the new line.

In March 1864, Gen. Gillem reported to Johnson: "I have just returned from the Northwestern Road, it is now progressing finely. I passed over 40 miles on the cars-and the track laying is going on well. An engine has gone to the other side of the road, and there is force enough to lay a mile daily of track."

The N&NWRR was completed on May 10th, 1864 and accepted as a U.S. military railroad on August 6th. At this point until the end of the war, W.R. Kingsley, who had served as division engineer since April 1864, was put in charge of maintaining the line.

McCallum estimated the cost of the railroad at $1,471,397 ($25 million in today's dollars).

Johnson told Stanton on Aug. 19th, 1864: "The importance of the Northwestern Railroad is now being seen and felt and our Army could not be sustained without it."

The Johnsonville depot comprised 90 acres and included an elevated trestle, guarded by a blockhouse and supplemented with a turntable, separating into north and south spurs leading to two large transfer buildings. The compound included civilian houses, soldier's huts and barracks, sawmills, horse corral, machine shops, telegraph offices,

Waverly fort controlled valley, railway

Fort Waverly, aka Fort Hill, was an earthen fort constructed by the 12th and 13th U.S. Colored Infantry to protect the Nashville and Northwestern Railroad. The 1st Kansas Artillery was stationed there. The 8th U.S. Iowa Cavalry and the 1st Kansas Battery also spent time as garrison troops stationed at Waverly.

This field fortification consisted of an irregular shaped redout that was constructed between 1863-1864. The redoubt was located on the military crest of the hill overlooking Waverly and a section of the original route of the Nashville and Northwestern Railroad.

The circumference of the walls was over 25 feet and the interior encompassed approximately one acre. The walls of the redoubt were surrounded by an outer ditch that measured seven feet to eight feet from the bottom of the ditch to the top of the wall. The interior of the fortification was about three feet lower than the top of the rampart wall with a slight rise in the center.

An irregular projection, or bastion, from the main work on the northeast corner of the redoubt would have been used for the placement of a cannon. Is it uncertain if the artillery piece would have been firing through an embrasure in the fort wall or over the crest of the parapet wall, called firing in barbette, that would allow the gun a wider range of fire.

An opening in the east wall of the fortification was used as the main entrance or sallyport in and out of the redoubt. The position high on the hill gave the artillery positions a commanding advantage over the valley below.

Two skirmishes took place in Waverly during the war. The first skirmish was on Oct. 22-25th, 1862, between a group of Napier's Confederate guerillas and a detachment of the 83rd Illinois Volunteers who, after three days of fighting, forced the Confederate guerillas to surrender. Another skirmish of Jan. 16th, 1863, involved a raid on Waverly that resulted in the capture of Confederate soldiers along with their horses and weapons. The town served as headquarters for Confederate guerilla forces during the early years of the war.

Fort Hill was the headquarters of the 13th USCT, led by Col. John A. Hottenstein, from the fall of 1863 to the end of the war.

Lt. James Nicholas Nolan of the 1st Kansas Battery returned to Waverly after the war, and in 1870 he built a house on the hill opposite the fort where he served. A successful businessmen, he later served as a city alderman and mayor of Waverly.

Fort Hill and the Nolan House are both listed on the National Register of Historic Places. The earthworks can be observed today adjacent to the Humphreys County Museum. The Nolan house is a bed-and-breakfast.

Raids Upon the Nashville & Northwestern Railroad - 1864

OFFICER ASST. INSPECTOR RAILROAD DEFENSES,
DEPARTMENT OF THE CUMBERLAND,
Eastern Section Nashville and N. W. Railroad, Section 20,
October 25, 1864.

SIR: In compliance with instructions received yesterday from your office, dated October 22, I have the honor to report the following particulars of the attack upon trains at section 36, Nashville and Northwestern Railroad, on the morning of the 18th instant; also, on the afternoon of the 21st instant:

The track repairers at section 36 were taken prisoners by McNary's gang (variously estimated at from 15 to 40 men, while some place the number at exactly 23) on the night of the 17th, about 12 o'clock, and held till late on the following morning, and made by McNary to draw the spikes from a rail and remove the fastenings at its end so as to be loose. The gang then drew back from observation, and in this condition of affairs the first a.m. train passed safely by them, except that a shower of bullets was poured in, which wounded a surgeon, Hogle, Engineer E. Andrews, and killed a boy, who was cook and brakeman, dead on the bunk, where he happened to by lying. The second a.m. train came to the loose rail and ran off; the engineer and fireman were wounded. Everybody was stripped of whatever money, watches, or valuables they had which pleased the fancy of the robbers. The locomotive was upset and slightly injured by cutting places with axes. One box-car was burned, but their efforts to burn the flat-cars loaded with iron, which composed the balance of the train, were not successful, and these were slightly injured. The third train, loaded with sawed timber from Ayres' saw-mill at section 29, ran up and was fired into. All hands jumped off and were robbed, except Engineer W.H. Stevens, who ran the train back to section 32, White Bluffs, in safety. Meantime the first train, Civil Conductor Charles White, arrived at Sneedville, and Colonel Murphy, who was on board, had the telegrapher, G.W. Leedon, send a dispatch to Lieutenant Orr, at White Bluff's, to come on with his cavalry. The dispatch was promptly obeyed, and Lieutenant Orr arrived with twenty-five men twenty minutes after the gang had taken their departure, and pursued them a short distance unsuccessfully, and his horses being tired and inferior he returned. A wrecking train was dispatched with hands from Gillem's Station, section 51, to clear the road, and Lieutenant Cox, with a detachment of Company B, One hundredth U. S. Colored Infantry, and Captain Frost, with a detachment from companies of the Twelfth U. S. Colored Infantry from Sullivan's Branch, were sent to section 36, and the road made clear on the following morning, 19th instant.

Again on the 21st instant, as the p.m. train for Johnsonville was passing section 36, it was signaled by the section foreman, whose cook had informed him she had seen men tearing up the track. Captain O.B. Simmons, military conductor, had the train stopped, and with his large train guard pursued the bushwhackers, whose numbers could not be ascertained, for a considerable distance, but as they were mounted the pursuit was unavailing. Civil Conductor Charles White fastened down the rail and the train passed on. Afterward the gang returned and burned the house and commissary of the section foreman, who lay in the bushes in sight. They also burned nearly all the negro and other dwellings along the railroad for two miles. Piles of wood at sections 38 and 39 were burned, and various estimates placed the loss in wood at from 3,000 to 15,000 cords. The wood being in several ranks close to the road many ties were burned at the ends, and the rails warped by the intense heat, so that the 3 o'clock train for Nashville could not pass. The telegraph operator at Sneedville called operator at White Bluffs, section 32, and while calling the line was cut before getting an answer. Captain J.W. Dickins, at Sneedville, went to the burning wood with part of this company, and arrived in time to hear the retreating bushwhackers laughing and talking, but was not able at that time (11 o'clock night) to do anything, and returned to Sneedville. On the 22nd Military Conductor Captain Van Skike, from Nashville, found out the condition of the road at sections 38 and 39, and took a detail up from White Bluffs and repaired the road as soon as possible so that trains ran through on the 23rd of October.

I have made no delay in gathering the materials from authentic sources for this report, and hope it may prove acceptable.

WILLIAM L. CLARK,
First Lieutenant, Twelfth U. S. Colored Infantry,
Division Inspector Eastern Section Nashville and Northwestern R. R.

Major JAMES R. WILLETT,
First U.S. Vet. Vol. Engrs., and Chief Insp. Railroad Defenses.

The larger south transfer building (Warehouse No. 2) showing the tracks and railcars next to the siding. The Tennessee River would be just to the right of this view. Under the tracks can be seen the channels used by the freight-hauling apparatus that pulled cargo up the inclined wharf to the warehouse. Loaded on the siding are dozens of stacked wheelbarrows. The double-roofed warehouse measured 600 x 90, the largest building at Johnsonville. (Library of Congress.)

blacksmith and wheelwright shops, freighthouses, and other structures, 190 in all. The long wharf was macadamized and sloped at 14 degrees.

W.W. Wright described the transfer of materiel from steamboat to freight train at Johnsonville:

"...two large transfer freight-houses were designed and built, one on each side of the railroad, with tracks starting from main line at the bluff and curving right and left until parallel with the buildings and river bank. The freight-house or shed on the north or lower side, 600 feet long by 30 feet wide, was hastily knocked up so as to bring it into immediate use, and the levee in front graded off to the water's edge with a slope of 9 degrees or about 16 feet rise in 100 feet horizontal. The freight-house on south side, 600 feet long and 90 feet wide, was a much more complete building. The floor was two feet and a half above high-water mark and the levee in front graded to a slope of 14 degrees, on which it was designed to lay railroad tracks from low-water mark to floor of freight-house. The plan for transferring freight from steam-boats to cars was to load from the boats onto small cars, which were hauled up the levee to the level of the freighthouse floor by a wire rope passing round a pulley or spool, which was dropped into or lifted out of gear with the main shaft by a lever. This main shaft was 500 feet long and passed through the center of the building immediately below the floor or platform and was operated by an engine located in the middle of the building. The freight was then passed directly through the building and loaded into cars on the opposite side. The levee was of sufficient length to allow at least four or five boats to unload at the same time, and the side tracks were so arranged that a whole train of cars could be loaded at once, and as soon as loaded could be moved away and another train run right alongside the house."

The Johnsonville depot included a turntable for rotating locomotive engines. The turntable was a large circular pit with a stone outer base. Located in the middle of the pit was a central pivot (made of iron and wooden timbers) that supported a single section of track called a bridge. As trains arrived empty from Nashville, rail cars were uncoupled and left at the supply warehouses located along the bank of the Tennessee River. The engineer advanced the locomotive onto the turntable bridge. A rail running around the floor of the pit supported the ends of the bridge when a locomotive entered or exited. Rotation of the track was accomplished by two men pushing on gears to swing the bridge on the central pivot. The engine was turned until the locomotive faced in the opposite direction for the return to Nashville. Constructed in late 1863, the outer stone base and pivot foundation are all that remain. In 1890, the Civil War era railroad bed was raised by the Nashville Division of the Nashville, Chattanooga, and St. Louis Railroad. This improvement made the right side of the turntable inoperable.

In May 1864, Assistant Quartermaster Charles A. Reynolds was named quartermaster at Johnsonville. When

Standing on a rail platform are members of Battery A, 2nd U.S. Colored Light Artillery in their uniforms and raingear. Arrayed in front of them are ten sets of 12-pounder howitzers and limbers, minus their horses. The horse corral can be seen in the distance beyond the rows of tents. The Tennessee River runs to the left with rail cars visible on a siding. The artillery crews are preparing to ship out to Nashville. (Library of Congress.)

he was reassigned elsewhere in July, he was replaced by Capt. Henry Howland, who shared quarters with his brother Walter, a quartermaster department lieutenant.

Once the railway was built, the 13th USCT guarded the western end, with its regimental headquarters at Waverly. The 12th concentrated at the eastern end, although one of its companies furnished the provost guard at Johnsonville. Companies of the 100th USCT guarded bridges and trestles in the central part of the line. Overall, 1,900 men were posted at 22 sites along 78 miles of track.

In mid-October, Colonel Charles R. Thompson, 24, a native of Maine and commander of the 12th USCT Regiment, arrived at Johnsonville and took command of all land forces. The garrison at the depot totaled 2,270 men and consisted of the 700 men of the 43rd Wisconsin; the 500 men of parts of the 12th, 13th, 40th, 41st, and 100th USCT; and 800 quartermaster employees (many of them loyal Southerners). Also there was the 11th Tennessee Mounted Infantry, 2nd Tennessee Mounted Infantry, the 1st Kansas battery, the 2nd US Colored Light artillery, and quartermaster department artillery.

Four "tinclad" gunboats — converted steamers that were lightly armored but heavily armed — were stationed at Johnsonville under the command of Acting Volunteer Lt. Edward M. King. They were the *USS Key West* (King's boat, No. 32), the *USS Tahwah* (29), the *USS Undine* (55), and the *USS Elfin* (52). Four hundred naval officers and sailors bunked at Johnsonville.

On October 16th, Lt. Gen. Richard Taylor ordered Major Gen. Nathan Bedford Forrest to disrupt Sherman's lines of communications as General Hood and the Army of Tennessee invaded the state, moving north toward Nashville.

On October 29th-30th, at Paris Landing, on the Tennessee River south of Fort Heiman, Forrest's cavalry, armed with field artillery, set a trap and captured four Federal vessels. The *Mazeppa* freighter was looted and burned, as was the *J.W. Cheeseman*. The *Venus* freighter was armed with two large Parrott rifles and along with the captured *USS Undine* gunboat formed an ersatz Confederate brown-water navy. The U.S. forces suffered eight killed, 11 wounded, and 43 captured, while one Confederate was wounded. Captain Frank M. Gracey of the 3rd Kentucky Cavalry, a former steamboat captain, was given command of the *Undine* and Lt. Colonel William A. Dawson assumed command of the *Venus* and was named "Commodore."

In mythology, an undine is a female sea creative who roamed the seas looking for a man (ship captain) to marry so she can gain a soul.

As the mounted Forrest watched from the riverbank, the *Undine's* flag draped over his saddle, the newly appointed "horse marines" hoisted the Confederate flag up the standard and practiced maneuvers on the river for two hours to the yelps and shouts of the delighted men. During this time, another freighter, the *Anna*, escaped the trap, fled downriver to Paducah, and warned U.S. Lt. Commander LeRoy Fitch of Forrest's activities. Fitch then quickly launched a flotilla of six tinclads upriver — the *USS Paw Paw, Fairy, Curlew, Brilliant, Victory,* and *Moose.*

On Nov. 1st, Forrest's 3,400 troops began a muddy 40-mile trek south down the west bank of the Tennessee River accompanied by the *Undine* and *Venus.*

At Johnsonville, Federal commander Thompson wired Thomas in Nashville that he feared Forrest would attack, and added, "I have not now, nor have had any idea of surrendering. Will fight to the last if attacked. I feel confident that I can hold the place."

By Nov. 2nd, Forrest was six miles from the riverbank directly across from Johnsonville. That afternoon, the

The Horse Marines by artist John Paul Strain depicts Forrest and his command following the capture of the Federal gunboat *USS Undine* on the Tennessee River. Artwork used with permission.

Venus steamed ahead of the column, came around a bend at 3:30 pm, and faced two Union gunboats, the *Tawah* and the *Key West*, at Green Bottom Bar near Reynoldsburg Island. In a 20-minute engagement, the *Venus* was grounded at shore, with the crew taking to the woods, not bothering to set it afire. The Confederates aboard the *Undine* pulled back from the action. The *Venus* with its two Parrott guns was recaptured by the Federal tinclads and arrived back at Johnsonville by 6:30 pm.

Lt. King, commander of the gunboat flotilla at Johnsonville and aboard the *USS Elfin*, wired Lt. Commander James Shirk at Paducah: "All anxious about this place. Please send up more gunboats at once. We won't allow this place to fall into enemy hands, if our forces can prevent, but please send up more gunboats."

On Nov. 3rd, Lt. Commander Fitch set out from Paducah with his six-vessel rescue flotilla. After shelling the Confederate position at Paris Landing, Fitch's gunboats approached the Reynoldsburg Island sector, where the Confederates had placed five batteries, most of them masked. The east channel past the island was too shoal for any boats to pass; in the deeper main channel the *Undine* sat waiting, bait for the trap. As Fitch moved south, King steamed north. Cautious, King held back his three tinclads, sensing an ambush. By this time, at noon, Forrest noted that the *Undine* was nearly out of coal for its steam-engine boilers. No other actions were undertaken that day.

That night, local lore has it that Captain Jack Hinson, a prolific guerilla sharpshooter, led Forrest's troopers through the Cypress Creek swamp to favorable artillery positions across the river from Johnsonville. According to historian Michael Bradley, Colonel Kelley knew that the steep banks of the river made it impossible for the guns aboard the U.S. boats to hit targets on top of the banks because the targets were so far above the water. When the boat crews elevated their guns to hit targets on top of the high banks, the shells went over the top and landed far inland. The guns in the U.S. redoubt behind and high above the depot were also ineffective. As Morton told Forrest: "The fort is so elevated that they can't depress their guns sufficiently to affect me, and the gunboats are so much more below in

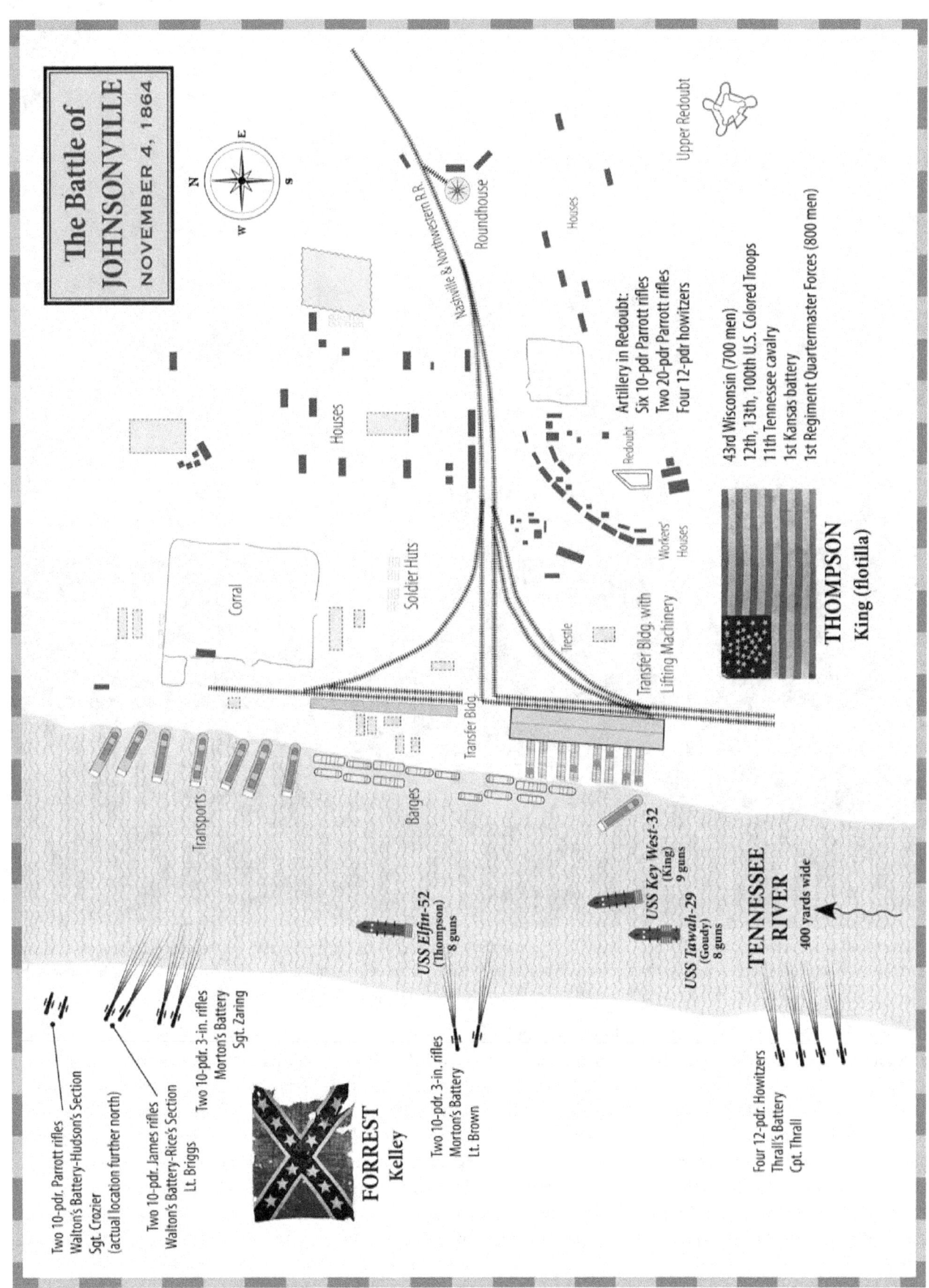

Looking southeast from the main railroad grade toward the south transfer-building railroad trestle, workers' barracks, and the lower redoubt where a solitary soldier in a rain poncho can be seen standing. Artillery pieces can be seen on the hilltop to the extreme right. In the middle distance is Sawmill No. 1 fronted by horse-drawn wagons inside of which a worker is guiding timber through a saw channel. (Library of Congress.)

the river that they will fire over me, and I'll be in an angle of comparative safety."

Meanwhile, on Nov. 4th, back at Reynoldsburg Island, Capt. Gracey aboard the *Undine* was caught between Fitch's flotilla to the north and King's gunboats to the south. Most of the Confederates had moved on to Johnsonville, so about 8:00 am Gracey guided their prized Yankee gunboat, nearly out of fuel, to shore, set the torch to it, and blew it up with a gunpowder charge. Abandoned in three feet of water 75 yards from Reynoldsburg Island, the *Undine* burned down to her waterline and then exploded spectacularly. During a melee with the remaining Confederate batteries of Hudson and Rice, the *Key West* was hit 10 times in the upperworks, seven through the decks, and two in the hull. The *Elfin* was also damaged. By mid-morning King's three gunboats returned to the depot. Assistant Quartermaster Henry Howland later reported: "The *Key West*, in advance, ran into a battery of heavy guns within two miles of Johnsonville and but a short distance above where the *Undine* was lying. She received 19 shots from 20-pounder guns, which passed entirely through her, before she could escape from this newly discovered battery."

North of the island, aboard the *USS*

This photo is a close-up of buildings featured in the previous photo, only from a different angle, more to the east. Again, the solitary soldier can be seen standing atop the hilltop redoubt. Foreground shows rails and crossties. In the middle distance can be seen an outhouse. It is obvious that the clearing of trees and vegetation has created muddy conditions. (Library of Congress.)

Moose, Fitch moved up at 2:30 pm after hearing cannon fire and engaged for 15 minutes with Hudson's guns, then fell back. For all day on the fourth, Fitch refused to run through the narrow channel gauntlet. Fitch wrote: "I thought it mere folly to attempt such a hazardous move, as I am almost confident that not over two boats out of six would have got through, and they never could have got back again." According to biographer Jack Smith: "The enemy batteries, narrowness of the channel, shoal water, and the possibility of a damaged vessel blocking the passage mitigated against any decision (by Fitch) to proceed (upriver)." Fitch's decision not to proceed was highly debatable, according to historian Edward Williams III, who claimed that Fitch was stopped by "only his imagination." Fitch's flotilla of six gunboats arrived back at Paris Landing on Nov. 4 at 10:30 pm.

Meanwhile, back near the depot, the Confederates were getting ready for action. Setting up his cannon, unnoticed by the enemy, chief artillerist Morton peered 400 yards across the river at the Johnsonville depot. He noted that "two gunboats with steam up were moored at the landing ... a third plied almost directly beneath the bluff." There were "a number of barges clustered around; negroes were loading them, officers and men were coming and going, and passengers could be seen strolling down to the wharf. The river banks for some distance back were lined with quantities of stores, and two freight trains were being made up. It was an animated scene, and one which wore an air of complete security."

The Battle of Johnsonville was "a convoluted mess," noted historian Wooten. Although warned of imminent attack, the Federal authorities at Johnsonville apparently did little to prepare for that eventuality.

At the depot were the three Union tinclads commanded by King, with a total of 25 guns. The Union redoubt, perched 100 feet above and behind the river depot, sported six 10-pounder Parrott rifles of the 1st Kansas Battery, two 12-pounder Napoleons of the 2nd USCT Artillery, two 12-pounder Napoleons of the 1st Regiment Quartermaster Forces, and the two 20-pounder Parrott rifles seized from the Venus. The Parrotts were capable of firing a 19-pound shell more than a mile at five degrees of elevation.

Across the river, the Confederate cavalry consisted of Colonel Mabry's brigade, Colonel Rucker and the 7th Alabama Cavalry, Lt. Col. Kelley and the 26th Tennessee Battalion, and Lt. Col. Logwood and the 15th Tennessee Cavalry. Forrest's artillery, arranged from north to south, consisted of the following:

- Sgt. Crozier commanded Hudson's section of Walton's Battery several miles north of Johnsonville, consisting of two 10-pounder Parrott rifles. These guns were not involved in the destruction of the depot.
- Lt. Brigg's section of Walton's Battery, consisting of two 10-pounder James rifles (U.S. guns captured at Brice's Crossroads)
- Sgt. Zarring of Morton's Battery, consisting of two 10-pounder 3-inch rifles
- Lt. Brown of Morton's Battery, consisting of two 10-pounder 3-inch rifles (directly across the river from the depot)
- Capt. Thrall's Battery of four 12-pounder howitzers

Precisely at 2:00 pm, ten field artillery pieces went off as one, aimed directly at the Federal gunboats. Howland testified that "…the cannonading was the most terrific I have ever witnessed," speculating that the rebels possessed up to 36 pieces of artillery. Within 40 minutes, the gunboats were abandoned. The *Key West* got underway but its paddlewheel fouled on a cable. The *Tawah* took the *Key West* in tow to a point upriver but eventually Commander King ordered all three boats abandoned and fired for fear of capture. Aboard the *Tawah*, the smallest of the three gunboats and known as "a miserable ship," one hundred shells exploded in a deafening cacophony. The gunners in the U.S. redoubt played almost no role in the battle. Inexperienced, they didn't respond to the attack because they couldn't see through the smoke.

Depot commander Thompson evidently assumed the Confederates would cross downriver and attack the depot by land. He ordered Quartermaster Howland to set fire to the barges at the wharf to prevent them from being captured. Fire from the barges spread to the depot warehouses and by 4:00 pm the entire wharf was an inferno. Howland said the warehouses ignited with the bursting of a rebel shell and "the intense heat of the burning boats (barges), which had been driven against the wharf by the strong wind, fired the stores in another place." He ordered the fires extinguished, but nobody obeyed his orders. The supplies in the wooden warehouses burned with a certain fury. A torrent of spilled whiskey ran ablaze from the warehouse down to the river.

Forrest said, "By night the wharf for nearly one mile up and down the river presented one solid sheet of flame."

Colonel Mussey of the 100th USCT reported that the behavior of the colored troops at Johnsonville was excellent … Some of the 13th USCT were upon the riverbank as sharpshooters, and armed with the Enfield rifle, and did good execution. The affair was slight, but it has gained credit for the colored troops."

Boat crews fled to the fort and interfered with the gun crews there. Four hundred Federals scrambled onto a freight train and chugged off eastward for Nashville. Twelve miles later, at Waverly, the railroad agent uncoupled the overloaded cars and sped to Nashville with just the locomotive. The abandoned Federal troops commenced to looting the railcars.

Back on the west bank, Forrest was in an unusually festive and feisty mood. Jostling with his gunners, he tried his hand at firing a cannon. "Elevate the breach of that gun lower," he commanded to his grimacing gunners. He said he wanted to rickety-shey a shell off the water. Needless to say, the gunners were not impressed with their commander's artillery skills.

That night Forrest moved his men six miles to the south. The Confederates did not cross the river. The next morning, he and Morton returned to the previous day's artillery positions to survey the damage. The depot was still smouldering and lay in ruins. The destruction totaled $1.5 million to $6.7 million, including the gunboats (more than $100 million in 2018 dollars). The Federals lost eight men killed and wounded.

Forrest told Morton: "There is no doubt we could soon whip old Sherman off the face of the earth, John, if they'd give me enough men and you enough guns."

Forrest later reported that the Federals lost four gunboats, 14 steamboats, 17 barges, quartermaster stores of 75,000 to 120,000 tons, and 150 prisoners. Forrest reported two killed, nine wounded, and two Parrott guns lost. On Nov. 7th, Forrest put 400 cavalrymen across the river at Perryville and soon linked up with Hood's army approaching Nashville.

After the Johnsonville debacle, rumors abounded throughout the North, including one that had Forrest approaching Chicago, where the local militia was called out. On Nov. 30th, 1864, the Johnsonville depot was abandoned and never used again, although the Federals did improve the upper redoubt on the ridge high above the river. The depot wreckage was not cleaned up for months.

News of the Confederate cavalry's exploits were received by Sherman, who on November 15th pushed off from Atlanta on his March to the Sea with the sardonic comment, "That devil Forrest was down around Johnsonville, making havoc among the gunboats and transports."

Commander King was court-martialed in May 1865 for burning the three tinclads, but he was acquitted of the charges. Witnesses testified that even if the gunboats had been scuttled in five feet of water (instead of burned) the Confederates could have salvaged them. Fitch reported, "The Key West, Tawah, and Elfin fought desperately and were handled in magnificent style, but it is impossible for boats of this class, with their batteries, to contend successfully against heavy rifled field batteries in a narrow river full of bars and shoals, no matter with what skill and desperation they may be fought."

Although Johnsonville was probably the most destructive raid of the war, and the result provoked near-hysteria in the North, the attack was too little too late to affect Sherman or any other Federal offensive operation.

As Hood approached Franklin and Nashville, people thronged the N&NWRR stations, trying to board eastbound trains headed to Nashville. At Johnsonville, the U.S. Colored Troops, including the 2nd U.S. Colored Light Artillery, were ordered to abandon the railroad, destroy surplus weapons, and return to defend Nashville. They arrived by December 7th, five days after Hood "besieged" Nashville. Two hundred soldiers of the 43rd Wisconsin withdrew from Johnsonville, marching to the Nashville & Chattanooga depot to sleep on the floor crowded with refugees, awakened at midnight by trains bearing 600 wounded from Franklin.

The Johnsonville depot used a railroad turntable that looked and functioned much like this one in Virginia.

The damage to the Nashville & Northwestern Railroad was inventoried following the battle at Nashville. Fifteen bridges had been destroyed and four others partially destroyed. All eleven water tanks along the route were damaged and two completely destroyed. Telegraph lines were cut in 19 places. All government and railroad buildings and warehouses were destroyed save for one blockhouse. One of the four sawmills along the route was destroyed.

The Federals rebuilt the N&NWRR bridges in January and February of 1865. Shortly after that, floods destroyed most of the bridges, which were then replaced by sturdier, prefabricated truss bridges.

The *Eclipse* was a 223-ton sternwheeler tied up at the old Johnsonville wharf, transporting the 9th Indiana Battery downriver. Capt. Vohris had complained to the quartermaster department that the boat's boilers were leaking and unsafe. At 6:00 am on Jan. 27th, 1865, the steamboat exploded at

The 1865 explosion aboard the Eclipse *killed 27 soldiers and wounded 78.*

the wharf, killing 27 men and wounding 78 others. General Thomas ordered two Ohio regiments and three Missouri batteries to Johnsonville in case the explosion had been sabotage.

In 1867, a 1,900-foot-long railroad bridge was constructed over the Tennessee River at Johnsonville, connecting Nashville with Memphis on the N&NWRR. The iron bridge used seven stone pillars and took ten months to construct.

In 1944, the federal government dammed the Tennessee River and created Kentucky Lake, the 25th largest in the nation. Along with Fort Henry, much of old Johnsonville (and several sunken gunboats) sits underwater. Johnsonville Historic State Park and the N.B. Forrest State Park (across the lake) offers a modern visitors center, interpretive signage, and trails exploring some of the finest extant earthworks and redoubts in the U.S.

Artifacts recovered from the steamboat wrecks during 1999-2010 are displayed at the visitors center. The boats were submerged in only 15 feet of water. The boats were the *USS Elfin*, *USS Tawah*, and two military barges, *Goody Friends* and *Chickamauga*.

Federal Garrison Towns in Mid-South Region

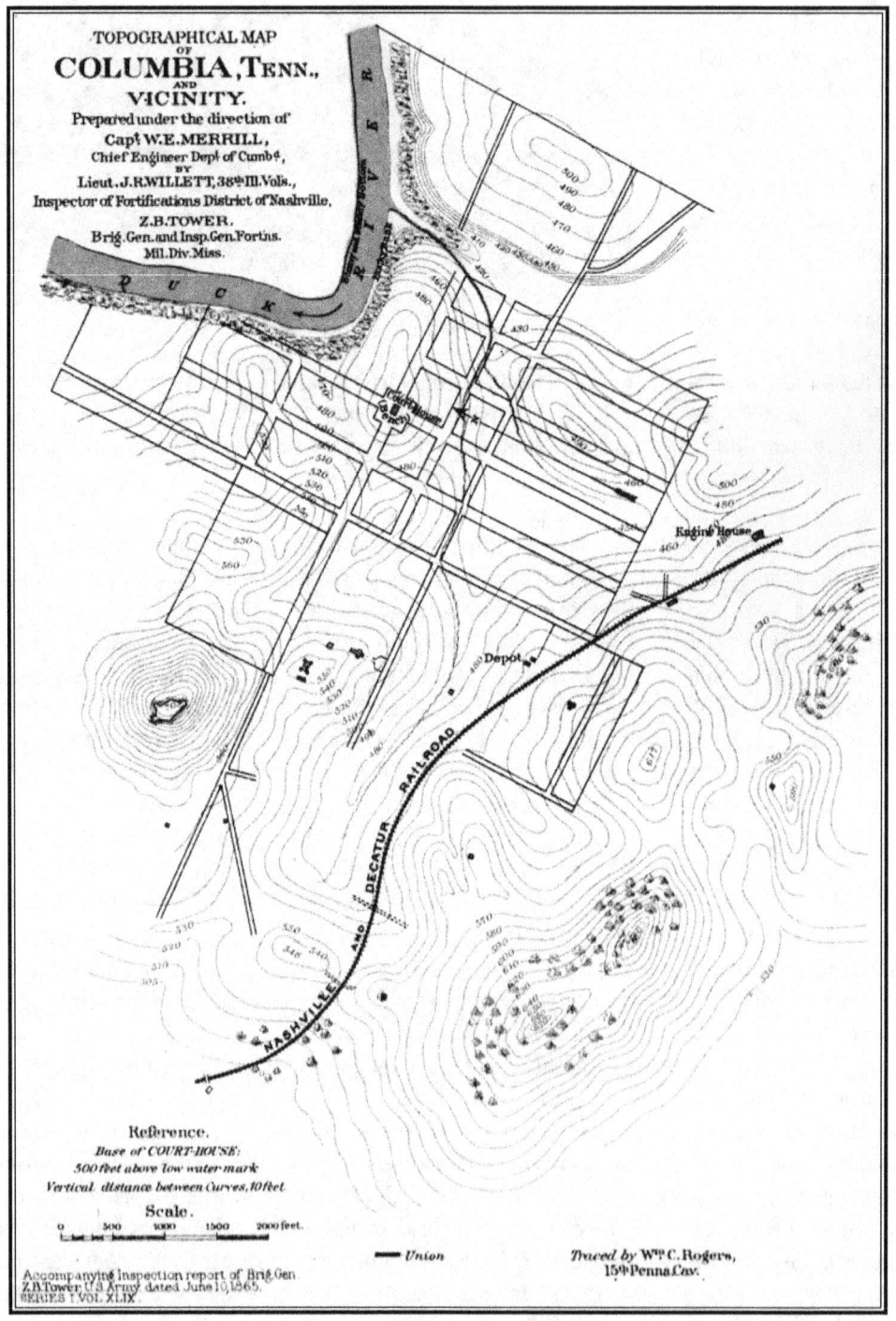

May 8, 1864 — From the *New York Times* correspondent in Nashville: "**Columbia**, a charming town about 40 miles south…, has been a notoriously disloyal town. The inhabitants…have taken oaths by the batch, yet still practice the most unheard of crimes, all arising from their ever-existing hatred to the Government. Something transpires in this Bedlam weekly of a distressing nature. On the 15th ult. two soldiers were found dead in the streets, one having a nail drove into his head…" Columbia was home to many fine mansions and the hub of a prosperous farming region, with many large plantations. At Columbia, Hood stole a march on Schofield during his 1864 invasion of Tennessee. The town on the Duck River was also a convenient spot to reform the Confederate rearguard during Hood's subsequent retreat following the Battle of Nashville.

Fortress Nashville: Pioneers, Engineers, Mechanics, Contrabands & U.S. Colored Troops

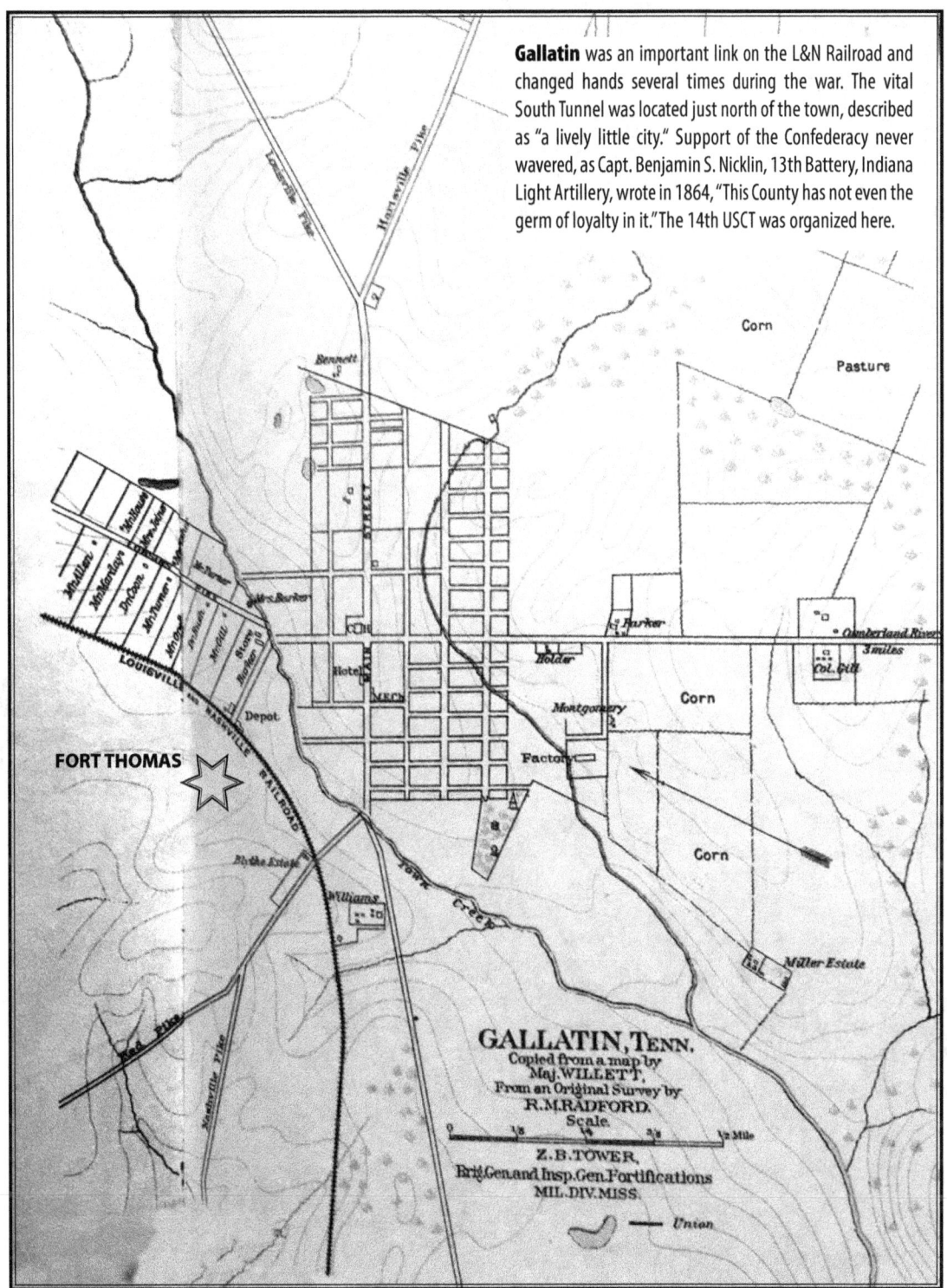

Gallatin was an important link on the L&N Railroad and changed hands several times during the war. The vital South Tunnel was located just north of the town, described as "a lively little city." Support of the Confederacy never wavered, as Capt. Benjamin S. Nicklin, 13th Battery, Indiana Light Artillery, wrote in 1864, "This County has not even the germ of loyalty in it." The 14th USCT was organized here.

All maps from *The Atlas to Accompany the Official Records of the Union and Confederate Armies*, unless otherwise noted.

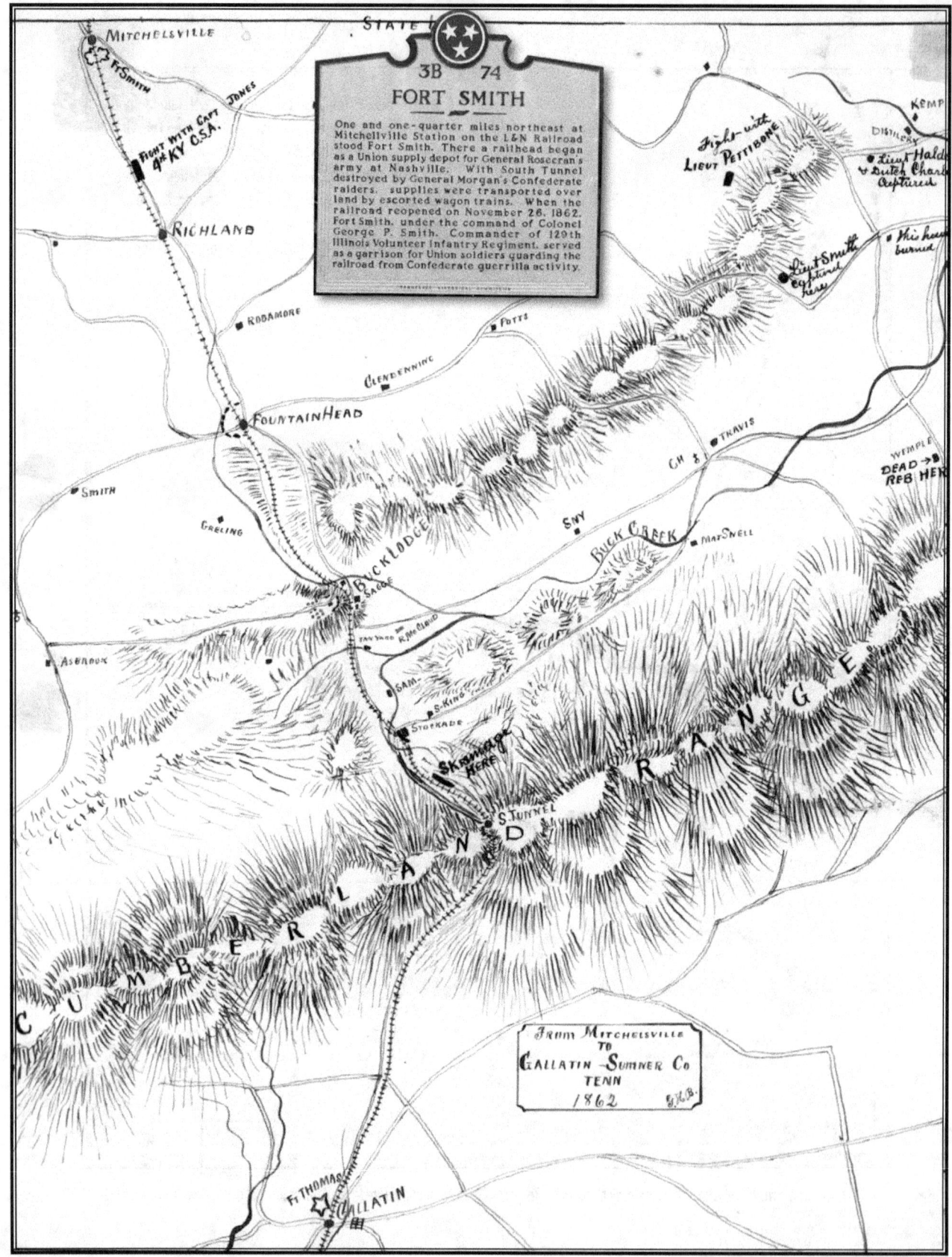

Sketch made in 1862 by George H. Blakeslee of the 129th Illinois Infantry Regiment, of South Tunnel through the ridgeline and the territory between Gallatin (Fort Thomas) and Mitchellville at the Kentucky state line. The next page shows drawings of various locations, including a diagram of Fort Smith at Mitchellville. (Library of Congress)

Sketch made in 1862 by George H. Blakeslee of the 129th Illinois Infantry Regiment, of various locations between Gallatin (Fort Thomas) and Mitchellville at the Kentucky state line, including a diagram of Fort Smith at Mitchellville. (Library of Congress)

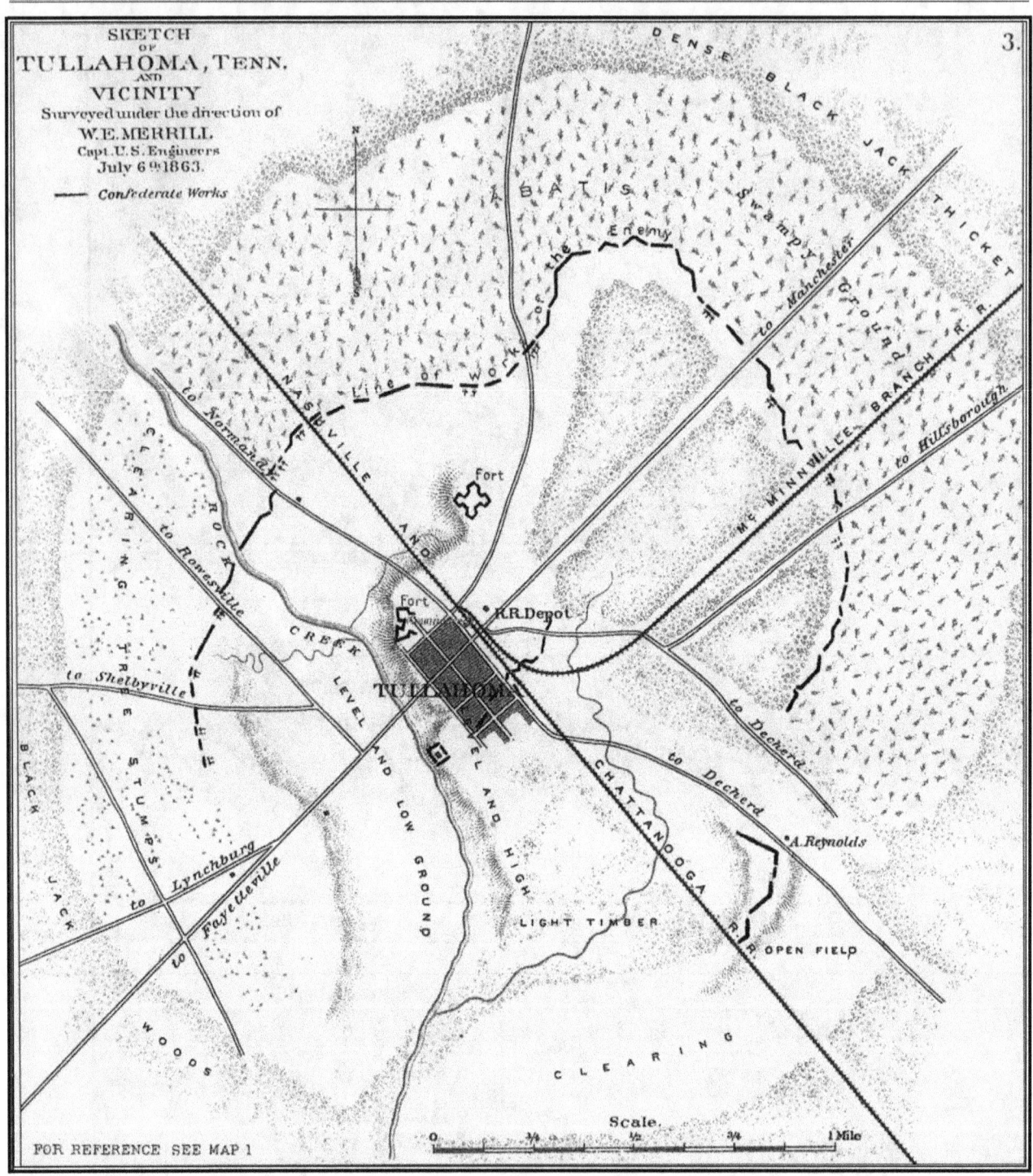

In January 1863, after the Battle of Stones River, Confederate Gen. Braxton Bragg's Army of Tennessee fortified **Tullahoma** to protect the supply depot and Bragg's headquarters. Tullahoma also served as the army's medical center, with divisional and general hospitals. The Nashville and Chattanooga Railroad was one of the most strategically important transportation corridors in the Western Theater. Tullahoma, the mid-point of the line, served both Confederate and Union armies as a logistics center. Federal troops arrived here in the spring of 1862, but the Confederates took back the town in the fall and supplied their army at Murfreesboro by rail. Fort Rains, a bastioned fort, was built by the Confederates in the early part of 1863 on the slight eminence along Fort Street between Ovoca Road and Forrest Drive. The fort was named for Brig. Gen. James E. Rains, who was killed at the Battle of Stones River. The fort was 125 yards across, surrounded by a ditch and a wall, held 12 cannon, and could be garrisoned by 500 men. Since no fighting occurred at Tullahoma, the fort was never used. It was leveled for homesites in the mid-20th Century. In June 1863, Union Gen. William S. Rosecrans made Tullahoma a major objective for his Army of the Cumberland. After the success of the campaign, the railroad became the Union army's lifeline, and massive earthworks were erected to protect it. Today the earthworks are almost gone, but the Confederate Cemetery, containing 407 graves, remains to recall the Civil War period.

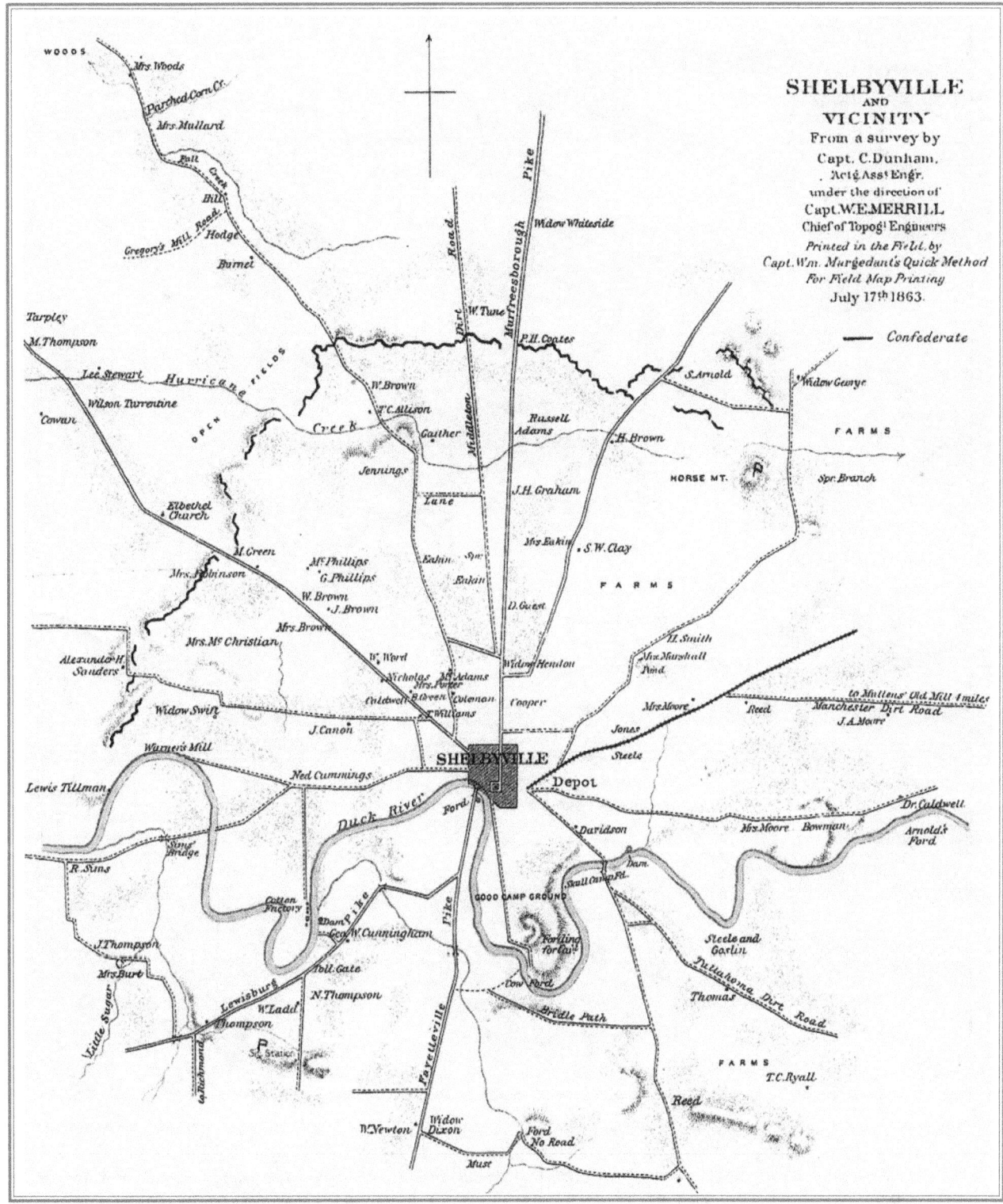

During the Civil War, Bedford County was divided in its loyalties and supplied nearly equal numbers of troops to the Confederate and Union armies. Although the pro-Union stance of **Shelbyville** earned that city the title of "Little Boston," one of the Confederacy's best-known generals, Nathan Bedford Forrest, was born in Bedford County in 1821. Extensive earthworks were built in early 1863 by Confederates under Edward Sayers, chief engineer of Polk's Corps. On June 27th, 1863, Confederate Gen. Braxton Bragg's army withdrew from Shelbyville and concentrated at Tullahoma. Union Gen. David S. Stanley's cavalry clashed outside town with Gen. Joseph Wheeler's troopers, who were screening the infantry's retreat. Wheeler's men temporarily occupied part of the defensive works on the outskirts of Shelbyville. Stanley's cavalrymen found a portion that was undefended, entered it, and moved along it until they struck the Confederate flank. Part of Wheeler's command fled through Shelbyville, where street fighting included both cavalry charges and artillery duels. Heavy fighting took place around the railroad depot, where the Confederates made a brief stand before continuing the retreat. The Union cavalry overwhelmed Wheeler's men. Some men and horses were trampled near the bridge or drowned in the river. For lack of infantry support, Stanley discontinued the pursuit of Wheeler. Over the next several days, Bragg's army made successive withdrawals to Tullahoma, Decherd, and Cowan, before the final retreat through the rugged mountains to Chattanooga on July 3rd.

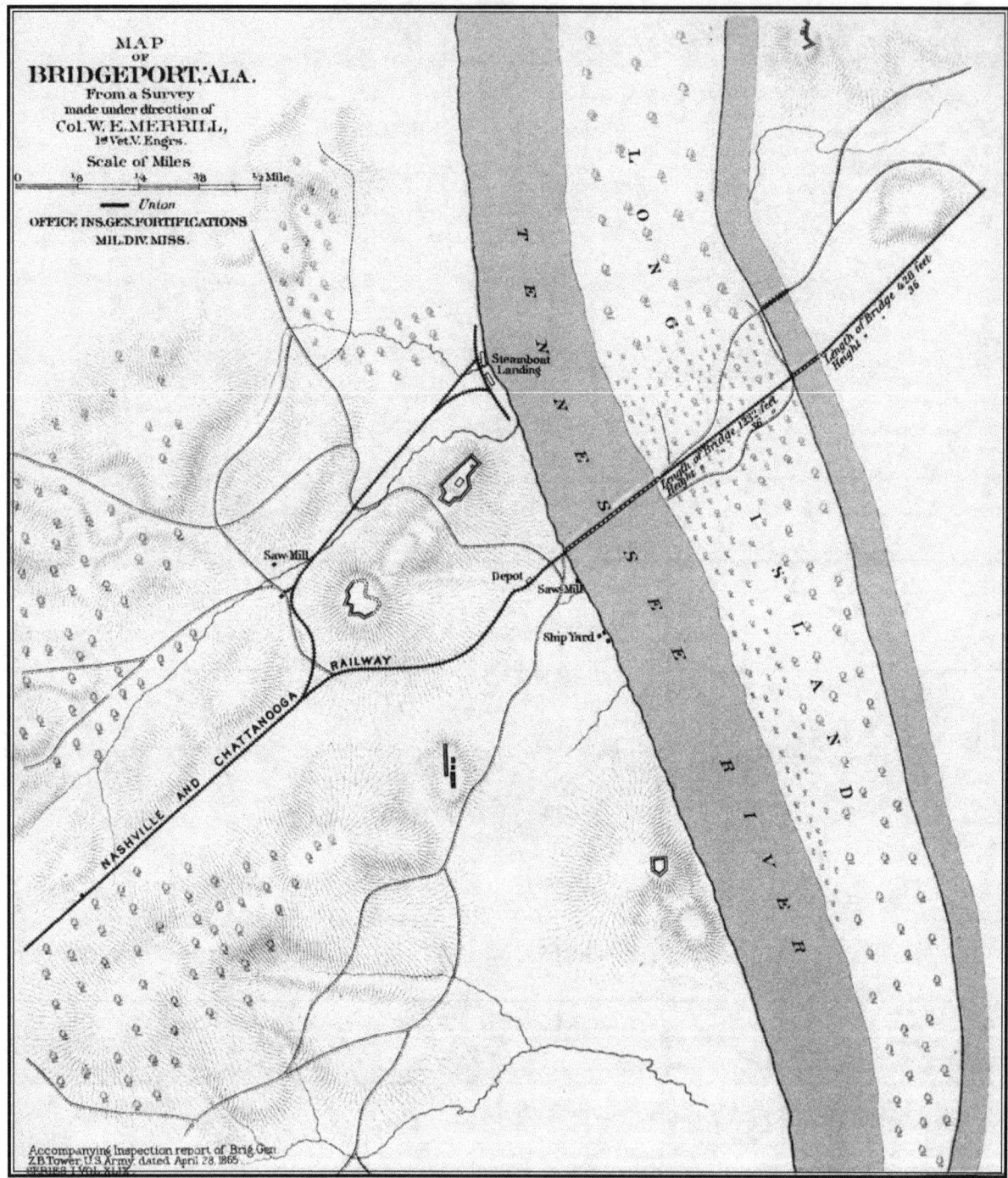

Bridgeport is located in the northeast corner of Alabama in Jackson County. As the end of usable railway from Nashville, the town became a key base of operations in the U.S. victory at Chickamauga and lifting the siege of Chattanooga. The vital Memphis-Charleston Railroad spanned the Tennessee River here. The bridge was burned several times in 1862-63. During the early part of the war, the Confederacy controled Bridgeport and its strategic bridge. Confederate Brigadier General Danville Leadbetter commanded 450 troops, defending the town at a fort situated on Battery Hill approximately 500 yards from the bridge. In April of 1862, Federal forces seized Bridgeport in a battle that lasted more than an hour. Union General Ormsby Mitchel led more than 5,000 troops into Bridgeport, forcing Leadbetter to retreat toward Chattanooga. Subsequently forced to relinquish the site, the Federals recaptured the town in July 1863 as Gen. William Rosecrans took Chattanooga (upriver). With the Union controling the bridge, Bridgeport became the major shipping center for troops and supplies going to General Sherman's "March to the Sea." The shipping route from Bridgeport to Chattanooga became known as the "Cracker Line." The town hosted a Union field hospital and served as the construction site for several Union gunboats, including the *USS Chattanooga* and *USS Bridgeport*.

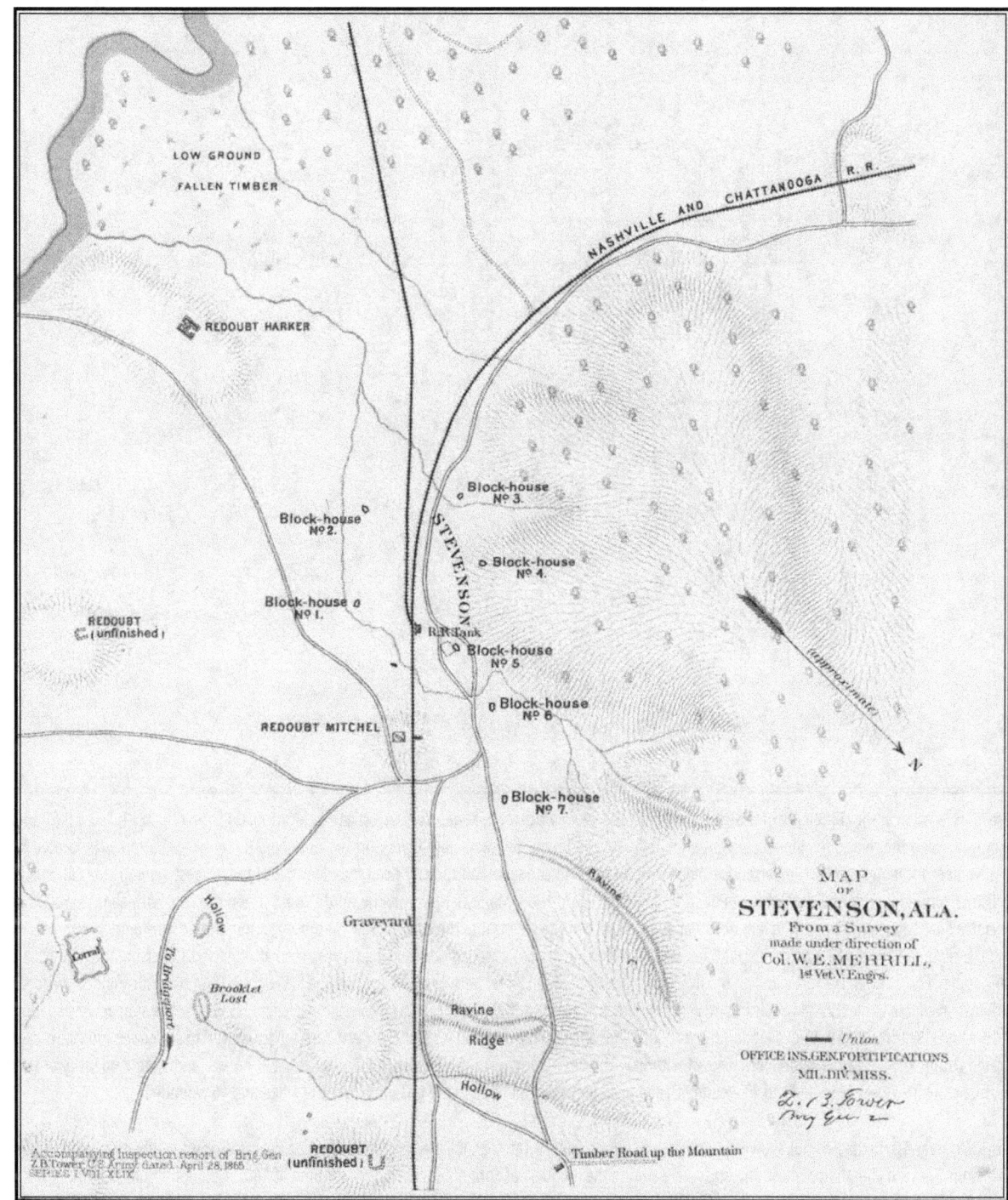

Stevenson, Alabama, was a major supply station and staging ground for several decisive Federal campaigns. Tens of thousands of soldiers, horses, wagons, prisoners of war, refugees, wounded men and others passed through Stevenson during the late summer and fall of 1863, before, during and after the Battles of Chattanooga and Chickamauga. *Harper's Weekly*, an influential newspaper of the time, noting the juncture of East-West and North-South rail lines here, called Stevenson "one of the seven most important cities in the South." Constructed in the summer of 1862 and expanded in 1864, using soldiers and freed slaves, Fort Harker was built on a broad hill a quarter mile east of town. It overlooked Crow Creek and was well within firing range of Stevenson's strategic railroad lines, supply depots, and warehouses. Fort Harker was an earthen redoubt, 150 feet square, with walls 14 feet high, surrounded by an 8-foot-deep dry moat. It contained seven cannon platforms, a bomb-proof powder magazine, a draw-bridge entrance, and an 8-sided wooden blockhouse at its center. Soldiers building the fort reported that "the soil is very hard, requiring the continual use of a pick." Despite that, Ft. Harker was critical to Union plans. The officer in charge was ordered by his commanding general "to work night and day" to complete the fort "as rapidly as possible." One other large fort, two smaller redoubts, and at least seven blockhouses were constructed along the railroad lines at Stevenson during the war.

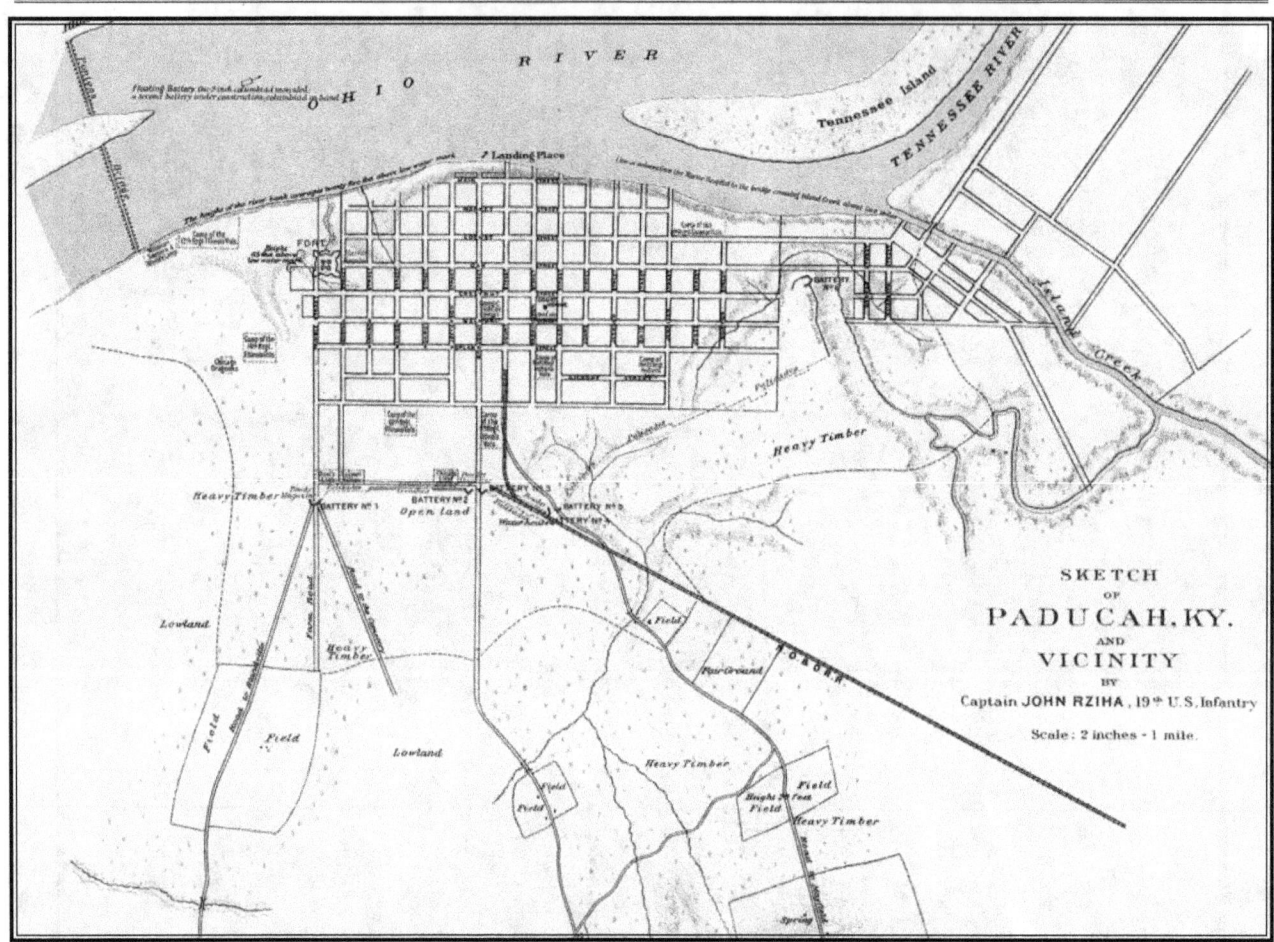

Because of its location at the confluence of the Ohio and Tennessee rivers, **Paducah**, Kentucky, was a strategic location. On Sept. 6th, 1861, Federal commander U.S. Grant occupied the river port immediately after Confederates occupied the Mississippi River port of Columbus. Union headquarters was located here, a more convenient location than Fort Anderson for those with business to conduct with the city's commander. One of the longest serving was Col. Stephen G. Hicks, district commander during the Battle of Paducah. On March 25th, 1864, Major Gen. Nathan Bedford Forrest seized the city. With about 2,500 men, the Confederates quickly overran six outlying redoubts that guarded each of the streets into the city. Col. A.P. Thompson, commanding the Kentucky Brigade, moved through the city pursuing the Federal troops, who retreated into Fort Anderson, a Mahan-style earthen fort named after Robert Anderson, the Kentuckian who had surrendered Fort Sumter. The fort was near the river at the west end of 4th Street between Park Avenue and Clay Street. Thompson surrounded the fort. Hicks had 665 men available in the 122nd Illinois Infantry, 16th Kentucky Cavalry, and the 8th U.S. Colored Artillery (Heavy), as well as the support of two gunboats in the Ohio River. Skirmishing continued around the fort until 4:30 pm, when a demand for surrender was sent. Hicks refused to surrender. After burning some confiscated cotton and a steamer in dry dock, Forrest withdrew the next morning to Mayfield with 50 prisoners, 400 horses ,and a large quantity of supplies. Confederate casualties were ten killed and 40 wounded. Federal casualties, in addition to the prisoners, were 14 killed and 46 wounded.

In June 1861, near **Springfield** (not shown) in Robertson County, Confederates established a major induction center, Camp Cheatham, named in honor of Gen. Benjamin Franklin Cheatham, whose ancestors were among the founders of Springfield. During the autumn of 1862, Gen. John Hunt Morgan's cavalry raid destroyed the Dead Horse Trestle near Ridgetop. The 9th Pennsylvania Cavalry occupied Springfield on March 25th, 1862. The Federals constructed a fortified camp to observe the city and protect the Edgefield and Kentucky Railroad. Earthworks were built from the fort's north side to Sulphur Fork Creek, and another line extended a mile from the south end of the camp. Union soldiers also manned railroad blockhouses at Baker's Station, the Ridgetop Trestle, the Dry Creek Trestle (Greenbrier), Sulphur Fork Trestle, and the Red River Blockhouse No.1 near Adams. The First Presbyterian Church on Locust Street was used as a stable. From August 1864 until the end of the war, Col. Thomas J. Downey and the 15th U.S. Colored Infantry guarded the town and its valuable railroad line.

Lebanon (not shown) in Wilson County escaped the worst of the war. Gen. Alvan C. Gillem reported in August 1864: "Wilson County shows but slight signs of the war. In Lebanon everything indicates peace. The houses have never been disturbed." Historic Cumberland University had its campus damaged by Federal troops in 1863. Late in 1864, Confederates burned it completely. General Alexander P. Stewart was a professor when the war began, and he returned to teach in 1867. Near dawn on May 5th, 1862, Col. John Hunt Morgan's 800 Confederate cavalry were camped around the Public Square and at the University, and were attacked by Gen. Ebenezer Dumont's Federal cavalry of 600 as they advanced from Murfreesboro in a torrential rain. After 90 minutes of charges and countercharges, Morgan and most of his men withdrew on the Rome and Trousdale Ferry Pikes pursued by the Federals. Meanwhile, barricaded in the Odd Fellows Hall on West Main Street, 60 to 70 of Morgan's men fired upon the Federals. Later these Confederates surrendered when Dumont threatened to burn the town. During the attack, the Federals lost 10 killed, 21 wounded, and five missing; the Confederates lost 60 killed and an unknown number wounded and missing.

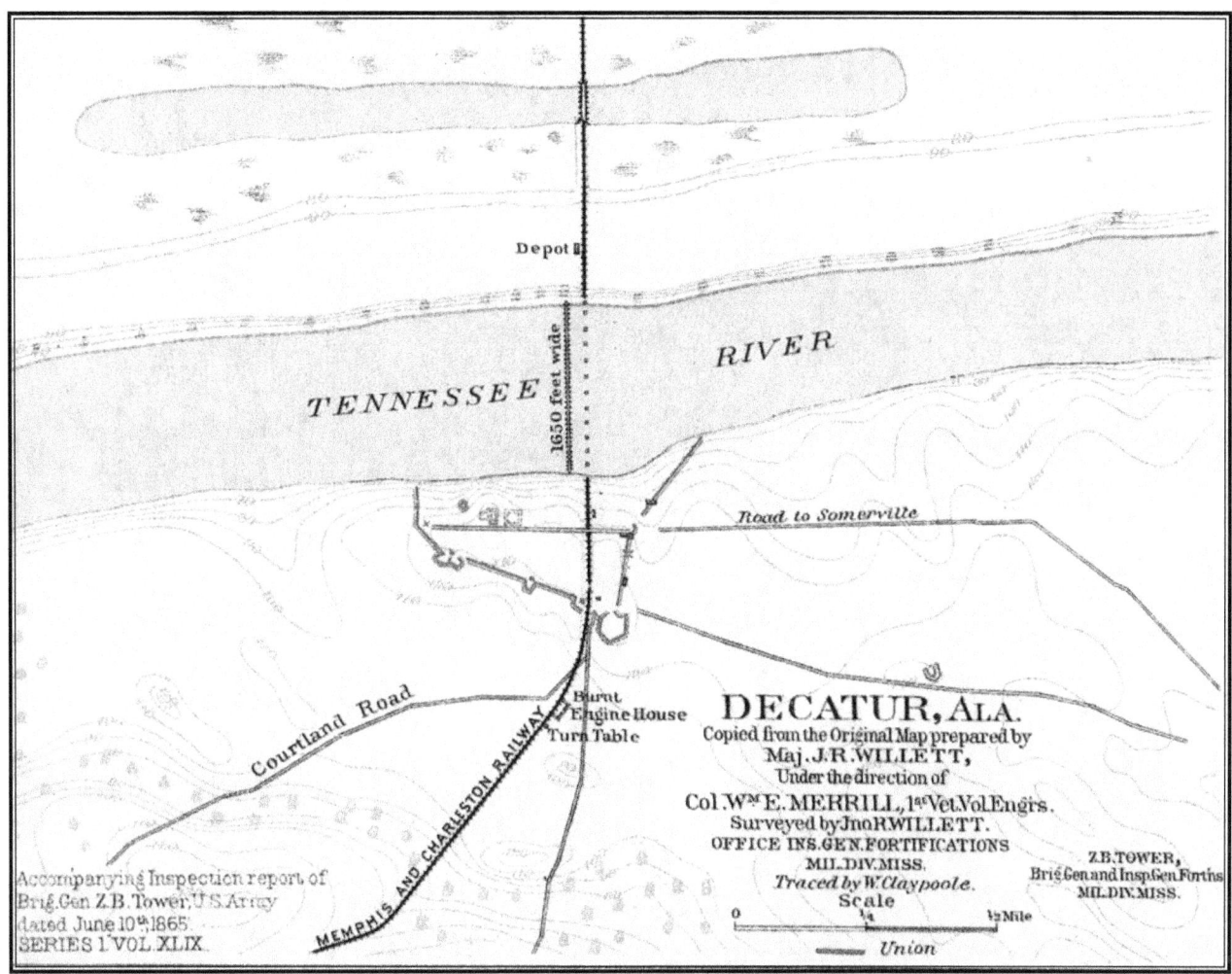

In 1860, the Memphis and Charleston Railroad was the only east-west route through the U.S. south of the Mason-Dixon Line. Maintaining control of this rail line was essential to Confederate strategy. Union Brigadier General Ormsby Mitchell occupied **Decatur**, Alabama (pop. 800) on April 13th, 1862. Confederate defenders attempted to destroy this bridge, but failed. Union troops would destroy the bridge themselves on April 27, 1862. The Federals occupied Decatur briefly in the summer of 1862 and in the fall of 1863, and returned permanently on March 8, 1864. Construction of a pontoon bridge on the site of the destroyed railroad bridge was immediately commenced. The pontoon bridge that crossed the Tennessee River into Decatur served as the logistical lifeline to the Federal garrison — all food, ammunition, medical supplies, mail, and reinforcements had to cross this bridge.

On Oct. 26-29th, 1864, Gen. John Bell Hood's Army of Tennessee tried to cross the river at Decatur but was repulsed. Federal officers reported a total loss of 113 officers and enlisted men killed, wounded, and captured. Estimates of Confederate casualties ranged from 500 to 1,500. A correspondent from Hood's army, writing to a Mobile, Alabama newspaper, stated, "We attempted to take Decatur, but found it a hard nut to crack..."

Once Hood appeared at Decatur, Gen. Robert S. Granger ordered forward regiments from throughout North Alabama and Tennessee to reinforce the garrison. Among these regiments was the 14th U.S. Colored Troops. Raised in 1863 at Gallatin, Tenn., the 14th USCT consisted primarily of freed slaves from the Middle Tennessee area. The town changed hands during the Civil War at least eight times, because of its strategic importance astride the junction of two railroads, and its location on the Tennessee River. Jefferson Davis passed through twice, once on his way to inauguration as the Confederacy's first and only President, and again on his way home after release from prison in 1867. Confederate Generals Albert Sidney Johnston, P.G.T. Beauregard, John Bell Hood, and Nathan Bedford Forrest also fought or gathered their troops here. Future U. S. president James Garfield visited here as a Colonel, along with Union Generals such as William T. Sherman, James B. McPherson, Robert S. Granger, James B. Steedman and Grenville M. Dodge. Both Confederate and Union regiments drawn from the surrounding countryside were organized at Decatur, and fought in the major battles of the war.

Because of the transportation routes running through **Bowling Green**, Ky., (next page) both armies recognized its strategic location during the war. The city was occupied briefly by Confederate troops, who used many of the surrounding hills for fortifications. Confederate troops took up this key position in the Southern defense line on Sept. 18, 1861. After Fort Henry fell and Fort Donelson was threatened, they evacuated Feb. 11-13, 1862. Gen. O. M. Mitchell and Federal troops entered Feb. 14, 1862, occupying the evacuated fort and securing the defense line for the North. For the remainder of the war, the Union Army used the city as a hospital center. Briefly during 1861, the city had served as the Confederate state capital. Before evacuating Bowling Green in mid-February 1861, the Confederate Army destroyed the Louisville & Nashville Railroad trestle. To prevent that kind of destruction from happening again, the Union Army constructed a defensive stockade to protect the railroad crossing.

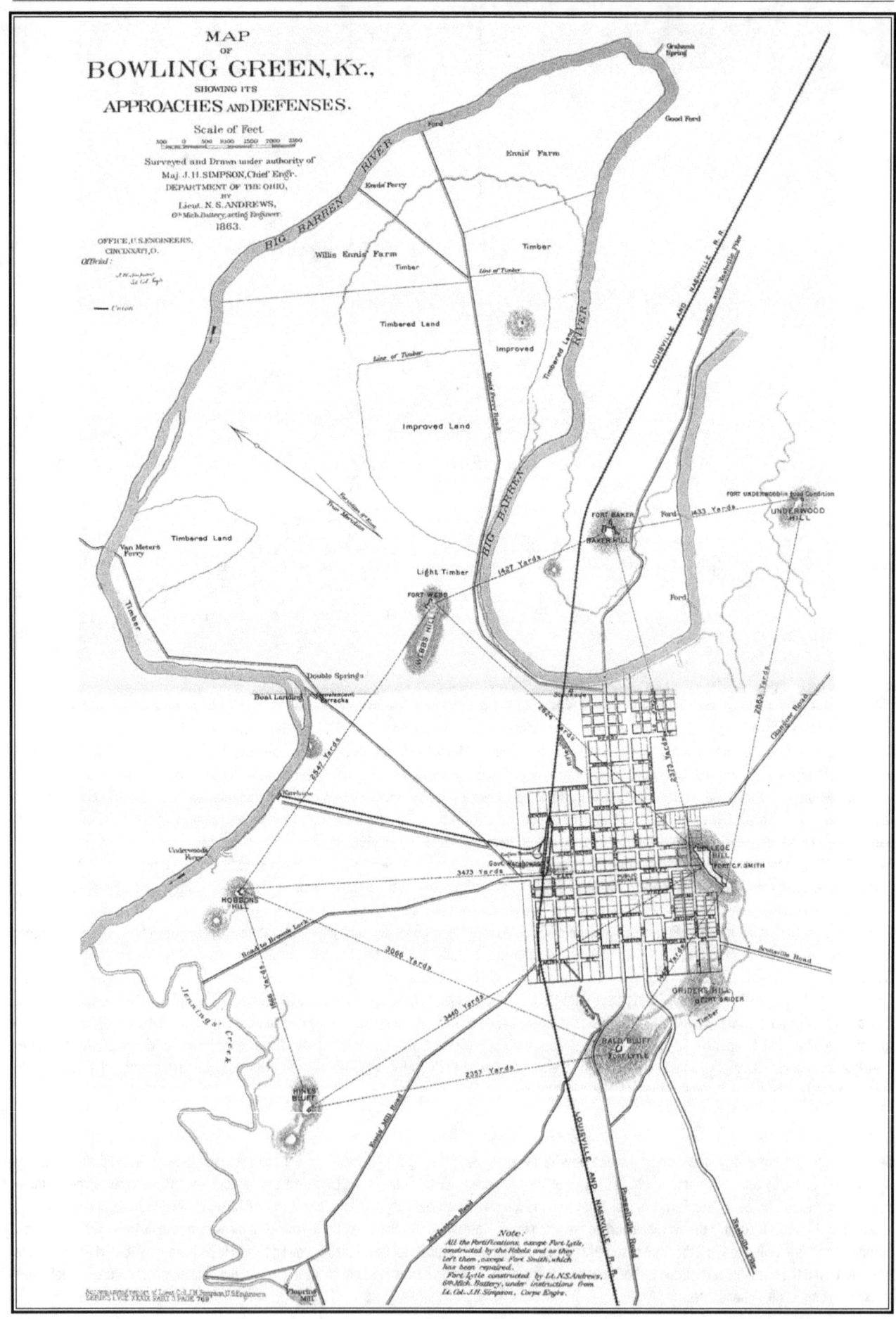

Fortress Nashville: Pioneers, Engineers, Mechanics, Contrabands & U.S. Colored Troops

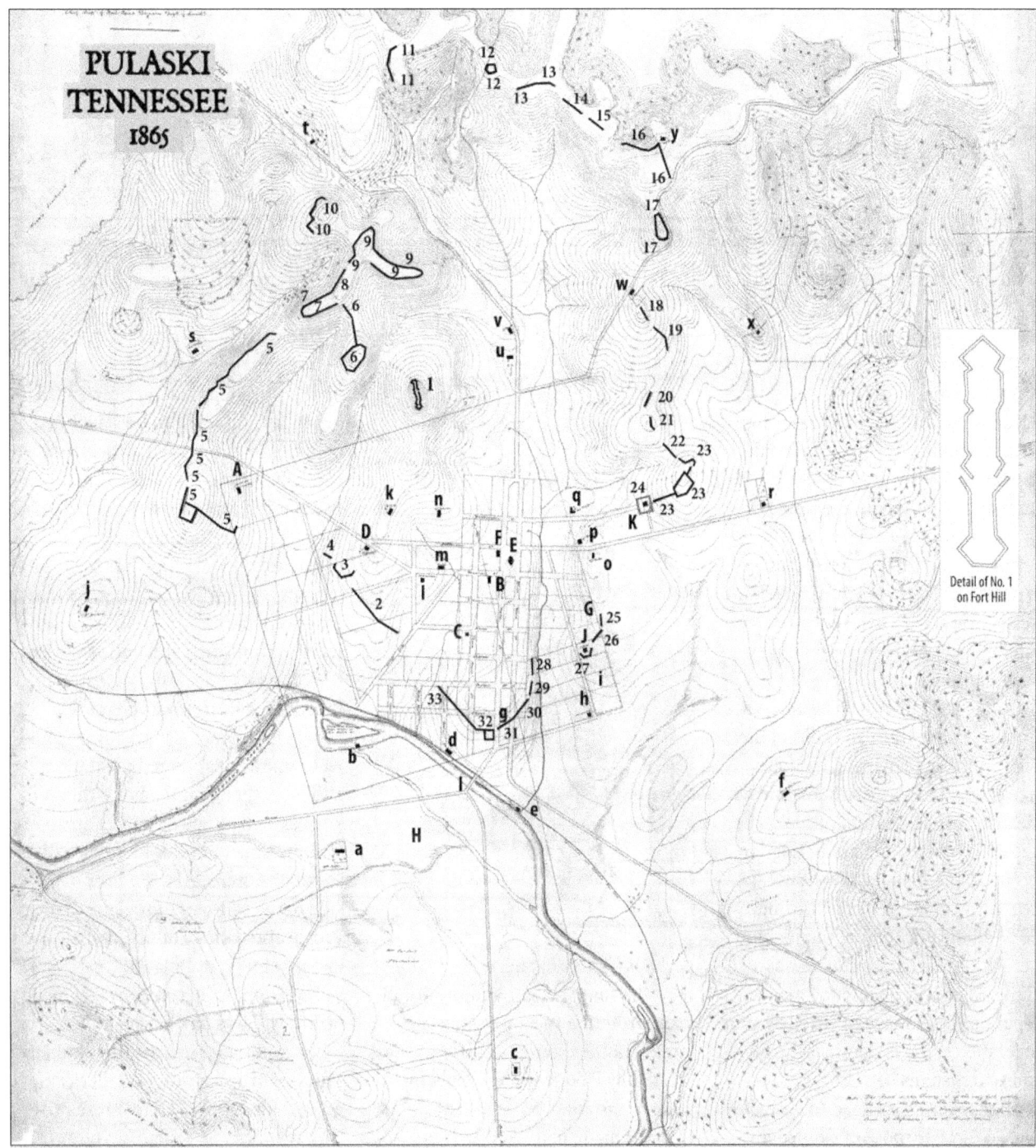

Topograpical Map of Pulaski, Tennessee and Federal Defenses. Surveyed and drawn by P.M. Radford, May 1865, under the direction of Col. Wm. E. Merrill, Chief Engineer, Dept. of the Cumberland, and Major James R. Willett, Chief Inspector of Railroad Defenses, Dept. of Cumberland. Showing courthouse, city streets, Richland Creek and numerous tributaries, Tennessee and Alabama Railroad. (National Archives). Earthworks and features identified by David J. Meagher © 1987. Researched by Ed Wheeler Jr.

LEGEND

a. Mrs. Carter's Plantation
b. Flour Mill
c. Dr. Tom White
d. Railroad Depot
e. Flour Mill
f. Mrs. Valentine's Plantation
g. Cemetery
h. Joe Dunlap
i. Maplewood Cemetery
j. Dr. Carter's Plantation
k. Mr. Rose
l. Mr. Simonton
m. Carter
n. Dr. Arteway
o. Dr. Batte
p. Rt. Gordon
q. Joe Childress
r. Mr C.E. Rose
s. J.T. Helms
t. Frank Lester
u. Factory
v. Wilkinson
w. J. Manning
x. McKinnian
y. James Ross

A. Smallpox Hospital
B. Jones Law Office-Original KKK founded here.
C. Thomas Martin House-2nd KKK home, for a weekend.
D. Carter House-Klan had rituals in basement of this ruined house.
E. Court House
F. Jail, where Sam Davis imprisoned.
G. Sam Davis hanged here. Site of museum.
H. Campground for soldiers
I. Covered Bridge
J. Giles College-Wartime Hospital.
K. Hawthorne Villa-Used by both sides as general officer lodging and command post.

Solid lines with numbers indicate rifle pits / trenches.

Federal General / Railroad Engineer Also Spymaster

Grenville Dodge was a Massachusetts native who eventually settled in Iowa and performed surveys for the Union Pacific railroad. At the outset of war, he was named colonel of the 4th Iowa Volunteer Infantry. He was noted for his performance in the battle of Pea Ridge, Arkansas, where he was wounded. One of the results of his experiences was an appreciation of intelligence networks. He tested certain aspects of such a network in Arkansas and further refined those aspects in Mississippi and West Tennessee. He served as Grant's intelligence chief during the Vicksburg Campaign.

Dodge used up to 100 agents at a time — people from all walks of life, men, women, teenagers, old folks, free slaves and those still in bondage. He assigned names and numbers but kept the information about them in his head and on a list of names he kept only on his person. Only four Federal officers knew about his spy network. At the time Dodge was also supervising a large force of men who repaired the damage inflicted by Confederates on the Federally controlled railroad systems.

The spies were trained to evaluate the size and purpose of Confederate units they encountered. All were pro-Union, and Southern, if possible. They usually carried mail addressed to real persons in the South. Thus, if stopped by rebel forces, they had a real reason for travel. Going through a Confederate line, they were usually sent to a commanding officer or provost marshal for a pass to proceed. On the way the agent observed the type of unit and its strength. The mail the agent carried sometimes contained hidden messages in the form of pinpricks or

General Grenville Dodge

perhaps a route cypher. (The Union used the route cypher and the South used a rotating message-wheel system.) Another source of information created by Dodge was an Alabama cavalry and infantry regiment (Union) made up of Southerners. At the front of the cavalry unit was a company-sized element wearing Confederate uniforms. These spies would contact real Confederate forces and, once accessing their strengths, ride back to the main Federal cavalry force and report.

As Federal forces moved from Mississippi northeast through Tennessee, Dodge's men began repairing the railroad systems there and in areas under Federal control. In November 1863 he moved his headquarters to Pulaski, Tenn., in Giles County. A sizable force of railroad workers and bridge and trestle constructors as well as sawmill operators arrived with him.

Due to the emancipation of slaves and the abandonment of plantations by their pro-Confederacy owners, a growing number of freed slaves fled to Federally controlled areas. Dodge's contraband solution was to create Federal military units of colored soldiers who were primarily construction workers. Thus the problems of feeding and housing the freed men were partially solved. They also received uniforms and basic training.

At the same time, the newly formed Freedmans Bureau was creating camps for former slaves so they could work on the abandoned plantations and have shelter and pay for their upkeep as well as education. Camps were built just north of the Tennessee River south of Athens, Ala.; on the north side of the railroad tunnel north of Prospect in Giles County; and also just northwest of Pulaski.

The Tunnel Hill camp, as it came to be known, was part of a complex. Due east about two miles on the Elkton to Pulaski turnpike was a smaller camp mainly for supply and to house pro-Union white war refugees. The two camps probably could signal each other. Dodge's trestle men built a substantial bridge over Richland Creek between the tunnel camp and the turnpike. Besides the large agricultural areas surrounding the camps, there were sawmills and woodyards around the rail tunnel area and south to Prospect and the two forts near the Elk River. In addition, Dodge designed a string of log blockhouses along the rail line from Athens to Columbia at the railroad trestles. The blockhouses

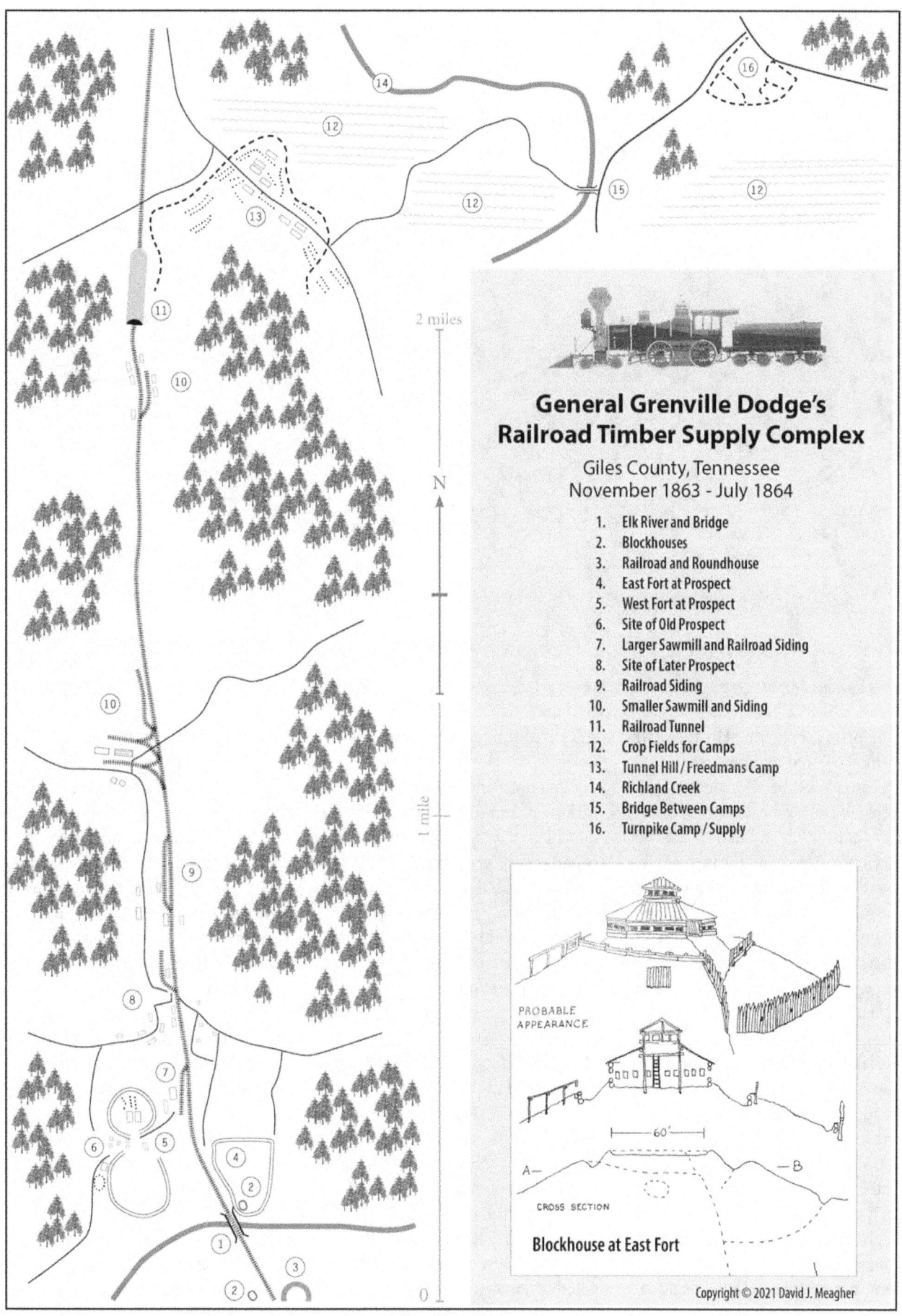

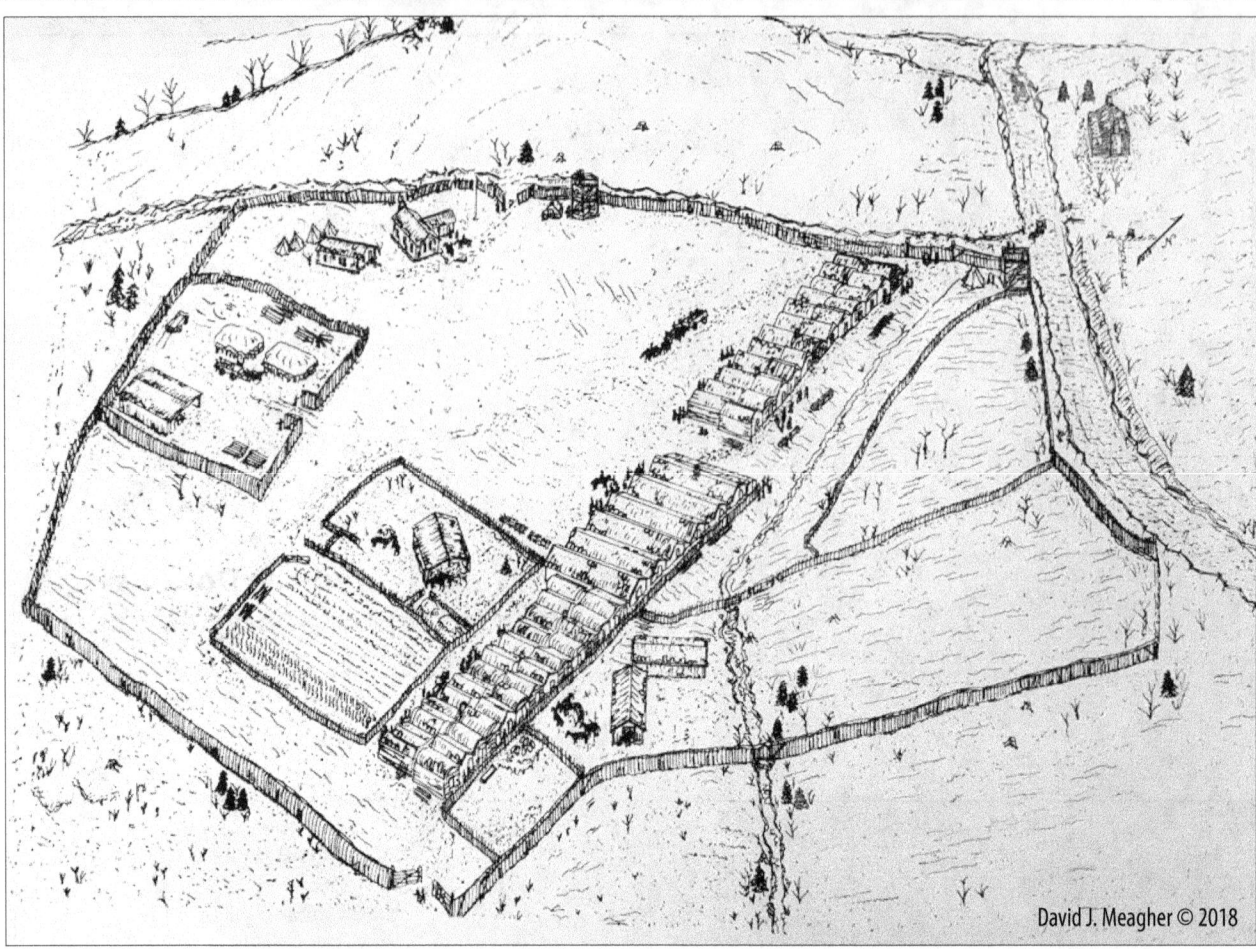

A conjectural depiction of the contraband / refugee camp at Conway, Giles County, Tenn., which was destroyed by Forrest's cavalry in September 1864.

were primarily manned by the newly formed colored regiments.

From November 1863 to May 1864, Gen. Dodge was headquartered at Hawthorne Villa in Pulaski. The military units trained and drilled while the sawmills turned out timbers for the railroad ties and trestles up and down the line. The black soldiers manned the blockhouses and cut firewood, which they sold to contractors who piled these cords beside the rail lines. In the camps, thousands worked the land and planted corn and cotton. The camps contained schools and hundreds of small huts built as shelters.

Dodge kept busy with his spy apparatus. In November 1863, just after his arrival in Giles County, a double agent of his and a Kansas unit captured a young Confederate courier carrying some papers that included detailed intelligence on Federal units in Nashville. Even though Sam Davis, a Coleman Scout, was technically a courier and not a spy, Dodge was rattled with suspicion that the source for the information was a real Confederate spy apparatus operating in his "military backyard." He attempted to pressure Davis into revealing his source of the intelligence by threatening to hang him. When Davis would not give up his source, Dodge then quickly hanged Davis before the possibility of an appeal. (Later in Missouri, Dodge would order 30-plus men hanged in the same way.)

At this time, Dodge was probably using the two agricultural camps as a contact point for agent debriefing. A Confederate doctor who lived nearby wrote (after the war) that Dodge and his officers frequented the white refugee camp. There were supplies and possibly females, "soiled doves," among the camp inhabitants for Dodge's soldiers, but the general was probably talking to people from the nearby railroad and tunnel camp and relaying a coded signal to the smaller installation. The meetings between handler and agents were usually innocuous and not meant to be noticed.

At one point, General Hurlbut in Memphis demanded that Dodge reveal his spies, but he refused and was backed up by Grant.

After the war, Dodge gained fame and fortune as chief engineer for the Union Pacific in charge of building the eastern portion of the transcontinental railroad and later as a Congressman from Iowa. He died in 1916, taking most of the secrets of his spy network to his grave.

Author's Note — Most of the information in this chapter comes from Giles County historian/artist David J. Meagher, who also supplied the accompanying maps and sketches.

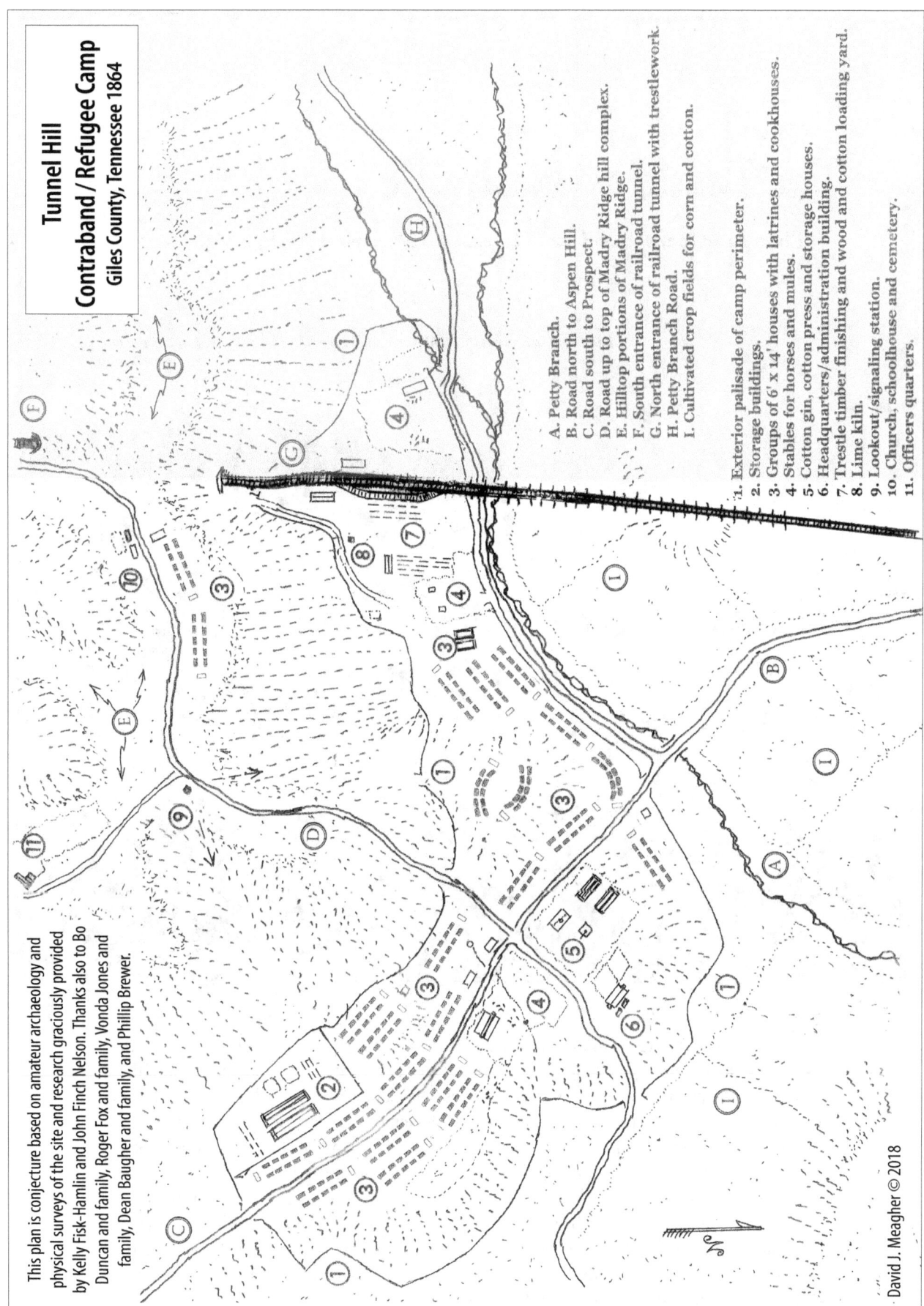

The Battle of Nashville
Thomas attacks Hood on Thursday, December 15th, 1864

Civilians watch events unfold from a vantage point on the outskirts of the city.
(Library of Congress)

Desperate times call for desperate measures. After Atlanta fell, Gen. John Bell Hood and his Army of Tennessee moved into northern Alabama, regrouped, and headed north toward Nashville, aiming to recapture the capital city occupied by the Federals 33 months before. Losing precious time waiting on supplies, the Confederates finally moved in three columns to Columbia, where Hood stole a march on Gen. John Schofield. At Spring Hill, the Federals slipped through the Confederates' grasp and the next day Hood launched a frontal assault just south of Franklin that crippled his army. Both sides moved on to Nashville, where Hood besieged the city. Or at least he tried to, but he didn't have enough men. The Confederates had to refuse their line and began to build five small forts or redoubts on their left flank. The going was rough as winter weather closed in. Then, in mid-December, the weather moderated, and Federal commander Gen. George Thomas made his move. The result was the decisive Battle of Nashville, Dec. 15-16th, 1864.

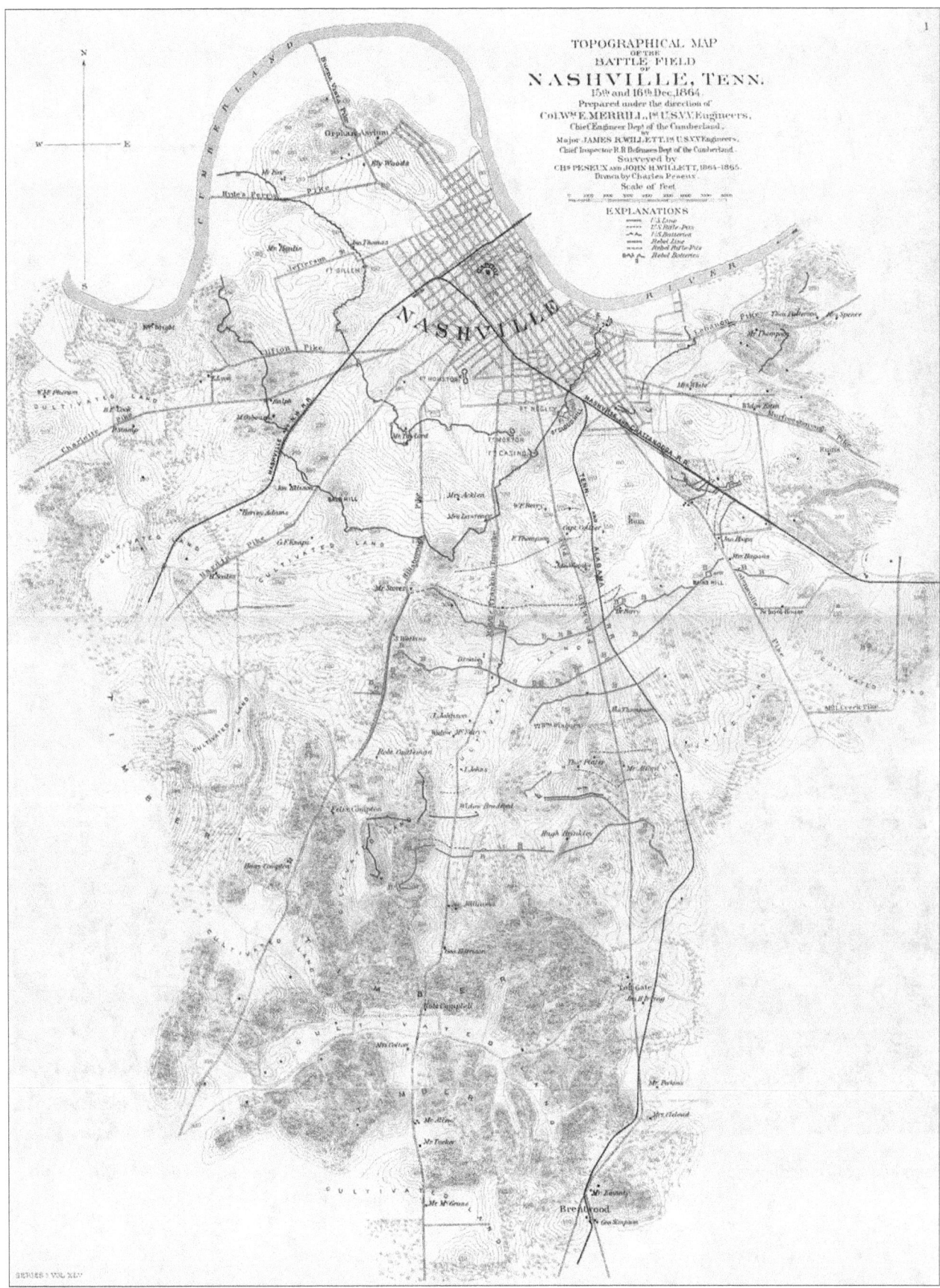

Map of Nashville showing the positions of the Federal defensive lines and the Confederate earthworks and redoubts prior to the Battle of Nashville and positions of both armies during the two-day battle December 15th and 16th, 1864. Map drawn by Federal Army Topographical Engineers.
(The Atlas to Accompany the Official Records of the Union and Confederate Armies)

Tents and soldier huts built along the Federal fortification line, with Fort Morton in the background and the Franklin Turnpike running perpendicular to the line. Only a few trees remain. (Library of Congress)

Another view of the Federal fortification line showing stacks of arms and soldiers camped next to it. (Library of Congress.)

(Library of Congress)

There is hardly any celebrated enterprise in War which was not achieved by endless exertion, pains, and privations; and as here the weakness of the physical and moral man is ever disposed to yield, only an immense force of will, which manifests itself in perseverance admired by present and future generations, can conduct us to our goal.
Carl von Clausewitz, On War

A Sight To See

The two-day Battle of Nashville was fought south of town, beyond the outer defensive line. Confederates in their works, just beyond the reach of the Federal forts' heavy guns, could easily see the State Capitol tower and the spires of downtown churches. Civilian spectators could watch the mass movements of infantry and troopers from vantage points atop tall buildings and higher elevations. Early morning fog obscured the action both days as did the pall of gunsmoke which hung over the battlefield. The roar of the artillery reverberated among the many hills and valleys. Afterwards began the grim task of policing the battle grounds and tending to the wounded and the dead, both man and beast.

Darkest of All Decembers. General John Bell Hood orders his men in Redoubt No. 4 to hold at all hazard. Original artwork by Rick Reeves. Used with permission.

The Confederate Army of Tennessee Redoubts

The Army of Tennessee besieged Nashville for almost two weeks in December 1864, waiting for the Federal forces under General George Henry Thomas to emerge from their fortifications and attack. Neither army sat idle during that time, however. Hood's line ran in a shallow arc from Granbury's Lunette in the east to the Hillsboro Pike in the west, his depleted army lacking the numbers to reach from river bend to river bend. At Hillsboro Pike, Hood's line under corps commander Lt. Gen. Alexander Peter "A.P." Stewart was refused or bent in a 90-degree angle to the south. Anchored at the salient and spaced along the pike for two miles were five small forts or redoubts (numbered 1 to 5 north to south), each consisting usually of four Napoleon artillery pieces, 85 artillerymen, and supported by 100 infantrymen. The redoubt was an earthwork reinforced with heavy timbers.

Redoubt No. 1 was on high ground, within sight of the Tennessee state capitol 3.75 miles to the northeast. This is one of only two of the five Confederate redoubts that exists today, a half-acre owned by the Battle of Nashville Trust (BONT) and located amidst a residential neighborhood.

A native Tennessean nicknamed Old Straight, Stewart was a highly competent general officer, historically under-rated due to his lack of flambouyance or quirky personality. One female admirer described him as "a man very charming in his manners, he is a perfect gentleman, and very entertaining and social, and one of the most modest and unassuming gentleman I ever saw." A reporter described him as "wide awake and constantly in the saddle." His aide Captain William D. Gale described him as "perfectly well bred and highly educated, very quiet and sedate, a thorough soldier and in all his actions, precise and military."

On the first day of battle, Dec. 15th, 1864, overwhelming Federal infantry and cavalry forces swept over the redoubts and forced Hood's army two miles to the south. At several redoubts, Confederates made courageous stands for as long as they could, usually outnumbered at least seven-to-one. Stewart's 5,321 men faced an enemy numbering 45,000.

The redoubts were built hastily under the supervision of Stewart's chief engineer, Major Wilbur Fisk Foster. None of the redoubts were finished by the eve of battle, harsh wintery weather hampering work beginning on December 8th. The thermometer dropped to ten degrees, with ice and snow. The ground was nearly impenetrable. The northern part of the line

(Continued on Page 262)

Federal attacks on Confederate Redoubts, December 15th, 1864

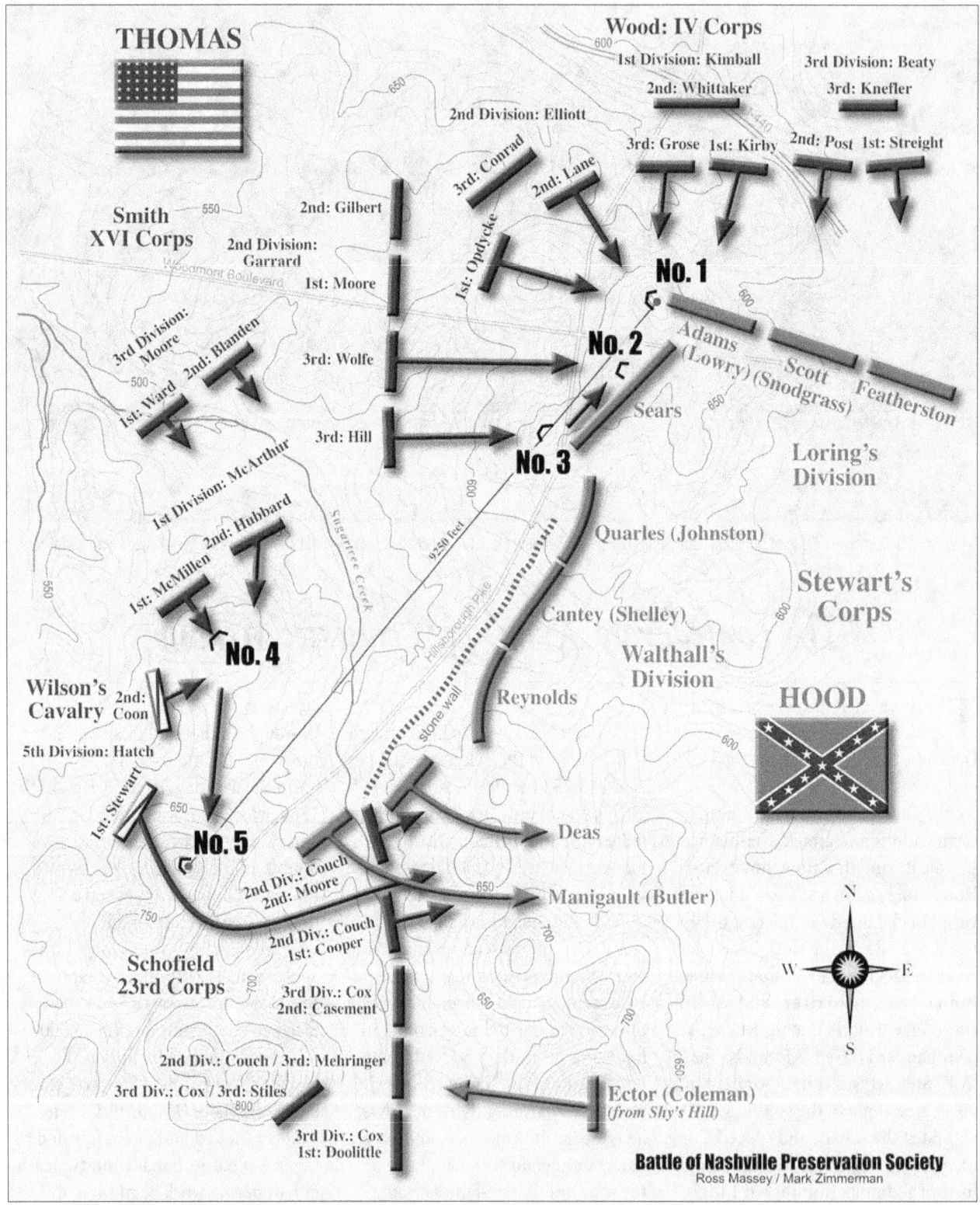

Locations of Confederate redoubts on modern Nashville map

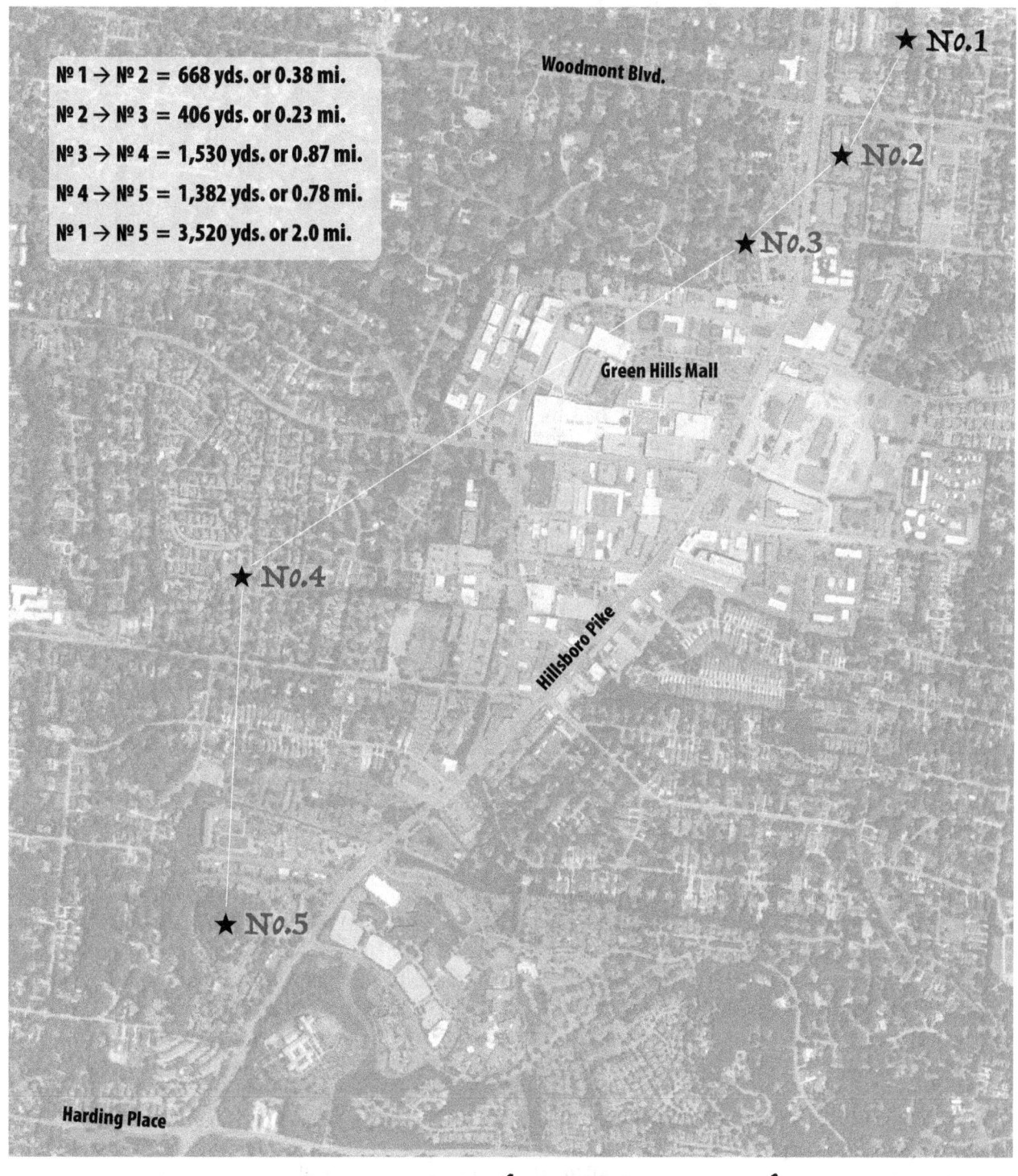

Redoubts:
- No. 1 — Preserved Historic Site - 3423 Benham Avenue
- No. 2 — Non Extant
- No. 3 — Earthworks Remnant - Calvary UMC
- No. 4 — Preserved site with marker - Foster Hill Road
- No. 5 — Non Extant

Confederate gunners in Redoubt No. 1 awaiting the inevitable battle. From this high ground, visible in the distance (to the right - a,b,c and d) are the State Capitol, First Presbyterian Church, Blockhouse Casino, and Fort Negley. Original artwork by Philip Duer, used with permission.

(Continued from Page 259) was under the command of division leader Major Gen. William Loring; the southern part under Major Gen. Edward C. Walthall.

While the officers slept in warm beds in their headquarters, usually consisting of well-appointed farmhouses, the rebel infantry, many without blankets or proper footwear, dug out the ground and slept in shallow trenches, warmed by small wood fires. Food was scarce and sanitary conditions non-existent.

On the foggy morning of December 15th, the Federal troops got a late start marching out of their fortifications. Following a diversionary attack on the eastern flank, the main forces, mostly under Generals Andrew J. Smith and John M. Schofield and supported by the cavalry of General James H. Wilson, swung in a lengthy arch and attacked the redoubts along the pike. Wilson's cavalry was actually mounted infantry, nearly all armed with repeating rifles or carbines.

The town came out to watch the fight, from a distance. One of the Federal officers recorded that the citizenry "came out of the city in droves. All the hills in our rear were black with human beings watching the battle, but silent. No army on the continent ever played on any field to so large and so sullen an audience." Thomas watched from high ground north of Montgomery Hill; Hood witnessed from Compton's Hill (now Shy's Hill).

An accounting of the capture of the redoubts comes from Ross Massey, a founder of BONT, and John Allyn, a BONT board member whose home is only a few hundred yards from Redoubts 2 and 3. A precise chronological account can be found in an accompanying article.

The main frontal assault targeted the refused Confederate line running south from Redoubt No. 1. Significant artillery fire and close combat erupted from the west, first at Redoubt No. 4 and then No.5, about one and a half miles to the south of Redoubt No. 1, and the flow of Union troops moved steadily northward to clash at Redoubt No. 3 and then No. 2. Much of this overwhelming attack could prob-

(Continued on Page 266)

Sight line from Redoubt No. 1 to State Capitol (3.72 miles)

1. Redoubt No. 1
2. State Capitol
3. RR Bridge over river
4. Fort Negley
5. Blockhouse Casino
6. Fort Morton
7. Fort Houston
8. Hill 210
9. Fort Gillem

Remnants of Confederate Redoubt No. 4 looking eastward, the earthwork facing to the north. Below, the same vantage point, looking westward. The earthworks would have been much more pronounced in 1864.

The granite marker at the bottom of Redoubt No. 4.
(Photos by Todd and Tom Lawrence, Battle of Nashville Trust, used with permission.)

Colonel Sylvester Hill is shot off his horse by a sharpshooter as his brigade takes Redoubt No. 3, a fanciful sketch by George H. Ellsbury published in January 1865 by *Harper's Weekly*. Hill was the highest ranking Federal officer killed at the Battle of Nashville. His horse, Dixie Bill, survived.

(Continued from Page 262)
ably be seen by the men defending the high salient of Redoubt No. 1, but that blue wave was not their only concern. The Union army was also closing in quickly on the fort from the north and northwest, having fought their way through thawing mud and skirmishers on Montgomery Hill during the day, and were preparing an assault up the steep and brush-strewn northern slope to the redoubt.

The IV Corps under Brig. Gen. Thomas J. Wood eventually moved into position below the redoubt to the north and northwest with three divisions, commanded by Brig. Gens. Washington Elliott, Nathan Kimball, and Samuel Beatty. Their progress during the day had been slowed considerably by the three Napoleons at Redoubt No. 1 (one of the battery's guns had been moved out of the redoubt earlier in the day to help defend against the Union forces charging from the west). In his report, Wood famously wrote that the redoubt guns "had been seriously annoying us all day." In response, he had two batteries of artillery, each consisting of six accurate rifled guns, brought up to pound the redoubt from converging angles at close range, doing considerable damage at the embrasures and killing and demoralizing the troops.

The final assault on the redoubt's steep grade began at approximately 4:00 in the afternoon of a dark, cloudy day. The day-long momentum of the Union forces, plus the gradual wearing down of the Confederate line, resulted in a reasonably quick capture by multiple components of both Wood's IV and Smith's XVI Corps. Because numerous Federal units essentially arrived within the works at the same time, field reports varied as each attempted to take credit for the capture.

Kimball's Division of Wood's Corps, charging uphill from the north, received the official recognition. However, various elements of Elliott's Division, charging from the northwest, claimed the win also. Elliott reported his division's victory of the redoubt, as a unit, in great detail, but one of his skirmishers, Lt. William Hall of the 36th Illinois, claimed that he had led a band of about two dozen men in a sneak attack along a stone wall on the west side of Hillsboro Pike and took the redoubt by surprise from the rear. Smith's XVI Corps, fresh from overtaking Redoubts 3 and 2, claimed a piece of the prize from the southwest side.

The confusion brought about by the simultaneous convergence of different units from multiple directions created this controversy, but by most accounts the Confederate line was already giving ground due to the overwhelming pressure of the assault.

As darkness came, the remainder of

Confederate Redoubt No. 1 today, looking westward from Benham Street. The vertical marker lists donors to the BONT preservation project. Below is a fanciful version of the capture of Redoubt No. 1, from *Harper's Illustrated*.

the Confederate line, following the collapse of Stewart's Corps on the left flank, fell back to the south to reestablish new positions primarily stretched along current Harding Place and Battery Lane, and anchored on the west by Compton's (now Shy's) Hill and on the east at Peach Orchard Hill. Following its capture, Redoubt No. 1 was used as a field hospital.

Colonel Sylvester Hill, an Iowan commanding the Third Brigade, First Division, XVI Army Corps, was killed by a sharpshooter during the attack on Redoubt No. 3, which lay on a tall hill west of the Hillsboro Pike. He was riding his famous horse Dixie Bill at the time (the horse survived the war). The redoubt contained three cannon and approximately 100 additional infantry supports. Colonel Hill, 44, was the highest ranking Federal officer killed in the battle.

Hill was born in Rhode Island and was trained as a cabinet maker. After moving to Cincinnati, Ohio, in 1849, he was swept up in the California Gold Rush and went west to seek his fortune. He was not successful. He ended up in Muscatine, Iowa, where he engaged in the lumber business. In the summer of 1862, Hill helped raise the 35th Iowa Infantry Regiment. Being a prominent Republican, he was appointed colonel of the regiment in September 1862. The regiment went on garrison duty in Illinois and Kentucky. In the spring of 1863 it joined the Army of the Tennessee besieging Vicksburg.

The Confederate troops manning Redoubt No. 3 were in the process of withdrawing when Hill's Brigade attacked; the loss of Redoubts No. 4 and No. 5 made the position untenable. However, two cannons were captured before they could be pulled out of the redoubt. The third gun had made it across Hillsboro Pike before capture. In the confusion following the assault, troops of the IV Corps were given credit for the capture. This error has been perpetuated on the historical marker located on Hillsboro Pike commemorating this action. The sharpshooter who killed Colonel Hill was firing from Redoubt No. 2, about a 300-yard shot.

The U.S. troops formed about 11:00 am along the ridge (now Estes Road) approximately one-half mile to the west of Redoubt No. 4. Meanwhile, more than two dozen U.S. field artillery pieces dueled with the redoubt's four smoothbore Napoleon cannons manned by the 48 men of Lumsden's Battery, supported by 148 men of the 29th Alabama Infantry under Capt. S. Abernethy (Cantey's Brigade under Brig. Gen. Charles M. Shelley). Capt. Charles Lumsden was an experienced artilleryman, a graduate of the Virginia Military Institute and commandant of cadets at Tuscaloosa's University of Alabama. They were ordered to hold their ground at all hazards.

In the late morning, the men of Ector's Brigade, commanded by Colonel David Coleman, fled eastward past the redoubt. Lumsden pleaded with Coleman to stop and reinforce the battery, but Coleman refused, stating that the Federal skirmishers, not to mention the regular infantry, outnumbered his men two or three to one.

According to Maxwell: "From 11 am, when the 24 guns opened on our four, to 2:00 p.m., or three full hours, we kept our four guns replying."

Firing at the redoubt were the gunners of Battery I, 1st Illinois Light Artillery, the 2nd Iowa Light Battery, and others. Just firing an artillery piece can be a dangerous job. An artilleryman of the 2nd Iowa lost both arms from a premature discharge (the gun was not properly swabbed). The artillery duel lasted about an hour when Federal Gen. Hatch ordered Col. Coon to charge the works. The Southerners began firing canister rounds and the infantrymen fired as fast as they could reload. The Federals were not greatly impeded. Coon said that "each regiment was competing with the others to reach the redoubt first." Slightly to the north, the infantry of the 95th Ohio, 10th Minnesota, and 114th Illinois and 93rd Indiana were closing in. The Confederates blasted canister at the Federals at one hundred yards. Sgt. James Maxwell served his gun well until the redoubt was overrun and the fighting became hand-to-hand. Lt. Cole Hargrove told the men to run right, which they did. Maxwell heard a shout and ducked just in time to avoid a double canister blast over his head. Maxwell fired a second artillery piece at th enemy twice before abandoning it (the man with the friction primers had fled to the rear). Lumsden then shouted, "Take care of yourselves, boys," and the rout was on. Maxwell, Lumsden, and Hargrove all made it to safety behind the Hillsboro Pike stone wall. Later that evening at dinner, Lumsden had to pick pieces of brain tissue (from a fellow cannoneer) out of his beard. Lumsden would survive the war only to die three years later in a lumber manufacturing accident.

The assault by 7,000 Federal infantrymen and dismounted cavalrymen, most armed with repeating rifles and carbines, began at 2:15 pm. The Alabama troops were overwhelmed about 3:00 pm and a number were captured. The Federals then crossed Hillsboro Pike and attacked the Confederate troops sheltered behind the stone walls.

First to reach Redoubt No. 4 was the 2nd Iowa Cavalry company of George W. Budd. No sooner had the Federals captured the four guns of Redoubt No. 4 but they were fired upon by the gunners in Redoubt No. 5 to the south.

Portions of the north wall of the last remaining redoubt, No. 4, can be seen at the top of a hill in the gated residential community of Abbotsford in South Nashville. Shortly after 2 o'clock in the afternoon, Redoubt No. 4 was

Timeline~Taking of the Redoubts

Thursday, December 15th, 1864 at 9:30 am: 10,551 troops of the U.S. 16th Corps moved out, supported by almost 10,000 cavalry. The 569 C.S. soldiers in Ector's Brigade realized they could not hold Harding Pike, and fell back toward Hillsboro Pike, passing a detached redoubt held by under 200 men. The men in the redoubt hoped Ector's Brigade would stop and help them make a stand. But they continued on, warning, "It can't be done; there's a whole army in your front."

10:00 am: Brig. Gen. Claudius W. Sears, in Redoubt 3, looked through his field glasses and saw a brigade in front of his pickets. He also advised his corps commander, Lt. General A.P. Stewart, "a heavy column of infantry is moving to our left." Sears was seeing the U.S. troops moving on Redoubt 4. Col. Robert Lowry, commanding the line around Redoubt 1, advised his division commander, Maj. Gen. William Loring, of the threat to the redoubts.

11:00 am: U.S. artillery batteries were in position along a ridge (today's Estes Rd.) where they began pounding Redoubt 4 for three hours. The 16th Corps troops began deploying to attack the redoubts.

2:15 pm: The 1st Division, supported by cavalry, attacked Redoubt 4. The three attacking brigades had about 7,000 men. Lumsden's Battery, supported by 100 infantrymen, were in Redoubt 4. They held until the last possible moment. Some defenders were able to escape.

2:30 pm: U.S. troops captured Redoubt 4. Two howitzers from Tarrant's Battery, in Redoubt 5, opened on them, making it too dangerous to stay put. As the 3rd Division was moving up in support, the 1st Division and the cavalry reformed to attack. Major Thomas E. Jameson commanded over 100 men in Redoubt 5, and fought until overwhelmed.

2:45 pm: During the attacks on the detached redoubts, both sides were bringing up reinforcements. The U.S. Army's 23rd Corps (almost 12,000 men) was brought around behind the 16th Corps positions. The 23rd Corps and a cavalry brigade swung past Redoubt 5 to attack Stewart's flank. Hood had ordered Johnson's Division, of Lee's Corps, to reinforce Stewart's flank.

3:00 pm: Two of Major General Edward Johnson's brigades were in place on Maj. Gen. Edward Walthall's Division flank. Deas' Brigade (628 men) and Manigault's Brigade (838 men) were facing most of the 23rd Corps' 2nd division, and a dismounted cavalry brigade, for a total attacking force of about 6,000 men.

3:15 pm: The U.S. Army's 4th Corps was aligned to strike Redoubt 1, and the line held by Loring's Division. The three divisions of the 4th Corps totaled over 15,000 men. Loring's Division of 2,524 men waited patiently, amidst artillery fire. About 200 men were in Redoubt 1.

3:30 pm: The ever-active Lt. Gen. Stewart sent a dispatch to Walthall, advising him that the brigades of Deas and Manigault had already fallen back. Stewart was still steady, and advised Walthall that reinforcements from Cheatham's Corps were coming (they did not make it).

4:00 pm: U.S. Col. John Mehringer advanced his brigade on the right of the 23rd Corps to the foot of Compton's Hill (Shy's Hill). Stewart's Corps was in a closing vise, except for Ector's Brigade. It had drifted south to Shy's Hill. Hood had been on hand and ordered them to "hold this hill regardless of what transpires around you." Ector's Brigade suddenly pounced on Mehringer's men. More brigades were rushed to Mehringer's support, causing Ector's Brigade to withdraw back up to Shy's Hill. Holding the hill enabled Hood to use it as the left salient point the next day.

4:15 pm: Redoubt 3 was attacked. Brig. Gen. Sears was in No. 3 with about 150 men. He knew Walthall's Division was outflanked, and he probably had received orders from Stewart to retreat. Attacking U.S. troops were under Col. Sylvester Hill, who was killed as No. 3 fell. Hill had just given the order to attack Redoubt 2.

4:30 pm: The U.S. 4th Corps attacked the salient, which included Redoubt 1. Stewart's orders to abandon the line had been received. As the attackers climbed the hill, they could see silhouettes of Stewart's men moving east toward Granny White Pike. Only a few men remained, hoping they could get their horses up to remove the cannon.

5:00 pm: It was past sunset and dark. Stewart's Corps had reformed along a line roughly parallel to Granny White Pike. Deas' and Manigault's men had been inclined to continue their retreat until Miss Mary Bradford had run from her home to beseech them to make a stand. General Sears was looking through his field glasses when a U.S. shell blew his left leg off and killed his horse. Rucker's Brigade had held their position across Charlotte Pike. During the night they rode to Hillsborough Pike.

Quartermaster Forces Manned Inner Trenches

The soldiers and civilians who worked for the Quartermaster's Department in Nashville — 7,000 strong — not only ran a massive logistics and transportation center, they manned the trenches during battle, and they were well-trained to do so. Under the direction of Quartermaster Colonel James Donaldson and under the overall command of Gen. James Steedman, eight regiments of Quartermaster Forces lined the interior works of Nashville from Fort Morton to Hill 210 (more than 2.5 miles), allowing regular forces to directly attack Hood's Army of Tennessee during the Battle of Nashville, Dec. 15-16, 1864. These units were issued uniforms, firearms, and flags, and they drilled on a regular basis, according to historian Greg Biggs. By September 1864, five regiments had been formed from the Nashville Depot numbered 1st through 5th Regiment Quartermaster Forces. On September 1st, Colonel Donaldson issued Special Orders No. 112 that stated that each regiment was to receive "one national and one regimental color, two camp colors…" The same day, the Quartermaster Forces Brigade paraded through the streets of Nashville. Captain J.C. Crane, of the Receipting and Disbursement Office, commanded the 2nd Regiment initially and later the Second Brigade. Overall, the three brigades were listed as being between 4,000 to 6,000 men strong.

Minnesota Monument on Shy's Hill.

overrun by attacking Federal ground troops, followed shortly thereafter by Redoubt No. 5 (which has been destroyed by residential development), allowing the surge of Federal forces attacking in a huge wheeling motion from the northwest to advance and begin to overrun the Confederate line near the Hillsboro Pike. The commander of Redoubt No. 5 was Major Thomas Jameson, who was shot in the thigh, captured and sent to prison at Camp Chase in Ohio. The other three redoubts fell shortly thereafter.

It is worth noting that historical accounts of this phase of the Battle of Nashville disagree as to which of the two southern-most redoubts, Nos. 4 and 5, fell first. The venerable book by Stanley F. Horn, *The Decisive Battle of Nashville* (1956), as well as the Tennessee Historical Society's engraved summary on its Redoubt No. 4 monument, show No. 5 falling first. James Lee McDonough, in a more recent book, *Nashville: The Western Confederacy's Final Gamble* (2004), interprets field reports as showing No. 4 falling first. Taking a general overview of the battle, this is a minor debate, considering the fact that thousands of Federal troops overwhelmed the few hundred Confederates holding these two fortifications within just minutes of each other.

The Confederate redoubt sites today:

• Redoubt No. 1 at 3423 Benham Road has limited parking space, interpretive signage, flags representing the combatants, examples of a 12-pound Napoleon and a 3-inch ordnance rifle, and magnificent views of the city.

• Redoubt No. 2 was destroyed during the building of a house in the 1920s. The site is now occupied by condominiums.

• Scant remnants of Redoubt No. 3 can be seen at the rear of Calvary United Methodist Church.

• The earthworks of Redoubt No. 4 have been preserved by the Tennessee Historical Society, which placed an interpretive monument at the base of the north face of the earthen wall, located at the south end of Foster Hill Road in the Abbottsford subdivision. The marker was placed through the Joseph B. Leu Memorial Fund of the society.

• Redoubt No. 5 near Summit Ridge Drive does not exist and there is no marker or signage (except along Hillsboro Pike).

Compton's Hill (Shy's Hill)

The Confederate Army of Tennessee would be routed by the Federals on the second day of battle. The precipitous attack occurred on the western flank at Compton's Hill, now known as Shy's Hill. During the night of Dec. 15-16th, Confederate engineers directed the placement of temporary field fortifications atop Compton's Hill. However, instead of placing them at the "military crest" of the hill, they were placed at the geographic crest, farther on up the slope. This enabled the attackers — men of Gen. Andrew Jackson Smith's XVI Corps — to reach nearly the top of the hill without being seen. The defenders were overrun in hand-to-hand fighting, and the hasty retreat was on, due in part to an engineering blunder.

U.S. Colored Troops
Former slaves fight, die for their freedom at Nashville

Freemen and fugitive slaves were recruited into the U.S. Colored Troops, supervised by white officers. By the end of the war, USCT comprised ten percent of the Union armies.

Freedom's Destiny

Our destiny is largely in our hands.
— Frederick Douglass

Our citizens yesterday saw, for the first time, a regiment of colored troops marching through the streets of Nashville. The novelty of armed negro troops elicited many remarks about the policy of the Administration in raising them—both pro and con," stated the *Nashville Daily Press* of October 3rd, 1863. Reactions among the local populace included shock and awe, fear and loathing, and pride and joy, depending upon one's experience with or perception of the Southern institution of slavery.

Most white Southerners were scared to death of slave insurrections and most Confederate soldiers abhorred the sight of a black man clad in a blue uniform and bearing a rifle.

Other than the black soldier, the most relevant and consequential reaction would be that of the Federal commanders themselves, many of them highly skeptical of the usefulness of Negro soldiers. William T. Sherman stated in June 1864, "I confess I would prefer 300 negroes armed with spades and axes than 1,000 as soldiers." He did not take any black soldiers with him on his March to the Sea. U.S. Grant was ambivalent but generally supportive. General George H. Thomas did not believe that the negro, especially a fugitive slave, had the discipline to make a good soldier.

Historian Robert G. Lambert noted, "Initial reservations about using black troops were twofold: one, prejudiced whites might simply refuse to fight side-by-side with blacks, and two, blacks were widely viewed as servile and cowardly."

Fugitive slaves, contrabands, and impressed freedmen laboring to build Fort Negley in 1862 begged their superiors to arm them against threats of Confederate cavalry attacks. They wished to protect themselves, their families, and the fort that they were building, but they were refused. So they armed themselves with axes, picks, and shovels. As it turned out, they didn't have to use them, but these Negro laborers represented perhaps the first black soldiers of the Civil War, noted Krista Castillo, museum director at Fort Negley Park.

Blacks had fought in American wars since the Revolutionary War, and enslaved men had been running away from their masters for decades. In Middle Tennessee during the Civil War, the one sure way for a fugitive to escape slavery was to enlist in the U.S. Colored Troops (USCT).

Would the black man have the discipline to make a good soldier? Would he obey orders? Would he fight and die for his freedom? These were questions that lingered throughout the second half of the war. Questions that were answered one day in December 1864 at a hill named for a peach orchard.

In 1860, one-fourth of the population in Nashville consisted of enslaved blacks, owned legally as property by a Southern aristocracy of planters (only a small percentage of whites actually owned slaves). Slaves were bought and sold routinely at the Public Square and the market at Cedar and 4th Avenue. There were also free blacks, about one thousand. During the war, due to a flood of refugees, the black population of the city swelled to 12,000. Eventually, Federal authorities established three contraband camps in Nashville—the northwest side of Fort Negley, Edgefield, and north of the city at Hendersonville.

Overall, 180,000 to 200,000 blacks served in the USCT infantry, along with 7,122 white officers, comprising nearly 10 percent of the army by the end of the war. USCT fought in 449 engagements, including 39 major battles. Approximately 37,300 lost their lives, 29,000 of those from disease, the remainder in combat. There were also 29,000 blacks in the Federal navy, and as many as 200,000 blacks were hired as laborers. Seventeen black soldiers and four sailors were awarded the Congressional Medal of Honor for their bravery.

The USCT comprised 170 infantry regiments, 13 heavy artillery units, and ten light artillery batteries. About 20,133 soldiers came from Tennessee (only Louisiana and Kentucky contributed more). Approximately 94,000 men in the USCT were ex-slaves from seceded states; 44,000 were ex-slaves or freemen from the border states;

The Emancipation Proclamation

All persons held as slaves within any State or designated part of a State, the people whereof shall then be in rebellion against the United States, shall be then, thenceforward, and forever free.
—January 1, 1863,
as issued by President Abraham Lincoln under his War Powers

the remainder were freedmen from the North.

Battery A of the 2nd US Colored Light Artillery fought at the Battle of Nashville, which involved more USCT than any other battle in the war. Each USCT light artillery battery comprised three white officers and 141 enlisted men. The six-gun battery, which included a forge and equipment wagon, was composed of three sections, each consisting of two platoons led by a lieutenant. Each platoon, led by a sergeant, included 20 to 30 men for each gun, a limber, and two caissons for storage of ammunition.

Blacks were promoted as non-commissioned officers — those who were literate could become corporals, leaders were appointed sergeants.

During the Civil War, which raged longer than anyone expected, there

A pair of USCT sergeants pose for a portrait.

An unidentified USCT private has his image immortalized in a tintype portrait.

Photography was very popular with most soldiers on both sides. (Library of Congress)

A life-sized statue of a USCT infantryman was dedicated at the Nashville National Cemetery in 2006. Sculpted by Roy Butler, the model for the statue was local USCT re-enactor Bill Radcliffe. The statue stands amidst rows of headstones of USCT soldiers, 2,133 in all, some unknown but to God. Sixteen thousand Civil War soldiers are buried at Nashville National Cemetery.

was an acute shortage of soldiers, manpower and labor. In fact, by the end of the war the Confederacy advocated putting blacks into uniform in exchange for their freedom.

In early 1863, President Lincoln expressed the nation's enthusiasm for the recruitment of black soldiers in a note to Andrew Johnson, military governor of Tennessee: "The bare sight of 50,000 armed and drilled black soldiers upon the banks of the Mississippi would end the rebellion at once." He summarized the contribution of negro soldiers and laborers in a letter of Sept. 12, 1864: "We can not spare the hundred and forty or fifty thousand now serving us as soldiers, seamen, and laborers. This is not a question of sentiment or taste, but one of physical force which may be measured and estimated as horse-power and steam-power are measured and estimated. Keep it and you can save the Union. Throw it away, and the Union goes with it."

The use of free blacks for menial labor was recognized early by the seceding states. In June 1861, the Tennessee legislature authorized the governor "at his discretion, to receive into the military service of the state all male free persons of color between the ages of 15 and 50." The enlisted blacks would receive $8 a month, and would be used for manual labor.

Nashville was captured by Federal forces in February 1862, but this did not result in any legal change in status for slaves. Those slaves who ran away from their masters to the protection of the Federal army were treated as contrabands or confiscated property.

In August 1861, the U.S. Congress passed the First Confiscation Act, stating that all enslaved persons fighting or working for the Confederate military were freed and relieved of obligations to their masters. In July of 1862, Congress expanded this legislation with the passage of both the Second Confiscation Act and the Militia Act. The Second Confiscation Act declared that the enslaved persons of Confederate civilians and military officers were to be "forever free." However, there was a catch — this act could only be enforced in Union-occupied areas of the South. The Militia

> **Headquarters Commissioner for the Organization of U.S. Colored Troops, Nashville, Tenn.**
>
> Colored men in the Department of the Cumberland will be enlisted into the service of the United States as soldiers, on the following terms:
>
> 1st. All freemen who will volunteer.
>
> 2d. All slaves of rebel or disloyal masters who will volunteer to enlist, will be free at the expiration of their term of service.
>
> 3d. All slaves of loyal citizens with the consent of their owners will be received into service of the United States; such slaves will be free on the expiration of their term of service.
>
> 4th. Loyal masters will receive a certificate of the enlistment of their slaves which will entitle them to payment of a sum not exceeding the bounty, now provided by law for the enlistment of white recruits.
>
> 5th. Colored soldiers will receive clothing, rations, and $10 per month pay. $3 per month will be deducted for clothing.
>
> Recruiting Stations are established at Nashville, Gallatin and Murfreesboro. Other stations will be advertised when established.
>
> GEORGE L. STEARNS,
> Major and A.A.G., U.S.V.,
> Commissioner of organization U.S. Colored Troops.

Officers of the 16th U.S. Colored Troops (USCT) Infantry Regiment atop Lookout Mtn, Tenn., July 4th, 1864; (left to right) Captain Samuel Galloway, and Lieutenants David H. Dickinson, Joseph H. Barbour, Jeremiah Chauncy, William Jones, Charles W. Seidel, and Lovett S. Rivenburg.

Act provided another limited step forward. The legislation allowed the President to receive blacks into the military, but it did not clear them for combat.

On Aug. 21st, 1862, the Confederate War Department issued an order directing that, should any officer engaged in organizing slaves for armed service be captured, "he shall not be regarded as a prisoner of war, but held in close confinement for execution as a felon."

On Aug. 25th, 1862, Secretary of War Edwin Stanton directed Brig. Gen. Rufus Saxton to recruit no more than 5,000 black troops in South Carolina. This was a turning point for the Lincoln administration. It was not long before colored troops were fighting against the Confederates in a series of raids along the coast.

On July 30, 1863, Confederate President Jefferson Davis announced that black soldiers of the USCT would be treated as escaped slaves and returned to their owners. Captured white USCT officers were declared to be inciting servile insurrection and subject to summary execution. This provision was not always followed. However, black troops were massacred at Fort Pillow early in 1864 and three USCT officers were executed (one survived) following the Battle of Nashville.

Each USCT regiment needed 35 white officers. Officers frequently volunteered for service with USCT units since promotion went with the assignment. William O. Stoddard, Lincoln's secretary, noted, "It was astonishing how large a number of second lieutenants of volunteers were willing to sacrifice themselves for the good of the service as majors and colonels." The officers were nearly all white, but the non-com ranks, such as corporal and sergeant, were available to blacks.

On March 26th, 1863, the Secretary of War issued an order directing Adjutant General Lorenzo Thomas to

The men of Battery A, 2nd U.S. Colored Light Artillery, drill as a unit while performing garrison duty at Nashville. (Chicago History Museum, ICHi-007774)

organize black regiments in the Mississippi Valley. On May 22nd, the War Department established a Bureau of Colored Troops to handle the recruitment, organization, and service of the newly organized black regiments. Reuben D. Mussey, Jr., a native of New Hampshire who served as captain of the 19th U.S. Infantry Regiment, helped recruit blacks into the USCT, and was named colonel of the 100th USCT on June 14, 1864. After the war, he served as President Andrew Johnson's private secretary.

Equal pay was a problem with USCT units. White privates received $13 a month, plus $3.50 for clothing, and sergeants got $21. USCT privates received $10 a month, $3 of which was assessed for clothing. For higher ranks, the difference was even greater.

Confiscation Act and Militia Act (excerpts)

Second Confiscation Act - July 17, 1862
SEC. 9. And be it further enacted, That all slaves of persons who shall hereafter be engaged in rebellion against the government of the United States, or who shall in any way give aid or comfort thereto, escaping from such persons and taking refuge within the lines of the army; and all slaves captured from such persons or deserted by them and coming under the control of the government of the United States; and all slaves of such person found on [or] being within any place occupied by rebel forces and afterwards occupied by the forces of the United States, shall be deemed captives of war, and shall be forever free of their servitude, and not again held as slaves.

Militia Act - July 17, 1862
SEC. 12. And be it further enacted, That the President be, and he is hereby, authorized to receive into the service of the United States, for the purpose of constructing intrenchments, or performing camp service or any other labor, or any military or naval service for which they may be found competent, persons of African descent, and such persons shall be enrolled and organized under such regulations, not inconsistent with the Constitution and laws, as the President may prescribe.

Selected Units of United States Colored Troops in Middle Tennessee

Battery A, 2nd United States Colored Light Artillery Regiment—Organized at Nashville, April 30, 1864, attached to Post and District of Nashville, Dept. of the Cumberland, to March, 1865. District of Middle Tenn., Dept. of the Cumberland, to Jan. 1866. Service: Garrison duty at Nashville, and in Middle Tenn. until January, 1866. Battle of Nashville Dec. 15-16, 1864. Mustered out January 13, 1866.

12th Regiment, USCT—Organized in Tenn. at large July 24 to August 14, 1863. Attached to Defenses of Nashville Camp; Northwestern Railroad, Dept. of the Cumberland, to October, 1864. 2nd Colored Brigade, District of the Etowah, Dept. of the Cumberland, to January, 1865. Defenses of Nashville Camp; N&NWRR, District of Middle Tenn., to May, 1865. 3rd Sub-District, District Middle Tenn., Dept. of the Cumberland, to January, 1866. Service: Railroad guard duty at various points in Tenn. and Alabama on line of the Nashville Camp; N&NWRR until Dec., 1864. Repulse of Hood's attack on Johnsonville, Nov. 2, 4 and 5. Action at Buford's Station, Section 37, Nashville Camp; N&NWRR, Nov. 24. March to Clarksville, and skirmish near that place Dec. 2. Battle of Nashville Dec. 15-16. Pursuit of Hood to the Tenn. River, Dec. 17-28. Action at Decatur, Ala., Dec. 27-28. Railroad guard and garrison duty in the Dept. of the Cumberland until January, 1866. Regiment lost during service 4 officers and 38 enlisted men killed and mortally wounded and 242 enlisted men by disease. Total 284.

13th Regiment, USCT—Organized at Nashville, Nov. 19, 1863. Attached to Defenses Nashville Camp; Northwestern Railroad, Dept. of the Cumberland, to Nov., 1864. 2nd Colored Brigade. District of the Etowah, Dept. of the Cumberland, to January, 1865. Defenses Nashville Camp; Northwestern Railroad, District Middle Tenn., Dept. of the Cumberland, to May, 1865. 3rd Sub-District, District Middle Tenn., Dept. of the Cumberland, to January, 1866. Service: Railroad guard duty in Tenn. and Alabama on line of Nashville Camp; Northwestern Railroad until Dec., 1864. Repulse of Hood's attack on Johnsonville, Tenn., September 25, and Nov. 4 and 5. Eddyville, Ky., October 17 (Detachment). Battle of Nashville Dec. 15-16. Pursuit of Hood to the Tenn. River Dec. 17-18. Railroad guard and garrison duty in the Dept. of the Cumberland until January, 1866. Mustered out Jan. 10, 1866. Regiment lost during service 4 officers and 86 enlisted men killed and mortally wounded and 265 enlisted men by disease. Total 355.

14th Regiment, USCT—Organized at Gallatin, Nov. 16, 1863, to Jan. 8, 1864. Attached to Post of Gallatin, to January,1864. Post of Chattanooga, Tenn., Dept. of the Cumberland, to Nov., 1864. Unattached, District of the Etowah, Dept. of the Cumberland, to Dec., 1864. 1st Colored Brigade, District of the Etowah, to May, 1865. District of East Tenn., to August, 1865. Dept. of the Tenn. and Dept. of Georgia until March, 1866.
Service: Garrison duty at Chattanooga, Tenn., until Nov., 1864. March to relief of Dalton, Ga., August 14. Action at Dalton August 14-15. Siege of Decatur, Ala., October 27-30. Battle of Nashville , Tenn., Dec. 15-16. Overton's Hill Dec. 16. Pursuit of Hood to the Tenn. River Dec. 17-28. Duty at Chattanooga and in District of East Tenn. until July, 1865. At Greeneville and Dept. of the Tenn. until March, 1866. Mustered out March 26, 1866.

15th Regiment, USCT—Organized at Nashville, Tenn., Dec. 2, 1863, to March 11, 1864. Attached to Post and District of Nashville, Dept. of the Cumberland, to August, 1864. Post of Springfield, District of Nashville, Dept. of the Cumberland, to March, 1865. 5th Sub-District, District of Middle Tenn., Dept. of the Cumberland, to April, 1866. Service: Garrison and guard duty at Nashville, Columbia and Pulaski, until June, 1864. Post duty at Springfield, and in District of Middle Tenn. until April, 1866. Mustered out April 7, 1866.

16th Regiment, USCT—Organized at Nashville, Dec. 4, 1863, to Feb. 13, 1864. Attached to Post of Chattanooga, Dept. of the Cumberland, to Nov., 1864. Unattached, District of the Etowah, Dept. of the Cumberland, to Dec., 1864. 1st Colored Brigade, District of the Etowah, Dept. of the Cumberland, to January, 1865. Unattached, District of the Etowah, to March, 1865. 1st Colored Brigade, Dept. of the Cumberland, to April, 1865. 5th Sub-District, District of Middle Tenn., to July, 1865. 2nd Brigade, 4th Division, District of East Tenn. and Dept. of the Cumberland. to April. Service: Duty at Chattanooga, Tenn., until Nov., 1864. Battle of Nashville, Tenn., Dec. 15-16. Overton Hill Dec. 16. Pursuit of Hood to the Tenn. River Dec. 17-28. Duty at Chattanooga and in Middle and East Tenn. until April, 1866. Mustered out April 30, 1866.

17th Regiment, USCT—Organized at Nashville, Dec. 12 to 21, 1863. Attached to Post of Murfreesboro, Tenn., Dept. of the Cumberland, to April, 1864. Post and District of Nashville, Tenn., Dept. of the Cumberland, to Dec., 1864. 1st Colored Brigade, District of the Etowah, Dept. of the Cumberland, to January, 1865. Post and District of Nashville, Tenn., Dept. of the Cumberland, to April, 1866.
Service: Duty at McMinnville and Murfreesboro, Tenn., until Nov., 1864. Battle of Nashville, Dec. 15-16. Overton Hill Dec. 16. Pursuit of Hood to the Tenn. River Dec. 17-27. Decatur Dec. 28-30. Duty at Post of Nashville, and in the Dept. of Tenn. until April, 1866. Mustered out April 25, 1866.

18th Regiment, USCT—Organized in Missouri at large Feb. 1 to September 28, 1864. Attached to District of St. Louis, Mo., Dept. of Missouri, to Dec., 1864. Unassigned, District of the Etowah, Dept. of the Cumberland, Dec., 1864. 1st Colored Brigade, District of the Etowah, Dept. of the Cumberland, to January, 1865. Unassigned, District of the Etowah, Dept. of the Cumberland, to March, 1865. 1st Colored Brigade, Dept. of the Cumberland, to July, 1865. 2nd Brigade, 4th Division, District of East Tenn. and Dept. of the Tenn., to Feb., 1866. Service: Duty in District of St. Louis, Mo., and at St. Louis until Nov., 1864. Ordered to Nashville, Tenn., Nov. 7. Moved to Paducah, Ky., Nov. 7-11, thence to Nashville, Tenn. Occupation of Nashville during Hood's investment Dec. 1-15. Battles of Nashville Dec. 15-16. Pursuit of Hood to the Tenn. River Dec. 17-28. At Bridgeport, Ala., guarding railroad until Feb., 1865. Action at Elrod's Tan Yard January 27. At Chattanooga, Tenn., and in District of East Tenn. until Feb., 1866. Mustered out Feb. 21, 1866.

44th Regiment, USCT—Organized at Chattanooga, Tenn., April 7, 1864. Attached to District of Chattanooga, Dept. of the Cumberland, to Nov., 1864. Unattached, District of the Etowah, Dept. of the Cumberland,

John Eaton

AAG George Luther Stearns

Gen. Lorenzo Thomas

It took two years to rectify this situation. Part of the problem was an interpretation of the Militia Act of 1862 which held that blacks should be paid as laborers and not as soldiers — USCT soldiers were getting the same pay as laborers.

In November 1862, U.S. Grant appointed John Eaton of New Hampshire as superintendent of freedmen. A year later, Eaton, who began the war as chaplain of the 27th Ohio Infantry, was named Superintendent of Negro Affairs for the Department of the Tennessee. Eaton subsequently supervised the establishment of 74 schools for negroes. By that time, he was colonel of the 63rd USCT.

On Jan. 1, 1863, Lincoln issued the Emancipation Proclamation freeing all slaves in the Confederate states (except Tennessee, southern Louisiana, and parts of Virginia) and announced the administration's intention to enlist black soldiers and sailors. By late spring, recruitment was underway throughout the North and in all the Federal-occupied Confederate states except Tennessee. In October 1863, the War Department ordered full-scale recruitment of black soldiers in Maryland, Missouri, and Tennessee, with compensation to loyal owners.

In the spring of 1863, General Lorenzo Thomas was appointed commissioner for the Organization of Colored Troops in Tennessee. He began actively raising black regiments in Memphis and had 3,000 troops by June. By war's end, Thomas had raised nearly 24,000 black troops from Tennessee and other states, filling 22 infantry regiments and eight artillery units.

George Luther Stearns, Assistant Adjutant General for the Recruitment of Colored Troops, was put in charge of USCT recruiting in Middle Tennessee. An abolitionist, Stearns was John Brown's biggest financial backer and even owned the rifles Brown had used at Harper's Ferry. He had recruited one of the first African-American regiments, the 54th Massachusetts, and would later become a leader in establishing the Freedmen's Bureau. Stearns quickly earned the wrath of Military Gov. Andrew Johnson, who wished to use military recruits to work on the Nashville & Northwestern Railroad and not used in combat.

On Sept. 16th, 1863, Stanton telegraphed Major Stearns, "You will not act contrary to the wishes of Gov. Johnson in relation to enlistments without express authority for so doing from this Dept." Three days later,

Selected Units of United States Colored Troops in Middle Tennessee (cont.)

to Dec., 1864. 1st Colored Brigade, District of the Etowah, Dept. of the Cumberland, to January, 1865. Unattached, District of the Etowah, to March, 1865. 1st Colored Brigade, Dept. of the Cumberland, to July, 1865. 2nd Brigade, 4th Division, District of East Tenn., July, 1865. Dept. of the Cumberland and Dept. of Georgia to April, 1866. Service: Post and garrison duty at Chattanooga, until Nov., 1864. Action at Dalton, Ga., October 13, 1864. Battle of Nashville, Tenn., Dec. 15-16. Pursuit of Hood to the Tenn. River Dec. 17-28. Post and garrison duty at Chattanooga, in District of East Tenn., and in the Dept. of Georgia until April, 1866. Mustered out April 30, 1866.

Source: Civil War Soldiers and Sailors System

100th Regiment, USCT—Organized in Kentucky at large May 3 to June 1, 1864, Attached to Defenses of Nashville & Northwestern Railroad, Dept. of the Cumberland, to Dec., 1864. 2nd Colored Brigade, District of the Etowah, Dept. of the Cumberland, to January, 1865. Defenses of N&NWRR, Dept. of the Cumberland, to January, 1865. Defenses of Nashville and N&NWRR, Dept. of the Cumberland, to Dec., 1865. Service: Guard duty on N&NWRR in Tenn. until Dec., 1864. Skirmish on N&NWRR September 4. Action at Johnsonville Nov. 4-5. Battle of Nashville, Tenn., Dec. 15-16. Overton Hill Dec. 16. Pursuit of Hood to the Tenn. River Dec. 17-28. Again assigned to guard duty on N&NWRR January 16, 1865, and so continued until Dec., 1865. Mustered out Dec. 26, 1865.

In October 2021, a bronze statue of a U.S. Colored Troops soldier, **"March to Freedom,"** sculpted by Joe F. Howard, was dedicated in front of the Williamson County Courthouse in Franklin on the same town square which has displayed the statue of a Confederate infantryman since 1899. Chains of bondage broken at his feet, the USCT soldier bears sergeant's stripes on his shoulder. Before the war, slaves were bought and sold at the courthouse. During the war, at least 300 black men from Williamson County enlisted in the USCT, and although none fought at the Battle of Franklin, many fought at the nearby Battle of Nashville. In recent years, five interpretive markers have been installed at the square telling "The Fuller Story" of the local black community. The statue and signs were financed with privately raised funds. The Franklin sculpture is one of six public statues in the U.S. recognizing the USCT, and the only one located in a public town square. Another is located at the Nashville National Cemetery.

Like A Man

This was the biggest thing that ever happened in my life. I felt like a man, with a uniform on and a gun in my hand…

— USCT Pvt. Freeman Thomas

Stanton further advised him: "You will conform your actions to his views. All dissension is to be avoided, and if there is any want of harmony between you, you had better leave Nashville and proceed to Cairo (Illinois) to await orders."

In Nashville, on Sept. 20th, 1863, educator John Berrien Lindsley wrote in his journal: "At African church a negro man shot down by the guards engaged in pressing. It is the custom of the Military authorities to go to the colored people's churches on Sunday when they wish to make a big haul of pressed men. The man died afterwards…"

In January 1864, planters in Rutherford County, Tenn., in an effort to retain their slaves, promised to pay them 10 cents per pound of cotton picked and to "take care of & support them as they have before done." This move did little to stem the tide of emancipation.

It was not until March 13th, 1865, a few weeks before the Confederacy collapsed, that the Confederate Congress enacted legislation authorizing the enrollment of blacks into the Confederate Army (Gen. P.R. Cleburne had proposed doing so in 1864 but was strongly rebuked).

USCT regiments were posted throughout Middle Tennessee. During the spring of 1864, the 12th and 13th USCT guarded the N&NWRR line, while the 14th and 16th USCT were sent to Chattanooga. The 15th was at Nashville, the 17th at Murfreesboro. During the summer, the 40th and 101st USCT from Tennessee and the 100th USCT from Kentucky were posted along the Louisville and Nashville Railroad. The 42nd and 44th USCT went to Chattanooga, the 1st U.S. Colored Artillery to Knoxville, and the 2nd and 3rd Alabama, renumbered the 110th and 111th USCT in June, at Pulaski and Athens, Ala.

The men of these USCT regiments sometimes served as laborers. Early in 1864, the quartermaster corps had expanded from 8,000 men to 15,000 (with 4,510 working on the railroads). Despite this increase, by fall of 1864, more workmen were needed to supply Sherman's armies at Atlanta. Quartermaster Donaldson, employing USCT troops as laborers, explained, "I did not want to do this, for I believe in Colored Troops and think they should take the Field and fight the same as White ones, but I knew there were Colored Regts. in the Dept. not yet fit for the Field and that, for obvious reasons, they had more work in them than I could get out of any other troops."

According to William A. Dobak's history of the USCT: "Later, Donaldson excused the men of the 15th and 17th USCT regiments from stevedore tasks and assigned them to the more military duty of guarding rail lines and quartermaster depots. The 15th USCT stretched out along the Edgefield and Kentucky Railroad, guarding bridges and trestles from the north bank of the Cumberland River opposite Nashville 40 miles northwest to the state line; the 17th concentrated at Nashville, with companies stationed up and down the river at sawmills and wood yards, and sometimes provided armed guards for riverboats."

An official reprieve came in the summer of 1864. On June 14th, the U.S. War Department issued a directive that stated in part: "Accordingly the practice which has hitherto prevailed, no doubt from necessity of requiring (black troops) to perform most of the labor of fortifications, and the labor of fatigue duties of permanent stations and camps, will cease, and they will only be required to take their fair share of fatigue duty with white troops."

The possibility that the negro suffrage proposition may shock popular prejudice at first sight, is not a conclusive argument against its wisdom and policy. No proposition ever met with more furious or general opposition than the one to enlist colored soldiers in the United States army. The opponents of the measure exclaimed on all hands that the negro was a coward; that he would not fight; that one white man, with a whip in his hand could put to flight a regiment of them; that the experiment would end in the utter rout and ruin of the Federal army. Yet the colored man has fought so well, on almost every occasion, that the rebel government is prevented, only by its fears and distrust of being able to force him to fight for slavery as well as he fights against it, from putting half a million of negroes into its ranks.

— Statement by Black Citizens at Union Convention, Nashville, Jan. 9, 1865

The memoirs of two USCT soldiers in Middle Tennessee illustrate the struggles and trevails experienced by the typical Negro soldier. As some have noted, fugitive slaves impressed by the Federals and then recruited into the Union army basically exchanged one master for another.

Ned Scruggs, 24, of Franklin, labored at Fort Negley. He was employed for five months at a rate of $7 per month. He was one of the relatively few men who actually received his wages of $35. On Sept. 24th, 1863 in Nashville, Scruggs enlisted in Company F of the 13th US Colored Infantry, the second Civil War-era federal black regiment of infantry soldiers in the United States. He was officially mustered into Co. F at Camp Rosencranz (Fortress Rosecrans) on November 19th. That day the men were presented with their regimental flag, described as a vibrant blue flag with a blazoned eagle and shield, marked "Thirteenth Regiment U.S. Colored Infantry" and "Presented by the colored ladies of Murfreesboro."

Scruggs was promptly promoted to corporal, indicating that he was literate. He signed up for a three-year term of service under the command of Colonel John A. Hottenstein. The 13th USCT guarded Johnsonville, Waverly, and other key points along the Nashville & Northwestern Railroad between May and December 1864, and again from mid-January 1865 to the end of the war. The record indicates that Co. F was summoned to Johnsonville in July 1864 from various points along the Nashville and Northwestern Railroad to help construct and garrison the Lower Redoubt at Johnsonville. In the course of that work, Scruggs "incurred injury of [his] back caused by lifting logs to build breastworks."

Ned Scruggs was admitted to Hospital 16 in Nashville with smallpox and returned to duty January 3rd, 1866 in Nashville. Just one week later, he was mustered out of service. He died Feb. 6, 1908 in Elkmont, Alabama, at age 72.

Freeman Thomas was born into slavery near Franklin. When he was about 16 years old he was impressed by a Federal army captain to work menial labor on Fort Granger in Franklin and then at Fort Negley in Nashville. He cut down trees, hauled brick, rock, soil, and other materials.

Thomas recalled, "I ran off from my master when I was about fifteen years old and joined the army. I was in the field shucking corn on the Murfreesboro Pike. All at once I heard a band playing. Everybody in the field broke and ran. Not a man was left on the place. We all went and joined the army. The captain asked what we wanted, and who our master was. We told him who our master was, and

This photograph of members of the 13th Regiment U.S. Colored Troops Living History Association was taken at Johnsonville with Civil War-era equipment and techniques. (Courtesy Bill Radcliffe.)

that we had come to join the army."

After working about three weeks at Fort Negley, Thomas was sent to Tullahoma for training. "This was the biggest thing that ever happened in my life. I felt like a man, with a uniform on and a gun in my hand." He enlisted in the 12th Infantry (Co. K) of the US Colored Troops on Aug. 12, 1863, along with 50 other men from Williamson County.

Originally his unit assisted in the construction of the N&NWRR, provided railroad guard duty at various points in Tennessee and Alabama on the line of the railroad until December 1864. They guarded the Elk River bridge on the Nashville & Decatur Railroad. After their time in Alabama, the 12th USCT was sent west to Johnsonville, where they were involved in repulsing Confederate General Forrest's attacks there.

The 12th USCT next saw action at Buford's Station, Section 37 of the Nashville & Northwestern Railroad, on Nov. 24th, 1864. Then they marched to Clarksville, and skirmished near there on December 2nd.

The 12th US Colored Infantry's most significant fighting was during the Battle of Nashville on Dec. 15-16th, 1864. Freeman Thomas was wounded in action there on Dec. 16th but survived. He was shot in the left ankle and was taken to Hospital No. 16 in Nashville where he was treated until February 1865. He rejoined his unit in camp near Nashville on the N&NWRR. At some point, the 12th US Colored Infantry was detailed to Murfreesboro to assist in burying the Union dead at Stones River Cemetery.

Thomas was granted a furlough and visited his old home in Franklin. "I went to see my mistress on my furlough, and she was glad to see me. She said, 'You remember when you were sick and I had to bring you to the house to nurse you?' and I told her, 'Yes'm, I remember.' And she said, 'And now you are fighting me!' I said, 'No'm, I ain't fighting you, I'm fighting to get free.'"

Freeman Thomas was honorably discharged on Jan. 16, 1866 with his regiment in Nashville, well after the end of the Civil War. He worked as a farm laborer near Franklin, married, and raised a family. On May 17, 1936, Freeman Thomas died at 91 years old.

The "Convention of Colored People of this State" met on Aug. 7th, 1865 at a chapel of the African Methodist Episcopal Church in Nashville. Twenty of the 116 delegates were soldiers in Tennessee regiments of the U.S. Colored Troops. Sgt. Henry J. Maxwell of Battery A, 2nd USCT Light Artillery, addressed them on the first day. "We want the rights guaranteed by the Infinite Architect," he told them. "We have gained one — the uniform is its badge. We want two more boxes, beside the cartridge box — the ballot box and the jury box."

Slavery was officially abolished in the United States on Dec. 6th, 1865, when the 13th Amendment to the Constitution was ratified. The 14th Amendment (July 9th, 1868) gave citizenship and due process of the law to freedmen. The 15th Amendment (Feb. 3rd, 1870) gave all qualified citizens the right to vote, not to be abridged by the states. Unfortunately, these new laws were not always uniformly enforced.

Shortly after the war, General Thomas ordered the establishment of a National Cemetery just north of Nashville, bisected by the Louisville & Nashville Railroad. Asked if the fallen soldiers should be segregated by state, Thomas replied to mix them up. He chose the location, he said, so that trainloads of citizens coming to Nashville from the North would have to witness the great cost suffered due to the war. Approximately 16,000 Civil War soldiers (all Union) are buried at Nashville, including 2,133 USCT.

Granbury's Lunette
Federals march into trap and are cut to pieces

On Dec. 13th, 1864, as the frigid winter weather eased a bit, Granbury's Brigade, the reserve brigade for Cleburne's (now Smith's) Division, was directed to build a four-gun lunette on a wooded hilltop about 400 feet northeast of the Nashville & Chattanooga Railroad on the extreme right flank of Hood's line. These men were part of Major Gen. Benjamin Franklin Cheatham's Corps (Cheatham himself would be inside the lunette on the first day of battle). The next day, the completed lunette was occupied by the 344 men of the brigade and the four smoothbore guns of Goldthwaite's Alabama Battery. The brigade, depleted at the Battle of Franklin, was commanded by Captain Edward T. Broughton, formerly of the 7th Texas Infantry. Broughton had been captured at Fort Donelson and also later at Raymond, Miss., spending captivity both times at Johnson's Island, Ohio, where he contracted smallpox and went partially blind. He took control of the brigade after Granbury was killed at Franklin.

Granbury's Texas brigade had gained a reputation as one of the hardest fighting units in the war on either side. Brig. Gen. Hiram B. Granbury, 33, had also been captured at Fort Donelson and was sent to Fort Warren Prison in Boston Harbor for five months. In 1863, following his release, his wife died of cancer. The 7th Texas fought at Raymond and Chickamauga and in the Atlanta Campaign. During Hood's invasion of Tennessee, Granbury was shot in the head and killed while charging the Federal works at Franklin. In 1913, a statue was erected in his honor at the courthouse in Granbury, Hood County, Texas.

Also on December 13th, the men of the 13th U.S. Colored Troops Regiment, part of the newly formed 2nd Colored Brigade, reconnoitered the Union left flank and skirmished with rebel troops. On the night of December 14th, on the eve of battle, Colonel Thomas Jefferson Morgan, commander of the 14th USCT, went out to

Sgt. Maj. Daniel W. Atwood, 100th USCT:

It was the first time in the memorable history of the Army of the Cumberland that the blood of black and white men flowed freely together for one common cause for a country's freedom and independence. Each was cheered on to victory by the cooperation of the other, and now, as the result, wherever the flag of our love goes, our hopes may advance, and we may, as a people, with propriety claim political equality with our white fellow-soldier and citizen; and every man that makes his home in our country may, whatever be his complexion or progeny, with propriety, exclaim to the world, 'I am an American citizen!' I ask, is there not something in this over which to rejoice and be proud?

General Hiram B. Granbury

Capt. Edward T. Broughton

Col. Thomas J. Morgan

examine the Confederate formations. On neither occasion did the Federals discover Granbury's Lunette.

Colonel John A. Hottenstein, commander of the 13th USCT, reported: "During the time from the 7th to the 13th [of December] this regiment was occupied in throwing up rifle-pits along the line and preparing for a campaign. The men were reclothed and refitted in everything necessary for a long campaign. On the 13th, the regiment was ordered out with the rest of the brigade on a reconnaissance near Rains' house, and had a lively skirmish during the afternoon, retiring at dusk. In this skirmish the regiment lost 1 man killed and 4 wounded. On the night of the 14th I received orders to be ready to move at 5 o'clock the following morning. Soon after daylight on the morning of the 15th we moved with the brigade and occupied the works thrown up on the right of the Chattanooga railroad and near the Nolensville pike. During the 15th, the regiment lay behind those breastworks, under a severe fire from a battery in our front, without sustaining any loss."

As the Federals prepared to launch

> ## They Lay Side by Side in Death: Colonel Morgan
>
> Thomas J. Morgan, Colonel of the 14th USCT and commander of the 1st Colored Brigade, gave the following account of his regiment:
>
> Nov. 29, 1864, in command of the 14th, 16th, and 44th Regiments U.S.C.I., I embarked on a railroad train at Chattanooga for Nashville. On December 1st, with the 16th and most of the 14th, I reached my destination, and was assigned to a place on the extreme left of General Thomas' army then concentrating for the defence of Nashville against Hood's threatened attack… Soon after taking our position in line at Nashville, we were closely besieged by Hood's army; and thus we lay facing each other for two weeks. Hood had suffered so terribly by his defeat under Schofield, at Franklin, that he was in no mood to assault us in our works, and Thomas needed more time to concentrate and reorganize his army, before he could safely take the offensive… About nine o'clock at night December 14th, 1864, I was summoned to General Steadman's [sic] headquarters. He told me what the plan of battle was, and said he wished me to open the fight by making a vigorous assault upon Hood's right flank. This, he explained, was to be a feint, intended to betray Hood into the belief that it was the real attack, and to lead him to support his right by weakening his left, where Thomas intended assaulting him in very deed. The General gave me [four colored regiments], … a provisional brigade of white troops…and a section of Artillery…of the 20th Indiana Battery…As soon as the fog lifted, the battle began in good earnest. [Thomas assailed] Hood's left flank, doubling it up, and capturing a large number of prisoners. Thus the first day's fight wore away. It had been for us a severe but glorious day. Over three hundred of my command had fallen, but everywhere our army was successful…General Steadman congratulated us, saying his only fear had been that we might fight too hard. We had done all he desired, and more. Colored soldiers had again fought side by side with white troops; they had mingled together in the charge; they had supported each other; they had assisted each other from the field when wounded, and they lay side by side in death. The survivors rejoiced together over a hard fought field, won by a common valor…"

their assault against Hood on Thurs., Dec. 15th, 1864, the Provisional Division of Brig. Gen. Charles Cruft relieved the troops of the Fourth and Twenty-third Army Corps, occupying their line of works and picketing the front of this line from the Acklen place to Fort Negley, and commanding the approaches to the city by the Granny White, Franklin, and Nolensville turnpikes. Post Commander Brig. Gen. John F. Miller's troops (142nd Indiana, 45th New York, and 176th, 179th, and 182nd Ohio) occupied the works from Fort Negley to the Lebanon pike, commanding the approaches to the city by the Mur-

freesboro, Chicken, and Lebanon turnpikes. Brig. Gen. J.L. Donaldson's trained quartermaster employees occupied the works from the right of General Cruft's command west to the Cumberland River, commanding the approach to the city by the Harding and Hillsboro turnpikes.

Federal commander Gen. George H. Thomas ordered Maj. Gen. James B. Steedman to attack the Confederate right flank at 5:00 am in a diversionary assault designed to confuse General Hood and draw rebel forces away from their left flank. Due partially to the previous protests of Col.

(Continued on Page 286)

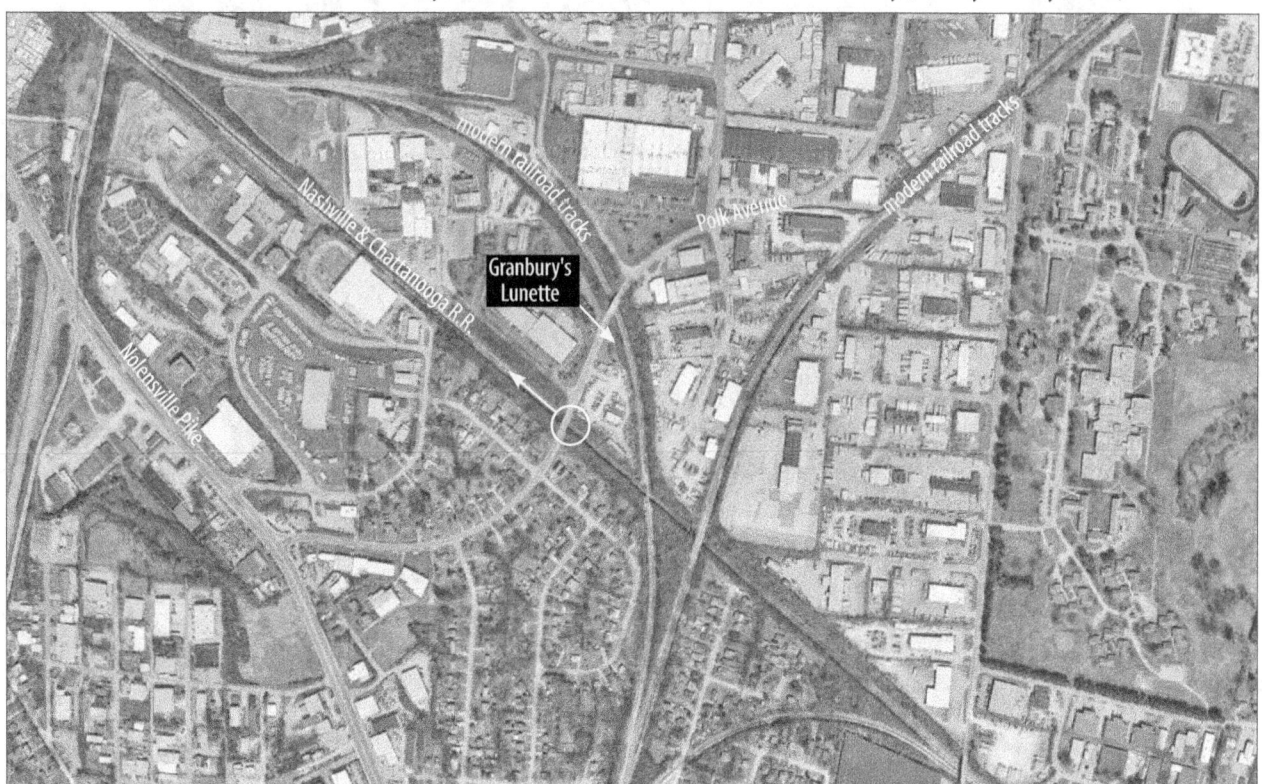

The Nashville & Chattanooga Railroad cut south of Granbury's Lunette, looking northwestward from the Polk Avenue overpass toward Nashville (see marker on map below). At the time of the Civil War, there would have been only one set of tracks and the cut would have been ten feet shallower. This railroad cut is not the one directly beside the lunette; that cut was made in the 20th Century. (Photo by John Allyn-BONT)

Eastern Sector - Battle of Nashville - Thurs., Dec. 15th, 1864

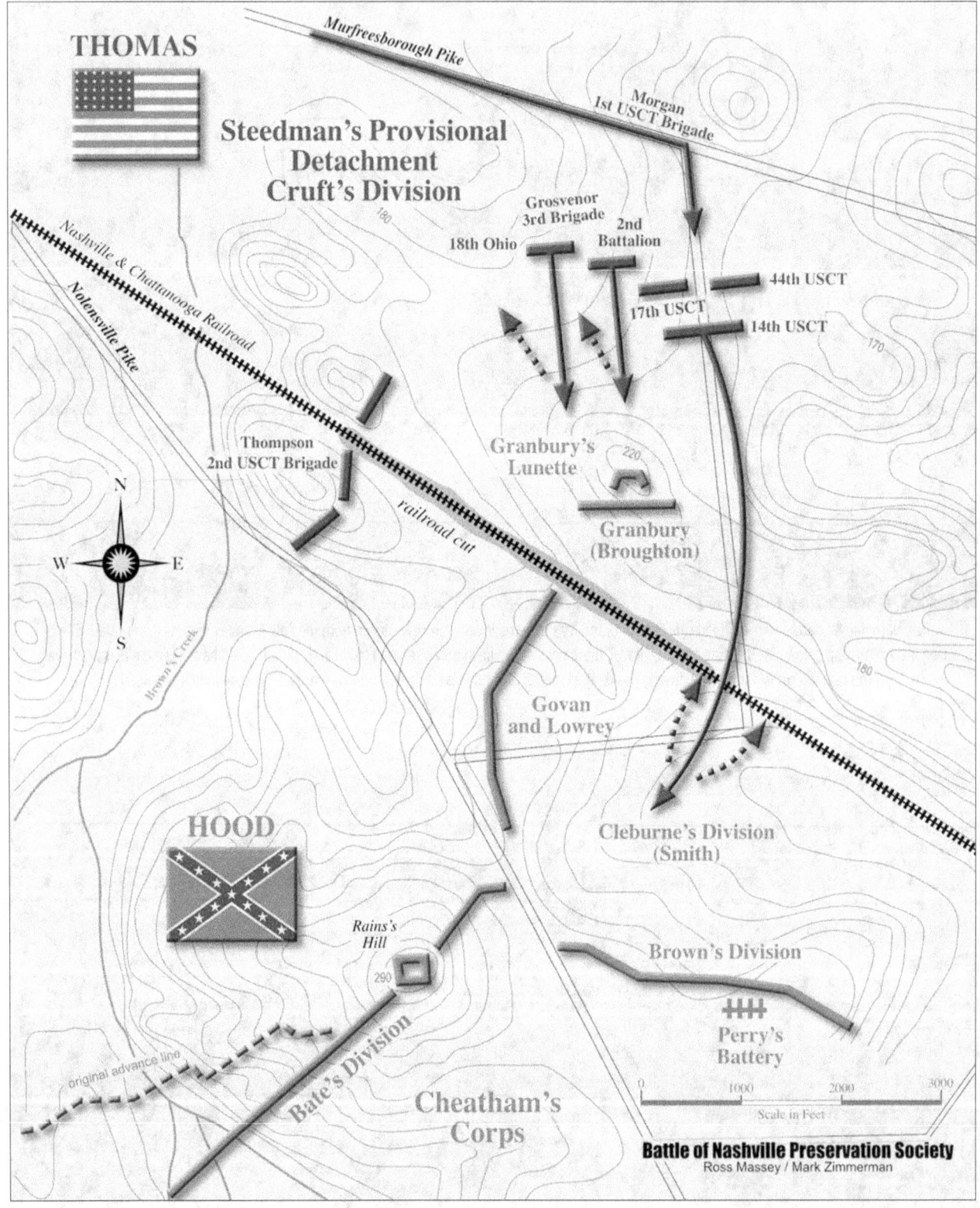

Granbury's Lunette

Above: Looking southeast along the railroad tracks (the modern railroad tracks that destroyed part of the lunette).

Left: The remains of Granbury's Lunette looking in a northeasterly direction from Polk Avenue.

Below: Looking northwest along the railroad tracks toward downtown Nashville.

(Photos by John Allyn-BONT)

Gen. James B. Steedman

Col. William R. Shafter

(Continued from Page 282)

Morgan, who desperately wanted to lead USCT men into combat, Steedman's command now consisted of two Negro brigades. Formerly the post commander at Chattanooga, Steedman now commanded the unique Provisional Detachment (District of Etowah). His lone division, under Cruft, consisted of two brigades of USCT and a white brigade composed of invalids, convalescents, untrained recruits, bounty jumpers, and others unable even to speak English.

The 1st Colored Brigade, commanded by Colonel Morgan, was composed of the following units:

• The 14th USCT of 610 officers and men, commanded by Lieutenant Colonel Henry C. Corbin. Raised in Gallatin, Tenn., the Fourteenth had been stationed in Chattanooga performing guard duty. Corbin and Morgan did not get along, dating back to an incident in which Morgan supposedly ordered his men to lie down and pray while Corbin thought they should be fighting. Morgan "early advocated [for] the organization of colored regiments." A deeply religious man and abolitionist from Indiana, Morgan would become a minister after the war and chief of the Bureau of Indian Affairs. Morgan stated in his report of the Nashville battle that Corbin "does not possess sufficient courage to command brave men." Corbin was subsequently tried before a general court martial on the charges of cowardice and misbehavior in the presence of the enemy. The court martial transcript makes it abundantly clear that it was the last step in a vendetta between the two men. Corbin was found not guilty and "was most honorably acquitted" by the court martial, which specifically condemned two of Morgan's witnesses for "misrepresentation" and Morgan himself for coaching witnesses.

• The 17th USCT, commanded by Colonel William R. Shafter of Michigan, approximately 600 officers and men. It had been raised in Murfreesboro and had previously served as commissary guards in Nashville, probably doing more warehouse work than actual guarding. According to historian John Allyn with the Battle of Nashville Trust, the 17th came to the 1st Colored Brigade more or less by accident. On the evening of Dec. 14th, 1864, the commander of the 16th wrangled a transfer from Morgan's brigade to the army's pontoon bridge train. Morgan, who apparently had a problem getting along with subordinates, described this transfer as an "unsoldierly process" and demanded a court of inquiry, which subsequently cleared the 16th's colonel of any misfeasance. Colonel Shafter volunteered his regiment to take the place of the 16th. Shafter was a career man and after the Civil War commanded Buffalo soldiers in the Indian Wars, acquiring the name of Pecos Bill. In the 1890s, he was awarded the Medal of Honor for conduct at the Battle of Fair Oaks/Seven Pines in 1862 (the legitimacy of the award is debatable). An obese man, Shafter was promoted to Major General and commanded the American expedition to Cuba during the Spanish-American War.

• The 18th USCT, 363 officers and men, commanded by Major Lewis D. Joy. This unit had been raised in Missouri in September 1864, and was sent to Nashville in late November.

• Two or three companies from the 44th USCT, commanded by Colonel Lewis Johnson, totaling 212 officers and men from Chattanooga. The bulk of this regiment had been captured when the garrison at Dalton, Georgia had surrendered to Hood's men on Oct. 13th, 1864.

The 2nd Colored Brigade, commanded by Colonel Charles R. Thompson, was composed of the following units:

• The 12th USCT, led by Lt. Col. William R. Sellon

• The 13th USCT, 576 men and officers, led by Col. J.A. Hottenstein

• The 100th USCT, led by Major Collin Ford

• The First Battery, Kansas Light Artillery, directed by Capt. Marcus D. Tenney

The 3rd Brigade (XIV Corps), commanded by Lt. Col. Charles H. Grosvenor of Ohio, included:

• 18th Ohio Veteran Volunteer Infantry, 325 officers and men commanded by Capt. Ebenezer Grosvenor, the brigade commander's brother. The Eighteenth Ohio was an 1861 unit

Lt. Col. Charles H. Grosvenor

which did not veteranize and had mustered out on November 9th. On Oct. 31st, however, a new unit – the Eighteenth Ohio Veteran Volunteer Infantry – came into being. It was composed of soldiers from five other Ohio regiments that had not veteranized: the 1st, the 2nd, the 18th, the 24th, and the 35th. Some of these soldiers had chosen to re-enlist, and others were soldiers whose terms of enlistment had not yet expired.

• The 68th Indiana Infantry, 255 officers and men, led by Lt. Col. Harvey J. Espy. This regiment was raised in August 1862. It was captured at Munfordville, Ky., in September 1862, was exchanged, and fought at Chickamauga and Missionary Ridge. Afterwards it served as part of the Chattanooga garrison.

• The 2nd Battalion, XIV Army Corps, 319 officers and men, led by Capt. D. H. Henderson, 121st Ohio Infantry. Henderson had been severely wounded while serving with his regiment at Jonesboro in late August. His command was a pick-up unit formed from casual troops nominally assigned to the XIV Army Corps of the Army of Georgia. It included Sgt. Kelsey's 59 recruits from the 2nd Minnesota. Many were unfit for duty, only partially recovered from illness; some untrained raw recruits; some European immigrants "unable to speak or understand the simplest words of our language."

• The 20th Battery, Indiana Light Artillery, 135 officers and men, commanded by Capt. Milton A. Osborne. The battery was another unit raised in the summer of 1862. For most of the war this battery had served in rear echelon garrisons with its only battle experience being in the Atlanta campaign.

Morgan was ready for a fight to prove that black men could be good soldiers and fight courageously. General Thomas told Morgan he believed that blacks would fight but only behind protective earthworks. Morgan countered that they would fight in the open field, but Thomas disagreed. Now was the time to find out.

The morning of Dec. 15th, 1864 was dark and somber, a wet fog delaying movement over the soggy winter ground until about 9:00 am. Morgan led his 1st Brigade down the Murfreesboro Pike; the plan was to march beyond Hood's right and then swing to the south and attack. Meanwhile, Thompson's 2nd Brigade was moving along the N&CRR tracks from the northwest. Together, the two units would close upon the Confederates like a vise. Grosvenor's 3rd Brigade was following Morgan's brigade down the pike in reserve.

Morgan's troops advanced, with the 14th USCT in the van as skirmishers, followed by the 17th USCT, and then the 44th USCT. Corbin waved his sword above his head to order the advance.

The Confederates knew what was transpiring; they would not be surprised. Some later reported that the black troops were being led with the white troops pointing bayonets at their backs, just in case. In any event, the Confederates were incensed to see blacks, many of them former slaves, in blue uniforms armed with rifles. They allowed the Federal troops to swing around and march to within a hundred yards of their hidden lunette. Then all hell broke loose—hundreds of muskets firing almost at once along with the belching of deadly canister shot from four artillery pieces.

Capt. Henry Romeyn of the 14th USCT recorded, "Pushing on, the right of the skirmish-line passing through an orchard and cornfield and the left through a field lately cleared of timber and thickly strewn with stumps and piles of brush, over the crest of the slope it had ascended, [the line of battle] found itself on a sloping field…and face-to-face with heavy earthworks on its opposite side, from which, came at once a heavy and deadly fire of both artillery and infantry."

The U.S. Colored Troops broke formation under the withering fire and scattered. In addition to not knowing about the lunette, they also did not know about the deep railroad cut to their left, twenty feet deep between rocky cliffs and several hundred feet long. Many of the colored troops jumped into the railroad cut to escape, only to suffer grievous injury and to be cut down by enemy fire. The Confederates gave no quarter. "The carnage was awful," said one graycoat. "It is doubtful if a single bullet missed." Another called the scene "a perfect slaughter pen." Yet another said the killing "was particularly sickening."

During this time, Grosvenor's men tramped through a soggy cornfield, virtually ignored until they got too close to the lunette. The 2nd Battalion panicked and ran to the rear, "behaving in the most cowardly and disgraceful manner." Now all the fire fell on the 18th Ohio, which eventually retreated. Their leader, Ebenezer Grosvenor, managed to gain the rebel works only to be shot dead, hit by two minié balls.

Meanwhile, Thompson's 2nd Brigade crossed the railroad north of the cut and occupied some works previously abandoned by the Confederates. There, they skirmished most of the day with the rebels, and the 20th Indiana Battery exchanged artillery fire. They sustained few, if any, casualties.

Morgan's men fell back to the Murfreesboro Pike, occupied the Rains house, and harassed the Confederates with sharpshooting the rest of the day. Shafter later noted it was an "awful battle." He reported 17 killed or mortally wounded and 67 wounded. Capt. Job Aldrich, Shafter's brother-in-law, was killed. The 14th USCT reported four killed, 41 wounded, and 20 missing. The 44th had four wounded. Some accounts reported "hundreds" of colored troops slain at the railroad cut.

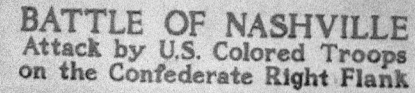

Although Steedman argued otherwise, his diversionary attack was not convincing, "a total failure," according to historian James Lee McDonough. Hood had not been fooled and had not sent any troops to shore up his right flank. Granbury's Brigade had performed brilliantly against overwhelming numbers. And although they had little chance of victory due to poor surveillance and leadership, the U.S. Colored Troops proved that they would march into combat and suffer the consequences.

Morgan wrote enthusiastically, "Colored soldiers had fought side-by-side with white troops. They had assisted each other from the field when wounded, and they lay side by side in death…A new chapter in the history of liberty has been written. It has been shown that marching under a flag of freedom, animated by a love of liberty, even a slave becomes a man and hero."

At 9:00 pm that night, Cheatham's Corps fell back to a secondary line on the extreme left flank of Hood's army. Hood's center corps, that of Lt. Gen. Stephen Dill Lee, fell back due south down the Franklin Pike to high ground known as Overton's Hill or Peach Orchard Hill.

Long after the Battle of Nashville, Granbury's Lunette was partially destroyed when a second railroad bed was cut through it. Granbury's Lunette Confederate Park and Memorial at 190 Polk Avenue was officially dedicated on Dec. 15th, 2001 by the Gen. Joseph E. Johnston SCV Camp 28. It is located on property owned by McCord Crane Co., which helped preserve the historic site. There is signage and a flag display but no other facilities.

US Colored Troops gained formal, public recognition on Oct. 15th, 2021, when the Metro (Nashville) Historical Commission erected a marker near Granbury's Lunette. The marker is the first-ever commemoration of the important and unique role played by African-American troops in the Battle of Nashville. Other markers will follow, including two in the planning stages by the Battle of Nashville Trust.

Close-up of the USCT statue erected in Nashville's National Cemetery.

Members of the 13th Regiment U.S. Colored Troops Living History Association proudly display the National banner. (Courtesy Bill Radcliffe.)

Peach Orchard Hill

Colored troops fought valiantly, died for their freedom

The 13th U.S. Colored Troops Regiment, led by Col. John A. Hottenstein, participated in a reconnaissance of the extreme right flank of Hood's Confederate lines a couple of days before the main battle and fell into a "lively skirmish" which resulted in one negro soldier killed and four wounded. A former Indian fighter, Hottenstein was a veteran of most of the major battles of the Western Theater. But his men were inexperienced. The 13th USCT regiment had been created only a year earlier and had worked on the railroad and performed garrison duty since then. They were at Johnsonville during Forrest's raid, but that was the only real action they had experienced. The men were eager to fight, eager to prove their worth.

The 13th USCT was part of the 2nd U.S. Colored Brigade led by Col. Charles R. Thompson, 24, a Maine native and St. Louis clothing salesman who had enlisted in October 1861 as a private in the Engineer Regiment of the West, Missouri Volunteers. In August 1862 he was appointed ordnance officer of the Army of the Mississippi. He became an aide-de-camp to Gen. Rosecrans, and by the end of the war was breveted brigadier general.

The 2nd US Colored Brigade, also consisting of the 12th and 100th USCT, had seen little action on the first day of the Battle of Nashville. On the morning of the 16th, Thompson sent skirmishers forward and discovered that the Confederates had abandoned their rifle pits and withdrawn to a new line. That morning, Steedman's Detachment would take five hours to move into position opposite a prominence occupied by the Confederate right flank known as Overton's Hill or Peach Orchard Hill. The hill was part of the 1,050-acre plantation of Col. John Overton II, whose house, Travel-

The 13th US Colored Troops attack breastworks atop Peach Orchard Hill. (Original artwork by Philip Duer. Used with permission.)

lers Rest, had been used by Gen. John Bell Hood as headquarters during the Nashville campaign. The gently sloping, 300-foot-high hill sported breastworks, and below that a row of abatis, felled trees stripped of small branches and leaves, with the larger branches sharpened to create a defensive obstacle. The approach to the hill, north and east, was wide open, consisting of a soggy, muddy cornfield and a singular thicket of dense woods, standing out as if an island.

Occupying the hill, bristling with gun muzzles, were the brigades of Brig. Gen. James Holtzclaw, Alabamians facing northward and centered on the Franklin Pike, and Brig. Gen. Marcellus A. Stovall, Georgians in a curve around the salient. These men belonged to the division of Major Gen. Henry D. Clayton, corps of Major Gen. Stephen D. Lee. They had missed the battle at Franklin and did not see much action the previous day at Nashville.

The Federals on the eastern flank comprised the IV Corps under Brig. Gen. Thomas John Wood, the scapegoat of the debacle at the Battle of Chickamauga in 1863. Two of his brigades had not performed well at Franklin. He had not seen much action on December 15th at Nashville. He was looking for some of the glory that was much anticipated at Nashville, the Federals outnumbering the Confederates almost three-to-one. Commander Thomas told Wood to demonstrate against Hood's right flank that day and take advantage of any opportune opening. That's all Wood needed to hear. After a brief reconnaissance of the flank, he prepared to launch an attack. Also raring to go

Color Made No Difference: General Steedman

General James B. Steedman, an old Breckinridge Democrat who had originally been opposed to the enlistment of Negro troops, was in command of the left wing of the Army of the Cumberland during the battle. He stated in his official report:

"The larger portion of these losses, amounting in the aggregate to fully 25 per cent of the men under my command who were taken into action, it will be observed fell upon the colored troops. The severe loss of this part of my troops was in their brilliant charge on the enemy's works on Overton Hill on Friday afternoon. I was unable to discover that color made any difference in the fighting of my troops. All, white and black, nobly did their duty as soldiers."

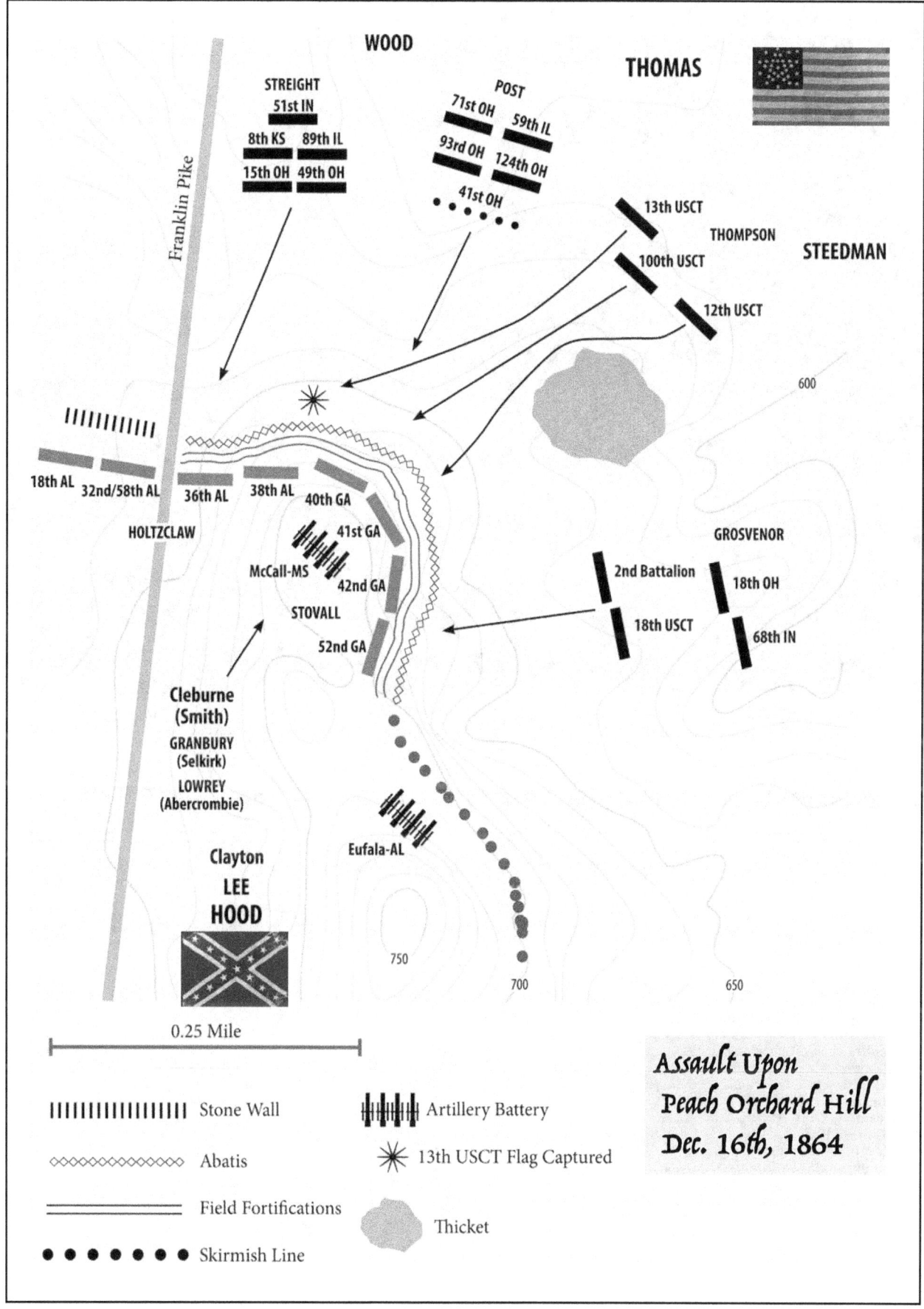

The Peach Orchard Hill battlefield has been obliterated by residential and commercial development and the construction of highways and railyards. The area bordered by the dashed line indicates where the fiercest fighting by the U.S. Colored Troops occurred.

Casualties-Battle of Nashville-Steedman's Provisional Detachment-USCT Units

	Killed Officers	Killed Men	Wounded Officers	Wounded Men	Missing Officers	Missing Men	Total Officers	Total Men	
Fourteenth U.S. Colored Infantry	......	4	......	41	......	20	......	65	⎫ Organized as the First Colored Brigade, Colonel T.J. Morgan, commanding.
Forty-fourth "	1	2	......	27	2	49	3	78	
Sixteenth "	......	1	......	2	......	......	......	3	
Eighteenth "	......	1	......	5	......	3	......	9	
Seventeenth "	7	14	4	64	......	......	6	78	⎭
Twelfth "	3	10	3	99	......	......	6	109	⎫ Organized as the Second Colored Brigade, Colonel C.K. Thompson, commanding.
Thirteenth "	4	51	4	161	......	1	8	213	
One Hundredth "	......	12	5	116	......	......	5	128	⎭
Eighteenth Ohio Infantry	2	9	2	38	......	9	4	56	⎫ Included in the Provisional Division, A.C., Brigadier-General Cruft, comd'g.
Sixty-eighth Indiana Infantry	......	1	......	7	......	......	......	8	
Provisional Division, A.C.	1	19	3	74	......	33	4	126	⎭
Twentieth Indiana Battery	......	......	2	6	......	......	2	6	Captain Osborn.
Aggregate	13	124	23	640	2	115	38	879	
								38	
Total								917	

was Col. Sidney Post, 2nd Brigade, 3rd Division. Wood advised Steedman of his plans, and Steedman agreed to also attack, protecting Wood's left flank.

Although the Confederates on Peach Orchard Hill were being relentlessly and heavily bombarded by Federal artillery batteries most of the day (and holding their own ammunition in reserve for the actual attack), their lofty position was formidable. Captain Henry V. Freeman of the 12th USCT opined, "It was probably their strongest position. The slope of the hill was obstructed by tree-tops. The approach was over a ploughed field, the heavy soil of which, clinging to the feet, greatly impeded progress." Facing the 12th, Freeman noted, "was a thicket of trees and underbrush so dense as to be almost impenetrable, constituting a kind of wooded island, in the midst of the cornfield."

Between 2:20 and 3:00 pm, the signal battery sounded and the attack was launched, first by Wood's men. Post's brigade, mostly Buckeyes, began their march due south along the Franklin Pike against Holtzclaw's regiments behind the works. Forty yards behind Post was the brigade of Col. Abel D. Streight of the infamous Jackass Brigade raid. They had not

Col. Sidney Post

Col. Charles R. Thompson

advanced far when the Confederates opened fire — hundreds of rifles and deadly canister from the four guns of Stanford's Mississippi Battery. Post was wounded and taken out of the battle when his horse was hit and killed by canister. He would be awarded the Congressional Medal of Honor. Some of Streight's troops managed to get close to Holtzclaw's works, but their advance could not be sustained.

To the east, Thompson's 2nd Colored Brigade moved out, the front line with the 100th USCT to the right, 12th USCT to the left, and the 13th USCT following behind. One of the odd aspects of Thompson's tactics was his positioning of the USCT troops according to the men's heights — tallest in the middle, then the shortest, and finally the men of middle height on the flanks of the regiment. The reasoning for this formation was never explained.

On their left flank was Grosvenor's rag-tag brigade, which also included that day the small 18th USCT Regiment.

Angered by the constant bombardment of their position, Holtzclaw's and Stovall's men were further enraged to discover themselves facing uniformed and armed negroes. Some USCT probably had been slaves owned by the Overtons, on whose farm they were now marching. The Confederates were determined to give no quarter to this enemy.

John Kendall of the 4th Louisiana noted: "A shout of derision promptly broke from our men. Instead of firing on those poor fellows, we began to jeer them, begging them to come on quickly, to black our boots, as some said; in order that none of the bullets should go astray, as others cried out. On the negroes came, and still we held our fire. It is a ticklish thing even for trained troops to advance against an enemy who does not shoot. The suspense of waiting seems rather more than human nerves can stand. On this occasion our dusky enemies approached to within 200 feet before our officers would permit us to press a trigger. Here they came to a halt. Cheered on by their white officers, however, the negroes took heart and moved forward. Then the jesting ceased. Our breastwork was suddenly belted with flame ... Two volleys were all that they could stand. Then they broke and fled."

Lucius W. Hull, 14, drum-major of the 18th Ohio, witnessed an odd event. He said he was within ten feet of Lt. Col. Benedict when the officer was shot in the mouth by a bullet glancing off the flagpole of the 14th USCT. He spit out a handful of teeth along with the spent ball. He could not speak so he was sent to the rear.

Captain D.E. Straight of the 100th USCT said some of the men, before moving out, had asked the officers to safeguard their papers and valuables. One man who was wounded during the assault asked his sergeant what he should do. The officer told the man to lie down. The wounded man went to ground, and eventually died. Straight testified that the "air seemed as full of the death-laden missiles as of hail in a driving hailstorm."

Historian Noah Andre Trudeau described some of the action. "Among the first to be hit was Private Alexander Helms. A friend caught him as he was spun around by the impact of the bullet, then gently lowered him to the ground. 'Lord have mercy,' Helms groaned as his comrades moved past. Nearby, Private William Smith was struck in the breast by a minié ball that tore through several layers of clothing before flattening itself, merely bruising the stunned soldier."

The 12th USCT, led by Lt. Col. William Sellon, trod forward through the mud but had to veer to the right to get around the thicket of trees. As the 100th USCT, led by Major Collin Ford, continued onward (somewhat disorganized from negotiating fallen trees), the men of the 12th increased their pace to the point that most men thought a double-time charge had been ordered. Thompson realized what had happened but decided that countermanding the "order to charge" would cause more confusion than doing nothing.

A Confederate gunner on the hill remarked, "On they came in splendid order, banners flying, mounted officers with drawn swords careering up and down in front of the lines. Then our artillery had its opportunity." The four guns of the Eufala Alabama battery joined in the fulisade, with cannon shot mowing down rows of blueclad negro soldiers. General S.D. Lee noted in his report that his troops had "reserved their fire until (Union troops) were within easy range, and then delivered it with terrible effect." Freeman noted, "They were so compact that every shot from Rebel muskets and cannon was telling with fearful effect."

"The rebel infantry blazed away at a fearful rate," noted *New York Times* correspondent Benjamin C. Truman, "and the artillery discharged sixteen shots of canister, which made the assaulting column reel, waver, and almost fall back."

Grosvenor's units to their left would be of no assistance that day. Disgraced the day before, the weak 2nd Battalion fled from the field and was not seen

Sgt. James Wilson's company muster roll with the 13th USCT denoting he was a colorbearer killed Dec. 16th at Peach Orchard Hill (Tennessee State Library and Archives).

for the rest of the campaign. With no reinforcements on their right, the two regiments of USCT failed to breach the parapet and eventually the assault withered and men began to fall back.

Then on came the 13th USCT, seemingly more determined than even the other bluecoats before. Col. Hottenstein remarked that he felt seeing the first line of Union troops lying down and taking cover from the Rebel fire would negatively affect the courage of his relatively raw regiment. But his men charged onward. Several infantrymen actually reached the parapet but were quickly struck down. Five color bearers were successively shot down trying to raise their regimental flag, which was eventually captured. Two of the color bearers were Sgt. Charles Rankin of Co. F and Sgt. James Wilson of Co. B.

"There were very few negroes who retreated in our front, and none were at their post when the firing ceased; for we fired as long as there was anything to shoot at," said a graycoat from Alabama. One USCT captain said the ground was "strewn with dead and wounded as thickly as a farmer's field with sheaves of a more peaceful reaper."

USCT Private John Beach, a big man at 200 pounds, was hit by a shell and injured his hip falling. Continuing on, he was shot in the head and face, fracturing his skull and knocking him senseless to the ground. He was aroused as his unit fell back, only to be hit again, this time in the side. He lumbered off the killing field and somehow managed to survive the war.

In the back parlor of nearby Glen Leven, home of the Thompson family (no relation to Col. Charles Thompson), the piano was used as an operating table for wounded USCT soldiers. Dozens were buried on the plantation, their remains later moved to the Nashville National Cemetery. In 2006, a life-sized bronze statue of a USCT infantryman was dedicated amongst the rows of USCT stone markers at the cemetery.

The 100th USCT lost 12 killed, 121 wounded; the 12th USCT had 10 killed, 104 wounded. But the 13th USCT suffered the worst.

Colonel Hottenstein's (partial) account of that day:

"At daylight on the morning of the 16th the regiment was under arms ready to move, and about sunrise I received orders from the colonel commanding to move across the Nolensville pike and feel the enemy in our front. I advanced my skirmishers to a piece of woods in our front, but the enemy had retired. I then received orders to move over to the Nolensville pike, where the remainder of the brigade then was, and to form my regiment as a reserve, in rear of the other two regiments of the brigade, and to regulate my movements by them. The brigade then moved to the right and front, and after considerable maneuvering joined the right to the left of the Third Division, Fourth Corps, where the men were ordered to lie down. In this position we were shelled considerably, by the enemy without any material damage. At about 2:30 I received notice that we would assault the works in our front, and in a few minutes afterward the order to advance was given. The regiment advanced with the brigade in good order, but before we arrived near the rebel works the troops in our front began to lie down, and skulk to the rear, which, of course, was not calculated to give much courage to men who never before had undergone an ordeal by fire. The fire of the enemy was terrific, but nevertheless the men, led by their officers, continued to advance to the very muzzles of the enemy's guns, but its numbers were too small, and after a protracted struggle they had to fall back, not for the want of courage or discipline, but because it was impossible to drive the enemy from his works by a direct assault. Before falling back, all the troops on our right had given way, and it was futile to continue the struggle any longer. The regiment reformed on the ground occupied just previous to the assault by the One Hundredth U.S. Colored Infantry, and was ready to again advance when a staff officer of the colonel commanding ordered me to take my regiment over to the left, where the remainder of the brigade was formed. I moved to the left, as ordered, and joined the brigade, which moved about four miles to the front and encamped for the night, in the meantime the enemy retiring toward Franklin. The regiment went into action on the morning of the 16th, 556 men and 20 commissioned officers, lost 4 commissioned officers and 55 enlisted men killed, and 4 commissioned and 165 enlisted men wounded; total loss, 220."

Colonel Thompson praised his men's performance: "The Thirteenth U.S. Colored Infantry, which was the second line of my command, pushed forward of the whole line, and some of the men mounted the parapet, but, having no support on the right, were forced to retire. These troops were here for the first time under such a fire as veterans dread, and yet, side by side with the veterans of Stones River, Missionary Ridge, and Atlanta, they assaulted probably the strongest works on the entire line, and though not successful, they vied with the old warriors in bravery, tenacity, and deeds of noble daring. The loss in the brigade was over twenty-five per cent of the number engaged, and the loss was sustained in less than thirty minutes."

"Better fighting was never done. Their chances were hopeless and they knew it. Still they showed courage and discipline." This from Ambrose Bierce, a Federal soldier on Gen. Beatty's staff who watched the USCT attack. Previously Bierce had applied to be an officer in charge of black troops but then declined because he didn't think blacks would fight. The struggle at Overton Hill changed his mind. Bierce would become a world-famous writer of fiction following the war.

Although the troops of the 13th USCT did not take the hill, they performed heroically and won the admiration of their superior officers and fellow soldiers. After the battle, Federal commander Thomas, who had doubts about black soldiers, witnessed the carnage and commented to his staff, "Gentlemen, the question is settled—Negroes will fight."

In an unprecedented action, Confederate General Holtzclaw noted the "gallant" efforts of the USCT units in his official after-action report. Noting the overwhelming odds against them, the general stated that "they came only to die."

"Many of these men escaped from bondage and enlisted, often without having ever handled a weapon, in order to end the practice of slavery once and for all in the United States. These soldiers may have been born into slavery, but they fought, and in many cases died, as free men," stated Andrew McMahan of the Tennessee State Library and Archives.

In an egregious decision during the battle that day which may have sealed his fate, Hood, worried about the artillery bombardment on his east flank, pulled two brigades from Cleburne's Division (Gen. James A. Smith) defending the southwest slope of Shy's Hill (Lowery's and Granbury's) to reinforce Peach Orchard Hill. The reinforcement was not needed at Peach Orchard Hill, and worse yet, their removal from Shy's Hill greatly aided the Federal encroachment in that sector.

As dusk was settling in, an amazing development swept over the field of battle. The Confederate left flank at Shy's Hill collapsed and many of the Confederate soldiers began an unorganized scramble towards the Franklin Pike. Lee's Corps evacuated Peach Orchard Hill, losing many of its 28 artillery pieces, and shortly thereafter made a stand on the pike that probably saved Hood's army from destruction. Nevertheless, Wood's Corps and the surviving USCT troops swarmed over Peach Orchard Hill and chased their foes a mile down the pike.

Following the two-day Battle of Nashville, the Federals chased the Army of Tennessee 100 miles south to the Tennessee River in northern Alabama. Steedman was ordered by Thomas to move his troops by railroad south to Alabama to intercept the fleeing Southerners but the Federal soldiers arrived too late to stop Hood's men from crossing the river to safety.

The Tennessee invasion was the last major campaign in the Western Theater. Five months later, the long war was finally over.

Twenty years later, Colonel Thomas J. Morgan of the 14th USCT wrote a report about his experiences, including this notation: "I cannot close this paper without expressing the conviction that history has not yet done justice to the share borne by colored soldiers in the war for the Union."

Battle of Nashville Trust

Great strides have been made in the past 30 years by concerned citizens and organizations to preserve what was remaining of the battlefield at Nashville. The Battle of Nashville Preservation Society, Inc. (BONPS) was formed in 1992 by a group of Nashvillians who recognized that the last remnants of the battlefield were quickly disappearing, and that the Battle of Nashville was poorly understood by both Nashvillians and visitors alike. The name was changed in 2020 to The Battle of Nashville Trust, Inc. (BONT).

Initially, its goal was to search for and identify unimproved ground that played important roles in the 1864 battle and find a way to protect it from development. The group's work and tenacity ultimately led to the ownership and protection of two iconic pieces of the battlefield – Confederate Redoubt No. 1, and the summit of Shy's Hill. Both have now been granted permanent protection from development through the Land Trust For Tennessee, and have been improved by BONT with the addition of interpretative kiosks and signage as well as reproduction field artillery pieces.

BONT has been instrumental in contributing to the preservation of numerous other portions of the battlefield, including efforts to protect and interpret Fort Negley, Kelley's Point, and numerous other smaller sites as well as many historic markers. It played a key role in the relocation of the Battle of Nashville Monument to Granny White Pike. BONT continues to assist in maintaining not only the sites it owns, but also in watching over and maintaining many others including the Monument. BONT is a tax-exempt, open-membership Section 501(c)(3) non-profit corporation which is managed by elected officers and a board of directors.

The difficulty that Nashville had with the U.S. occupation, which did not end until 1876, combined with the losses at Franklin and Nashville, made battlefield preservation a topic of little, if any, discussion. The Federal army had destroyed the countryside by digging miles of trenches and cutting down every tree within the Nashville city limits. Nashville wanted to forget what had happened, and consequently no preservation occurred. There was a movement in Congress in the 1880s to make Nashville and Franklin federal battlefields but that failed. On the 50th anniversary of the battle in 1914, 20 signs were placed at strategic points explaining the first day of the battle; these were accessible to the streetcar service from downtown. Trees had re-grown and the U.S. forts were either in ruin or had been demolished, but the earthworks were still present and many Nashvillians deemed them "scars" of the war. The battlefield for the most part was lost in history.

In 1923, Mrs. James Caldwell, a prominent Nashville historian, started the Nashville Battlefield Association and some money was raised, but no sites were saved. The monument was erected in 1927. The Great Depression then hit Nashville. But for battlefield preservation, it was a positive. In 1934, the Works Progress Administration rebuilt Fort Negley. It quickly fell into ruins again as it was "Yankee fort" and most of Nashville did not care about it. This trend continued for another 60 years.

By the end of World War II the farms that comprised the second day of the Battle of Nashville were starting to be developed and the land became expensive – $500 per acre after platting. In 1954, the developer of the Shy's Hill property donated the top 3.5 acres to the Tennessee Historical Commission, the only notable battlefield preservation until BONPS was formed in 1992.

Since 1992, BONPS has taken the premier leadership role in saving Redoubt No.1 and Shy's Hill. Due to a generous loan from its first president, Wes Shofner, Redoubt No.1 and the lot on Shy's Hill were purchased and the debt later paid. BONPS was instrumental in having Kelley's Point, a site of the naval engagement on the Cumberland River before the Battle of Nashville, made into a park. Mayor Bill Purcell was approached by BONPS in the mid 1990s to restore Fort Negley and build a visitors center. This $2.5 million project was accomplished and the visitor's center was opened in 2007. Additional battlefield signs have been placed in south Nashville and negotiations are underway to establish a one-acre park on the eastern flank near Peach Orchard Hill. Realistically, the future of additional preservation in Nashville is bleak due to the fact that Nashville is a boom town and land prices have sky-rocketed to such a degree that any core site, if it can be found, cannot be purchased. Most of the second day of the fighting in south Nashville is on property that is valued at $1.2 million or more per acre. A marker commemorating the USCT has been placed near Granbury's Lunette.

Addendum A: Glossary of Fortification Terms

Abatis: A line of felled trees with their branches sharpened, tangled together, and facing toward the enemy. It strengthened fortifications by preventing surprise and delaying an attacking enemy once within the defenders' range.

Advanced Works: Entrenched positions within supporting range in front of the main line of earthworks. They included rifle pits, picket lines, and vidette posts. They served as observation points and a first line of defensive positions.

Angle: Point where two faces of a fortification met. A reentrant angle pointed away from the enemy, a salient angle pointed toward the enemy.

Apex: Angle in a fortification closest to the enemy position.

Approach: Trench dug toward the enemy position.

Banquette: A raised step leading up to the rampart that served as a firing platform for defenders. The top was called the tread and the inclined plane leading up to the tread was called the slope. The banquette allowed defenders to fire and then step back to a covered position to reload.

Barbette: Raised platform or mound allowing an artillery piece to be fired over a fortification's walls.

Bastion: A fortification projecting outward from the curtain. Bastions were designed to prevent attackers gaining shelter from the defenders' fire.

Berm: Small horizontal space between the top of the ditch and the bottom of the parapet. It was designed to prevent earthwork from sliding back into its ditch. After completion of an earthwork's construction, some engineers chose to minimize the berm's size to prevent attackers from using it as a foothold while attempting to scale the wall.

Blockhouse: A log structure built to withstand attack from any direction, typically used to protect railroad bridges and depots. Blockhouses incorporated elements of fortification design and could have small ditches dug around them with the dirt piled against the outer log wall for additional structural support. The walls had loopholes and embrasures to allow the garrison to fire artillery and small arms in its defense.

Bombproof: A portion of the fortification designed to protect the garrison from enemy artillery fire. Bombproofs were built with heavy timbers and their roofs were covered with dirt.

Breach: A large gap in a fortification's walls or embankments created by artillery fire or mine. This exposed the inside of the fortification to assault.

Breaching Battery: A designated artillery position constructed during siege operations to fire upon a vulnerable position in the enemy line, opening it up for assault. Breaching batteries were placed on the parallel lines closer to the enemy position.

Breastworks: Fortifications made of piled material (logs, fence rails, stones) usually built up to breast height. Typically converted to a rampart if used long-term.

Entrenchment Profile

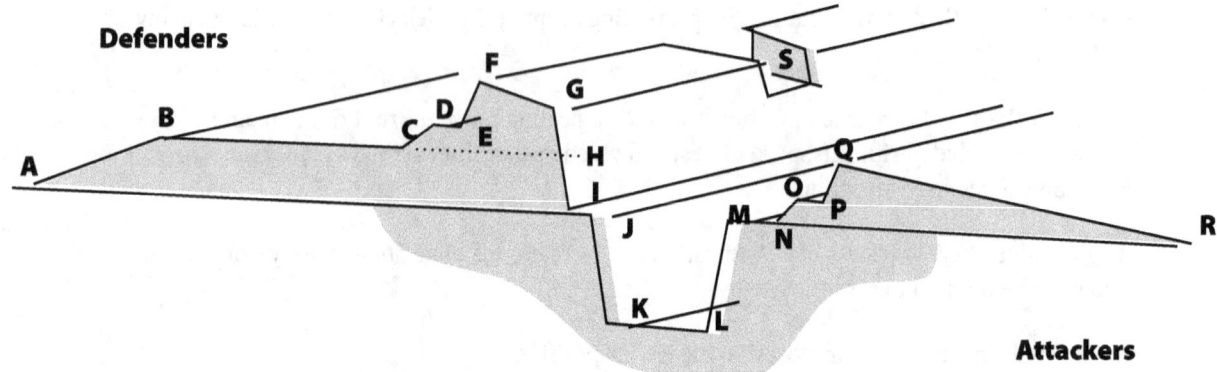

ABHI	Rampart or Bulwark	OP	Banquette
CDEFGH	Parapet	PQ	Interior Slope
JKLM	Ditch	QR	Glacis Slope
NOPQR	Glacis	S	Embrasure
AB	Parade or Slope		
BC	Terreplein		
CD	Banquette Slope		
DE	Tread of the Banquette or simply Banquette		
EF	Interior Slope		
FG	Superior Slope		
GI	Exterior Slope (if no rampart, GH)		
IJ	Berm		
JK	Scarp Wall		
KL	Bottom of the Ditch		
LM	Counterscarp Wall		
MN	Coverd Way		
NO	Glacis Banquette Slope		

High Points or Crest:
- F............ Interior Crest
- G............ Exterior Crest
- J............ Scarp Crest
- M........... Counterscarp
- Q............ Glacis Crest

Low Points or Foot:
- C............ Foot of Banquette Slope
- E............ Foot of Interior Slope
- I............. Foot of Exterior Slope (if no rampart, H)
- K............ Foot of Scarp
- L............ Foot of Counterscarp
- R............ Foot of Glacis

Fortification Forms

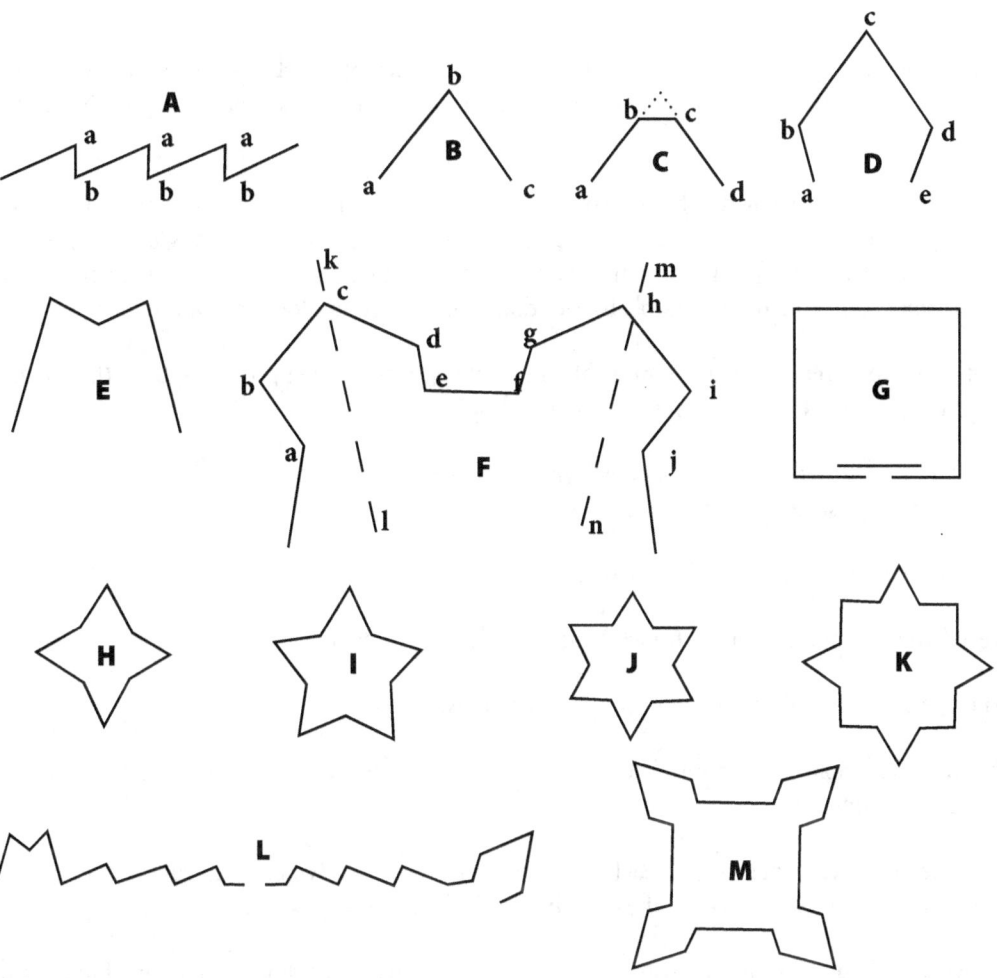

A	Cremaillere or Indented Line a - Salients b - Re-Enterings	**F**	Bastioned Fort ef - Curtain abcde - Lunette fghij - Lunette kl, mn - Capitals
B	Redan ab - Face bc - Face ac - Gorge	**G**	Redoubt (in this case a Square) Traverse protects Outlet of Gorge
C	Redan with Pan Coupe (bc)	**H-K**	Forms of Star Forts
D	Lunette bc, cd - Faces ab, de - Flanks	**L**	Plan of indented line between Priest-Cap and Lunette Salients
E	Priest Cap or Swallow Tail	**M**	Plan of Bastion Fort drawn from a square

Addendum A: Glossary of Fortification Terms

Casemate: A sturdily-built, arched masonry chamber enclosed by a fortification's ramparts or walls. Casemates were often used to protect gun positions, powder magazines, storerooms, or living quarters.

Chevaux-de-Frise: A defensive obstacle constructed by using a long, horizontal wooden beam (usually a log) with sharp wooden lances inserted as spokes at forty-five-degree angles to a present a spiked wooden fence. They were constructed before entrenching to use at the beginning of earthwork construction and provided a removable barrier that facilitated future forward movements.

Communication Trench: Smaller entrenchments that connected larger positions along the fortifications. These allowed the movement of troops and supplies.

Corduroy: A hardened surface created by laying parallel logs. A corduroy road, inside or outside of a fortification, was usable in wet weather.

Counterscarp: Outer sloped wall of the ditch.

Covered Way: A communication trench built to conceal movement.

Curtain: A line of fortifications connecting two bastions.

Defilade: Depressions in the natural contours of a landscape that allowed an attacking force to seek shelter from enemy fire.

Ditch: The deep trench dug around each earthwork. The ditch was typically in front of the fortification, but some advanced works had the ditch built behind the raised surface.

Earthwork: A field fortification constructed out of dirt. An earthwork could be a mound but typically consisted of a ditch and a parapet.

Embrasure: An opening or hole through the earthworks through which artillery was fired.

Enfilade: To fire along the length of an enemy's battle line. Fortifications were frequently designed to maximize the potential for enfilading fire against an attacking force.

Entrenchments: Long cuts (trenches) dug out of the earth with the dirt piled up into a mound in front. They enabled a defending army to fight with advantage because it sheltered them from enemy fire, posed an obstacle to the enemy's approach, and provided the means for defenders to effectively use their weapons. Name applied to all fieldworks. Frequently spelled historically as intrenchments.

Exterior Slope: The part of the rampart facing the enemy.

Fascine: Small branches tied into a bundle by wire or rope. The defensive purpose of fascines was to construct revetments, field magazines, and blinds or to reinforce earthworks, trenches or lunettes. They could also be used on the offensive to fill in a ditch.

Addendum A: Glossary of Fortification Terms

Field Fortifications: Temporary entrenchments built to last for a short period (the operations of a single campaign). Typically built from dirt and wood. Also known as fieldworks.

Flank: The end, or side, of a military position. An unprotected flank was considered "in the air", while a protected flank was referred to as being "refused."

Fort: A fully enclosed earthwork.

Fortification: A man-made structure or portion of the natural terrain that made a defensive position stronger. Man-made fortifications were permanent (mortar and stone) or temporary (wood and soil). Natural fortifications included waterways, forests, hills, and swamps.

Fraise: Stakes or palisades placed horizontally along the berm or at the top of the counterscarp to stop or slow a climbing attacker. They prevented the earthworks from being taken by surprise or sudden assault.

Gabion: A cylindrical basket of woven sticks made in advance for quick use in building or repairing a parapet. Gabion were frequently filled with earth once placed into a fortification.

Glacis: Gentle slope leading up to the ditch in front of the fortification. The glacis was created to prevent attacking soldiers from taking cover while approaching the ditch.

Headlog: Wooden beam placed on top of parapet with a small amount of space underneath, providing cover for marksmen and allowing defending infantry to fire without exposing themselves.

Interior Slope: The part of the rampart on the side where the defenders are located.

Loophole: An opening in the fortification through which small arms could be fired.

Lunette: A fortification similar to a redan but consisting of two faces and two flanks. Like a redan, its rear is open.

Magazine: A fortified location, similar to a bombproof, where powder and supplies were stored.

Military Crest: The highest location on the slope of a hill that still allowed defenders to observe and fire upon the base of the hill. It was located below the topographic crest.

Palisade: A line of sharpened sticks angled toward the attacker to stop or slow their movement.

Parallel: A series of parapets connected by saps and constructed in sequence toward the enemy. Used when advancing by regular approaches in siege operations.

Parallel Fire: Musketry and artillery fire directed straight across the front of an entrenched position.

Parapet: The top of the rampart. Sometimes the term was used interchangeably with rampart.

Permanent Fortifications: Durable artificial defenses designed to last an extended amount of time, typically to defend cities, garrisons, or other fixed strategic positions.

Addendum A: Glossary of Fortification Terms

Picket: Soldiers posted on guard ahead of a main force.

Priest Cap: Two redans placed adjacent to one another to provide enfilading fire.

Profile: A cross-section of a fortification, containing the rampart and ditch.

Quaker Guns: Large logs painted to resemble artillery piece. They were used to produce the impression that a fortification was of greater strength.

Rampart: A broad earthen mound surrounding a fortified place to protect it from artillery fire and infantry assault.

Redan: A fortification consisting of two faces jutting out past the rest of the defensive line of works who unite at a salient angle toward the enemy. Redans are open to the rear.

Redoubt: An enclosed fortification constructed to defend a position from attack from any direction.

Regular Approaches: Digging toward an enemy by use of saps, parallel and breaching batteries.

Revetment: Support for the embankment to protect against erosion, often made of wood, sandbags, gabion, or masonry.

Rifle Pit: A set of small fortifications (usually in advance of the main line) containing a short ditch with a low earthen wall in front.

Salient: The portion of a fortification jutting out toward the enemy. Can be vulnerable points because they can be attacked from multiple direction. They are nevertheless constructed to link already entrenched positions, to provide enfilading fire, or to follow the natural terrain contours.

Sally Port: An opening left in a fortification during construction to allow passage to facilitate movement to the advanced works and toward the enemy.

Sap: An approach trench built to connect parallel trenches with each other that is used when employing regular approaches toward an entrenched enemy position. Saps could be built directly toward the target or in a zigzag.

Sap Roller: A large gabion placed in front of working soldiers to shield them from enemy fire. Sap rollers were used when initially constructing fortifications and when attempting to build approach trenches toward the enemy position.

Scarp: Inner sloped wall of the ditch.

Sentry: A soldier standing guard, used interchangeably with picket.

Siege: A military strategy with the objective of blocking the supply lines and escape routes of a city or encampment in order to force its surrender.

Addendum A: Glossary of Fortification Terms

Siege Operations: The tactical use of regular approaches (saps, parallels, and breaching batteries) to advance toward an enemy position. Though lengthy in time, this forward movement would eventually reach the enemy ditch through digging rather than risking frontal assault.

Skid: Series of logs running perpendicular to the earthwork and used only when headlogs were employed. If an enemy shell knocked the headlog back, it would roll down the skids instead of crushing the soldiers in the trench.

Stockade: A line of tall stout posts securely set into the ground.

Traverse: Small rampart perpendicular to the parapet to protect against flanking fire and limit a successful attacker from expanding any breech.

Vidette: The sentry (usually mounted) closest to the enemy position.

Adapted from the American Battlefield Trust website (www.battlefields.org)

Civil War Fortification Study Group
Earthworks Classification System

Class One: Prepared Artillery Fortifications
Forts, redoubts, bastions, lunettes, redans, batteries, blockhouses

Class Two: Prepared Infantry Fortifications
Siegeworks, main lines, parallels, connecting curtains, blockhouses

Class Three: Rapid Artillery Fortifications
Breastworks, minor artillery lunettes or demi-lunes

Class Four: Rapid Infantry Fortifications
Breastworks, rifle trenches

Class Five: Communication and Supply
Communication trenches, covered ways, entrenched military roads

Class Six: Internal Works
Magazines, bombproofs, bunkers, traverses, associated with enclosed or semi-enclosed artillery fortifications

Class Seven: Personal Field Shelter
Discrete fox holes, picket or skirmish holes, command holes, slit trenches, rifle pits

Class Eight: Defenses of Convenience
Stone walls, piled stone breastworks, sunken roads, railroad cuts/fills, often enhanced by digging

Dennis Hart Mahan: Master of Fortifications

Professor of Military and Civil Engineering
US Military Academy, West Point

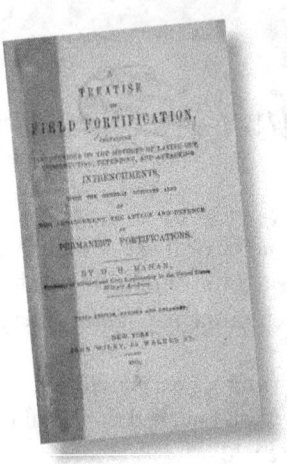

Principles that regulate the plan and profile of intrenchments:

  I. A flanked disposition should be the basis of the plan of all intrenchments.
  II. Every angle of defense should be 90°.
  III. A line of defense should not exceed 160 yards.
  IV. A salient angle should not be less than 60°.
  V. A strong profile is essential to a vigorous defense.VI.
  VI. The bayonet should be chiefly relied on to repel the enemy.
  VII. Intrenchments should be arranged to facilitate sorties.
  VIII. Intrenchments should contain a reserve proportioned to their importance.
  IX. Intrenchments should be defended to the last extremity.

As a field fort must rely entirely on its own strength, it should be constructed with such care that the enemy will be forced to abandon an attempt to storm it, and be obliged to resort to the method of regular approaches used in the attack of permanent works. To effect this, all the ground around the fort, within the range of cannon, should offer no shelter to the enemy from its fire; the ditches should be flanked throughout; and the relief be so great as to preclude any attempt at scaling the work.

As a general rule, the following dimensions may be taken for the parapets and other coverings for field-works:—

Brick wall of one brick,	
Stone ditto, 6 inches,	
White Pine fence, 12 inches,	Musket proof.
Yellow ditto, 9 "	
Oak (seasoned), 4 "	
Earth, three to four feet,	
Earthen parapet against field-pieces, from 9 to 12 feet.	

The star fort takes its name from the form of the polygonal figure of its plan. It is an enclosed work, with salient and re-entering angles; the object of this arrangement being to remedy the defects observed in redoubts. This, however, is only partially effected in the star fort: for, if the polygon is a regular figure, it will be found, that, except in the case of a fort with eight salients, the fire of the faces does not protect the salient; and that in all cases there are dead angles and all the re-enterings. The star fort has, moreover, the essential defect, that occupying the same space as a redoubt, its interior capacity will be much less, and the length of its interior crest much greater, then in the redoubt: it will, therefore, require more men than the redoubt for its defense, whilst the interior space required for their accommodation is diminished. These defects, together with the time and labor required to throw ups at your work, have lead engineers to proscribe it, except in cases where they are compelled by the nature of the site to resort to it.
To plan a star fort, its salients should not be less than 60°, and its faces may vary from 30 to 60 yards.

The bastion fort satisfies more fully the conditions of a good defense, than any other work; but, owing to the time and labor required for its construction, it should be applied only to sites of great importance, which demand the presence of troops during the operations of a campaign.
The bastion of a fort may consist of a polygon of any number of sides; but for a field fort, the square and pentagon are generally preferred, owing to the labor and construction.

Addendum B: Timeline of Events

Eons: Millions of years ago, this territory lay at the bottom of a vast sea close to the equator. Over eons of time, the land rose above the sea in a dome which somehow settled into what is now called the Nashville Basin, much of which consists of solid limestone lying close to the surface. It is this limestone which supports large bluegrass tracts feeding thoroughbred horses and caverns and springs producing pure water suited for the production of whiskey. The indigenous limestone has been used by humans to build all types of structures, from farm wall fences to iron furnace stacks to the State Capitol, and eventually Federal fortresses.

Prehistory: Paleo-Indians hunt game in the area. Large settlements are established near the river, depositing layers of consumed shellfish marking their locations. Ceremonial and burial mounds are built throughout the region. One thousand years ago, natives built what is now known as Old Stone Fort. The Middle Tennessee area became a common area for hunting, trapping and trading by Indians and European long hunters until settlement in the late 18th Century.

Before 1492: Native American tribes identify the Salt Lick near the Cumberland River and began to use the area of the Central Basin as common hunting ground. The Cherokee and Chickasaw drive the Shawnee out of Middle Tennessee.

1710: Frenchman Charles Charleville establishes a trading post, called French Lick, near the river.

1769: French-Canadian fur trader Jacques-Timothe (Timothy) Demonbreun camps near French Lick, sometimes living in a cave.

1770s: Long hunters such as Kasper Mansker and Bigfoot Spencer (who sometimes hides in a hollow tree trunk) roam the territory.

1779-80: James Robertson leads settlers westward from the Watauga Settlement to create a new settlement on the bluffs of the Cumberland River near the French Lick. They are joined the next year by John Donelson's party, arriving in flatboats after navigating the Holston, Tennessee, Ohio, and Cumberland Rivers.

1780: The eligible citizens (256 white male landowners) sign the Cumberland Compact, the first constitutional government in the settlement. Eight stations or forts are built in the region; most are quickly abandoned due to conflicts with natives.

1781: Battle of the Bluffs near what becomes known as Fort Nashborough.

1783: North Carolina creates Davidson County.

1788: The North Carolina General Assembly establishes Mero District for the Cumberland Settlements.

1792: Battle at Buchanan's Station, followed by siege.

1794: Treaty of Holston ends most hostilities between Native-American tribes and European settlers.

1796: Tennessee becomes the 16th state admitted to the United States on June 1st.

1806: The City of Nashville is incorporated by the state's General Assembly. Joseph Coleman is elected first Mayor of Nashville.

1812-1815: Through his military service at Natchez, at Horseshoe Bend, and in New Orleans, Andew Jackson catapults to the national stage.

1819: The *General Jackson,* owned by William Carroll, becomes the first steamboat to arrive in Nashville.

1822: The Nashville City Cemetery opens near Saint Cloud Hill.

1843: The Tennessee General Assembly selects Nashville as the state's permanent capital. The capitol cornerstone is laid in 1845.

1850: Designed by architect Adolphus Heiman, the Adelphi Theatre opens, with the second-largest stage in the country. That same year, the first steam locomotive arrives in Nashville by steamboat.

1859: The city and county celebrate the completion of the Louisville and Nashville Railroad, which provides the area with a north-south rail line.

Addendum B: Timeline of Events

Nov. 6, 1860 — Abraham Lincoln elected President. Nashvillian John Bell, Constitutional Union Party, carries Nashville, Davidson Co., and Tenn.

Dec. 20, 1860 — South Carolina votes to secede from the Union.

Feb. 11, 1861 — Tennessee votes against holding a secession convention. Memphis and Nashville elect Union candidates by overwhelming majorities.

April 12, 1861 — Fort Sumter in Charleston, S.C. harbor bombarded by Confederate artillery.

April 15, 1861 — As Fort Sumter surrenders, Lincoln declares a state of insurrection and issues call for 75,000 troops. Recruitment and military training begin in earnest.

April 17, 1861 — Tennessee Governor Isham Harris notifies Secretary of War Cameron that the state will not honor Lincoln's demand for two regiments of Tennessee Militia. The Rock City Guards drill in Nashville.

May 6-7, 1861 — The Tennessee General Assembly approves secession, subject to ratification. Tennessee enters into a "military league" with the Confederate government.

May 15, 1861 — The General Assembly authorizes the governor to call up 25,000 men into immediate service, with a reserve corps of 30,000, and issues $5 million in state bonds.

May 16, 1861 — Kentucky State Legislature declares neutrality but later votes to remain in the Union.

June 8, 1861 — The citizens of Tennessee vote 105,000 to 47,000 to secede from the Union. In Nashville, the vote is 3,033 for separation, 249 against.

June 21, 1861 — Battle of Bull Run (Manassas) in Virginia is a massive Federal defeat.

June 28, 1861 — The Tennessee General Assembly authorizes a draft of free black men into the Confederate army. Most free black men will manage to evade both the Confederate draft and the local sheriffs compelled to enforce it.

July 3, 1861 — Tennessee takes control of the Nashville end of the L&N Railroad, which will remain owned by stockholders for the duration of the war.

Aug. 6, 1861 — Lincoln signs the First Confiscation Act, authorizing Union seizure of Rebel property and ordering Union officers not to return escaped or confiscated slaves who are working or fighting for the Rebel forces.

Sept. 3, 1861 — Confederate General Polk occupies Columbus, officially breaking Kentucky's neutrality.

Sept. 6, 1861 — Grant occupies Paducah and Smithland, Ky., at the mouths of the Tennessee and Cumberland rivers.

Nov. 9, 1861 — US Major Gen. Henry Halleck is given command of the states east of the Mississippi, and Brig. Gen. Don Carlos Buell is put in command of eastern Kentucky and Tennessee.

Feb. 6, 1862 — Fort Henry on Tennessee River falls to ironclad gunboats under Foote.

Feb. 14, 1862 — Confederate gunners at Fort Donelson defeat Foote's ironclad gunboats. The next day, Fort Donelson is surrendered to US Grant.

Feb. 16, 1862 — News of capture of Fort Donelson sparks widespread panic in Nashville.

Feb. 19, 1862 — Clarksville on the Cumberland River occupied by Federal troops.

Feb. 25, 1862 — Nashville occupied by Federal forces under Buell. It is the first Confederate state capital to surrender, and will remain under Federal control for the remainder of the war.

Addendum B: Timeline of Events

April 6-7, 1862   Battle of Shiloh (Pittsburg Landing).

May 5, 1862   Military action at Lebanon.

July 13, 1862   Forrest captures U.S. garrison at Murfreesboro

July 17, 1862   Congress passes Second Confiscation Act and Militia Act.

Aug. 6, 1862   Buell orders Morton to Nashville to build fortifications.

Aug. 12-13, 1862   Morgan's cavalry captures US garrison at Gallatin, disables South Tunnel on L&N Railroad.

Aug. 18, 1862   Clarksville recaptured by Confederates.

Aug. 25, 1862   Second Battle of Dover, Woodward's cavalry attacks 71st Ohio and is repulsed.

Sept. 7, 1862   Clarksville recaptured by the Federals.

Sept. 15, 1862   "Siege of Nashville" begins as Buell moves into Kentucky.

Oct. 8, 1862   Battle of Perryville. Buell withdraws from Kentucky.

October 1862   Numerous skirmishes at Murfreesboro, Lavergne, and throughout Middle Tennessee.

Oct. 24, 1862   Rosecrans named commander of Army of the Cumberland.

Nov. 3, 1862   Rosecrans orders formation of Pioneer Brigade.

Nov. 5, 1862   Forrest moves on Nashville as Morgan attacks Edgefield.

Dec. 7, 1862   Work on Fort Negley, the largest Union fort west of Washington, D.C., completed.

Dec. 25, 1862   Col. Sanders Bruce occupies Clarksville with full brigade.

Dec. 29, 1862   Military Governor Andrew Johnson shuts down Nashville newspapers.

Dec. 31, 1862   The Battle of Stones River begins at Murfreesboro. Pioneer Brigade holds off Rebel assaults.

Jan. 1, 1863   1st Michigan Engineers fight Rebel cavalry at Lavergne.

Jan. 1, 1863   Lincoln signs the Emancipation Proclamation; Tennessee exempted.

Jan. 2, 1863   Battle of Stones River (Murfreesboro) ends, Bragg retreats.

Jan. 13, 1863   Wheeler's cavalry attacks Federal shipping at Harpeth Shoals; Forrest's cavalry does same at Palmyra

Feb. 1, 1863   Franklin is occupied by Federal troops under Col. Robert Johnson.

Feb. 3, 1863   Third Battle of Dover, as Wheeler and Forrest unsuccessfully attack 83rd Illinois.

Feb. 4-7, 1863   Series of skirmishes near Murfreesboro as Wheeler, Forrest, and Starnes attack Federal infrastructure.

Feb. 8, 1863   Federal troops enter Lebanon and capture 600 prisoners.

Addendum B: Timeline of Events

March 3, 1863   The Conscription Act / Enrollment Act requires all able-bodied men to enroll in the Union Army. They can pay $300 for an exemption or send a substitute. Only 46,347 of the 776,892 men receiving draft notices will actually wear a uniform.

March 4-5, 1863   Skirmishes at and near Chapel Hill, Unionville, Spring Hill, and Thompson's Station.

March 5, 1863   Steedman begins work on Triune earthworks.

Spring 1863   Fort Granger built on Harpeth River near railroad in Franklin.

Spring 1863   Fortress Rosecrans built on Stones River, N&C RR near Murfreesboro.

March 25, 1863   Forrest, Wheeler, and Wharton, cross Harpeth River six miles above Franklin and attack Federal forces at Brentwood, with 300 Federal soldiers taken prisoner.

April 3, 1863   Woodward's cavalry attacks Federal transports at Palmyra. Three days later, Federal gunboats under Fitch destroy the town.

April 7-11, 1863   Wheeler's raid on Nashville and Chattanooga Railroad.

April 10, 1863   Affair at Antioch Station. Engagement at Franklin, where Van Dorn attacks General Granger. Rebels attack passenger train on the N&CRR near Lavergne, killing a number of guards, and destroying the train.

May 6, 1863   As Gen. R.S. Granger assumes command of Nashville, a number of Confederate sympathizers are sent south, among them a former governor of Tennessee.

May 22, 1863   The War Department establishes a "Bureau of Colored Troops" to facilitate the recruitment of black soldiers into the Union Army.

June 7-9, 1863   Skirmishes at Triune and Spring Hill.

June 11, 1863   Forrest attacks Gen. Mitchell's troops at Triune.

June 24-July 3, 1863   Tullahoma Campaign drives Confederate troops out of Tennessee.

July 3, 1863   Battle of Gettysburg. Lee's Army of Northern Virginia defeated.

July 4, 1863   Confederates surrender Vicksburg on Mississippi River to US Grant.

Aug. 10, 1863   Col. Innes named military railroads superintendent.

Sept. 10, 1863   Bureau of U.S. Colored Troops opens in Nashville, with George Luther Stearns in charge.

Sept. 18, 1863   Battle of Chickamauga begins; ends on Sept. 19th with Federal army driven back into Chattanooga.

Sept. 23, 1863   Forrest's cavalry raids the Nashville & Decatur R.R. through Sept. 26th.

Sept. 28, 1863   Five thousand wounded brought to Nashville from Chickamauga. Thirteen hundred prisoners pass through city on way to Northern prison camps.

Oct. 5-6, 1863   Skirmishes near Readyville. Confederate troops destroy a large railroad bridge south of Murfreesboro.

Nov. 1, 1863   Major Morgan organizes 14th USCT at Gallatin.

Nov. 25, 1863   Battle of Misssionary Ridge.

Dec. 10, 1863   Nashville's hospitals are filled with soldiers wounded at Chattanooga

Addendum B: Timeline of Events

January 1864	Smallpox epidemic hits Nashville; subsides in March.
Feb. 21, 1864	Report from Missouri states that over 12,000 blacks have enlisted in the Federal Army in Tennessee.
March 7, 1864	Local election in Nashville favors Unionist candidates, with 1,100 votes cast.
March 17, 1864	Sherman arrives in Nashville, to replace Grant as commander in the West.
March 31, 1864	Enlistment of USCT soldiers continues to go well in Middle Tennessee – 5,000 men at Shelbyville and Lebanon are ready for the field.
April 10, 1864	The new powder magazine at Nashville is nearly completed. It will be the largest and most advanced in the country, with many modern safety features.
April 12, 1864	Fort Pillow captured by Forrest, massacre of black troops reported.
May 1, 1864	Gen. Sherman, in Nashville, issues orders concerning what may and may not be published in newspapers.
May 4, 1864	Collision at Gallatin between construction train and passenger train kills at least three men of Tenth Indiana Cavalry.
May 7, 1864	Sherman begins campaign through northern Georgia to capture Atlanta.
May 10, 1864	Nashville & Northwestern Railroad to Johnsonville depot completed.
June 27, 1864	Battle of Kennesaw Mountain, Ga. Generals Charles Harker and Dan McCook killed.
Aug. 3-4, 1864	Skirmishes at Triune.
Sept. 10, 1864	Pioneer Brigade disbanded.
Sept. 26, 1864	Forrest raids Nashville-Decatur Railroad. Skirmish at Richland Creek near Pulaski between Forrest and Rousseau.
Oct. 11, 1864	Skirmish near Fort Donelson between Confederate troops under Lawry and USCT soldiers under Weaver.
Oct. 29, 1864	Forrest captures Union gunboats at Paris Landing, Tenn.
Nov. 4, 1864	Forrest destroys shipping at Johnsonville depot on Tennessee River.
Nov. 4, 1864	Lincoln re-elected President, with Andrew Johnson as Vice-President. Nashville gives Lincoln 1,317 votes versus 25 for McClellan.
Nov. 15, 1864	Sherman leaves Atlanta and begins his March to the Sea.
Nov. 21, 1864	Hood marches north from Florence, Ala. to invade Tennessee.
Nov. 30, 1864	Battle of Franklin. Army of Tennessee badly mauled.
Dec. 7, 1864	Battle of the Cedars at Murfreesboro.
Dec. 15, 1864	Battle of Nashville. USCT debacle at Granbury's Lunette. Capture of Confederate redoubts.
Dec. 16, 1864	Battle of Nashville: USCT attacks Peach Orchard Hill; Shy's Hill rout. Army of Tennessee retreats.
Dec. 21, 1864	Sherman marches into Savannah, Ga. on the seacoast.
Dec. 25, 1864	Hood's army retreats across the Tennessee River.

Addendum B: Timeline of Events

1865:   Nashville's black population nearly triples during war years, growing from 4,000 to more than 11,000.

April 9, 1865   Lee surrenders to Grant at Appomattox, Va.

April 14, 1865   Lincoln shot at Ford's Theater in Washington, DC; dies the next day. Andrew Johnson becomes President.

April 26, 1865   Joseph Johnston surrenders to Sherman at Bentonville, N.C.

Jan. 9, 1866:   Fisk Free School established in former army barracks to provide education for former slaves. It is incorporated as Fisk University in 1867 to train teachers.

July 24, 1866:   Tennessee becomes first state readmitted into the Union. General Assembly ratifies 14th amendment.

1870:   Randall Brown is elected Davidson County Commissioner, the first black man to hold elected office in the state.

Feb. 2, 1928   House committee hearing on proposal to establish national military park at Fort Negley site.

1930s   WPA crews reconstruct Fort Negley; neglected during WWII.

June 4, 1975   St. Cloud Hill placed on National Register of Historic Places

Dec. 10, 2004   Fort Negley Park opens to the public for first time in 60 years. Interpretive Center opens in 2006.

Note—Much of this information derived from Tennessee State Library and Archives and Metro Nashville Government.

Addendum C: Inspection Reports on Federal Defenses

Barnett Report - December 1862

Headquarters Fourteenth Army Corps, Department of the Cumberland, Nashville, Tenn., December 5, 1862.

GENERAL: Below is a report of the number and caliber of guns, mounted and dismounted, at Nashville, which were captured from the enemy:
Number 1 — 24-pounder iron gun, mounted on bank of river near reservoir.
Number 2 — 32-pounder iron gun (Parrott), mounted on corner of reservoir.
Number 3 — 24-pounder iron gun (smooth bore), mounted on Lebanon pike.
Number 4 — 32-pounder iron gun (Parrott), mounted on end of Summer street.
Number 5 — 32-pdr iron gun (Parrott), mounted at Gen. Palmer's headquarters.
Number 6 — 24-pdr iron gun (smooth bore), mounted under Saint Cloud Hill.
Numbers 7 and 8 — 24-pdr iron guns (smooth bore), mounted on Fort Negley.
Number 9 — 24-pounder iron gun (smooth bore), mounted at railroad tunnel.
Number 10 — 24-pounder iron gun (smooth bore), dismounted at Fort Negley.
Number 11 — 32-pounder howitzer (iron), mounted at old Lunatic Asylum.
Number 12 — 32-pounder iron Parrott, mounted on floating bridge.
Dismounted at ordnance depot: one 100-pounder columbiad; two 32-pounder rifled iron guns, five 24-pounder carronades, and twelve 6-pounder iron guns, unserviceable, spiked; three 24-pounder iron smooth bores and one 18-pounder iron smooth bore, serviceable, and four 6-pounder iron guns, unserviceable.
Of the guns at the ordnance depot there are but three 24-pounders and one 18-pounder iron smooth bores that are considered safe.
Very respectfully,
JAMES BARNETT
Colonel, and Chief of Artillery Fourteenth Army Corps
Major General W.S. ROSECRANS,
Commanding Fourteenth Army Corps.

Dana Report - September 1863

Report on Forts Negley and Morton in Nashville, Sept. 8, [1863]-7 p.m.

I have spent the afternoon in examining the fortifications for the defense of this place. The principal works are three in number, all on the southern side of the town. One of these, the easternmost, named Fort Negley, is finished, or nearly so, and armed. It is a work of very intricate design, and requires about a thousand men for its garrison. The central work, known as Fort Morton, is scarcely yet commenced. Simpler in design and more powerful when done than Negley. It is situated on a hill of hard limestone, and the very extensive excavations required must all be done by blasting. At the present rate of progress it will take two years to finish it. A part of it, namely, the demilune in its front, is partly done, so far in fact that its parapet might be used as a rifle-pit and might afford some protection to field guns. This work will require a garrison of from 1,500 to 2,000 men. The two redoubts and barracks connecting them, of which its main body consists, will be altogether 700 feet long. The third and westernmost fort is precisely the same in plan as Morton, but is on land that can be easily dug. This fort is about one-quarter done, and can be completed with comparative rapidity and cheapness. The cost of Morton must be heavy. Nothing new from the front. Judicious men here think there will be no battle, and that Bragg has only the shadow of a force at Chattanooga to delay Rosecrans' advance.
[C.A. DANA.] [Hon. E.M. STANTON, Secretary of War.]

Mendenhall Report - January 1864

HDQRS. CHIEF OF ARTY., DEPT. OF THE CUMBERLAND,
Chattanooga, Tenn., January 14, 1864.
Brig. Gen. J.M. BRANNAN,
Chief of Artillery, Department of the Cumberland:
GENERAL: I have the honor to submit the following written report of my inspection of a portion of the artillery in this department, between the 25th of December, 1863, and 9th of January, 1864; in addition to which I submit a regular inspection report:

FORT DONELSON

The fort is in good condition, except the curtain on the river side, the scarp and exterior slope of which are giving way, but it is being repaired by the garrison. The magazine is large and in good condition, frequently aired, and the ammunition well looked to and in good order.

The fort is armed with four 22-pounder sea-coast and two 12-pounder iron guns, and one 8-inch siege howitzer. There is one old 6-pounder iron gun, on a broken carriage, lying near the fort.

The men understand the drill very well, and the guns and implements are well taken care of; military appearance, discipline, and police, good. The men are in comfortable huts.

Battery O, Second Illinois Artillery, stationed here, has four James rifles. This battery is in very good condition, everything neat and well cared for; horses in excellent condition; stables not very good, but expect to make new ones soon. Men are in comfortable huts. Garrison consists of left wing of Eighty-third Illinois Volunteer Infantry, under command of Lieutenant-Colonel Brott.

CLARKSVILLE

The fort is in very nice order. The magazine is slightly damp overhead, but the ammunition is in good condition, being frequently examined and aired.

The fort is armed with two 24-pounder siege guns, which are kept in good order, and the men drill very well. There are two 6-pounder field guns in the fort belonging to Battery C, Second Illinois (with carriages, limbers, and caissons complete), and also one iron 6-pounder taken out of the river and not mounted. Battery H, Second Illinois, stationed here, has two 6-pounder guns and four James rifles; drill at manual pretty well. Battery well taken care of; very comfortable stables, and horses in fine condition. Military appearance and police very good. Men in comfortable huts. The garrison consists of the right wing Eighty-third Illinois Infantry, Colonel Smith commanding.

GALLATIN

The fort is in good condition and the magazine in good order.

The Thirteenth Indiana Battery, having one 6-pounder, one 12-pounder howitzer, and four 3-inch guns, are at the fort (the guns inside). There are also three rebel field guns, with carriages, limbers, and caissons, in the fort, viz: One 6-pounder bronze, one 3-inch (not U.S.), and one howitzer, iron, probably a 12-pounder. Drill at manual very good; military appearance, discipline, police, care of guns and battery very good. The horses are in very good condition, in a good stable, well stacked with fodder. Men in comfortable quarters.

A lieutenant and 13 men from this battery are at CARTHAGE in charge of two 3-inch guns. The guns do not belong to any particular battery. Garrison at Gallatin, Seventy-first and One hundred and sixth Ohio, General Paine commanding.

FRANKLIN

Fort is in very good condition. The magazine is large and leaks badly, but a shed was being put over it to try to keep it dry. The ammunition did not seem to be damaged from dampness, it being frequently taken out and aired. The magazine is used for a commissary store-house as well as to keep ammunition.

The fort is armed with one 30-pounder Parrott, two rifled 24-pounders, and three 8-inch siege howitzers; another 8-inch howitzer is in a small work a few hundred yards northwest of the main work. The men are in comfortable huts inside the fort; they drill well. Military appearance, care of guns and implements, and police very good. The garrison, consisting of two companies of the Fourteenth Michigan Infantry, are also quartered inside the fort. The lieutenant colonel commands the post.

COLUMBIA

Lieutenant Gifford, with a detachment from the Fourteenth Michigan, has charge of a section, one 6-pounder gun and one 12-pounder Wiard gun, with limbers, caissons, horses, and implements completed. Garrison, Fourteenth Michigan, Colonel Mizner commanding.

Wiard gun

NASHVILLE

Fort Negley seems to be in good condition. The magazines (of which there are two) are in good order, and the ammunition is well looked to. The fort is armed with one 30-pounder, rifled (on a barbette carriage with a circular platform); three 24-pounder siege guns, two 24-pounder howitzers (field), and two 6-pounder field guns, manned by the Twelfth Indiana Battery. The guns and implements are well taken care of, and the men drill very well indeed. The men's quarters are not first rate. Tents are old and lumber very scarce. Military appearance, discipline, and police very good.

The Seventy-third Indiana Infantry has charge of all the other guns that are in position at Nashville. At the capitol there are four 30-pounder and two 20-pounder Parrotts. At Fort Houston one 24-pounder siege gun and four 6-pounder field guns (this work is unfinished). At Fort Morton one 30-pounder Parrott, one 32-pounder sea-coast, and one 24-pounder siege gun (the last two are mounted on carriages like casemate carriages without the chassis). There is a 24-pounder siege gun at the termination of Broad street, one 100-pounder Parrott between termination of Broad street and officers' hospital (in the camp of the One hundred and twenty-ninth Illinois Regiment), one 30-pounder Parrott near officers' hospital, one 24-pounder siege gun near Lebanon pike, one 100-pounder Parrott at water-works, and one 24-pounder siege on river bank. These isolated guns are mounted like the two before mentioned at Fort Morton, carriage of casemate gun without chassis.

At the capitol the magazine is a portion of the basement. The ammunition keeps well, though complaint has been made that it is too damp. Such is perhaps the case in summer rather than in winter. The magazines at Forts Morton and Houston are in good order.

At each isolated gun there is a small magazine capable of holding about 100 rounds of ammunition. They are very indifferent, but so far the powder has kept pretty well by being taken out and aired as often as the weather would permit. There were from 80 to 100 rounds with each gun, and in one or two, ammunition belonging to other guns. I directed

20-pounder Parrott rifle

Captain White, chief of artillery, to leave 50 rounds per gun and send the balance to one of the large magazines. The men are generally very comfortable, some in tents and some in huts. A squad of about 10 men are with each isolated gun. The military appearance, police, and discipline very fair; drill, good; guard duty seemingly well performed.

Battery E, First Michigan, belongs to the artillery of the post. Battery drill, general appearance of the battery, care of guns and carriages, packing ammunition, &c., very good; camp not in first rate order, and police horses not in good condition as they should be, having been at Nashville since August; no stable for animals. I told Captain White that he must see that Captain Ely had stables built, even if he had to get out the material with his men.

Captain White is a young, energetic officer; he also has charge of the Twentieth Indiana Battery, which is stationed, one section at Stockade No. 1, 6 miles from Nashville, one section at La Vergne, and one section at Stewart's Creek. This battery has two 6-pounder guns and four James rifles. Captain White reports this battery to be in first-rate condition. I did not inspect it.

Col. James Barnett is at Nashville with six batteries of the reserve artillery. His men are comfortably hutted in a nice camp, and their stables were about half completed on the 31st December. The men got posts, rafters, &c;, and timber, which they split into boards, from the woods. He had drawn no horses yet, he was waiting till the stables were finished. Four of his batteries had drawn their guns. Captain Stokes has four reserve batteries and three batteries belonging to the Eleventh Corps under his command; the latter are in good condition, horses in good order, and ready for the field if necessary. Captain Stokes had not yet established his camp for the winter. One of his reserve batteries had drawn guns and two of them had drawn horses. He expected to get lumber from the quartermaster in a few days to build his stables.

Captain Stokes has drawn no mules for his caissons yet, and does not like the idea of using them in the place of horses.

Colonel Barnett also very much dislikes to draw mules for his caissons, and while speaking upon the subject he called my attention to the fact that a great number of Government wagons in Nashville were drawn by fine horses, which would be excellent for the artillery, and at the time a Government wagon was passing drawn by as fine horses as I nearly ever saw, and while I was at Nashville I saw quite a number of Government wagons drawn by fine horses.

MURFREESBOROUGH

The fort is manned by the First Kentucky Battery and about 800 convalescent officers and soldiers, all under the command of Major Houghtaling, First Illinois Light Artillery.

The guns are divided into batteries of from three to nine guns each, under the charge of a commissioned officer, and from 60 to 108 enlisted men present.

Battery Mitchell is commanded by Lieutenant Irwin, of the First Kentucky Battery, and is armed with one 12-pounder and one 6-pounder field gun, and two 8-inch siege howitzers.

Battery at Lunette Palmer, by First Lieutenant Jones, Seventy Ninth Indiana Volunteers, armed with four 6-pounder Parrott field guns and one 8-inch siege howitzer.

8-inch siege howitzer

Battery at Lunette McCook, by Capt. J.R. Fiscus, Seventeenth Indiana, armed with one 24-pounder, rifled, four 6-pounder Parrott field guns, and two 8-inch siege howitzers.

Battery at Lunette Negley, by Capt. D.M. Roberts, Seventy-fifth Illinois, armed with two 6-pounders, one 3-inch, one 6-pounder James rifle field-guns, and one 8-inch siege howitzer.

Battery at Lunettes Rousseau, Sheridan, and Reynolds, by Capt. W.A. Gregory, Twenty-second Illinois, armed with three 6-pounder field guns, one 24-pounder, rifled, and one 8-inch siege howitzer.

Battery at Lunettes Granger and Crittenden, by Capt. W.N. Doughty, Thirty-seventh Indiana, armed with one 6-pounder and one 3-inch gun, and one 12-pounder field howitzer.

Battery at Redoubt Johnson, by Lieut. William Pool, Eighty seventh Indiana, armed with four 24-pounders, rifled.

Battery at Redoubt Schofield, by First Lieut. William H. Leamy, Nineteenth U. S. Infantry, armed with one 30-pounder Parrott, four 24-pounders, siege, and one 6-pounder, field guns.

Battery at Redoubt Wood, armed with four 24-pounders, rifled. Battery at Redoubt Brannan, by Second Lieut. J.D. Williams, Ninth Michigan, armed with three 30-pounder Parrotts, two 12-pounder field guns, and one 8-inch siege howitzer.

The First Kentucky Battery (Captain Thomasson), besides its own guns, two 6-pounders, one 3-inch and two 6-pounder James rifled field guns, has charge of one 24-pounder, rifled, and three 8-inch siege howitzers. Each battery, except the one at Lunettes Granger and Crittenden, has a magazine, all of which are in good condition. A little dampness can be seen in three or four of them after a long, heavy rain. At Redoubt Schofield the magazine leaked slightly, but will be fixed as soon as the weather will permit. Major Houghtaling told me that this magazine had heretofore been considered the best in the fort.

I could not get into the large magazine, the man who had the key could not be found. General Van Cleve and Major Houghtaling, who were in it a few days before, said that it was in a very good condition, but that there was slight dampness after long rains. The gallery around this magazine has fallen in in two places, but can be easily repaired.

The military appearance, discipline, drill, police, care of guns and implements, very good. The men are in comfortable huts. Mess arrangements good.

The First Kentucky Battery, 54 horses, some of which are convalescent horses, and the others are such as could be bought in the country around Murfreesborough. They are not in good condition, and are generally too light for artillery. The battery is kept in good order and the horses have a very good stable.

The scarps of the redoubts are giving way badly where galleries were to have been made. Some of the traverses are also falling down; one of them has been almost entirely rebuilt by the garrison.

The block-houses all leak badly, and are therefore little used even for store-houses.

Garrison, One hundred and fifteenth Ohio and Twenty-second and Thirty-first Wisconsin.

TULLAHOMA

The Ninth Ohio Battery is at the fort, and has four 12-pounders (light) and two 3-inch guns. There are two 24-pounders (rifled) on barbette carriages, one 24-pounder (smooth-bore) not mounted, and one 3-inch rebel gun with carriage and caisson, also under charge of Ninth Ohio Battery. The magazine is reported to be in good condition; the lock was filled within and it could not be unlocked.

Military appearance, discipline, and police only tolerable; men in comfortable huts and tents; guns and harness well cared for; horses in very fair condition and under good shelter. The work seemed to be in good condition. Garrison, Twenty-seventh Indiana and four detached companies, division and corps headquarters.

ELK RIVER AND DECHERD

The Second Kentucky Battery has two 3-inch guns in the fort at Elk River and two 3-inch guns in a little redoubt at Decherd.

The magazine at Elk River is in good condition (no magazine at Decherd). The horses, most of which are with the section at Decherd, are not in good condition. They have very good stables. Military appearance, drill, discipline, and police fair. Care of guns, harness, and ammunition good. Men in comfortable huts.

The work at Elk River is in good condition; that at Decherd small and falling to pieces. Garrison at Elk River, Second Massachusetts Regiment, seven companies of which were to start home on furlough next day (the 8th instant); at Decherd, the Forty-sixth Pennsylvania Regiment.

STEVENSON

Battery F, Fourth Artillery, six Napoleon guns; horses in bad condition for want of rough forage, otherwise the battery is in very good condition. Military appearance, discipline, police, very good. Care of guns, implements, and harness good. Men in comfortable log-huts. Horses under good shelter. Garrison, Colonel Ireland's brigade, Brigadier-General Geary's division.

BRIDGEPORT

Battery K, Fifth Artillery, four Napoleon guns in very good condition. Horses in pretty good order. No stables, but will have soon. Guns, implements, and harness well cared for. Men in comfortable huts. Military appearance, discipline, and police very good.

Battery M, First New York, four 10-pounder Parrotts, in fair condition. Horses in pretty good order; building stables. Guns and harness well cared for. Men in comfortable huts. Police, &c., not first-rate. The guns are in a small work near the river. Magazine dry and small, nothing in it. This battery has charge of two 3-inch guns and one 12-pounder howitzer, which were shipped down from Murfreesborough about the 1st of October. Four guns were shipped but only three are at Bridgeport; the missing one is a 3-inch ordnance.

Major Lawrence is at Bridgeport with two batteries of reserve artillery. He has the men in comfortable huts, and they are building stables for the horses. Garrison, General Geary's division, two brigades.

I am, very respectfully, sir, your obedient servant,
JOHN MENDENHALL,
Assistant Chief of Artillery.

Tower Report - October 1864

In October 1864, Brig. General Zealous B. Tower, who was in charge of the Nashville defenses, issued the following report. It should be noted that his recommendations were being made only two months before the Battle of Nashville occurred.

First, Forts Morton and Houston should be completed.
Second, Fort Gillem should be modified.
Third, the lines from the reservoir over University Hill strengthened by batteries.
Fourth, work on hill 210 built.
Fifth, work to right toward river built.
Sixth, Fort Negley strengthened.
Seventh, Casino Hill defended.

10-pounder Parrott rifle

These salient points occupied, the defensive lines would be strongly held. This might, perhaps, be accomplished with an expenditure of $200,000. If I can secure a black regiment, some 200 men, which have been promised, it will be a great gain. It is proper to remark that this place has been a depot for engineer property used in the Department of the Cumberland. Much labor in the workshops for other points, and in receiving, storing, and forwarding material, &c., has been paid for on the Nashville rolls. All the iron-bound tanks for the block-houses on the different railroads, and the large reservoir tanks for Chattanooga, have been made or are nearly completed in the engineer workshops here. Several block-houses have been prepared in Nashville for near points on the railroads, and labor rolls at other points have been paid here. A canvas pontoon train was gotten up at this point in part by engineer labor. Gen. Morton prepared a small steamer as a gun-boat, cleared much ground on the opposite side of the river, and that in the hills selected as sites of forts, pulled down houses, dismantled the suspension bridge, prepared pontoon bridges at two or three different times for crossing the Cumberland, with the labor charged upon his pay-rolls; and during the siege of Nashville, lasting between two and three months, nearly all his force, over 1,000 strong, was employed upon temporary structures. The engineer department has built a grand depot magazine, the largest and best devised that I have ever seen. Its interior measurement is 150 feet by 60, high, airy, and well ventilated, solidly constructed, and lighted at either end by locomotive reflectors placed in small masonry rooms. The structure is covered with earth to a depth of eight feet. A covered roadway with stone masonry side-walls passes through the embankment and communicates with the magazine entrance. A solid trestle-work branch railroad from the main track has been built into the magazine yard, and a long building erected to receive the large quantities of fixed ammunition in transit. Had it not been an absolute requirement of the department to construct this magazine, I think Forts Negley, Morton, and Houston would have been completed, or, at least, available, so that with the aid of temporary batteries and rifle-pits Nashville might be looked upon as a fortified place. I make the above statements to show in part why the defenses of Nashville have not progressed more rapidly, and to account for large expenditures which have been applied to the forts. The forts planned were entirely too large to be speedily built. When Gen. Morton commenced on the defense of Nashville great numbers of blacks could be obtained at small wages. It is probable that he expected to carry out his system by cheap labor. The enlisting of the negroes broke up this arrangement, and an uncommon mortality among them interfered with the progress of the system. Both Gen. Morton and Capt. Burroughs, while in charge of the Nashville works, were frequently required elsewhere. Col. Merrill, in charge of engineering operations in the Department of the Cumberland, has scarcely had time to inspect this post. Gen. Morton selected for the defense of Nashville a line extending from the reservoir over University Hill, crossing the railroad to Fort Negley, Morton, Houston, and thence along the edge of the City to the river. The first portion of the line was simply an intrenchment or rifle-pit, probably supported by field batteries. The next was to consist of three large, strong works, of a somewhat permanent character and capable of resisting a siege after the City had been captured. The third portion was a simple intrenchment, supported by the intrenched and stockaded capital. This was the weakest portion of the line. The selection was natural at that time, and with the exception of the third portion, is yet the best defensive line. But the three works devised were unnecessarily large, and would have involved immense expenditures. Fort Negley, the least of the three, has been essentially completed. It requires, however, some extensive changes to give it more offensive strength. Fort Morton, after an expenditure of $15,000 at least, was abandoned by direction of Col. Merrill, Engineers, when he took charge of the Department of the Cumberland. Fort Morton, as now being constructed, is a simple polygon, sufficient for the purpose intended. Fort Houston has in part been constructed according to the original plan, which like that of Fort Morton, is a double bastioned Choumara work. It has already involved large masses of embankment. The most expensive portion may be omitted and the modified work completed within more rational limits of expenditure. A small blockhouse has been constructed on Casino Hill in advance of Fort Negley and Morton, which is insufficient to hold the position. I propose, for the entire defense of Nashville, including the advanced portions of barracks, hospitals, store-houses, and corrals the following batteries and small works:

First. A battery at the reservoir.

Second. A small work held by a strong block-house on University Hill.

Third. A battery on the next rise toward Fort Negley, to sweep the railroad and turnpike.

12-pounder field howitzer

Fourth. Fort Negley to be strengthened by an interior, double-cased blockhouse, with a parapet on the top, each star salient to be arranged so that the gun may be covered. It is now entirely exposed. The re-entering to be strengthened by some obstacle, as abatis, chevaux-de-frise, or palisade, in a sufficient excavation to be covered from enemy's fire. Thus modified Fort Negley will be a strong work.

Fifth. A strong, double-cased block-house for Casino Hill, covered against direct fire from the high ground to the west by a parapet or battery for guns, the battery to be protected by external obstacles, and connected with the blockhouse by a palisade. The block-house to be a bomb-proof, surmounted by a parapet.

Sixth. Fort Morton to be completed as now being built. The rear parapet will, however, be reduced to the minimum. It may be necessary to pile up rock and earth on the exterior for a glacis, and some exterior obstacles, as the work is neither flanked nor has a ditch, and the ground near the fort is not seen from the parapet. The interior block-house, covered by the parapet against direct fire, will serve as a keep and bomb-proof.

Seventh. Fort Houston to be completed as cheaply as possible. Instead of the line of casemated bomb-proof connecting the two polygons, I propose a double caponiere. The parapets of these works will be made the minimum on the rear line, and the northern one left much lower than the plans indicated. The immense traverse bomb-proof will be omitted, and perhaps a small blockhouse bomb-proof put in their place. Of course, the independent scarp will not be constructed.

Eighth. Fort Gillem was built by Gen. Gillem; it is about 100 feet

12-pounder James rifle

square; it is not defiladed from the near hill (210 ref.) to the southwest, has no bomb-proof nor magazine, but has a deep ditch, walled with dry stone. The emplacements for eight guns are a barbette. I propose to defilade this work, to throw up merlons for the protection of the guns and gunners, to build a small magazine and block-house, bomb-proof.

Ninth. Gen. Cullum proposed a work for hill (210 ref.) in advance of Fort Gillem, and that one or two batteries to the left toward Fort Houston should be constructed to aid in the defense of that portion of the line. Hill (210 ref.) should doubtless be held to prevent an enemy from seizing it, and to give flank fire upon approach, by the valley, to the left. As this hill is very rocky, a work upon it will be expensive. I propose a polygonal work with exterior obstacles, and a double-cased, strong block-house, bomb-proof, for interior defense, covered against direct fire from the hills in the vicinity by the parapet of the works, and surmounted by an earth parapet. The batteries to the left may be simple batteries, defended either by contiguous rifle-pits or by a stockade inclosure.

Tenth. Gen. Cullum in his conversation with me expressed an opinion in favor of continuing the line of defense, from hill (210 ref.) to the sharp bend of the river near old Fort Zollicoffer. I think this would press the defensive line too far forward, and too near the range of high hills; and would not cover the ground in front so well as a more retired line.

I therefore propose to go to the river near Hyde's Ferry, indicated on the map. For the defense of this line two works will be required, and two batteries; the first battery on the knoll to the right of Fort Gillem may be a simple structure with contiguous rifle-pits; the next point will require an inclosed work, with bomb-proof block-house and magazines. The next battery, protected by obstacles and stockade; the work on the river-bank strong, with necessary bomb-proofs for its garrison. The line thus constructed will effectually guard Nashville, and will cover the advanced structures and corrals, and will require 3,000 men for garrison in time of attack, supported by 2,000 movable troops and by quartermaster employes, who can throw up rifle-pits; 30,000 men could not probably take Nashville thus defended. The cost of these works will be at least $300,000 at present prices, as nearly as I can judge, aided by the opinion of those who have been superintending works at this point. It is proper to remark that the rocky nature of the soil on the hills makes the works built there very costly. The parapets of Fort Morton, as far as completed, have been built of loose stones covered to a depth of three or four feet with earth; the earth is hauled up the hill. The crown of the hill has been removed by blasting to give the requisite reference. Magazines can only be constructed and drained by heavy blasting operations. I think these works must cost three or four times as much as they would were it possible to construct them, as in the vicinity of Washington, by simply excavating a ditch and throwing up the earth for a parapet. The difficulty is inherent to the limestone formation of this locality. Some of the knolls that I propose to occupy by batteries will probably give sufficient earth for their construction so as to avoid blasting. It will be very difficult in this vicinity to get abatis or material suitable for gabions. You must expect that the works will be costly. I propose to push Forts Morton and Houston to completion, and to modify Fort Gillem. The City can thus be defended by the aid of these forts, and the temporary constructions which the troops can erect against any large raiding party. In the mean time I hope to arrange plans for the proposed structures along the whole line.

Brig. Gen. Zealous B Tower

Tower Report - April 1865

Office of the Inspector General of Fortifications
Military Division of the Mississippi, Nashville, Tenn., April 28, 1865
Major General George H. Thomas,
Comdg. Mil. Div. of the Miss. West of the Alleghany Mountains:
General: I have the honor to submit the following report of the defenses of Bridgeport and of the railroad line thence to Nashville:

BRIDGEPORT

The Tennessee River at Bridgeport is divided into two branches by an island, and is spanned by two railroad truss bridges respectively 1,850 and 650 feet long. These important structures required special protection, as their destruction would have involved in a serious delay, at least, of the Atlanta campaign. It was the most important point on the line of communication, not excepting Chattanooga. Fortunately its approaches from the south bank of the Tennessee were very difficult for a large raiding party with field pieces, and probably impracticable for heavier guns. These difficulties doubtless saved the place from attack in that direction. An attack from the north could only be effected by crossing the Tennessee at distant points, and by long marches which would have given time to the various detachments in Middle Tennessee to concentrate and cover Bridgeport, or it least relieve it. This vital position was thus well protected by natural obstacles. Its defenses, however, though not yet finished, received the early attention of the engineers and of the commanding general. Two large artillery and infantry block-houses, in the form of a cross, were erected—one on the island near the abutment of the short bridge, the other on the south bank near the other abutment. A battery on the hill to the east, half a mile distant, strengthened by a small single block-house, was intended to prevent the enemy from taking possession of this position, from which he might have seriously annoyed the defenders of the bridge below him. When inspecting, March 7, I directed that the flanks of this battery should be prolonged to the bluffs, so as to make it an inclosed work. It required a magazine and embrasures for a full field battery, and that the scarp in places should be made more difficult. If the hill was to be occupied it was necessary that it should be held by a redoubt, too strong to be carried by assault when defended by its proper garrison, and then it should be prepared for guns superior to the possible artillery of an attacking party. It is quite probable that the two block-houses would have proved sufficient to protect the bridge against a raiding party coming to the south bank of the Tennessee. Yet it was a proper precaution to hold the hill on which the battery was constructed. On the northwest bank of the Tennessee are three redoubts. Redoubt No.

12-pounder Napoleon smoothbore

3, on a knoll to the west of the railroad, is finished and armed with two 3-inch Rodman guns. It has sand-bag embrasures, badly constructed, and is defended by a small block-house in the gorge. The ditch of this redoubt is not a serious obstacle. It requires deepening. The block-house keep, however, is its safety against assault. It has a good distant fire, but does not see all the ground within canister range, as a portion of the elevation on which it stands is abrupt and convex. It covers the naval shops. Fort No. 2 stands on the northwest end of the hill, near the north abutment of the long bridge. This is a star fort with a stockade gorge. The two flanks that should connect with the gorge were unfinished when I inspected the work on March 7. It is not a strong work and seems to have been designed simply as a battery. Finishing the flanks, deepening the ditches, and building an interior block-house would give this redoubt sufficient strength. On the south end of the same hill is an inclosed polygonal redoubt 500 feet long. It is unnecessarily large. The parapet, magazine, and embrasures, and ditches required much labor to finish them at the date of my inspection. As this work has a large block-house keep when completed as directed it will be strong, but will require a large garrison. The guns of these works see well upon the surrounding country, but the steep hill slopes within canister-range are not well swept by them. Forts in such positions are more readily carried than when placed on level ground or in slight elevations. When practicable the ditches should be at least seven feet deep, and so steep that no soldier could scale them without much assistance. They should also be strengthened by a bomb-proof keep, and their guns should fire through embrasures with the least width of throat. It would have been a proper precaution to have placed small picket stockades at the abutments of the bridges. The defenses of Bridgeport have grown up much like those of other important stations in the department, under different engineer officers and different commanding officers, and it could not be expected that they would be the best possible. Moreover, the labor required to fortify so large an extent of territory is immense, and soldiers are not willing to overtask themselves, except when necessary for protection against threatened attack. The positions selected for defending Bridgeport are well chosen. The post, however, has not received as much attention as Knoxville and Chattanooga. The natural obstacles to its approach, together with the large garrison at the post, have probably prevented any large expedition attempting the destruction of the bridges. The gun-boats, though simply musket proof, would have given some assistance to this post had it been seriously attacked.

STEVENSON

Is ten miles distant from Bridgeport. It lies at the junction of the railroad to Huntsville and Decatur with that to Nashville. Its seizure by an enemy would not seriously affect war movements except for the time being, as would the holding any point of the railroad. It was, however, a suitable position for a garrison to cover Bridgeport, and has some importance as a railroad station. Hence it required defenses. These defenses are quite ample, consisting of two redoubts and seven block-houses. Fort Mitchel, just south of the depot, is a small redoubt about 100 feet square, with a magazine and a small block-house keep. It has a barbette platform at each angle, and shows some attempt at imperfect embrasures, or rather to cover the gunners with sand-bag merlons. Fort Harker, half a mile distant, is a similar redoubt, about 150 feet square, with barbette platforms for seven guns, a magazine, and interior bomb-proof keep. The block-houses are mostly distributed to the east of the railroad, near the foot of the abrupt hills overlooking the depot. Other forts were commenced by General Granger when he held the place, during Hood's invasion of Middle Tennessee, but they were afterward abandoned as unnecessary. The accompanying sketch shows the relative positions of the defenses of Stevenson.

DECHERD

Is thirty miles from Stevenson and eighty-two from Nashville and about seven miles from the dividing ridge through which the tunnel passes. The country from Stevenson is closed in by high hills and almost without inhabitants. Decherd is the principal intermediate stopping place between Nashville and Chattanooga, but has no military importance further than that which arises from the necessity of distributing forces at intervals along the line of railroad. One redoubt with a block-house keep would have been sufficient for this place. Its defenses consist of two polygonal breast-high inclosures, respectively 20 feet and 100 feet in diameter, and of a square stockade. These structures are not entitled to the appellation of redoubts. Decherd requires no additional works now.

ELK RIVER,

Five miles from Decherd, the largest stream between Bridgeport and Nashville, is spanned by a bridge 480 feet long, resting upon four stone piers and four wooden trestles. The bridge is protected by two double-cased block-houses, which are sufficient. On a hill about 800 feet distant is a large redoubt with good ditches, built by the soldiers. It has no keep, however, and unless strongly garrisoned would be rather prejudicial than otherwise to the defenses of the position. Although this bridge could be quickly replaced if destroyed, much inconvenience would have resulted from two days' delay during the Chattanooga campaign. It was necessary, therefore, to protect so large a bridge against raiding detachments and guerrilla bands.

TULLAHOMA

Is six miles from Elk River and sixty-nine from Nashville. Being a large village, a garrison was necessary to control it and the guerrillas of the vicinity. It also covered to some degree the crossings of Elk and Duck Rivers, a few miles distant on either side. Near the station is a small stockade, and half a mile distant is a large bastion fort, nearly 300 feet square on the curtain lines, built by the rebels. This fort stands on the general level of the table-land. It has no bomb-proof keep, and its magazine was badly constructed. At each salient and each shoulder angle there is a gun platform, and on the parapet merlons have been raised to cover the gunners. With an interior block-house it would have been a very strong work.

DUCK RIVER.

Across Duck River is a bridge 353 feet long resting on twelve trestles. It is protected by a double-cased block-house. For greater security to this important bridge another block-house was commenced last winter. From Tullahoma to Murfreesborough the road required protection from the numerous guerrillas that infested the country. Small garrisons at the stations and in the blockhouses at the numerous river crossings guarded the road. The towns being small, no forts were built to control them.

MURFREESBOROUGH.

The city of Murfreesborough is situated about one mile and a half southeast of Stone's River. The country round about is generally level, and was formerly populous. One large fort near the city and depot, garrisoned by a regiment, would have controlled the place and neighborhood. A double-cased blockhouse would have been sufficient to protect the trestle bridge across Stone's River, 218 feet long. While Gen. Rosecrans' army was encamped in the vicinity, Fortress Rosecrans, inclosing 200 acres on either side of Stone's River, was constructed under the direction of Gen. St. Clair Morton, of the Corps of Engineers. This large work is composed of a series of bastion fronts, with small, irregular bastions and broken curtains; or more properly it may be described as consisting of lunettes connected by indented lines, having in the interior four rectangular redoubts, and one lunette as keeps to the position. In large permanent works, with high scarps, the ditches are swept by guns in the flanks, because the depression of the guns prevent the canister-balls from rising above the parapet. In field forts, with ditches only six feet deep and long curtains, opposite flanks cannot fire in the same manner as in permanent works without risk to the defenders; but by breaking the curtain line the ditches are swept by close musketry. This is the manner of flanking the ditches of Fortress Rosecrans. Its lines give powerful cross fires and direct fires, both of artillery and infantry, on all the approaches. Placed on the crests of the elevations, they not only command the distant country, but effectually sweep the gentle slopes within canister-range. This fortress could not be taken except by siege, if properly garrisoned and well defended. The parapets have high commands and when built were well revetted with fascines. The work has many traverses, covering against ricochet fire. Most of the guns are in embrasures, made with gabions. Lunettes Thomas and McCook and the four interior redoubts have large block-houses in the form of a cross. The magazines, except in Fort Brannan, are small. That in Lunette Mitchell is subject to being flooded, and is consequently useless in the wet season. The ditches of the redoubts are not so well preserved as those of the main lines. In fact the exterior slopes of the parapets and the scarps have taken the natural slopes, about 45 degrees. These redoubts, however, are strong against attack, being defended by large keeps, which deliver their fire upon every part of the interior. It requires much labor to keep so large a work in repair; small portions of the parapets have sloughed off, due to frosts and heavy rains. These effects were especially noticeable in Lunettes Mitchell and McCook. Some thirty feet of the parapet revetment of Lunette Thomas had fallen down, when I inspected March 10. Parts of the revetted traverses in Lunette Negley were badly broken down, and I have been informed that the heavy and uncommon rains since have caused some further damage. Temporary field-works are liable to frequent injury by storms. The garrison should keep them in order. Those that have been built for two or three years, of perishable material, must necessarily require repairs; gabions, fascines, boards, and nails, in contact with wet earth and exposed to the air, will decay rapidly, and in consequence parapets and embrasures crumble down and magazines leak. This large work, originally built as a refuge for the army in the event of disaster, is not needed in the present condition of the rebellion. The interior redoubts ought to be kept in order. A small garrison sufficient to hold them will control the neighborhood. At the date of my inspection Fortress Rosecrans was occupied by three artillery companies and mounted fifty-seven guns. The city was held by infantry. The depots were not within the fort. The accompanying drawing is well executed, and shows the positions and lines better than they can be described.

LAVERGNE

Is fifteen miles and a half from Nashville. It has a redoubt which has not been garrisoned for a long period. In truth the town is desolate and requires no defenses.

BLOCK-HOUSES.

Before Hood's invasion there were seven block-houses between Nashville and Murfreesborough to protect the railroad bridges across the streams; six of these were abandoned to avoid the capture of the garrisons, and were in consequence burned by the enemy; the seventh, at Overall's Creek, stood a heavy attack until the enemy were driven away by a sortie from the garrison of Fortress Rosecrans. Between Murfreesborough and Bridgeport there are twenty-nine railroad bridges protected by block-houses. These are mostly double-cased. Two large artillery block-houses defend the south bridge over the Tennessee, and ten have been erected to protect the bridges between Bridgeport and Chattanooga. Thus in the line between Nashville and Chattanooga the bridges and

3-inch ordnance rifle

trestle-works, whose preservation was essential to the running of the road, have been effectually protected against guerrillas and raiding parties of cavalry by forty-seven block-houses, mostly double-cased. These block-houses always resist and drive off the infantry. Field pieces, unless in numbers, and of the caliber of 12-pounders, cannot reduce them. They have performed a most important service, and it was a very happy application of the double-cased block-house. Had they not been used it would have been necessary to have built small redoubts with single block-houses inside as keeps. The rectangular form of the block-house is defective, as the fire on the capital is a single musket. Those now in process of construction are octagonal. No new defensive works are required on this line. Drawings of Bridgeport, Stevenson, and Murfreesborough accompany this report.

Very respectfully, your obedient servant,
Z.B. TOWER
Inspector-General of Fortifications, Mil. Div. of the Mississippi

Tower Report - May 1865

An inspection report dated 25 May 1865 by Brigadier General Zealous B. Tower, Inspector General of Fortifications, Military Division of the Mississippi:

Nashville was first occupied by the U. S. army in March, 1862. Gen. Morton, then captain, U.S. Corps of Engineers, commenced fortifying the position soon after its occupation. His plan was to hold Morton and Houston Hills and that on which Fort Negley stands by three large works controlling Casino Hill by a block-house and the fire of the two forts in rear. He also built defenses around the capitol, which is situated on a high hill within the city. It is presumed that these works were to be connected by an entrenched line when the necessity should arise. Forts Morton and Houston were designed as very large works, the double bastions of Choumara with a demilune, and were to be built in a permanent manner, with detached stone scarps. I have been informed that he expected these works to hold out after the city had been taken, and therefore devised them with interior capacity for the defensive materials and provisions for resisting a siege in the event the lines around the city could not be maintained. The magnitude of these works prevented the carrying out of his views. They would have required more labor than building all the necessary redoubts to completely inclose the city.

Fort Negley (now called Fort Harker).-This large work was nearly completed by Gen. Morton, assisted by Capt. Burroughs, Corps of Engineers. It is a complex fort. Within stands a square stockade twelve feet high, with flanking projections on each face. It is surrounded by a redoubt essentially square, with redan projections on the east and west sides. Its parapets are heavy, and the scarps were walled with dry stone, over which, however, the earth of the embankment falls, so as to give a continuous slope. On the south are two bastions, the flanks of which join to the south face of the main work, as a curtain, thus forming a bastion front. Each bastion has two interior entrenchments rising in stages, which are themselves small bastion fronts, the bastions small bomb-proofs loop-holed, flanking the interior ditch, and with infantry and artillery fire to the exterior. These small bomb-proofs are surrounded nearly to the height of the loop-holes by a parapet with low, dry stone scarps. Immediately below the main parapet to the east and west are outer parapets about nine feet thick, apparently for infantry, with sharp salients and dry stone scarps. They connect on the north side with the main work and on the south with the bastion front. Near the entrance in one of the salients is a bomb-proof, loop holed, which flanks the gateway front, serves as a guardhouse, and as a keep to the east star-shaped outwork. The main work connects with each of the outworks by two open passages without gates, wide enough for artillery. Within this work are two casemates of timber, covered on the slope toward the enemy with railroad iron and made bomb-proof with earth. The other guns, four in number, are en barbette. No embrasures were prepared either in the upper or lower parapets. A strong work against assault, its power to resist siege is weakened by uncovered dry stone walls and exposed woodwork. In some measure it throws away the advantages of a simple earthen redoubt in an effort to gain security against coup de main. It is, however, a very imposing fort, and its appearance alone would keep an enemy at a good distance. Its offensive power would be much increased by excavating the interior of the east outwork and placing guns there in embrasure. The terreplein of the western outwork is sufficiently low; guns could be placed in embrasure there also, as well as in the main work. If Casino Hill were strongly held, Fort Negley could only be attacked from much lower ground than its own site, and the emplacements for the attacking batteries would be distant. The hill slope is too rocky for the construction of trenches. Nothing has been done to this work under my direction further than the arrangement of the lower parapets on the western front for placing two guns in embrasure. The accompanying drawing explains this complex work.

Casino Hill is half a mile distant from Fort Negley and one-third mile from Morton and is ten feet higher than this last fort. Gen. Morton placed on this hill a single-cased block-house in the form of a cross, relying upon the combined fire of Morton and Negley to drive an enemy from the position should he attempt to build batteries there. Had Fort Morton been finished of the magnitude originally intended, its powerful armament might have accomplished that object by deluging the hill by its fire. I designed for this position a simple battery, with a deep ditch and eight-foot rock scarps. The two faces were directed upon Morton and Negley, so as to expose the hill to the fire of these forts. The forge line, simply a stockade closing on the block-house, leaves the interior open to fire from the works in the rear, so that no enemy could hold the battery, should he succeed in carrying it. Lack of men and the urgent necessity for forwarding more exposed points on the defensive line prevented the commencement of this battery. The hill is limestone rock with scarcely any soil, and steep on the line of approach.

Fort Morton.-This work had made some progress, according to the original plans, when Col. Merrill (captain, Engineer Corps), foreseeing that it would never be finished, directed its abandonment and the substitution therefor of a polygonal redoubt, with guns en barbette and an interior block-house. When I assumed general direction of the Defenses of Nashville this fort was not half finished. I modified it slightly by increasing the number of guns and placing them in embrasure, diminishing the parapets unnecessarily thick, introducing two service magazines, which would serve also as traverses, and reducing the block-house from 120 to 80 feet in length. It was my intention also to build a glacis around the work, revest the scarps with dry stone, and put flanks in the redan, so as to sweep the ditches of the fronts of attack; this has in part been done. The accompanying sketch shows these arrangements. The rocky character of the site of Fort Morton, its position on a high hill, the necessity for blasting the terre-plein and for the magazines, and for hauling earth from a much lower level, and the large keep have made this work expensive and retarded its progress. Fort Morton is nearly finished.

Fort Houston (now called Fort Dan McCook).-More labor has been expended on this fort than would have been required to build a large bastion work. In November, 1864, it was in a very unfinished condition. It progressed very rapidly for the period of three weeks, by the hands of a large number of workmen, mostly from the quartermaster's department. It was made ready for twenty-six guns at the time of the battles of Nashville, though the polygons were not inclosed. A small force has been employed upon Fort Houston since December last. Nearly all the gabion embrasures have been constructed, and entrances walled, and the works inclosed. Much labor is required to finish it. Its dimensions are so great that a small number of workmen make slow progress upon it. When completed it will mount thirty-five guns for direct fire and ten flanking guns. The original design was very costly, involving independent scarp walls, an immense traverse, and bomb-proof store-houses.

All these structures have been omitted in the modified plans. The north polygon, not being inclosed, was reduced in size, to avoid heavy embankments, and the reference of the interior crest dropped. The accompanying sketch will show the magnitude and character of this fortification. The almost unprecedented rains of December, January, February, and March have greatly retarded progress upon all the forts about Nashville.

Capitol Hill.-Gen. Morton built some earth parapets and stockades around the capitol building large enough to mount fifteen guns and to give room for a regiment of infantry. The position has a good command over the country around, and, thus strengthened, was a good keep for the north portion of the city. No longer needed, the stockade is being removed at the request of the Legislature and by direction of the commanding general. Gen. Morton's line of defense successfully resisted Morgan's and Forrest's attacks during Buell's march into Kentucky. Afterward Nashville became a great depot, and public buildings, as hospitals, store-houses, and corrals, extended far beyond the limits of the city and necessitated a much longer defensive line.

Fort Gillem (now called Fort Sill).-Gen. Gillem, while in command of the Tenth Tennessee Regiment, built this fort. It was a redoubt about 120 feet square, with narrow ditches, walled with dry stone, six feet high, having emplacements for eight guns in barbette, but without magazines or bombproof, and not defiladed from hill (ref. 210) looking into it. It was neatly constructed and was a good redoubt. I modified its interior arrangements with a view to increased strength and protection to its defenders. The parapet toward hill (210) has been raised two feet for defilement; two service magazines, which also serve as traverses, constructed on the faces, which would naturally be subject to ricochet from attacking batteries; thirteen embrasures, finished mostly with gabions, and a block-house keep set up. This structure has not been covered for lack of timber. Much blasting was required for the magazines, the drains, and of the terre-plein to prepare the site of the block-house. It is proposed to finish this block-house, set up a gate at the entrance, and build a suitable bridge across the ditch. When thus completed the work will be ready for a small garrison and should be kept in repair.

Hyde Ferry, Fort Garesche.-As Fort Gillem is nearly one mile and three-quarters distant from the Cumberland River, it became necessary to close this space by one strong redoubt, at least. Having therefore obtained from the commanding general the aid of the One hundred and eighty-second Ohio Volunteers November last, they were set at work building a strong redoubt on the knoll crossed by the Hyde Ferry road about three-quarters of a mile distant from the ferry and one mile north of Fort Gillem. This position had a good command over the approaches in every direction. Rapid progress was made, so that the fort was prepared to mount a battery at the time of the battles of Nashville. The regiment was called upon to do military duty after the battles, resuming labor upon the work in strength about the middle of January. The ditches and parapet have been finished, and the latter mostly sodded; three magazines, serving as traverses, completed and also sodded. Gabion embrasures have been formed for fourteen guns and twelve platforms laid.

The large block-house keep with flanking redans is set up and covered with timber. This covering, after being made water-proof, will be loaded with its parapet. The gateway has yet to be completed. This fort when finished will be very strong and a good specimen of polygonal redoubt. Its angles are made open so that the guns of the faces fire parallel to the capitals. It should be garrisoned and preserved. Were the scarps revetted, it would be easily kept in order.

Redoubt Donaldson (now called W.D. Whipple) is situated midway between Hyde Ferry Fort and Gillem. It is a small battery with seven exterior and two interior embrasures. On the gorge, closed by a stockade, is a little octagonal block-house of ten feet sides, made bomb-proof. This small redoubt, intended for a six-gun field battery, covers the ground between Gillem and Hyde Ferry Fort, and is supported by infantry entrenchments on either side. I devised it for a model battery. The faces form angles of 144 degrees, while the embrasures open 40 degrees, so that the guns on each face can fire parallel to the contiguous capitals. By this arrangement there are no sectors without fire; in fact, the fire on the bisecting line of the angles is equal to that in any other direction. Such batteries, placed at intervals of 600 yards along infantry entrenchments, constitute a good defensive line for inclosing a city. Key points should be occupied by redoubts as large as Hyde Ferry Fort. Within this inclosing line should be built one or more strong redoubts to serve as citadels or keeps to the outer line, and arranged to fire into the gorges of the batteries, which, being simple stockades, would not shelter the enemy should he succeed in acquiring temporary possession. Battery Donaldson, commenced while Hood's army was approaching Nashville, is completed. For its preservation the exterior slopes have been sodded by the soldiers of the field battery stationed near.

Defenses north bank of Cumberland River.-At my request the Thirteenth U.S. Infantry, Capt. La Motte commanding, commenced an octagonal redoubt about three-quarters of a mile from the railroad bridge, at bend of track, where there is usually a large collection of cars. The work would cover approaches to the bridge. The ditch was excavated, parapet raised and revetted with openings left for embrasures. Little has been done to this work since the battles. It is not necessary to complete it.

Hill 210 is situated half a mile west and beyond Fort Gillem and is higher than that redoubt. From its crest an enemy could fire at long range into the suburbs of the city and could make Cumberland Hospital and the large store-houses on the Northwestern Railroad untenable. I therefore planned a redoubt in October last to hold this hill. It was not commenced for the lack of means. When, however, Hood commenced his movement on Nashville, a large battery of two bastion fronts for fifteen guns, supported on either side by rifle-pits, was built, by the aid of employes from the quartermaster's department. The 30th of November, by my request, the commanding general directed large forces of the quartermaster and railroad departments to report to me for constructing an infantry line around the city. This line was built before the battles. It commenced at the reservoir and passed over Cemetery Hill to the railroad track, and was continued thence by Gen. Schofield to Casino Hill. From Fort Morton it passed around the Taylor barn, and thence north in rear of the Ellison house, to Hill 210. Most of the line from Hill 210 to the Cumberland River, touching at Gillem, Donaldson, and Hyde Ferry Forts, was a rifle-pit. This line was supported by twenty batteries, constructed with embrasures. The entrenchment is seven miles long; no shorter line, however, would inclose the store-houses and hospitals. The high range of hills, distant about three miles from the city, was entrenched by the army occupying them while Gen. Steedman threw up lines in front of the south suburbs of the city. Thus Nashville was doubly entrenched. The line of the hills was the best army line. It in part rested on Forts Negley, Morton, and Casino Hill, but received no support from Houston, Gillem, Donaldson, and Hyde Ferry Forts, and could not, therefore, be held except by an army. The interior line, while serving as a reserve to the exterior, would enable the usual garrison of Nashville, aided by the quartermaster employes, to hold the city against ordinary attacks from large raiding parties, under such generals as Forrest and Wheeler. Had the war continued it was my intention to put a redoubt on Hill 210 and support the two batteries to the left by block-houses. The battery at the Taylor barn would have been converted into a redoubt with a block-house keep. One small block-house between Morton and the Taylor house, and two between Negley and the reservoir would have completed the line of defense, and made it amply secure. These block-houses have all been prepared by a detachment of the One hundred and eighty-second Ohio Volunteers from timber cut down in the vicinity of Johnsonville. The spring floods destroyed the bridges on the Northwestern Road, and prevented the transportation of this material to the city. It is useless now to build these structures. As Nashville will probably have a garrison for one year at least, if not for a much longer period, I propose to complete Forts Morton, Houston, Gillem, and Hyde Ferry, almost finished, by the aid of soldiers. Negley and Donaldson are finished. Capt. Burroughs, U.S. Corps of Engineers, up to October, 1864, had charge of the works around Nashville, mostly under the direction of Gen. Morton. Maj. Willett, then lieutenant, also assisted Gen. Morton, and built the magazine. Col. Merrill gave little attention to the defenses of this depot, being principally occupied with those at Chattanooga. For so important a place, held so long by our troops, the Nashville defenses certainly were not pushed forward as much as they should have been. Little aid is given by commanding officers of posts when those posts are not in the front or constantly exposed. In such positions, building redoubts is the first operation, while far back on the line of communication it is very difficult to get a detail to throw up lines. Every other labor takes precedence. Capt. Barlow took immediate charge of the works around Nashville the 13th of November, under my direction, and has performed his duties faithfully and intelligently. Capt. Jenney gave me much assistance, superintending at Forts Houston and Gillem and upon the lines.

Maj.'s Dickson, Powell, and Willett assisted in the construction of the entrenchments around the city, which were mostly executed the first week in December, 1864, by the quartermaster and railroad employes. These departments also assisted at the same time on Forts Morton and Houston, and furnished lumber for gun platforms. In reviewing the works of the Department of the Cumberland it is due to Gen. Morton, of the Corps of Engineers, who was chief engineer for a long period with the army, to say that his work is visible along the Louisville road at this place, at Murfreesborough, Elk River, Bridgeport, and Chattanooga. His constructions are generally very well executed. He used the block-house in the form of a cross for interior keeps, and built some of the same model to defend bridges. Col. Merrill, of the Corps of Engineers, has doubtless the merit of applying the double-cased block-house for bridge defenses. He has given much attention to the study of this defensive structure. The railroad defenses of this department certainly deserve notice and commendation, and I doubt if in any other department such lines have been so thoroughly guarded against surprise or raiding parties. They do credit to Col. Merritt and Maj. Willett, and the other officers of the regiments engaged upon them. The posts of the Department of the Cumberland have been fortified principally by the labor of the soldiers.

I inclose drawings of Forts Negley, Morton, Houston, Gillem, Donaldson, and Hyde Ferry, and plans of redoubts on north bank of Cumberland, and for Hill (210); also of battery for Casino Hill and of the defenses inclosing the capitol. A general plan of the city and vicinity shows the defensive line and the position of the forts and batteries. There are eleven drawings accompanying this report. Since writing the above some names of forts have been changed, as indicated in red ink.

Wright Report on Railroads - April 1866

Office of Chief Engineer, U. S. Military Railroads
Washington, D. C., April 24, 1866

General D. C. McCallum
Director and General Manager Military Railroads U. S.

General,

I have the honor to submit the following final report showing the amount and cost of work done for construction and maintenance of way on the several military railroads in what was the Military Division of the Mississippi, and also on the military railroads in the Department of North Carolina. This report only embraces the operations on these roads subsequent to the time they were placed in your charge. There are no means at my command of ascertaining the amount of work done or its cost previous to that time.

The railroads included in this report *[redacted for this publication]* in the Military Division of the Mississippi are the Nashville & Chattanooga, Nashville & Decatur, Nashville & Northwestern, and the Nashville and Clarksville.

* * * * * *

On the 19th of December, 1863, I received your order to accompany you "to Chattanooga, Tenn., with such portion of the construction force as could be spared from the front" in Virginia.

One division of the Construction Corps, numbering about 285 men, was taken, and we arrived in the Military Division of the Mississippi on the 1st of January, 1864. At the time of our arrival the **Nashville & Chattanooga Railroad** (151 miles long, extending from Nashville to Chattanooga) was being operated between Nashville and Bridgeport, and the Tennessee River and Running Water bridges were building. Our construction force was at once put to work between Bridgeport and Chattanooga, the bridge builders to assist in the completion of the Running Water and other bridges, and the track layers to repair the track and relay the portion that had been destroyed. This work was completed and the first train run into Chattanooga on the 14th of January, some three weeks sooner than was deemed possible previous to our taking charge of the work. Although this road was now completed, it was not in condition to sustain the heavy traffic that would necessarily be thrown upon it when General Sherman's whole army would have to be supplied over it. The superstructure was old and much worn and had never been of first-class character. The rail used was light and of the U-pattern and laid on longitudinal stringers, which were so much decayed in many places that they would not hold the spikes. Accordingly orders were given to relay the track over the whole road with T-rail in the best manner. For this work, and that to be done on the other lines which were to be opened up, a large additional force was required, and arrangements were at once made for an abundant supply of men. The work of relaying the track was prosecuted steadily until completion, though necessarily at a great disadvantage in consequence of the large number of trains constantly on the road. When turned over to the company the road was in every respect in excellent condition. The following statement embraces the whole construction work done on this line, with the exception of some small pieces of track rebuilt, which had been destroyed by guerilllas, and of which no account was kept.

Track

	Miles
Rebuilt in first instance	115
Rebuilt after Wheeler's raid in 1864	7
Rebuilt after Hood's invasion	7 3/4
Total main track	129 3/4

Side-tracks

Location	Length Feet	Location	Length Feet
Nashville	38,628	Tunnel	264
Barracks	1,630	Tantalon	1,500
Glen Cliff Station	2,368	Condit	2,000
Antioch	990	Anderson	354
La Vergne	895	Stevenson	1,673
Smyrna	2,260	Bolivar	1,640
Stone's River	1,660	Bridgeport	9,472
Winsted	2,408	Carpenter's	1,037
Christiana	1,500	Alley's Spur	159
Fosterville	775	Whiteside	850
Normandy	929	Hooker	350
Tullahoma	609	Chattanooga	10,072
Estill Springs	1,582	Total	100,2767
Decherd	13,732		
Cowan	970		

	Miles
Main track	129 3/4
Side track, 100,277 feet, or	19
Total track laid by Government	148 3/4

Nashville & Chattanooga Railroad Bridges

No.	Location	Hgt.-Feet	Lgth.-Feet	Remarks
1	Mill Creek, No. 1	16	260	Rebuilt five times
2	Mill Creek, No. 2	18	250	Rebuilt four times
3	Mill Creek, No. 3	16	256	"
4	Hurricane			
5	Smyrna	20	120	Rebuilt three times
6	Stewart's Creek	29	183	"
7	Overall's Creek	20	160	"
8	Stone's River	22	420	"
9	Lytle's Creek	10	135	"
10	Murfreesborough	9	140	Not destroyed
	"	6	40	
	"	6	50	
	Creek Branch	7	50	
11	Stone's River (East Fork)	22	270	Rebuilt
12	Christiana		73	"
13	Bellbuckle	7	82	"
	Bragg's Bridge	9	128	Not destroyed
14	Wartrace	14	241	Rebuilt
15	Garrison's Fork	24	178	Rebuilt twice
16	Duck River	30	350	"
17	Poorhouse Creek	13	100	"
18	Elk River	60	470	Rebuilt
19	Cowan Creek	26	160	"
20	Crow Creek (South Fork)	17	160	Rebuilt twice
21	Dry Trestle, No. 1	12	84	Rebuilt
22	Dry Trestle, No. 2	10	75	"
23	Crow Creek, No. 1	15	225	Not destroyed
24	Crow Creek, No. 2	11	225	"
25	Crow Creek, No. 3	19	348	Rebuilt
26	Crow Creek, No. 4	16	254	"
27	Crow Creek, No. 5	11	160	"
28	Crow Creek, No. 6	8	100	Not destroyed
29	Crow Creek, No. 7	12	156	Rebuilt
30	Crow Creek, No. 8	18	143	"
31	Crow Creek, No. 9	11	234	Not destroyed
32	Crow Creek, No. 10	21	240	Rebuilt
33	Crow Creek, No. 11		225	"
34	Tennessee River		1,520	
35	Ben's Creek	10	100	
36	Widow's Creek	24	127	"
37	Dry Creek, No. 1	22	140	
38	Nickajack	34	200	Rebuilt twice
39	Dry Creek, No. 2	34	203	Rebuilt
40	Dry Trestle	16	301	
41	Running Water	120	789	
42	Lookout Creek	36	155	Rebuilt twice
42	Chattanooga	38	263	Rebuilt

	Lineal feet
Total bridging	10,543
Amount rebuilt	12,236
Total length of bridging on this line	22,779
Bridges not destroyed	1,052
Total built by Government	21,727 Or 4 miles 607 feet

Nashville & Chattanooga Railroad Water Stations

Where built	Number of tanks	Where built	Number of tanks
Nashville	5	Murfreesborough	2
Florence	1	Fosterville	2
Christiana	2	Garrison's Fork	2
Bellbuckle	2	Decherd	2
Normandy	3	Tantalon	2
Cowan	2	Stevenson	2
Anderson	4	Chattanooga	2
Poison Hollow	1	Total	35
Antioch	1		

The Nashville & Decatur Railroad

Extends from Nashville to the Memphis & Charleston Railroad at a point near Decatur, Ala., and is 120 miles long. The repairs were completed and the road opened in March, 1864. Much of the work in opening it the first time was done by soldiers, and I have no account of the cost of what they did. General Dodge was in command of the force employed on this work. The following statements show the amount of work done and the cost of that done by the Military Railroad Department:

Track

	Miles
Main track rebuilt in first instance	2
Main track rebuilt after Forrest's raid	7 1/2
Main track rebuilt after Wheeler's raids	22
Total main track	31 1/2

Sidings

Location	Feet
Eaton Depot	1,000
Nashville Junction	8,025
Brentwood	300
Franklin	290
Columbia	1,150
Prospect	600
Athens	1,480
Decatur Junction	1,170
	14,015
Add main track rebuilt	163,680
Total	177,695
	Or 34 miles 815 feet

Water stations

Where built	Number of tanks	Where built	Number of tanks
Little Harpeth	1	Carter's Creek	2
West Harpeth	1	Lynnville	2
Lytle's Creek	2	Near Tunnel	1
Pulaski	1	McDonald's	1
Elkmont	2	Total	15
Franklin	2		

Nashville & Decatur Railroad Bridges

No.	Location	Height Feet	Length Feet	Remarks	Rebuilt Feet
1	Brown's Creek	12	38	Not destroyed	
2	Little Harpeth	14	74		
3	Spencer's Creek	17	38		
4	Big Harpeth	38	187	Rebuilt twice and partly rebuilt twice	454
5	West Harpeth	13	58		
6	Near Springs Hill	12	53		
7	Spring Creek	15	21		
8	Carter's Creek, No. 1	18	112	Rebuilt twice and partly rebuilt twice	285
9	Carter's Creek, No. 2	21	184	"	470
10	Carter's Creek, No. 3	20	94	"	235
11	Carter's Creek, No. 4	20	94	Rebuilt twice and partly rebuilt once	228
12	Carter's Creek, No. 5	30	235	Rebuilt twice and partly rebuilt twice	587
13	Rutherford's Creek, No. 1	26	130	Rebuilt three times and partly rebuilt twice	455
14	Rutherford's Creek, No. 2	27	265	Rebuilt twice and partly rebuilt three times	723
15	Rutherford's Creek, No. 3	30	295	"	811
16	Rutherford's Creek, No. 4	50	270	Rebuilt twice and partly rebuilt twice	676
17	Duck River	72	627	Rebuilt twice	1,254
18	Lytle's Creek	14	22		
19	Hurricane Creek	14	23		
20	Harris Trestle	29	232		
21	Kalioka Trestle	37	1,130	Rebuilt	1,130
22	Grace Trestle	42	637	"	637
23	Robinson's Forks	18	126	"	126
24	Richland Creek, No. 1	32	160	Rebuilt twice	320
25	Richland Creek, No. 2	37	180	"	360
26	Richland Creek, No. 3	35	180	"	360
27	Pigeon Roost Creek	12	50	Rebuilt	50
28	Richland Creek, No. 4	41	315	Rebuilt twice	630
29	Tunnel Trestle	38	822	Rebuilt	822
30	Elk River	40	625	Rebuilt three times	1,875
31		10	48		
32	Mill Creek	30	330	Rebuilt	330
33	White Sulphur	71	570	"	570
34	Mud Creek	5	62	"	62
35	"	9	102	"	102
36	Athens Creek	10	134	"	134
37	"	11	64	"	64
38	Swan Creek	11	340	"	340
39	"	11	129	"	129
40	Black Creek	6	225	"	225
41	Junction Trestle	16	<u>275</u>	"	<u>275</u>
	Total		9,555		14,120

	Feet
Total bridging	9,555
Amount rebuilt	<u>14,720</u>
Total built by Government	24,475
	Or 4 miles 3,155 feet

The Nashville & Northwestern Railroad

Is seventy-eight miles long and extends from Nashville to the Tennessee River at Johnsonville. It was partly built before the war. On the 22d of October, 1863, the Secretary of War ordered this road to be constructed for "military purposes," and placed it in charge of Andrew Johnson, then Military Governor of Tennessee, who was empowered to "employ an engineer and other officers and workmen necessary to complete it without delay." Col. W. P. Innes was acting as engineer at the time the railroads in this military division were taken charge of by the U. S. Military Railroad Department, and had a considerable forces of soldiers and civilian laborers employed on the road. But as the work was not progressing to the satisfaction of the general commanding, he relieved Colonel Innes and placed the construction of the road in your charge. This order of General Grant's was given on the 17th of February, 1864, and on the 25th of the same month I received your order directing me to adopt the most energetic means at my command to complete the Nashville & Northwestern Railroad. I at once made an examination of the work to be done and found it to consist of a rather formidable amount of grading, bridging, track laying, and other work incident to the construction of a new railroad, and proceeded to take the necessary steps to complete the work as directed. I appointed Lieut. Col. John Clark engineer of construction, and by General Grant's direction sent North for 2,000 mechanics and laborers in addition to the force then on the road. Some time after we had got fairly under way Governor Johnson, claiming the right under the above-mentioned order of the Secretary of War to appoint an engineer, also selected Colonel Clark, who then filled this double position until the work of construction was so far completed that the track was connected through, an event which took place on the 10th day of May, 1864. Governor Johnson continued to exercise semi-control over the operations on this road until it was formally taken possession of by General Sherman and placed absolutely under the control of the general manager of military railroads, in accordance with the order of the President of the United States dated August 6, 1864. The Transportation Department then took charge of the movements of trains, and the maintenance of way, together with construction work, remained in my department.

On the 20th of August I appointed W. R. Kingsley, esq. (who had been connected with the road as division engineer since April), engineer in charge of construction and maintenance of way. He continued to perform the duties of this position faithfully and satisfactorily until the 1st of April, 1865, when, all construction work being done, the maintenance of way was turned over to the transportation department. The line of this road as originally located crossed the Tennessee River nearly perpendicular to the course of that stream and at an elevation of fifty-two feet above low water and nine feet above high water. The approach to the river was an embankment seventeen feet high above the surface of the ground on the river bank. The object of making this a military railroad being the transportation of army supplies from the Tennessee River to Nashville, it became necessary to construct ample and convenient arrangements for the transfer of freight from steam-boats to cars. Accordingly two large transfer freight-houses were designed and built, one on each side of the railroad, with tracks starting from main line at the bluff and curving right and left until parallel with the buildings and river bank. The freight-house or shed on the north or lower side, 600 feet long by 30 feet wide, was hastily knocked up so as to bring it into immediate use, and the levee in front graded off to the water's edge with a slope of 9 degrees or about 16 feet rise in 100 feet horizontal. The freight-house on south side, 600 feet long and 90 feet wide, was a much more complete building. The floor was two feet and a half above high-water mark and the levee in front graded to a slope of 14 degrees, on which it was designed to lay railroad tracks from low-water mark to floor of freight-house. The plan for transferring freight from steam-boats to cars was to load from the boats onto small cars, which were hauled up the levee to the level of the freight-house floor by a wire rope passing round a pulley or spool, which was dropped into or lifted out of gear with the main shaft by a lever. This main shaft was 500 feet long and passed through the center of the building immediately below the floor or platform and was operated by an engine located in the middle of the building. The freight was then passed directly through the building and loaded into cars on the opposite side. The levee was of sufficient length to allow at least four or five boats to unload at the same time, and the side tracks were so arranged that a whole train of cars could be loaded at once, and as soon as loaded could be moved away and another train run right alongside the house. This plan would undoubtedly have enable us to handle a large amount of freight with great rapidity and ease, but we had not the opportunity of bringing it to a practical test, for just as everything was about completed Hood's invasion of Tennessee took place and Johnsonville was evacuated by our troops, and during their absence the freight-house was burned, as is supposed, by rebel sympathizers in the neighborhood. However, the engine and all the most valuable parts of the machinery were saved by being taken to Nashville.

All could have been saved if we had had sufficient transportation for it. Although the road was opened through to Johnsonville after Hood's defeat at Nashville, but little work was done in rebuilding the houses and platforms at that point. Grading off the levee involved considerable work; about 30,000 cubic yards of earth had to be moved. It was designed to pave it, or put on a covering of broken stone, but owing to the delay in furnishing gun-boat protection to our boats, which were to bring stone down the river for this purpose, the work was but partially carried out. A row of piles were to have been driven at the edge of the water to protect the levee and prevent its washing away at time of floods, but the pile driver for this purpose never reached Johnsonville. It is but proper for me to state here that the work on the buildings and levee at Johnsonville was much delayed by the confusion and embarrassment caused by the conflict of authority incident to a divided control of the work.

*****

The following is a statement of the work done on this road:

Graduation

The amount of grading was very considerable, but I am unable to give the number of cubic yards moved, because when we took charge of this road I had no time to measure it, and I had no assistants to do it for me. By the time I procured the requisite assistance much of the work had been done. Thorough cuts of as much as forty and fifty feet in depth and 800 feet in length were taken out and high embankments made. Even where the grading had been done previously much labor was required to dress up the embankments and clean out the cuts.

Nashville & Northwestern Railroad Superstructure

The total length of track laid was:

	Miles
Main line	46 1/2
Sidings	4 1/4
Total	50 3/4

Seven different patterns of rails were used in the track; the amount of each kind is given below. With the exception of No. 1 and the U-rail, the iron was purchased by the Government. No. 1 pattern is the fish-joint bar belonging to this road, and the U-rail was taken from the Nashville & Chattanooga. Railroad.

Pattern	Weight per yard Pounds	Amount Tons
No. 1	56 1/4	1,315.61
No. 2	49 3/4	149.70
No. 3	45	382.11
No. 4	45	40.04
No. 5	60	1,096.84
No. 6	56 1/4	1,469.48
U	48	23.50
total		4,477.28
Deduct No. 1 pattern		1,315.61
Balance furnished by Government		5,161.67

One hundred and seven thousand cross-ties were used in laying the track. A considerable number was found on the line of this road, but we had to make the greater part.

Water Stations

Fourteen of these were built and located, as shown in the following table, 63,700 feet, B.M., of lumber:

Distance from Nashville Miles	Capacity	Remarks
7 1/2	One tank	
16 1/2	"	Discontinued
17 1/2	"	Destroyed and rebuilt
24 1/2	"	Destroyed
27 1/2	"	
28	"	
45	"	
53 1/2	Two tanks	
59 1/4	One tank	"
66 1/4	"	"
71 3/4	"	
75 1/4	"	Destroyed and rebuilt
77 1/2	"	"
78	Two tanks	"

Statement of bridges and trestles on the Nashville & Northwestern Railroad

Distance from Nashville Miles	Name	Number of spans or bents	Height Feet	Length Feet	Remarks
0	Nashville Trestle	170	21-28	2,151	
4.57	Richland Creek, No.1	2	15	76	Rebuilt once
5.36	Richland Creek, No. 2	1	9	66	Rebuilt twice
5.75	Richland Creek, No. 3	1	9	65	"
6.53	Branch Richland Creek	2	10	35	
6.72		1	8	17	
	"	1	8	26	
7.52	Over road	1	10	32	
8.91	Trestle over road	6	20	75	Rebuilt five times
13.39	Harpeth River, No. 1	2	34	87	"
13.94	Harpeth River, No. 2	2	35	201.5	Rebuilt four times
15.31	Harpeth River, No. 3	1	42	180	"
17.43	Harpeth River, No. 4	2	38	201.4	"
21.21	Harpeth River, No. 5	1	32	236.9	"
23.14	Harpeth River, No. 6	12	24	180	
	"	2	24	201.8	Rebuilt once
23.56	Harpeth River, No. 7	10	24-30	180	
	"	2	33	203.3	"
24.66	Turnbull River	43	12	516	
24.66	"	2	27	259	
24.66	"	20	20-25	270	
25.37	Trestle	66	20-32	792	
25.66	Sullivan's Branch	3	16	39	Rebuilt twice
25.66	"	1	48	89.7	
25.66	"	102	26-46	1,326	
26.44	Trestle	17	36-26	306	
27.18	"	21	38-24	262	
40.95	"	18	17-28	238	
41.71	"	17	14-25	225	
47.53	"	75	20-33	1,087	
49.49	"	30	19	442	
52	"	58	7-13	837	
52.38	"	8	10	145	
53	"	37	13-12	470	
53.44	"	2	18	40	
54.19	"	62	30-48	910	
55.79	"	70	40-72	980	
56.18	"	11	30-39	180	
	Branch Trace Creek	1	7	24	
60.05	"	2	8	47	
63.56	Trace Creek	2	14	216	Slightly injured and repaired
64.01		1	10	20	
64.61	Flood Creek	2	3	30	
66.51	"	3	4	35	
71.44		22	18	272	
73.08	Trace Creek	1	25	114	Rebuilt four times
74.44	Trestle	3	15	66	
78	Trestle at Johnsville	121	12-18	<u>1,525</u>	
	Total			15,956	Or 3 miles and 114 feet.

Add to this amount rebuilt, 5,366 feet, and we have a total of four miles and 200 feet of bridging and trestle on this road built by the Government. The lumber consumed in these structures amounted to 4,098,509 feet, B. M. 5,000

Nashville & Northwestern Railroad

The following table shows the location of and amount of lumber in the buildings on this road:

For what purpose	Location	Lumber Feet, B. M.	Shingles	Remarks
House for trackmen	Nashville	8,000	5,000	
House for switchmen	"	1,500		
Tool-house	"	3,000	5,000	
House for trackmen	Section 3	7,863	5,000	
"	Section 6	5,000	5,000	
Telegraph office	Section 18	8,000		Destroyed
House for trackmen	Section 20	5,728	5,000	
"	Section 24	15,037		"
Telegraph office	"	8,500	3,500	Destroyed and rebuilt
Blacksmith shop	"	5,000		
Outbuildings	"	800		"
House for trackmen	Section 29	10,162		
Telegraphic office	Section 32	5,000	3,500	
"	Section 42	11,000	3,500	"
"	Section 50	11,000	3,500	"
House for trainmen	"	4,000		Destroyed
House for trackmen	"	2,800		"
Telegraph office	Section 57	4,800		"
"	Section 66	6,800		"
House for trackmen	Section 77	6,800		"
House for yardmen	Johnsonville	18,200	4,200	"
House for engineers and firemen	"	25,200	22,000	"
house for station agent	"	28,900	21,000	"
Outbuildings	"	1,000		"
Wheelwright shop	"	5,570		"
Blacksmith shop	"	5,000		"
Saw-mill	"	6,656		Destroyed and rebuilt
House for carpenters	"	11,800		Destroyed
Depot	"	175,000	90,000	Destroyed and rebuilt
House for railroad purposes	"	110,400		
House for track hands	"	6,540		
House for mill hands	"	20,000		
Upper freight-house	"	1,097,600	566,000	Destroyed
Lower freight-house	"	165,000		Destroyed and partly rebuilt
Total		1,805,656	742,200	

The Nashville and Clarksville Railroad

Extends from Nashville to Clarksville, and is sixty-one miles long. It is composed of three links: First, the Louisville & Nashville Railroad from Nashville to Edgefield Junction, ten miles; second, the Edgefield & Kentucky Railroad to the State line, thirty-seven miles; and third, the Memphis, Clarksville Louisville Railroad to Paris, fourteen miles. On the 4th of August, 1864, I received General Sherman's order directing this road to be opened so as to provide another avenue of supply to the depot at Nashville. Having made the necessary arrangements for carrying on the work at the front during my absence, I took the First Division of the Construction Corps, under L. H. Eicholtz, division engineer, and proceeded to Springfield, where we arrived on the 11th of August. I found the road had been repaired and put it running order from Edgefield Junction to this station by Capt. C. H. Irvin, assistant quartermaster, who was using it to haul lumber from his numerous saw-mills to Nashville. The portion from State line to Clarksville was in running order and being operated by the Louisville & Nashville Railroad Company. Putting the construction force to work at once, I made an examination of the line between Springfield and State line and found the work to be done consisted principally of bridging; the track had not been much damaged. Some of the cuts were so filled up that it required the removal of a good deal of material to clear the track. The bridges destroyed were of considerable magnitude and all the timber for their reconstruction had to be cut and prepared. The work was completed and the road opened through to Clarksville on the 16th of September. The construction force remained on the road until October 16, employed in getting out bridge timber and cross-ties, and grading and laying a track with sidings 6,765 feet long from main line to the levee at Clarksville. On the 25th of October I appointed W. R. Kingsley, division engineer, engineer of construction and repairs, and he continued to occupy this position while we held and operated the road. The cross-ties were badly decayed in places and many had to be taken out and replaced with new ones. On the 4th of March a freshet carried away the Red River bridge and it was rebuilt by the 25th of same month. Another freshet on the 7th of April again carried away this bridge and it was not rebuilt. Still another freshet occurred on the 20th of May, destroyed the Sulphur Fork bridge and doing much additional damage to the road, all of which, however, was quickly repaired. The following tabular statement shows the amount of bridging and trestle-work on this road:

Name	Height Feet	Length Feet	Remarks
Springfield	44	410	
Sulphur Fork	60	433	Partially destroyed and rebuilt, 150 feet
Spring Creek	60	560	
Red River	85	680	Partially destroyed and rebuilt, 300 feet
Clarksville extension	6-20	900	
Total		2,983	

Add to this the amount rebuilt, 450 feet, and we have a total of 3,433 feet bridging and trestle on this road built by the Government. The lumber consumed in these structures amounted to 890,000 feet, B. M.

Track laid

	Feet, linear
On Edgefield & Kentucky Railroad	2,484
On Clarksville extension	6,055
On side-tracks, Clarksville extension	700
	9,249

Or 1 mile 3,969 feet.

Cross-ties

About 15,000 cross-ties were cut by the Construction Corps on the line of this road.

I have the honor to be, very respectfully, your obedient servant,
W. W. Wright
Chief Engineer Military Railroads United States

Acknowledgements and Notes on Sources

First, I would like to thank certain individuals who have supported me in my quest for knowledge about the Civil War and my efforts in writing and publishing books. Many are involved with the Battle of Nashville Trust (BONT) and the Civil War Fortification Study Group. These folks include the late Ed Bearss, Jim Kay, Fred Prouty, Jerry Wooten, Ross Massey, Wes Shofner, Sam Huffman, Stacey Allen, Phil Shiman, David Lowe, Philip Duer, Doug Jones, Bob Henderson, Jack Smith, James Hoobler, Greg Biggs, J.T. Thompson, Thomas Cartwright, Cody Engdahl, and many others. Special thanks go to my newspaper pals, David Rush Jr. and John Cannon.

I would like to thank the talented artists who contributed artwork for this project, including Andy Thomas, John Paul Strain, Rick Reeves, Philip Duer, and David Meagher. Their talents are unlimited and their generosity greatly appreciated. Please visit their websites (andythomas.com; www.johnpaulstrain.com; https://rickreeves.site) to view their products or contact them by email (pwduer@gmail.com).

I want to thank Krista Castillo, Jerry Wooten, Greg Biggs, Fred Prouty, David Meagher, Jim Kay, Bill Radcliffe and Susan Hawkins—all were gracious enough with their time to read all or parts of my draft manuscript and offer valuable feedback and comments.

I would especially like to thank Krista Castillo and her staff at the Fort Negley Visitors Center, including Tracy Harris and Howard Mann. They were extremely helpful in explaining the history of Fort Negley and Civil War Nashville. The members of BONT graciously allowed me to use photographs and text from their website, including photos by Tom Lawrence and research by John Allyn. Greg Biggs told me about Clarksville and Fort Defiance; Jerry Wooten about Johnsonville. Fred Prouty helped me with various war sites and archaeology. Trent Hanner, Jennifer Core, and the staffers at the Tennessee State Library and Archives were happy to help me in researching Nashville; they benefit from working in a magnificent new facility. The staff at the Nashville Room at the Metro Nashville Library provided research on fortifications, as did Sarah Arntz and other staff at Metro Nashville Archives. Much of the material on frontier stations comes from *Chronicles of the Cumberland Settlements*, a magnificent reference book by Paul Clements. Many of the photos and diagrams of forts come from the Library of Congress online files and illustrations from the *Atlas to Accompany the Official Records of the Rebellion*. Also helpful was the *Annals of the Army of the Cumberland* by John Fitch.

The American Battlefield Trust is a tremendous source of online information about numerous Civil War topics. Their maps are especially helpful. Please consider donating to this well-run non-profit organization that preserves hallowed battlefields.

David Meagher of Giles County, a talented and prolific marine illustrator, went the extra mile in providing me with lots of artwork and information on the Pulaski area and Johnsonville. I am indebted to him.

Zada Law of MTSU was gracious with her time in explaining the history of Fort Negley.

Much information was gained by visiting the National Park Service sites and sites operated by the State of Tennessee and various municipalities.

Books by James Hoobler and the late Walter Durham were very informative about Nashville during the war. Books by Earl Hess explore the logistics of Civil War armies.

James Lee McDonough wrote the most recent reference book on the Battle of Nashville and has disspelled some myths. Retired General John Scales wrote the book about the comings and goings of the cavalry chieftain Nathan Bedford Forrest.

Historians who proved vital to this project include Myron "Jack" Smith, Benjamin F. Cooling, Ross Massey, Stanley Horn, Larry Daniel, Ed Bearss, and Peter Cozzens.

Extremely helpful in providing background historical information was the *Tennessee Encyclopedia of History and Culture*, available online and in print.

Information on the Pioneer Brigade and the First Michigan Engineers and Mechanics derives from Geoffrey Blankenmeyer and Mark Hoffman, respectively, who wrote the books on those subjects. Their research was invaluable.

Material on the U.S. Colored Troops comes from many sources, including Bill Radcliffe, Dr. Bobby Lovett, Tina Cahalan Jones, James McRae, Noah Andre Trudeau, Dr. Learotha Williams Jr., Jennifer Edwards-ring, Roger Cunningham, Gary Burke, and Walter Dobak.

M. Todd Cathey and Ricky W. Robnett recently wrote a book detailing the history of the river gun batteries at Fort Donelson.

Last but not least, I'd like to thank my faithful readers and supporters, family and friends, who have sustained me in my humble writing and publishing ventures through the years. I am especially grateful for the friendship and guidance offered to me the past four decades by one of my mentors, Dennis R. Trisler, who passed away recently.

To God goes all the glory.

Bibliography and Suggested Reading

Abdill, George B., *Civil War Railroads, A Pictorial Story of the War Between the States, 1861-65,* Indiana University Press, 1961.

American Battlefield Trust, www.battlefields.org website.

Army, Thomas F. Jr, "Engineering Victory: The Ingenuity, Proficiency, and Versatility of Union Citizen Soldiers in Determining the Outcome of the Civil War" (2014). Doctoral Dissertations. https://doi.org/10.7275/j7nx-8q44 https://scholarworks.umass.edu/dissertations_2/50

Atlas to Accompany the Official Records of the Union and Confederate Armies, Washington-Government Printing Office, 1891-95.

Bailey, Joe R., "Union Lifeline in Tennessee: A Military History of the Nashville and Northwestern Railroad," *Tennessee Historical Quarterly,* Vol. LXVII, No. 2, Summer 2008.

Battle of Nashville Trust, https://www.battleofnashvilletrust.org.

Bearss, Edwin C., *The Construction of Fort Henry and Fort Donelson,* West Tennessee Historical Society Papers.

————, *The History of Fortress Rosecrans,* Stones River Military Park, December 1960.

Beckwith, H.W., Genealogy Trails History Group, Montgomery County, Indiana, http://genealogytrails.com/ind/montgomery/men-killed-in-civilwar.htm.

Biggs, Greg, "The Great Question Was One of Supplies: Sherman's Logistics in the Atlanta Campaign," Nashville Civil War Roundtable, Feb. 15, 2022.

————, "Stopping Hood: The Navy in the Tennessee Campaign," Battle of Nashville 153rd Anniversary Symposium, Dec. 16, 2017.

Blankenmeyer, Geoffrey L., *The Pioneer Brigade,* https://web.archive.org/web/20170702193619/http://www.thecivilwargroup.com/pioneer.html.

Brent, Maria C. and Brent, Joseph E., *Ready to Die for Liberty: Tennessee's United States Colored Troops in the Civil War,* Tennessee Wars Commission, 2013.

Brown, Lenard E., "Fortress Rosecrans: A History, 1865-1990," *Tennessee Historical Quarterly,* Vol. L, No. 3, Fall 1991.

Cammack, William letters, photocopies at www.flickr.com/people/fortnegley/; transcription available at Fort Negley Archives, transcribed by Howard Mann.

Castillo, Krista, interview with Allen Forkum, *The Nashville Retrospect,* nashvilleretrospect.com/nrc01.

————, "The Story of Jerry Jackson, Fort Negley Laborer/12th USCT Soldier," presentation, Nashville Civil War Round Table, Sept. 21, 2021.

————, personal interviews, July 8 and July 22, 2021.

Cathey, M. Todd and Robnett, Ricky W., *The River Batteries at Fort Donelson, Construction, Armament and Battles, 1861-62,* McFarland & Co., Inc., 2021.

Clark, Donald A., *The Nortorious "Bull" Nelson: Murdered Civil War General,* Southern Illinois University Press, 2011.

Bibliography and Suggested Reading

Clements, Paul, *Chronicles of the Cumberland Settlements, 1779-96*, 2012.

Cunningham, Roger D., "Black Artillerymen from the Civil War through World War I," *Army History*, U.S. Army Center of Military History, No. 58, Spring 2003.

Connelly, Thomas Lawrence, *Army of the Heartland: The Army of Tennessee 1861-62*, Louisiana State University Press, 1967.

————, *Civil War Tennessee*, University of Tennessee Press, 1979.

Cooling, Benjamin Franklin, *Forts Henry and Donelson: The Key to the Confederate Heartland*, University of Tennessee Press, 1987.

————, *Fort Donelson's Legacy: War & Society in Kentucky & Tennessee, 1862-1863*, Knoxville: UT Press, 1997.

Cozzens, Peter, *No Better Place to Die: The Battle of Stones River*, University of Illinois Press, 1989.

————, *This Terrible Sound: The Battle of Chickamauga*, University of Illinois Press, 1992.

Creighton, Wilbur F., *The Life of Major Wilbur Fisk Foster: A Civil Engineer, Condederate Soldier, Builder, Churchman and Free Mason*, Ambrose Printing Co., 1961.

Daniel, Larry J., *Battle of Stones River: The Forgotten Conflict between the Confederate Army of Tennessee and the Union Army of the Cumberland*, LSU Press, 2012.

————, *Days of Glory: The Army of the Cumberland, 1861-65*, LSU Press, 2004.

Dobak, William A., *Freedom by the Sword: The Official Army History of the U.S. Colored Troops in the Civil War 1862-1867*, U.S. Army Center of Military History, 2011.

Douglas, Byrd, *Steamboatin' on the Cumberland*, Tennessee Book Co., 1961.

Duane, Captain James C., *Manual for Engineer Troops*, D. Van Nostrand, 1862; https://books.google.com.

Durham, Walter, *Reluctant Partners: Nashville and the Union*, Tennessee Historical Society, 1987.

————, *Nashville: The Occupied City*, Tennessee Historical Society, 1985.

Edwards-ring, Jennifer, "Chattel, Soldier, Citizens: the United States Colored Troops at the Battle of Nashville," 140th Anniversary Battle of Nashville Symposium, Nashville Public Library, Dec. 10-11, 2004.

Eggleston, Michael A., *President Lincoln's Recruiter*, McFarland & Company, 2013.

Encore Interpretive Design LLC, *Fort Nashborough Interpretive Master Plan* (Nov. 2013), et.al. for Metro Nashville Parks Dept.

Ferguson, John Hill, *On To Atlanta*, edited by Janet Correll Ellison and Mark A. Weitz, University of Nebraska Press, 2001.

Fitch, John, *Annals of the Army of the Cumberland*, Stackpole Books, 1864.

Flagel, Thomas R., "The Fortress War: The Effect of Union Fortifications in the Western Theater of the American Civil War," dissertation, Middle Tennessee State University, May 2016.

Bibliography and Suggested Reading

Fort Negley Descendants Project, http://ftnegley.digitalprojects.network.

Foster, Wilbur F., *The Building of Forts Henry and Donelson,* Morningside Bookshop, Dayton, Ohio, 1978.

Gilfillan, Kelly, "Battle of Brentwood recalled on its 150th anniversary," www.williamsonhomepage.com/brentwood/news/battle-of-brentwood-recalled-on-its-150th-anniversary/article_f8f4724a-86d0-5fa3-8574-0f1c4e6d836b.html, Sept. 25, 2019.

Gott, Kendall D., *Where the South Lost the War: An Analysis of the Fort Henry-Fort Donelson Campaign,* February 1862, Stackpole Books, 2003.

Hagerman, Edward, *The American Civil War and the Origins of Modern Warfare,* Indiana University Press, 1988.

Hamlin, Kelly, Sources on Giles County's Contraband Camps (unpublished).

Harding, Cody J., "Crossed Hatchets and Detached Service: The Creation of the Pioneer Brigade," https://www.libertyrifles.org/research/regiments/pioneer-brigade.

Hess, Earl J., *Civil War Logistics: A Study of Military Transportation,* LSU Press, 2017.

―――, *Civil War Supply and Strategy: Feeding Men and Moving Armies,* LSU Press, 2020.

Historical Marker Database, www.hmdb.org.

Hoffman, Mark, *My Brave Mechanics: First Michigan Engineers and Their Civil War,* Wayne State Univeristy Press, 2007.

Hoobler, James A., *Cities Under the Gun: Images of Occupied Nashville and Chattanooga,* Rutledge Hill Press, 1986.

Horn, Stanley F., *The Decisive Battle of Nashville,* Louisiana State University Press, 1956.

Hulette, Brandon, "Old Story, New Tricks: New Tools to Explore the Battlefield of Nashville," Nashville Civil War Roundtable presentation, Feb. 15, 2022.

Jobe, James, "The Battles for Fort Henry and Donelson," *Blue & Gray Magazine,* Vol. XXVIII No. 4, 2011.

_____, "Forts Heiman, Henry, and Donelson," Civil War Trust, www.civilwar.org/battlefields/fortdonelson/fort-donelson-history-articles/donelsonjobe.html

Jones, James B., "Municipal Vice: The Management of Prostitution in Tennessee's Urban Experience," *Tennessee Historical Quarterly,* Vol. 50, No. 1, Spring 1991.

Jones, Tina Cahalan, "From Slaves to Soldiers and Beyond: Williamson County, Tennessee's African American History," http://usctwillcotn.blogspot.com.

Kemmerly, Phillip R., "Rivers, Rails, and Rebels: Logistics and Struggle to Supply U.S. Army Depot at Nashville, 1862-1865," *The Journal of Military History,* Vol. 84, No. 3, July 2020.

Kirkpatrick, Jim, "Boom Town: Cairo's Strategic Location in the Civil War," *The Southern,* Oct. 18, 2010, thesouthern.com/progress/section3/boom-town-cairo-s-strategic-location-in-the-civil-war/article_html

Bibliography and Suggested Reading

Law, Zada, Director of the Fullerton Laboratory for Spatial Studies at Middle Tennessee State University, interview with Mark Fraley Podcast (July 12, 2017) online.

———— and Rowe, David, *The Construction of Fort Negley: The Civil War Era*, Middle Tennessee State University, May 8, 2009.

Lewis, James Park Ranger, Stones River National Battlefield video, "Pioneer Brigade and Chicago Board of Trade Battery," Sept. 11, 2014, https://www.youtube.com/watch?v=wSAQH_ZShIY.

Library of Congress, Civil war photographs, 1861-1865, Prints and Photographs Division, www.loc.gov.

Lovett, Bobby L., *The Negro's Civil War in Tennessee, 1861-1865*, The Journal of Negro History, Vol. 61, No. 1, pp. 36-50, 1976.

————, "Nashville's Fort Negley: A Symbol of Blacks' Involvement with the Union Army," *Tennessee in the Civil War Vol 1*, Tennessee Historical Society, 2011.

Mackey, Robert R., *The Uncivil War: Irregular Warfare in the Upper South 1861-65*, University of Oklahoma Press, 2004

Mahan, Dennis Hart, *A Treatise on Field Fortification*, John Wiley, New York, 1860.

Massey, Ross, *Nashville Battlefield Guide*, Tenth Amendment Publishing, 2007.

Matlock, J.W.L., "The Battle of the Bluffs: From the Journal of John Cotten," *Tennessee Historical Quarterly*, Vol. 18, No. 3, Sept. 1959.

McCallum, David C., "Report of Brig. Gen. D. C. McCallum, Director and General Manager, Military Railroads, United States, from 1862 to 1866," Washington [D.C.] : Government Printing Office, 1866.

McDonough, James Lee, *Nashville: The Western Confederacy's Final Gamble*, The University of Tennessee Press, 2004.

————, keynote speaker, 140th Anniversary Battle of Nashville Symposium, Nashville Public Library, Dec. 10-11, 2004.

McKee, John Miller, *The Great Panic*, Elder-Sherbourne, 1977.

McKenney, Tom C., *Jack Hinson's One-Man War: A Civil War Sniper*, Pelican Publishing Co., 2014.

McMahan, Andrew, "Bravery, Tenacity, and Deeds of Noble Daring: The 13th United States Colored Troops in the Battle of Nashville," Tennessee State Library and Archives website, June 2019.

McPherson, James M.. *The Negro's Civil War* (Vintage Civil War Library), Knopf Doubleday Publishing Group. Kindle Edition.

McRae, James S., "I Don't Want to Be in a Hotter Place: The 12th USCT at the Battle of Nashville or David G. Cooke Joins the USCT," 140th Anniversary Battle of Nashville Symposium, Nashville Public Library, Dec. 10-11, 2004.

Merritt, Dixon, *Sons of Martha*, "Construction of the City of Old Hickory," Mason & Hanger Co., Inc., 1928.t

National Park Service, various National Battlefield websites.

National Tribune, Washington, D.C., "Fort Negley Was Silent," Sept. 24, 1903, www.newspapers.com/image/46297384.

Bibliography and Suggested Reading

Naval History and Heritage Command, "River Warfare During the Civil War,"
www.history.navy.mil/library/online/riverine.htm

Neal, Dr. W.A., *An Illustrated History of the Missouri Engineer and the 25th Infantry Regiments,* Donohue and Henneberry, 1889.

Perry, John C., "Forrest's Navy: The Raid on Johnsonville," *Confederate Veteran,* Jan.-Feb. 1993.

Plaisance, Aloysius F., and Leo F. Schelver. "Federal Military Hospitals in Nashville, May and June, 1863." *Tennessee Historical Quarterly,* Vol. 29, No. 2, 1970, pp. 166-175.

Powell, David A. and Wittenberg, Eric J., *Tullahoma: The Forgotten Campaign That Changed the Course of the Civil War,* Savas Beatie, 2020.

————, Chickamauga Blog, https://chickamaugablog.wordpress.com/2011/08/28/who-where-those-guys-part-one/

Prouty, Fred, Military Sites Preservation Specialist/Director of Programs, Tennessee Wars Commission (retired), personal interview, Feb. 2, 2022.

Quertermous, Grant, "Summary of Excavations at Fort Star: Archaeology of a Union Civil War Fortification," *Ohio Valley Historical Archaeology,* Vol. 14, 1999.

Roberts, Robert B., *Encyclopedia of Historic Forts: The Military, Pioneer, and Trading Posts of the United States,* Macmillan, New York, 1988, 10th printing.

Robinson, William M. Jr., "The Confederate Engineers," *The Military Engineer,* Vol 22 No 124, July-Aug 1930, pp297-305, Society of American Military Engineers.

Scales, John R., *The Battles and Campaigns of Confederate General Nathan Bedford Forrest,* Savas Beatie, 2017.

Shockley, Gary C., "The Union Legal Response to Hood's Invasion of Tennessee," 140th Anniversary Battle of Nashville Symposium, Nashville Public Library, Dec. 10-11, 2004.

Shiman, Philip L., "Engineering and Command: The Case of General William S. Rosecrans 1862-1863," in *The Art of Command in the Civil War,* ed. Steven E. Woodworth (Lincoln: University of Nebraska Press, 1998), 88.

Smith, Myron J. "Jack", *Le Roy Fitch: The Civil War Career of a Union River Gunboat Commander,* McFarland & Co., Inc., 2007.

———, *The Tinclads in the Civil War: Union Light-Draught Gunboat Operations on Western Waters 1862-65,* McFarland & Co., Inc., 2010.

Smith, Samuel D., Prouty, Fred M., Nance, Benjamin C., *A Survey of Civil War Period Military Sites in Middle Tennessee,* Tennessee Dept. of Conservation, Division of Archaeology, Report of Investigations No. 7, 1990.

Smith, Timothy B., *Grant Invades Tennessee: The 1862 Battles for Forts Henry and Donelson,* University Press of Kansas, 2016.

Stuart, Reginald C., "Cavalry Raids in the West: Case Studies of Civil War Cavalry Raids, Nathan Bedford Forrest and the Confederate Cavalry," *Tennessee in the Civil War,* Vol. 5, Tennessee Historical Quarterly 2013.

Bibliography and Suggested Reading

Taylor, Lenette S., *The Supply for Tomorrow Must Not Fail: The Civil War of Captain Simon Perkins Jr., a Union Quartermaster*, Kent State University Press, 2004.

Temple, Wayne C., editor, "Fort Donelson in October 1862," *Lincoln Herald*, Lincoln Memorial University Press, Spring-Winter 1967.

Tennessee: The Civil War Years, Tennessee 200 Inc., 1996.

Tennessee Division of Archaeology, https://www.tn.gov/environment/program-areas/arch-archaeology.html.

Tennessee Historical Society, *Tennessee Encyclopedia of History and Culture*, https://tennesseeencyclopedia.net.

Tennessee State Library and Archives, https://sos.tn.gov/tsla/history.

Trudeau, Noah Andre, *Like Men of War: Black Troops in the Civil War, 1862-65*, Little, Brown & Co., 1998.

War of the Rebellion: a Compilation of the Official Records of the Union and Confederate Armies, Cornell University Library, http://collections.library.cornell.edu/moa_new/waro.html.

Warwick, Rick, *Triune: Two Centuries at the Crossroads*, Williamson County Historical Society, 2004.

West, Mike, "Fabulous military career cut short" *Murfreesboro Post*, https://www.murfreesboropost.com/community/fabulous-military-career-cut-short/article_6816761b-2c61-540e-9c1f-c3265e65aadf.html

Wetherington, Mark V., Ku Klux Klan, *Tennessee Encyclopedia of History and Culture*, https://tennesseeencyclopedia.net/entries/ku-klux-klan/.

Williams, Max R., "Jermey Francis Gilmer," *Dictionary of North Carolina Biography*, 6 volumes, edited by William S. Powell, University of North Carolina Press, 1979-96.

Wooten, Jerry T., *Johnsonville: Union Supply Operations on the Tennessee River and the Battle of Johnsonville, November 4-5, 1864*, Savas Beatie, 2019; and personal interview, Aug. 5, 2021.

————, "Johnsonville: Union Supply Operations on the Tennessee River and the Battle of Johnsonville," November 4-5, 1864, Nashville Civil War Roundtable presentation, Aug. 17, 2021.

Wright, David Russell, *Civil War Field Fortifications : An Analysis of Theory and Practical Application*, thesis (MA), Middle Tennessee State University, 1982.

Zimmerman, Mark, *Guide to Civil War Nashville-2nd Edition*, Zimco Publications LLC, 2020.

————, *Iron Maidens and the Devil's Daughters*, Zimco Publications LLC, 2014.

About the Author

Mark Zimmerman is a retired newspaperman and publications manager who resides in Nashville, Tennessee. In addition to operating Zimco Publications LLC, he is a member of the American Battlefield Trust, Battle of Nashville Trust, Nashville Civil War Roundtable, Save the Franklin Battlefield, Civil War Fortification Study Group, Tennessee Squires Association and other historical preservation groups. He enjoys traveling to historic sites and museums, boogie music, a good joke, and collecting old bottles of whiskey.

Members of the Tennessee Trails Association on a 2021 tour of Fort Negley led by the author (second from right). Photo by Ron Jenkins.

Books for the Historically Curious
Website: zimcopubs.com
Contact: info@zimcopubs.com

Mud, Blood & Cold Steel:
The Retreat From Nashville - December 1864
ISBN: 978-0-9858692-6-7
Retail: $19.95 / 6 x 9 / 184 pages
Ebook: $9.95 / ASIN : B0892R2KBY

Takes a fresh look, with 16 maps, at the unprecedented and brutal pursuit of the Army of Tennessee by Federal troops following the Battle of Nashville. The winter campaign covered 10 days and 100 miles of cavalry-rearguard action.

Guide to Civil War Nashville:
2nd Edition (revised, updated, expanded)
ISBN: 978-0-9858692-2-9
Retail: $19.95 / 8.5 x 11 / 92 pages

The classic illustrated tour guide to Nashville during the Civil War and the decisive 1864 Battle of Nashville, with battle maps and orders of battle, period and modern photography, GPS coordinates, and driving tour map of 25 sites.

Iron Maidens and the Devil's Daughters:
US Navy Gunboats versus Confederate Gunners and Cavalry on the Tennessee and Cumberland Rivers, 1861-65
ISBN: 978-0-9858692-5-0
Retail: $24.95 / 8.5 x 11 / 184 pages

The fascinating story of the Federal naval invasion of Middle Tennessee on the Tennessee and Cumberland rivers and the Confederate response via river gun batteries and cavalry armed with artillery. Well-illustrated, with 26 maps, including 14 battle maps.

God, Guns, Guitars & Whiskey:
An Illustrated Guide to Historic Nashville, Tennessee - 2nd Edition
ISBN: 978-0-9858692-3-6
Retail: $19.95 / 8.5 x 11 / 164 pages

Spotlights more than 170 historic sites and monuments of Nashville, the capital of Tennessee and Music City USA. Includes downtown walking tour, 265 photographs, 17 maps, historical markers, and in-depth articles on local attractions.

Land of the Free, Home of the Brave:
Our Founding Documents & Concise History of the USA
ISBN: 978-0-9858692-8-1
Retail: $19.95 / 8.5 x 11 / 178 pages

The Declaration of Independence, the US Constitution, and the Bill of Rights, with wording straight from the National Archives. The history of the United States is laid out in concise, readable form. Plus many articles on the Presidents, the States, battlefields, and historic sites.

Gone Under: Historic Cemeteries of Nashville, Tennessee-2nd Edition
ISBN: 978-0985869243
Retail: $19.95 / 8.5 x 11 / 68 pages

An illustrated guide to the gravesites of the famous personages in Nashville history, including Presidents Jackson and Polk, famous pioneers and generals, and Nashville country music and Grand Ole Opry stars! Includes biographies, 128 photos, and 11 maps.

Bookstore owners, gift shop managers, other retailers order online from Ingram at 55% discount/returnable.

www.ingramcontent.com/pod-product-compliance
Lightning Source LLC
Chambersburg PA
CBHW051207290426
44109CB00021B/2372